Creative Conservation

Creative Conservation

Interactive management of wild and captive animals

Edited by

P.J.S. Olney
Federation of Zoological Gardens of Great Britain and Ireland
Zoological Gardens
London
UK

G.M. Mace
Institute of Zoology
Zoological Society of London
London
Uk

and

A.T.C. Feistner
Jersey Wildlife Preservation Trust
Trinity
Jersey
Channel Islands

CHAPMAN & HALL
London · Glasgow · New York · Tokyo · Melbourne · Madras

Published by Chapman & Hall, 2–6 Boundary Row, London SE1 8HN, UK

Chapman & Hall, 2–6 Boundary Row, London SE1 8HN, UK

Blackie Academic & Professional, Wester Cleddens Road, Bishopbriggs, Glasgow G64 2NZ, UK

Chapman & Hall Inc., One Penn Plaza, 41st Floor, New York NY10119, USA

Chapman & Hall Japan, Thomson Publishing Japan, Hirakawacho Nemoto Building, 6F, 1-7-11 Hirakawa-cho, Chiyoda-ku, Tokyo 102, Japan

Chapman & Hall Australia, Thomas Nelson Australia, 102 Dodds Street, South Melbourne, Victoria 3205, Australia

Chapman & Hall India, R. Seshadri, 32 Second Main Road, CIT East, Madras 600 035, India

---

First edition 1994

Typeset in 10/12pt Palatino by Acorn Bookwork, Salisbury

Printed in Great Britain by St Edmundsbury Press, Bury St Edmunds, Suffolk

ISBN 0 412 49570 8

A catalogue record for this book is available from the British Library

Library of Congress Cataloging-in-Publication data available

∞ Printed on permanent acid-free text paper, manufactured in accordance with ANSI/NISO Z39.48-1992 and ANSI/NISO Z39.48-1984 (Permanence of Paper).

CLAUDIO SILLERO ZUBIRI

Contents

Contributors

Adler, H.J.,
Allwetterzoo,
Sentruper Strasse 315,
D-4400, Münster, Germany.

Andrews, C.,
IUCN Freshwater Fish Specialist Group,
Director of Captive Propagation and Acting Director of Husbandry,
National Aquarium in Baltimore,
Pier 3, 501 East Pratt Street,
Baltimore,
MD 21202, USA.

Balen, B. van
ICBP Indonesia Programme,
P.O. Box 310/Boo,
Bogor 16003, Indonesia.

Banko, P.C.,
Wildlife Science Group
College of Forest Resources AR-10,
University of Washington,
Seattle,
WA 98195, USA.
Present affiliation: US Fish and Wildlife Service,
Patuxent Wildlife Research Center,
Laurel,
MD 20708, USA
(Stationed at:
US Fish and Wildlife Service
Hawaii Research Group,
P.O. Box 44,
Hawaii National Park,
HI 96718, USA.)

Beck, B.B.,
National Zoological Park,
Smithsonian Institution,
Washington, DC, 20008, USA.

Biggins, D.,
US Fish and Wildlife Service,
National Ecology Research Center,
4512 McMurry Avenue,
Fort Collins,
CO 80525, USA.

Black, J.M.,
The Wildfowl and Wetlands Trust,
Slimbridge,
Gloucester GL2 7BT, UK.

Boer, L.E.M. de,
National Foundation for Research in Zoological
Gardens and EEP Executive Office,
c/o Amsterdam Zoo,
P.O. Box 20164,
1000 HD Amsterdam, The Netherlands.

Bruford, M.W.,
Institute of Zoology,
Zoological Society of London,
Regent's Park,
London NW1 4RY, UK.

Butler, D.,
Threatened Species Unit,
Department of Conservation,
Box 10420,
Wellington, New Zealand.
(Present address
c/o New Zealand High Commission,
Apia, Western Samoa.)

Carroll, J.B.,
Jersey Wildlife Preservation Trust,
Les Augres Manor
Trinity,
Jersey JE3 5BF, Channel Islands.

Craig, J.L.,
Centre of Conservation Biology,
School of Biological Sciences,
University of Auckland,
Private Bag 92019,
Auckland, New Zealand.

Diaz, M.,
Fundación Nacional de Parques Zoológicos y Acuarios,
Aspartado 68387,
Caracas 1062-A, Venezuela.

Dietz, J.M.,
Department of Zoology,
University of Maryland,
College Park,
MD 20742, USA.

Dietz, L.A.,
World Wildlife Fund-US,
1250 24th Street, N.W.,
Washington, DC 20037, USA.

Dixon, A.M.,
Institute of Zoology,
Zoological Society of London,
Regent's Park,
London NW1 4RY, UK.

Durant, S.M.,
Institute of Zoology,
Zoological Society of London,
Regent's Park,
London NW1 4RY, UK.

Durbin, J.,
Durrell Institute of Conservation and Ecology,
University of Kent,
Canterbury CT2 7NX, UK.

Durrell, L.,
Jersey Wildlife Preservation Trust,
Les Augres Manor,
Trinity,
Jersey JE3 5BF, Channel Islands.

Ellis, S.,
IUCN/SSC Captive Breeding Specialist Group,
12101 Johny Cake Ridge Road,
Apple Valley,
MN 55124, USA.

Feistner, A.T.C.,
Jersey Wildlife Preservation Trust,
Les Augres Manor,
Trinity,
Jersey JE3 5BF, Channel Islands.

Foose, T.J.,
IUCN/SSC Captive Breeding Specialist Group,
12101 Johny Cake Ridge Road,
Apple Valley,
MN 55124, USA.

Gadsby, E.L.,
Pandrillus,
Housing Estate P.O. Box 107,
Calabar, Nigeria.

Garland, P.,
Orana Park Wildlife Trust,
Box 1530,
Christchurch, New Zealand.

Gepak, V.H.,
Kebun Binatang,
Jalan Setail 1,
Surabaya, Indonesia.

Girman, D.,
Department of Biology,
UCLA,
Los Angeles,
CA 90024, USA.

Ginsberg, J.R.,
Institute of Zoology,
Zoological Society of London,
Regent's Park,
London NW1 4RY, UK.

Groombridge, B.,
World Conservation Monitoring Centre,
219 Huntingdon Road,
Cambridge CB3 0DL, UK.

Hanebury, L.
US Fish and Wildlife Service,
National Ecology Research Center,
4512 McMurry Avenue,
Fort Collins,
CO 80525, USA.

Holt, W.V.
Institute of Zoology,
Zoological Society of London,
Regent's Park,
London NW1 4RY, UK.

Jenkins, M.,
World Conservation Monitoring Centre,
219 Huntingdon Road,
Cambridge CB3 0DL, UK.

Jenkins, P.D. Jr,
Pandrillus,
Housing Estate P.O. Box 107,
Calabar, Nigeria.

Johnson, T.H.,
World Conservation Monitoring Centre,
219 Huntingdon Road,
Cambridge CB3 0DL, UK.

Kaufman, L.,
IUCN Captive Breeding Specialist Group (Aquatic Section),
New England Aquarium,
Boston,
MA 02110, USA.

Kleiman, D.G.,
Department of Zoological Research,
National Zoological Park,
Smithsonian Institution,
Washington, DC 20008, USA.

Mace, G.M.,
Institute of Zoology,
Zoological Society of London,
Regent's Park,
London NW1 4RY, UK.

Magin, C.D.
World Conservation Monitoring Centre,
219 Huntingdon Road,
Cambridge CB3 ODL, UK.

Martin, R.D.,
Anthropological Institute,
University of Zürich,
Winterthurerstrasse 190
8057 Zürich, Switzerland.

Mickleburgh, S.,
Fauna and Flora Preservation Society,
1 Kensington Gore,
London SW7 2AR, UK.

Miller, B.,
Centro de Ecologia,
Universidad Nacional Autonoma de Mexico,
Apartado Postal 70-275,
Mexico D.F., 04510 Mexico.

Mujica, E.,
Fundación Nacional de Parques Zoológicos y Acuarios,
Aspartado 68387,
Caracas 1062-A, Venezuela.

Nagagata, E.Y.,
Department of Park and Recreational Resources,
Michigan State University,
East Lansing,
MI 48823, USA.

Oliver, W.L.R.,
IUCN/SSC Pigs and Peccaries Specialist Group,
Park End,
28A Eaton Road,
Norwich NR4 6PZ, UK.

Pearce-Kelly, P.,
Invertebrate Conservation Centre,
Zoological Society of London,
Regent's Park,
London NW1 4RY, UK.

Rakotonindrina, R.,
Association Nationale pour la Gestion des Aires Protégées,
B.P. 1424,
Antananarivo, Madagascar.

Rapaport, L.G.,
Department of Anthropology,
University of New Mexico,
Albuquerque,
NM 87131, USA.

Rebholz, W.E.R.,
Institute of Zoology,
Zoological Society of London,
Regent's Park,
London NW1 4RY, UK.

Reid, D.,
Jersey Wildlife Preservation Trust,
Les Augres Manor,
Trinity,
Jersey JE3 5BF, Channel Islands.

Rossiter, P.B.,
c/o Kenya Agricultural Research Institute,
Muguga,
P.O. Box 32,
Kikuya, Kenya.

Seal, U.S.,
IUCN/SSC Captive Breeding Specialist Group,
12101 Johny Cake Ridge Road,
Apple Valley,
MN 55124, USA.

Serena, M.,
Healesville Sanctuary,
P.O. Box 248,
Healesville,
Victoria 3777, Australia,
and
Chicago Zoological Society
Brookfield Zoo,
Brookfield,
IL 60513, USA.

Shepherdson, D.,
Metro Washington Park Zoo,
4001 S.W. Canyon Road,
Portland,
OR 97221, USA.

Smith, H.,
World Conservation Monitoring Centre,
219 Huntingdon Road,
Cambridge CB3 0DL, UK.

Soderquist, T.R.,
Department of Ecology and Evolutionary Biology,
Monash University,
Clayton,
Victoria 3168, Australia.

Stanley Price, M.R.,
IUCN/SSC Re-introduction Specialist Group,
c/o African Wildlife Foundation,
Box 48117,
Nairobi, Kenya.

Sunnucks, P.,
Institute of Zoology,
Zoological Society of London,
Regent's Park,
London NW1 4RY, UK.

Taylor, A.C.,
School of Biological Sciences,
University of New South Wales,
New South Wales,
Australia.

Toone, W.D.,
San Diego Wild Animal Park,
15500 San Pasqual Valley Road,
Escondido,
CA 92027-9614, USA.

Trebbau, P.,
Fundación Nacional de Parques Zoológicos y Acuarios,
Aspartado 68387,
Caracas 1062-A, Venezuela.

Vargas, A.
US Fish and Wildlife Service,
Cooperative Fish and Wildlife Research Unit,
Department of Zoology and Physiology,
University of Wyoming,
Box 3166, University Station,
Laramie,
WY 82070, USA.

Vàsàrhelyi, K.,
Anthropological Institute,
University of Zürich,
Winterthurerstrasse 190
Zürich, Switzerland.

Wallace, M.P.,
Los Angeles Zoo,
5333 Zoo Drive,
Los Angeles,
CA 90027, USA.

Waugh, D.R.,
Jersey Wildlife Preservation Trust,
Les Augres Manor,
Trinity,
Jersey JE3 5BF, Channel Islands.
(Present address:
Apdo 16324,
Candelaria,
Caracas 1011-A, Venezuala.)

Wayne, R.K.,
Institute of Zoology,
Zoological Society of London,
Regent's Park,
London NW1 4RY, UK.

Wemmer, C.,
Conservation and Research Centre,
Front Royal,
VA 22630, USA.

Wheater, R.J.,
Royal Zoological Society of Scotland,
Murrayfield,
Edinburgh EH12 6TS, UK.

Wilson, A.C.,
IUCN/SSC Re-introduction Specialist Group,
Villa Les Sapins 1,
Route de Basse-Ruche,
CH 1264 St Cerque,
Switzerland.

Wirth, R.,
IUCN/SSC,
F. Senn Strasse 14,
D-8000, München 70, Germany.

Woodford, M.H.,
IUCN/SSC/Veterinary Specialist Group,
Apt B-709, 500 23rd Street, N.W.,
Washington, DC 20037, USA.

Foreword
Past progress and future challenges

R.J. Wheater
Royal Zoological Society of Scotland, Edinburgh, UK.

In the past two decades much has been achieved in the sphere of breeding endangered species, and we should be pleased that our co-operative efforts have already borne so much fruit. However, on balance and despite the best efforts of conservationists, the position of wildlife in the wild places where they are best conserved has become worse, often dramatically worse.

Before returning to the United Kingdom in 1972, I was in Uganda for 16 years, most of which time was spent as Chief Warden of Murchison Falls National Park. Our main problem was that an over-population of large mammals was having a devastating impact on the habitat. Devastation was being wrought on woodland areas by the arrival of large numbers of elephants into the sanctuary of the Park, following changes in land use in the areas outside the Park. These changes were in response to the requirements of an ever-expanding human population. This was happening at a time when, all over the world, land set aside for wildlife was coming under similar pressure. Even the large reserves were fast becoming islands in a sea of intensive human activity, the protected areas often overburdened with additional animals, the species isolated once the reserves were surrounded. This was also the time when the zoo world (with one or two honourable exceptions) was regarded by conservationists at best as an irrelevance and at worst as a consumer of wildlife.

When I left Uganda in 1972 the parks were being well managed, and some culling of elephants and hippopotami had taken place in pursuit of the management aim of maintaining diversity. Coincidentally, but most importantly, the parks were providing foreign-exchange income through a successful park-based tourist industry. We were, however, aware in late 1971 of a rapidly deteriorating situation in the administration of the country, particularly in law and order. There were at that time an estimated 32 000 elephant and 200 each of black rhinoceros and northern white rhinoceros, as well as a wide range of other species, within the parks and reserves of Uganda. I returned to Uganda in 1980, just eight years after my departure, following the removal of the military dictator Idi Amin Dada. I found that the species that had prospered most was the vulture, for fewer than 1600 elephant remained, the black rhino population was reduced to two animals and the white rhino had been totally eliminated. In the 32 years that I have been involved directly in conservation an estimated population of 5000 white rhinoceros located in Uganda, Southern Sudan and Zaire has been reduced to fewer than 30 individuals, all located in the Garamba National Park in eastern Zaire. This one example illustrates the speed at which a major species can disappear, and how human pressures world-wide are isolating remnant populations. Things have improved greatly in Uganda in recent years and perhaps at some time in the future it will be possible to return white rhino, although not I fear, the northern race of that species.

Of course all is not gloom and doom and we must recognize this. For me, a magic moment came in 1990 when I was able to stand in the Yalooni Desert in Oman looking at a small, mixed-age group of Arabian oryx, back where they belong in the wild. Captive breeding efforts of zoos and some early activities of the Fauna and Flora Preservation Society had made this possible. Most importantly, the Omani conservation authorities had identified the oryx as a significant element of Oman's natural heritage and had invested substantial time and money in its return to the wild (Stanley Price, 1989). The fact that we had made such progress in captive breeding programmes led to the inclusion of these activities in IUCN's document *Caring for the Earth* (IUCN, 1990), and means that the task of implementing such an enterprise has received due recognition. This has led to the development of scientifically based cooperative programmes worldwide. In 1977 the Anthropoid Ape Advisory Panel was formed by zoos in the UK to develop good cooperative management of the UK population of anthropoid apes. Also in the UK the Joint Management of Species Group was created in 1978. Meanwhile the Species Survival Programme (SSP) was being developed by the American Association of Zoological Parks and Aquariums (AAZPA) in North America, with the efficiency and enthusiasm so characteristic of the New World. Australasia was developing a similar programme in response to a need to cooperatively manage all species because of the difficulty of importing exotic animals into that region. In more recent times we have seen the development of the European Endangered Species Programme (EEP) in Europe, a continent which in the past three years has changed so dramatically and which now provides such a great challenge to develop its new potential. Across the globe we hear of the development and the acceleration of such programmes in India, South-East Asia, Japan and South America, all responding to the needs of species conservation.

The supporting organizations have also grown dramatically during the same period. The IUCN/Species Survival Commission(SSC)/Captive Breeding Specialist Group, under its dynamic leadership, provides the overview and backup so essential to long-term success. The supporting databases operated worldwide, the International Species Information System (ISIS), assist an ever-increasing number of studbook keepers and species coordinators in the task of developing plans to secure the future of species. At the same time the scientific underpinning necessary to ensure that limited resources of animals, space, expertise and finance are properly and efficiently used is developing rapidly.

The development of organizational structures may seem a rather dull subject when compared with the more exciting breeding and reintroduction programmes. However, I believe that the successful structuring of organizations is fundamental to long-term efficient utilization of resources. There has been recent, and I believe positive, movement in reforming existing organizations and in the development of new ones – which will help to create mutual confidence and support. The world zoo body, the International Union of Directors of Zoological Gardens (IUDZG) as recently as 1991, following a number of years of debate, radically changed its Constitution to allow for wider membership of individual institutions, as well as of those representing national and regional organizations. This combination of individual, national and regional organizations now ensures that the very smallest collection, as a member of a represented national organization, can make its contribution on the world stage. A number of regional organizations are currently being developed in South America, South-East Asia and Africa, joining those longer established ones. Most recent of all has been the formation of an organization that embraces all of Europe, the European Association of Zoos and Aquaria (EAZA), which has risen in response to the changes in Europe and the recognition that such a structure is essential to the combining of efforts in coordinated programmes.

At the same time it must be recognized that individual zoos and regional and international organizations have a tremendous responsibility to maintain high standards of animal welfare, and to place an ever-increasing emphasis on wildlife and environmental issues through their role in public education.

One subject which does need to be addressed in a more mature and measured way is the question of surplus animals. We have failed to deal with this subject for long enough, and it is inconsistent that we regard surplus animals in such a negative way. We should accept the need to reduce surplus stock, and rejoice in the happy circumstance of having too many animals rather than too few. This is not a matter only of captive breeding, but one that must also be addressed in the wild. Generally speaking the inevitable outcome of breeding success will be a surplus, and we need to recognize that fact, as do those who express concern every time action has to be taken to reduce or remove the surplus. I really do despair that otherwise intelligent, and sometimes delightful, people are so removed from the realities of nature that they not only fail to recognize the need to manage animal populations, but actively promote disenchantment of such conservation action by their use of images of beautiful, often baby animals, which in species terms it may be necessary to cull if we are to use our limited resources wisely. One can only presume that such people would prefer to see a species go extinct rather than have it assisted by humans to survive. This attitude is particularly galling when, in most circumstances, it is the human species which has created the risk in the first place. I do so agree with the recent statement of Michael Hutchins that 'A short-sighted focus on individual animals could prove disastrous for long-term conservation efforts'.

So together we have much to do to improve our contribution to conservation breeding, environmental understanding and scientific study. Our ability to provide for the many species at risk depends on the availability of space and resources, both in the wild and captivity. This will call for more and more twin-financing between those who have the funds or other resources and those who have not. The international and regional organizations, with their widening membership, have a very real challenge to support those with potential but with few resources to develop that potential. Support may take a number of forms: meeting training needs or providing development advice, special expertise, equipment and funds – already a number of such initiatives are in operation. I am told that the global zoo community has approximately half a billion visitors annually, giving both the resources to carry on its work, and the challenge and opportunity to educate those who visit. Ideally the visitors of the future will be those who come not only because they enjoy the experience of seeing living animals but also because they understand that these animals form part of a wider process of species survival; and they will add their voices and their financial support to such activities.

There is no doubt that spreading the conservation message to individuals, to the board rooms of industry and the offices of government is a vital element of our work.

There is now a wider realization that truly global problems are facing us: global warming, acid rain, major habitat loss – all directly attributable to a combination of growth of the human population (now projected to reach 12.5 billion by 2050) and an ever-increasing demand by all of us for a better quality of life. We must not lose sight of our specific conservation activities, for when efforts to solve the major problems have succeeded, and succeed they must, our more finely focused activities will have helped to ensure that the brave new world continues to be populated by creatures which have fascinated us, provided for us and shared this planet with us over the millennia.

REFERENCES

IUCN (1990) *Caring for the Earth*, IUCN, Gland, Switzerland.

Stanley Price, M. (1989) *Animal Re-introductions: The Arabian Oryx in Oman*, Cambridge University Press, Cambridge, UK.

Preface

From the text of the Opening Address to the Sixth World Conference on Breeding Endangered Species by Her Royal Highness, the Princess Royal 4–6 May 1992, Jersey

The First World Conference on Breeding Endangered Species was held in 1972 in Jersey, twenty years ago. In the same year I became the Patron of the Jersey Wildlife Preservation Trust.

On my first visit to Jersey Zoo the atmosphere and the people I met were so friendly. It was rather like a glorified backyard . . . tremendous personal contact, a little bit higgledy-piggledy. It was what you might expect from the boy who turned into the man who wrote *My Family and Other Animals*, always torn between the fascination of watching creatures in the wild and taking them home to study more closely.

Even in those days Jersey's conservation objectives had already given the zoo a unique atmosphere, which was perhaps less to do with what they were doing, than why they were doing it.

People's memories of the first time they went to a zoo are often rather clouded and mixed. I certainly know that I was taken to London Zoo and to Whipsnade. I have to admit there's probably a slight confusion between one and the other. Looking back I suppose I thought the animals were there for my benefit – not, I think, as entertainers, but as something to go and learn about; for my benefit, yes, and for those other millions whose privilege it was to see them in their extraordinary variety, variety of movement, being alive and of smell – which is still something television cannot yet recreate.

If there were also more esoteric reasons for the zoo, such as for the advancement of science, I have to admit that they were probably lost on me at the time. I don't recall that the purpose of my visit was to contribute towards the conservation of species. But that might have been because, in those days, such a need had yet to be recognized by more than a handful of people. I'm sure, however, that those early impressions did help to contribute to my interest in the conservation and breeding of endangered species today.

It is most zoos' hope that a visit to the zoo, at least for some, develops an appreciation for wildlife and an interest in its protection and its habitat. This, of course, is a very difficult thing to prove or disprove and unfortunately we no longer have the luxury of time on our side to await a conclusion. If the primary role of zoos of the future is the conservation of species and if this is explained to the public, then I believe that the special atmosphere that I encountered on my first visit to Jersey will become less unique.

Having been enlightened about the significance of F1 and F2, perhaps we should ascribe this special atmosphere to F3 – the Fauna First Factor. I have now visited other zoos and had more than a whiff of F3. Zoos like Marwell, in Hampshire, which has built an admirable reputation for conservation and breeding, often in association with the Institute of Zoology (part of the Zoological Society of London), and in Edinburgh, where much emphasis is placed on conservation education, and at Calgary where I was especially interested in the free-ranging herd of Przewalski horses on the open plains of Alberta. I'm also very aware of the considerable resources and energy exhibited by some zoos of the United States in conservation projects in America and overseas.

Perhaps though, I have the strongest sense of F3 in some of the most unlikely places. In 1984, I was very pleased to open the International Training Centre in Jersey, since when I have had the opportunity to meet some of the graduates of the training scheme at zoos and breeding stations in their own parts of the world. These institutions were very varied.

I don't know how many of you have visited the zoo in Belize. It doesn't fit into the average definition of a zoo. Its creation was very much a result of the strength of personality of the individual concerned who adopted a cast of animal film extras and created a small zoo to teach Belizeans about their wildlife. In Hong Kong you find an example of creating the right atmosphere, particularly for captive breeding, in a concrete jungle – no mean achievement. In Jamaica there exists a battle of minds; a traditional zoo, which had been largely ignored, being transformed by individuals who have been well-trained and know what it is that they have to do. In Rio de Janeiro, there are lessons to be learned about the reintroduction of captive-bred tamarins into the wild. In Mauritius there are contrasts: between the kestrel, a classic example, whose numbers have grown dramatically from a tiny founder stock, and the parakeet which seems to have a death wish but is now, finally, being marginally more co-operative. And in Bahrain, where you have to say that the site of the zoo is extremely unpromising, but the people who are there now have the training and the knowledge to create something special for the good of Bahrain certainly and possibly southern Arabia as well.

Everyone I met described the importance of their institutions, not only as attractions for the people of the area, but as an opportunity to demonstrate the problems that face their wildlife.

These zoos and breeding stations, often poorly resourced, have a very strong sense of F3, of conservation interests being put first. You know you are contributing to much more than food, fuel and professional fees. You feel privileged to be there. Could it be that the ideal zoos of the future, those that concentrate on species of their own region, their breeding and their conservation by public education, will be found in places like Central America, the Caribbean and the Indian Ocean? And does that mean that the multi-species collections of Europe and North America will begin to seem less relevant in an atmosphere of species loss and the need for intensive breeding programmes?

I think the answer will be no if we in the so-called West recognize that our zoos are a means, not an end in themselves. A means to accumulate and export information, expertise, resources, funds, as well as animals, in support of the conservation programmes, breeding stations and zoos in those countries where biodiversity is under greatest threat.

I know that for the survival and success of the zoo itself, and therefore ultimately its contribution to conservation, a balance has to be struck between its serious work and being a day out for the family. But I don't think we should underestimate a growing public awareness, certainly in the West, that all is not right with the world and its wildlife.

From my work with The Save the Children Fund it is quite obvious to me that sustainable economy depends on a balance with nature. The root cause behind the undernourished millions is the same as that which is behind the depletion of the animal kingdom. It's a situation which bodes ill for our planet and one which will require all of our skill and energy if life, human or otherwise, is to be sustained.

Since Gerald Durrell first opened his zoo, in 1959, he has consistently said he wishes he could close it down. Well, I actually don't think his staff need fear for their jobs. I'm afraid their jobs, at least for the time being, are part of a growth industry.

Mr Durrell, as you would expect, is accomplished in the use of irony. When you ask why it is that he wishes to close down his beloved zoo, you realise that his explanation encapsulates what I call the Fauna First Factor and which I have encountered in a growing number of zoological establishments around the world.

Papers given in Jersey at the Sixth World Conference on Breeding

Endangered Species in May 1992 emphasized the interactive management of wild and captive animals and form the basis of this book, *Creative Conservation*. I thoroughly recommend it, not only as a good read, but as an essential background for anyone interested in the management of populations of animals that are threatened with extinction.

Anne

Her Royal Highness, the Princess Royal

Part One

General Issues

1

Species extinctions, endangerment and captive breeding

C.D. Magin, T.H. Johnson, B. Groombridge, M. Jenkins and H. Smith
World Conservation Monitoring Centre, Cambridge, UK.

1.1 INTRODUCTION

During the last four hundred years some 490 described species of animal are known to have become extinct. In this chapter the patterns of animal extinction are analysed by taxonomic group, against time, and between islands and continental land masses. From species–area curves based upon extrapolation of the rate of habitat loss (particularly tropical forests) estimates have been produced of the future rates of species extinctions. These estimates vary considerably from less than 5% of all animal species between 1985 and the year 2025 to more than 25% if rates of forest clearance accelerate (WCMC, 1992).

As estimated rates of extinction continue to rise, albeit probably more slowly than most of the more speculative predictions, the identification of species threatened with extinction will become ever more necessary to determine conservation priorities. IUCN (1990a) lists approximately 2500 vertebrate species and 2000 invertebrate species as threatened or

Creative Conservation: Interactive management of wild and captive animals.
Edited by P.J.S. Olney, G.M. Mace and A.T.C. Feistner.
Published in 1994 by Chapman & Hall, London.
ISBN 0 412 49570 8

possibly threatened. These figures are analysed by taxonomic group, biogeographical region and degree of threat.

Captive breeding, to preserve species at risk of extinction in the wild, and to supply animals for reintroduction projects, is increasingly recognized and accepted as part of the repertoire of creative conservation. The chapter concludes with a review of the current efforts, successes and shortcomings of captive breeding with regard to threatened species.

1.2 SPECIES EXTINCTIONS

1.2.1 The current crisis

Species are not immutable: they appear, persist for a time, and then disappear, a process to which the fossil record bears ample testimony. Species can be lost either through extinction, in which all individuals die out leaving no progeny, or through pseudoextinction, in which a lineage evolves over time or diverges into two or more daughter lineages. The relative frequency of these two types of extinction is not known, but probably fewer than one in a thousand of all the species that have ever existed are alive today (WCMC, 1992). Extinction is therefore an entirely natural occurrence.

However, the world appears to be experiencing an extinction event comparable in magnitude to many of those indicated in the fossil record. Current and predicted future extinction rates are much higher than the estimated natural or background levels, and most extinctions appear to be caused either directly or indirectly by human activity. Conservationists therefore have sound scientific reasons for concern in addition to the moral, ethical and aesthetic grounds on which the loss of any species may be regarded as a tragedy.

1.2.2 Past extinction events

The fossil record indicates that extinction rates have not been constant. Around 60% of extinctions occurred in a few, relatively short (as measured on the geological time-scale) intervals, known as 'extinction events'. From the start of the Cambrian period, 580 million years ago, to the present day, five major extinction events have been documented, in the late Ordovician, Devonian, Permian, Triassic and Cretaceous periods. The most severe was in the late Permian, 245 million years ago, when the number of marine animal families decreased by 54% and marine animal species may have declined by up to 96% (Erwin, 1990).

The frequency and severity of extinction events since the Cambrian varies between taxonomic groups: thus fish appear to have experienced eight such events, tetrapods six. Whilst some extinction events coincided among groups, others did not. Even when coincidence did occur, the extent to which different groups were affected varied considerably. For example, in the late Cretaceous tetrapods suffered more extinctions than other groups: 36 out of 89 families (40%) in the fossil record disappeared. These families were all members of three major groups which became completely extinct (dinosaurs, plesiosaurs and pterosaurs) while other vertebrate taxa were almost completely unaffected (WCMC, 1992).

Most of these extinction events occurred over periods lasting millions of years (for example the late Permian event probably lasted for 5–8 million years). There is however some evidence that the late Cretaceous extinction event was associated with an extra-terrestrial impact, an instantaneous event in geological terms (Alvarez *et al.*, 1984). During extinction events, extinction rates would have been many times higher than normal.

1.2.3 Background extinction rates

A very coarse estimate of the 'natural' or background rate of animal extinction prior to human influence is 2.5 species per year, calculated by dividing the estimated total number of animal species (approximately 10 million described and undescribed vertebrates and invertebrates) by the average persistence time (around four million years) of species in the fossil record (WCMC, 1992). However, by its very nature the fossil record is probably heavily biased towards successful, wide-ranging species, which undoubtedly have a far longer than average persistence time. If most species persist for less than four million years, background extinction rates will be correspondingly higher than estimated above, perhaps by as much as a factor of 10 (WCMC, 1992). A more realistic estimate of the background extinction rate is therefore 25 species per year for all animals, or roughly 3 species per year for the 1.3 million described species.

1.2.4 Recent extinction rates: documented and estimated

One problem in assessing recent extinction rates is the poor quality of information available from any period more than just a few hundred years ago. Adequate scientific sense can be made of records only from around AD 1600 to the present day. Since then, 491 animal species are known to have become extinct (WCMC, 1992). This is undoubtedly an underestimate of actual extinctions, for several reasons. First, the majority of animal species (almost certainly over 90%) have not yet been described by science. Many species, particularly invertebrates, will therefore have disappeared undocumented before being collected and described. Second, extinction is registered only when a species has not been recorded, despite significant efforts, for a considerable length of time – arbitrarily taken to be 50 years by both the World Conservation Union (IUCN) and the Convention on International Trade in Endangered Species of Wild Fauna and Flora (CITES). Accurate information on current status and abundance is available for only a tiny fraction of the world's described species, so many are certain to have become extinct without our knowledge.

Known extinctions since AD 1600 are concentrated amongst certain taxa. Although vertebrates constitute only approximately 3.5% of described animal species, they account for around 50% of known extinctions (Table 1.1). The majority of these are from the two scientifically best-known taxa, mammals (60 species) and birds (122 species).

Far more extinctions have occurred among birds and mammals than predicted from the estimated background extinction rate. The rate of documented extinctions among described animal species since AD 1600 is about 1.25 species per year, or only 40% of the estimated background extinction rate of three species per year. However, if background extinctions are spread uniformly across all animal species, the expected number of extinctions amongst the 4300 or so living mammals is around four species every 400 years. Amongst the 9700 or so birds only nine extinctions every 400 years would be predicted. Recent documented extinctions in these two taxa are therefore 10–15 times more than expected. If these rates of extinction were to continue, all birds and mammals would be lost in just 30 000 years.

Extinctions among other vertebrate taxa are not as conspicuously high as for birds and mammals, but as a group vertebrates have still experienced five times more extinctions than predicted from the estimated background extinction rate. Conversely, for all invertebrate groups except molluscs fewer extinctions have been documented than predicted.

There are two possible explanations for this finding. Vertebrates may genuinely be more prone to extinction than invertebrates, perhaps because they are typically much larger, and occur at lower densities. There are however a number of counter-arguments which suggest the reverse: e.g. many invertebrates have extremely small ranges which may render them vulnerable to extinction by habitat loss, natural disaster or disease. A widely accepted alternative explanation is that rates of extinction are similar between all taxa, and the documentation of a large number of vertebrate extinctions simply reflects their relative ease of observation and a bias of interest (particularly towards birds and mammals) among biologists.

If the latter is the case, and rates of extinction among all animals are actually similar to those documented among birds and mammals, then up to 18 000 (1.4%) **described** species, most of them invertebrates, may have become extinct since AD 1600. If the **total** number of animal species, including undescribed species, is considered, then up to 140 000 species may have been lost. Although the actual numbers involved will never be known exactly, what is certain is that for the past 400 years the world has been experiencing an extinction event which, if recent trends continue, will be comparable in magnitude to the most dramatic extinction events documented in the fossil record.

1.2.5 Recent extinctions: taxonomic distribution

Historical records of extinctions since AD 1600 are heavily biased, both taxonomically and geographically. Information on extinctions of terrestrial snails, birds, and mammals is good owing to the scientific attention these groups have received, but that for most other taxa is poor. Data for Europe and North America (including Hawaii) are much better than for the rest of the world. These biases make interpretation of recent extinction patterns problematic.

Molluscs are the group with the highest number of documented extinctions (191 species), followed by birds (122 species), mammals (60 species) and insects (59 species). Together these account for roughly 88% of all known animal extinctions. Fewer than 30 extinctions have been recorded for any other group, including reptiles and fish (Table 1.1).

Some 1.4% of recognized mammal species and 1.3% of birds are known to have become extinct, compared with only 0.3% of molluscs and 0.01% of insects (Table 1.1). In total, 0.5% of described vertebrates and 0.02% of invertebrates have become extinct since AD 1600. Thus although there are actually fewer known vertebrate extinctions (236 species) than invertebrate extinctions (255 species), vertebrates have experienced 25 times more extinctions on a proportional basis. A major contribution to this difference is the comparative lack of scientific attention that invertebrates have received.

1.2.6 Recent extinctions: geographic distribution

Most recently extinct species had restricted distributions: 96% were endemic to single countries (Figure 1.1) and the majority (75%) were restricted to islands or island groups. For the three best-documented groups, the proportion of island extinctions varies from 90% for birds to 58% for mammals, with molluscs intermediate at 80%.

Of the continental extinctions, at least 66% can be classed as aquatic species. Very few extinctions have been recorded from continental tropical forest ecosystems, although based on rates of habitat destruc-

Table 1.1 Taxonomy of known animal extinctions since AD 1600: current threat status, and numbers of described species[a]

Phylum (or sub-phylum)	*Class or group*	*Number of species extinct since 1600*	*Number of species threatened*	*Number of species Endangered*	*Number of described species*
Vertebrata	Mammalia	60	507	140	4327
	Aves	122	1029	132	9672
	Reptilia	23	169	38	6547
	Amphibia	2	57	8	4014
	Pisces	29	713	368	21 732
Vertebrate total		236	2475	686	46 292

Ciliophora		0	1	0	?
Cnidaria		1	154	0	9000
Platyhelminthes		0	4	2	12 700
Nemertea		0	10	0	650
Mollusca		191	409	85	70 000
Annelida		0	139	2	8700
Arthropoda	Insects	59	1083	56	950 000
	Merostomata	0	4	0	4
	Arachnida	0	18	1	75 000
	Crustacea	4	126	3	40 000
Onchyophora		0	27	0	65
Echinodermata		0	2	0	6000
Invertebrate total		255	1977	149	*c.* 1 300 000
Animal total		491	4452	835	*c.* 1 300 000

[a]Numbers of threatened and Endangered animal species obtained from IUCN (1990a). Threatened species are defined as those assigned any IUCN threat category except Extinct. Table includes all vertebrate groups, but only those invertebrate groups with one or more threatened species. Numbers of species extinct since AD 1600 obtained from WCMC (1992). Numbers of described species obtained from the following sources (? = data unavailable): amphibians, Frost (1983); birds, Sibley and Monroe (1990); mammals, Corbet and Hill (1991); reptiles, Halliday, Adler and O'Toole (1986); fish, Nelson (1984); Cnidaria, Platyhelminthes, Annelida and Echinodermata, Barnes (1989); molluscs, crustaceans, arachnids, insects (including all hexapods), invertebrate and animal totals, WCMC (1992). The estimated number of species for the whole of the world's biota is 12.5 million, of which very approximately 10 million are animal species (WCMC, 1992).

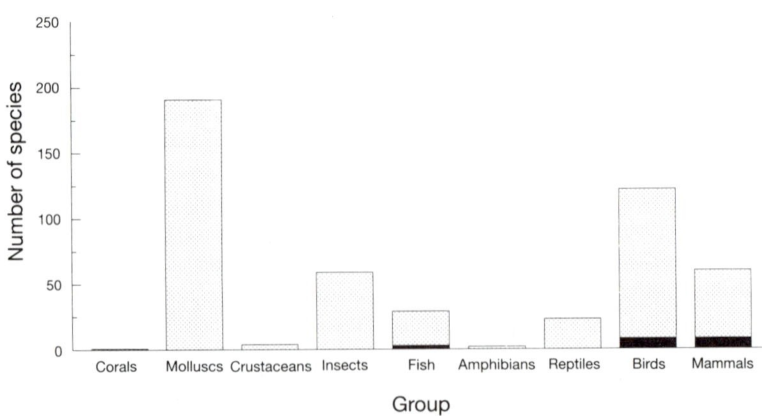

Figure 1.1 Documented animal species extinctions since AD 1600 (source WCMC, 1992).
Key: ■ multiple country □ single country

tion these are where most scientists predict species losses will be concentrated in the future.

There are several reasons why species confined to real or ecological islands (such as isolated inland waters) show high rates of extinction. Their restricted distribution may limit population size, and allow adverse factors to operate on the entire population. Island species are less likely to benefit from immigration or recolonization from neighbouring populations since they are not part of interlinked 'metapopulations'. They may also have evolved ecological adaptations, such as flightlessness and reduced reproductive rates, which render them vulnerable to predation by, or competition from, humans or introduced species. Finally, taxonomists have often classified island populations as full species when they should perhaps more reasonably be regarded as subspecies of species on adjacent islands or on the mainland, thus artificially inflating numbers of species extinctions from islands.

1.2.7 Recent extinctions: causes

In this study, the cause of extinction has not been examined for 55% of recent animal extinctions (most of which are invertebrates). Human activity has contributed to the overwhelming majority of the remaining 45% (Figure 1.2). Most extinctions can be attributed to more than one factor. Introduced animals and direct habitat destruction by man are the two most important causal agents of extinction (39% and 36% of

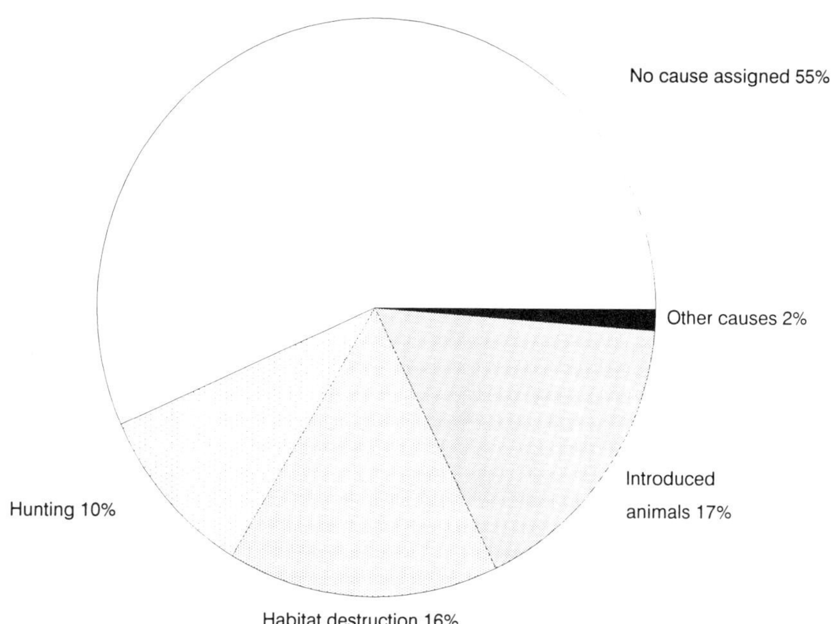

Figure 1.2 Causes of documented animal extinctions since AD 1600 (source WCMC, 1992). Figures were compiled by giving each species a score of 1 in the appropriate category if there was only one cause of extinction, 0.5 in each if there were two, etc.

extinctions with known cause, respectively) followed by hunting and deliberate extermination (23% of extinctions with known cause).

1.2.8 Recent extinctions: time series

Most species cannot be assigned an accurate date of extinction. Two trends are apparent from those for which the extinction date (as measured by the date of last record or estimated date of extinction) is known to within a decade. First, documented extinctions on islands began almost two centuries before those on continents; second, both island and continental extinctions have increased rapidly from early or mid-nineteenth century to the mid-twentieth century (Figure 1.3).

This increase has been more pronounced for continental species, although there are more island than continental extinctions in all periods. The highest number of known extinctions occurred in the late nineteenth century, reflecting the large number of mollusc extinctions on islands during this period.

The apparent decline in rate for both continental and island extinctions for the last 30-year interval, 1960–1989 (Figure 1.3), is probably

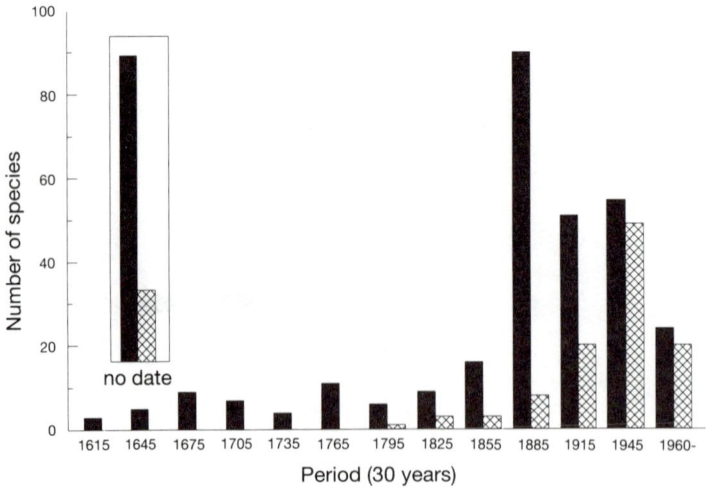

Figure 1.3 Time series of documented and dated extinctions in 30-year intervals since AD 1600 on islands and continents (source WCMC, 1992).
Key: ■ islands ▨ continents

attributable to two causes. First, there is an expected time-lag in documenting extinctions, because species are more likely to be regarded as extinct the longer the elapsed interval since their last record. In fact no species recorded since 1960 could yet be accepted as extinct with strict adherence to the 50-year rule of CITES and IUCN. WCMC (1992) adopted a more flexible approach in compiling its list of extinct species, and those regarded with a high degree of certainty as extinct (for example the Caribbean Monk Seal (*Monachus tropicalis*), last recorded in 1962) were included despite the time rule. Nevertheless a significant number of species have probably become extinct recently without yet being documented as such.

A second, more optimistic, possibility is that the great increase in conservation action over the past 30 years has actually reduced extinction rates. During this time, attention has focused largely on saving well-known species under imminent threat of extinction, and most efforts have succeeded, at least in the short or medium term. Some of these programmes have involved the maintenance of *ex situ* populations of threatened species by the captive breeding community.

1.2.9 Future extinction rates

A number of authors have made predictions about present and future extinction rates, based on extrapolation from estimates of habitat loss

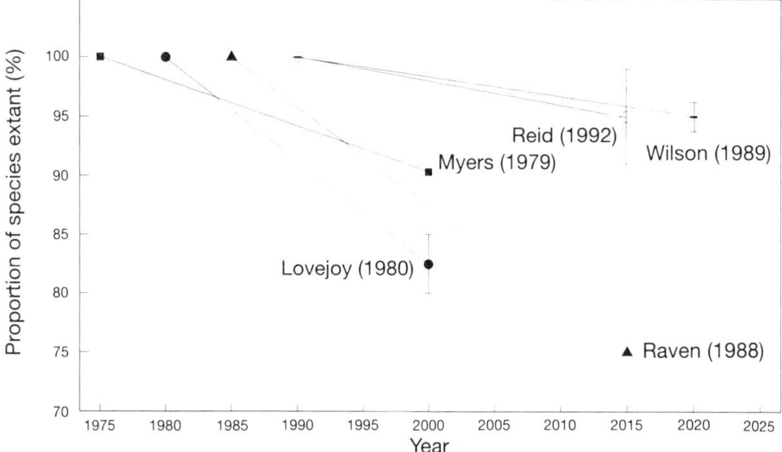

Figure 1.4 Estimated rates of future species extinctions. Most authors give a range for predicted species losses, e.g. 15–20% of species lost between 1980 and 2000. Points plotted show the medians of these ranges; error bars show the ranges themselves. Actual rates of loss may not be linear. Source: WCMC (1992).

coupled with assumptions derived from biogeography. In these predictions, numbers of species remaining in a reduced area of habitat are calculated from the species–area (Arrhenius) relation, in which $S = cz^A$, where S = number of species, A = area, and c and z are constants. Values for z commonly used vary between 0.15 and 0.40. The most widely quoted generalization is that the loss of 90% of a habitat results in the loss of half the species present (WCMC, 1992).

The majority of estimates of future species extinctions focus on tropical moist forest species, which account for most of the world's species diversity. Recent predictions have generally been more optimistic than those made in the early 1980s (Figure 1.4), but Reid (1992) still estimates that between 2% and 8% of all animal species will be lost between 1990 and 2015. Reid stresses (and this applies to other predictions too) that this is the percentage of species 'committed' to eventual extinction as a result of deforestation, not the percentage that will actually become extinct during this time. In many cases there will be a delay between the reduction in the area of habitat and the extinction of species dependent on it, especially for longer-lived species.

The overall trend in extinctions for the two best-studied taxa, birds and mammals (which also have the highest extinction rates of any animal groups), is upwards despite the apparent decline for the period 1960–1989 (Figure 1.5). However, predictions based on these documented extinctions are not as pessimistic as those based on habitat loss.

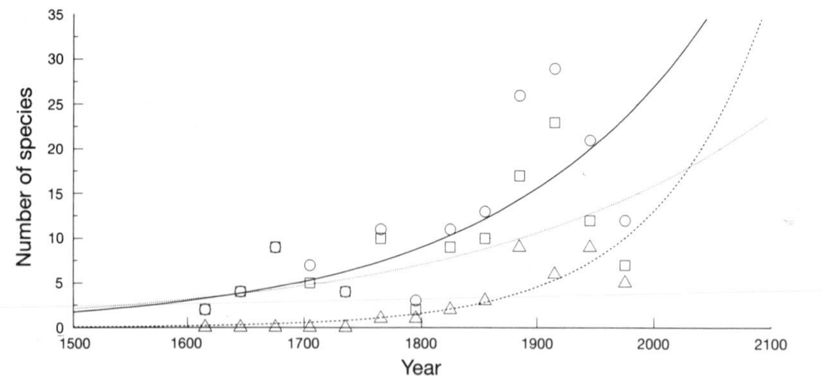

Figure 1.5 Documented bird and mammal extinctions in 30-year intervals since AD 1600 and extrapolated future trends. Lines plotted are exponential curves, which provide a better fit to the data than linear or logarithmic functions. Source: WCMC (1992).

Key: Birds ·□·· Mammals ------△----- Combined ——○——

Extrapolation from the curve predicts that only around 30 species of birds and mammals combined (0.2% of the described total) will be lost between 1975 and 2005. This is just one-tenth of the extinctions predicted if Reid's (1992) minimum estimate of species loss applied to all taxa equally.

There are many theoretical problems with models that predict extinctions based on the species–area curve (see also Vàsàrhelyi and Martin, Chapter 6) and rates of habitat loss, and their predictions should therefore be treated with caution (Heywood and Stuart, 1992). Many treat the loss of equal-sized areas of forest in different locations as having equivalent effects on species extinctions. Since species richness is not evenly distributed throughout the world's tropical forests this assumption is unlikely to be true: for example various authors have identified countries or regions which possess particularly high species diversity (e.g. Mittermeier and Werner, 1990; Myers, 1988; 1990). The proportion of endemic or restricted-range species in a fauna or flora is also highly variable, and the loss of an area of habitat containing a large number of unique species will cause more extinctions than one containing mainly widespread species.

Rates of habitat loss or conversion vary from country to country, and the type of modification may have implications for species extinctions. Thus selective logging may have less effect on a forest's biota than clear-felling and conversion to farmland. The ecology of most tropical forest species is poorly known, so that it is impossible to generalize the extent to which reduction in area or modification of a forest habitat will affect the species present.

Furthermore, most models do not consider the degree of habitat fragmentation, despite the well-known prediction from island biogeography theory that fragmented habitats will suffer more extinctions than expected on the straightforward basis of area loss. Finally, estimates of extinction rate cannot take into account the impact of large-scale changes in environmental conditions, such as climate change, which are likely to profoundly influence species survival (WCMC, 1992). Although actual rates of species extinction are therefore open to debate, all authors are agreed that the current extinction crisis will continue to worsen, at least for the foreseeable future.

1.3 THREATENED SPECIES

1.3.1 The *IUCN Red List of Threatened Animals*

The growth in public awareness of the problem of depletion and possible extinction of species is largely attributable to the development of the Red Data Book concept, pioneered by Sir Peter Scott during the 1960s. Red Data Books are now complemented by the Action Plans produced by the Specialist Groups of IUCN's Species Survival Commission (SSC), and by the *IUCN Red List of Threatened Animals* (IUCN, 1990a). The latter is the only accepted global list of threatened animal species.

The Red List is based on information provided by scientists, naturalists, and conservationists, much of it collated by IUCN SSC Specialist Groups. Information on threatened birds is coordinated by Birdlife International, formerly known as the International Council for Bird Preservation (ICBP).

The Red List has been compiled since 1986 by the World Conservation Monitoring Centre. Each taxon included is assigned to one of eight threat categories determined by a review of its conservation status throughout its range. Key factors examined include changes in distribution or numbers, degree and type of threat, and population biology. In the following analysis and discussion, the term 'threatened' includes taxa listed as Insufficiently Known (K) which are suspected but not definitely known to be threatened.

1.3.2 IUCN threatened species: what do we know?

A total of 4452 extant species are listed as threatened in the 1990 Red List, (*c.* 0.5% of described animal species). In addition, some extinct species and many threatened subspecies are listed. The two groups with the greatest number of threatened species are birds, with 1029, and

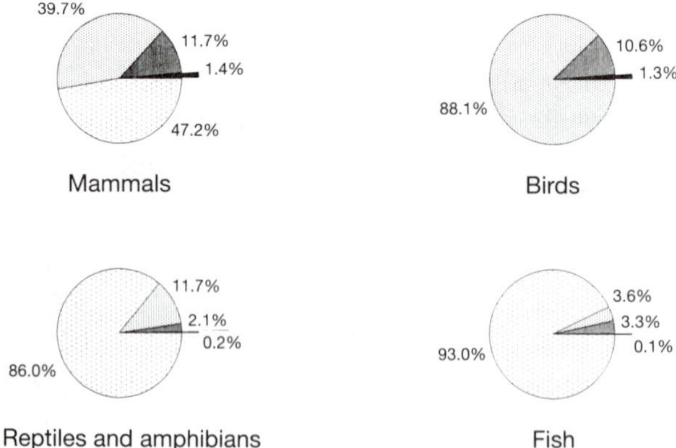

Figure 1.6 Assessment of the conservation status of vertebrate species. Proportions of IUCN threatened species and extinct species were calculated from data in Table 1.1; proportions of each group assessed for threatened status are WCMC estimates.
Key: ■ Extinct since AD 1600 ▩ Threatened □ Assessed but non-threatened □ Not assessed

insects with 1083 (Table 1.1). Clearly, the number of threatened species in a taxon is not directly proportional to the overall number of species: some taxa, particularly vertebrates, have relatively more than others. Mammals are the most threatened vertebrate group, with 11.7% of species threatened (Figure 1.6). Around 5.3% of all vertebrates are threatened.

In comparison, only some 0.15% of described invertebrates are threatened. The large number of insects listed represents just 0.1% of the world's total, and many invertebrate groups are not included in the Red List at all. Thus although only 1.25 times more vertebrate than invertebrate species are actually listed, on a proportional basis 35 times as many are threatened.

The dichotomy between vertebrate and invertebrate listings is apparent not only in the number of listings but also in the categories of threat assigned. Each of the five main vertebrate groups has a higher percentage of Endangered species (the most at risk of the seven categories of threat) than any of the invertebrate taxa (Figure 1.7). In total, 4.6 times more vertebrates than invertebrates are listed as Endangered, or 130 times as many on a proportional basis.

1.3.3 Threatened species: fact or bias?

Just as for documented extinctions, the huge differences between the proportions of vertebrate and invertebrate taxa listed as threatened

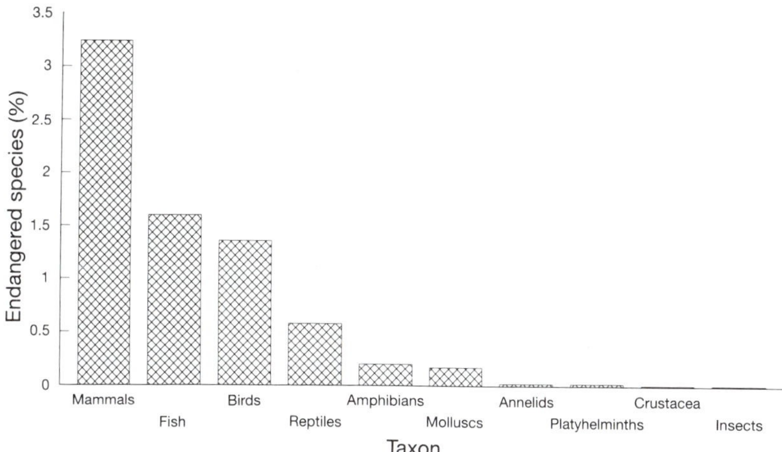

Figure 1.7 Proportion of species in various taxa listed as Endangered by IUCN (1990a). Calculated from data in Table 1.1.

could have two explanations. Either the allocation of threat categories accurately reflects the real biological situation, or it merely mirrors scientific interests and priorities.

Threat categories are usually determined or influenced by IUCN SSC Specialist Groups. The number of members in these groups can be taken as an approximate measure of conservation interest in each taxon (although it should be noted that membership policies may vary between groups, and some are more inclusive than others). Figure 1.8 suggests that taxa with lower proportions of threatened species are simply those that have aroused the least attention among conservationists. This hypothesis is supported by a recent estimate that there is one mammalogist for every seven mammal species, but only one entomologist for every 425 described insect species (Wheeler, 1990, cited in Lewin, 1991).

1.3.4 IUCN threatened species: what do we need to know?

The IUCN Red List is designed to be continually updated as new information becomes available. The number of taxa (i.e. species and subspecies) assigned Red List threat categories has grown by 60% between 1986 and 1990 (Figure 1.9) and is expected to continue increasing as scientific knowledge of the conservation status of the world's fauna improves.

Many scientists and conservationists regard the existing IUCN threat category definitions as largely subjective, and believe that, as a result, categorizations made by different authorities vary and may not actually

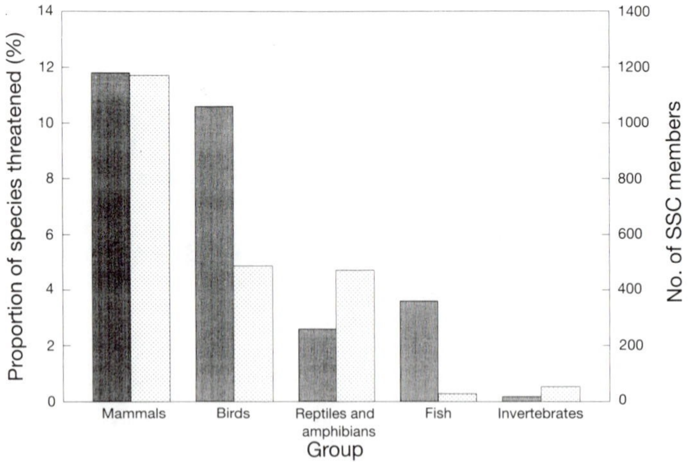

Figure 1.8 Comparison of proportion of threatened species with IUCN/SSC Specialist Group membership for various taxa. Proportions of threatened species calculated from Table 1.1, numbers of SSC members obtained from IUCN (1990b).
Key: ▩ % Threatened □ No. SSC members

reflect real extinction risks. Mace and Lande (1991) have recently proposed an alternative, but not yet universally accepted, system based on quantitative population viability analysis. If the Mace–Lande criteria are adopted, the number of species considered threatened may increase substantially. For example, 19 of the world's 234 waterfowl taxa are listed under the present IUCN classification, but 77 would be considered

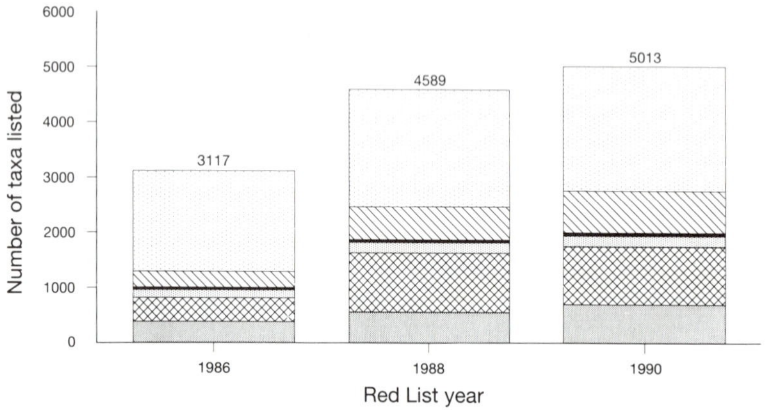

Figure 1.9 Recent growth in numbers of taxa assigned Threatened status in IUCN Red Lists (IUCN, 1986, 1988 and 1990a).
Key: ▩ Mammals ⊠ Birds ▤ Reptiles ■ Amphibians ⧅ Fish □ Invertebrates

threatened under the Mace–Lande criteria (Ellis-Joseph, Hewston and Greene, 1992).

Factors such as the uneven global distribution of species, conservation activity, and field work affect the degree to which the Red List has been completed for different taxa. With the exception of birds, for which ICBP instituted a world-wide review of the conservation status of all species, no vertebrate group has been fully assessed for inclusion in the Red List (Figure 1.6). The number of invertebrate species reviewed for inclusion is not known, but is undoubtedly an extremely small percentage of those described.

Two interdependent biases may be involved in the review process. There is apparently a tendency for 'sizeism': larger animals evoke more non-scientific interest, are more easily surveyed, and are more often regarded as flagship or keystone species. As an illustration, mammalian orders consisting mainly of larger species, such as the Cetacea and Artiodactyla, have been comprehensively reviewed for threatened status while those consisting of mainly smaller species, such as Rodentia, Insectivora and Chiroptera have not (Table 1.2).

In addition to body size, the level of scientific interest in taxa may affect assessment for potential threatened status, as appears to be the case for the proportions of taxa that are actually listed (see above). The most poorly assessed mammalian orders tend to be those with the fewest Specialist Group members (Figure 1.10), although a plausible alternative explanation is that they are also the orders with the largest numbers of species.

Progress in reviewing the conservation status of vertebrate taxa is being made by the SSC via a series of Conservation Assessment and

Table 1.2 Large mammalian orders not fully assessed for inclusion in the 1990 IUCN red list[a]

Mammalian order	*Number of species*		
	Threatened	*Assessed*	*World total*
Marsupialia	25	*c.* 200	*c.* 282
Insectivora	79	*c.* 230	*c.* 365
Chiroptera	45	*c.* 150	*c.* 977
Rodentia	54	*c.* 800	*c.* 1793
Mammalian total	507	*c.* 2280	*c.* 4327

[a]Numbers of threatened species obtained from IUCN (1990a). Threatened species are defined as those assigned any IUCN threat category except Extinct. World species totals for orders of mammals obtained from Corbet and Hill (1991). Numbers of species assessed for potential threatened status are WCMC estimates.

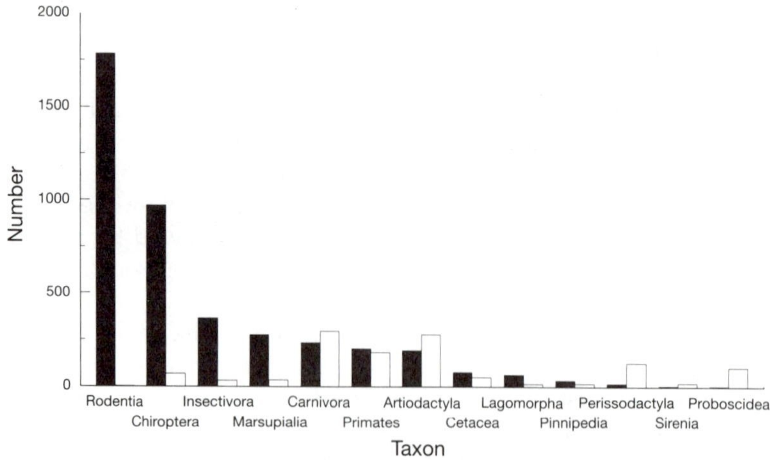

Figure 1.10 Relation between numbers of species in selected mammalian orders and SSC Specialist Group membership. Sources: Corbet and Hill (1991), IUCN (1990b).
Key: ■ Total no. of species □ No. of SSC members

Management Plan (CAMP) workshops, involving collaboration between the Captive Breeding Specialist Group (CBSG) and appropriate taxa-based Specialist Groups (Seal, Foose and Ellis-Joseph, Chapter 16). These aim to assess the status of selected taxa using the Mace–Lande criteria, in preference to the existing IUCN threat categories. To date CAMP workshops have been conducted for several groups, including parrots and waterfowl.

In summary, the Red List is the most comprehensive listing of globally threatened species available. It is a valuable tool for conservationists, and efforts should be made both to completely assess all higher vertebrate taxa for potential threatened status, and to increase the proportion of invertebrate species reviewed.

1.4 CAPTIVE BREEDING

1.4.1 Historical and future roles of zoos and aquaria

The principal institutions holding *ex situ* populations of animal species for captive breeding purposes are zoos and aquaria. Although most originated as menageries for public entertainment, and to some extent education, they are increasingly turning their attention to conservation. Captive populations of animals can play a significant conservation role as demographic and genetic reservoirs from which infusions of 'new blood' may be obtained for wild populations, as sources from which new populations can be founded, and as last redoubts for species which have no immediate chance of survival in the wild. If the captive breeding community is to succeed in its stated education and conservation aims it is important that efforts should be directed on both a geographic and taxonomic basis towards areas of need. In this section, past successes and the allocation of resources within the captive breeding sector are examined, and recommendations made for the future conservation priorities of zoos and aquaria.

1.4.2 Geographic distribution of zoos and aquaria

There are approximately 1150 zoos, aquaria and affiliated captive facilities worldwide (Foose, pers. comm.) of which 878 reported their addresses to the International Zoo Yearbook (IZY) (Olney and Ellis, 1991). At least 83 countries possess one or more of these institutions.

The overall geographic distribution of zoos and aquaria is very uneven: 573 (or 65%) are located in the developed world. Europe has

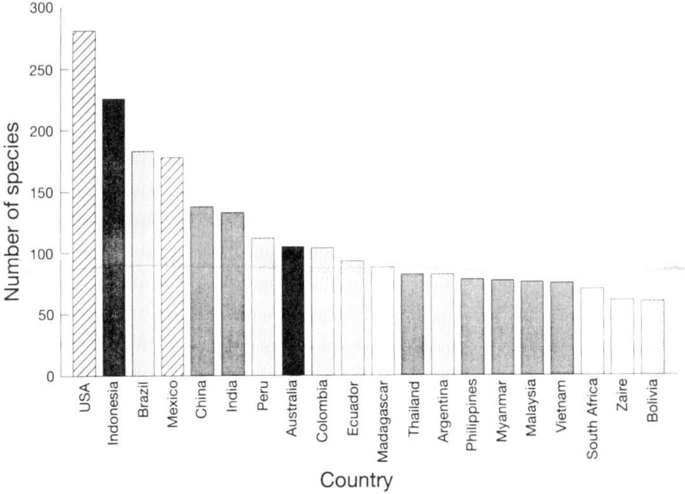

Figure 1.11 Top 20 countries containing the highest numbers of IUCN threatened vertebrate species (source WCMC, 1992). USA includes Hawaii but not dependent territories. Marine and extinct species, and about 252 species of Lake Victoria cichlids (many of which are believed to be severely threatened) are excluded. Threatened bird species are only attributed to countries within their breeding and wintering ranges. Figures for USA and Mexico are inflated by large numbers of threatened fish species (164 and 98, respectively): the ichthyofaunas of other countries may not have been as comprehensively reviewed for threatened status.
Key: ▨ Central and North America ▩ Asia ⬚ South America ■ Oceania □ Africa

298 institutions, the USA 160, Canada 24, Australia 17, New Zealand 8 and Japan 66. With the notable exceptions of the USA, with over 250 threatened vertebrate species (of which 164 are fish), and Australia (with over 100), these developed countries are areas of low species richness with relatively low numbers of threatened species, although these may comprise a high proportion of the total fauna. In contrast, those tropical countries with generally high species diversity and large numbers of threatened species (Figure 1.11) have few or no zoos or aquaria: the whole continent of Africa has just 32 institutions, South America 29, Central America 16, and Asia (excluding China and Japan) 55.

Surprisingly, this divide is not reflected in the sizes of zoological collections. The developing countries together hold 32% of all vertebrate specimens, in 35% of the reporting institutions. However, most of these institutions are poorly developed and underfunded.

Captive breeding institutions can play an important role in mobilizing public awareness of conservation issues, and the fact that so few are located in the developing world where major species losses are expected is therefore of concern. Given adequate funding, local institutions may be better placed than overseas ones to breed and maintain native species with particular ecological requirements, and to provide environmentally adapted specimens for eventual reintroduction projects. Support for local captive breeding facilities in developing countries should therefore be increased, possibly through collaborative ventures with captive breeding programmes in the developed world.

1.4.3 Captive conservation: successful preservation

During the past 30 years some of the species most at risk of extinction have been saved, at least in the short term, thanks to direct manipulative intervention. *Ex situ* preservation is just one of many options open to wildlife managers, but often represents the only hope for species near extinction in the wild. As an example, several projects have taken some of the last wild individuals of gravely threatened species into captivity (e.g. Arabian oryx [*Oryx leucoryx*] and Californian condor [*Gymnogyps californianus*]) and built up numbers until conditions were suitable for the re-establishment of free-living populations.

However, as yet the number of species that has actually been saved from extinction by captive breeding programmes is small (Table 1.3). Since AD 1600, 427 mollusc and vertebrate species have become extinct, 156 of them this century. Only 25 species have been preserved in captivity following extinction in the wild (a mere 14% of potential extinctions this century), and only seven of these (4%) have been successfully preserved and then reintroduced to areas of their former range.

Table 1.3 Captive breeding successes[a]

Group	*No. of threatened species*	*No. of threatened species held in captivity*	*Threatened species surviving only in captivity*	*Species extinct in the wild but reintroduced from captive bred populations*
Molluscs	409	?	11 *Partula* spp.	None
Fish	713	?	*Cyprinodon alvarezi* *Megupsilon aporus* *Xiphophorus couchianus* *Skiffia francesae*	None
Amphibians	57	5	None	None
Reptiles	169	34	None	None
Birds	1029	32	*Zenaida graysoni*	*Gymnogyps californianus* *Rallus owstoni*
Mammals	507	174	*Equus ferus* *Bos taurus*	*Mustela nigripes* *Canis rufus* *Bison bonasus* *Oryx leucoryx* *Elaphurus davidianus*

[a]Only fully threatened species are considered, numbers of which were obtained from IUCN (1990a). Numbers of species extinct in the wild but preserved in captivity or preserved and reintroduced taken from WCMC (1992). Data for numbers of threatened species held in captivity taken from the IZY Census of Rare Animals in Captivity 1989/1990 (Olney and Ellis, 1991); data for mammals supplemented by International Species Inventory System (ISIS) records for 31 December 1991 (Flesness, pers. comm.). The IZY Census underestimates the numbers of threatened species held, since some mammals, birds, and reptiles considered threatened in their natural habitats are common in captivity, breed regularly, and are therefore omitted from the census. The IZY Census does not cover rare fish or molluscs.

There are, though, a number of other species, such as the Bali starling (*Leucopsar rothschildi*), for which captive breeding efforts have contributed significantly to the viability of the remaining wild population (see van Balen and Gepak, Chapter 24). Both this number and the number of species maintained solely in captivity are likely to continue growing in the foreseeable future. For example, Heywood and Stuart (1992) estimate that by the year 2015 nearly half (450) of the world's 1029 threatened bird species will be 'committed to extinction' (i.e. will face inevitable extinction in the wild unless major conservation action such as habitat restoration, reintroduction, elimination of introduced predators, captive breeding, etc. is undertaken). The maintenance of viable captive populations of most, if not all, threatened species should therefore be an important component of the strategy to preserve the world's biodiversity.

1.4.4 Analysis of threatened species held in captivity

Institutions which reported their specimen numbers to the International Zoo Yearbook (Olney and Ellis, 1991), which maintains the most complete global record available, collectively held approximately 1 232 000 vertebrate specimens as of 31 December 1989. Nearly half (584 000) of these were fish. The numbers of other taxonomic groupings held were mammals (202 000 specimens), birds (351 000), reptiles (74 000) and amphibians (21 000).

However, only a small percentage of these specimens actually contribute to the *ex situ* conservation of biological diversity, since surprisingly few threatened species are held. Of the world's threatened tetrapod species only around 9% of amphibians, 20% of reptiles, 3% of birds and 34% of mammals are held in captivity (Table 1.3). Zoos and aquaria undoubtedly have considerable capacity to preserve threatened species, but to date efforts have been relatively limited.

1.4.5 Analysis of threatened mammals held in captivity

The situation for captive mammals can be analysed further, since more data are available than for other vertebrate groups. The following analysis is based on a combination of data from the IZY Census of Rare Animals in Captivity 1989/1990 (Olney and Ellis, 1991), and ISIS (International Species Information System) records for 31 December 1991 (Flesness, pers. comm.). Where the two systems report different numbers of a threatened taxon held or bred in captivity, the higher figure is taken in each case. Approximately 30 300 specimens of threatened mammal species are kept in zoos, of which at least 21 350 (70%) are captive-bred. Only some 15% of zoo 'mammal spaces' are therefore

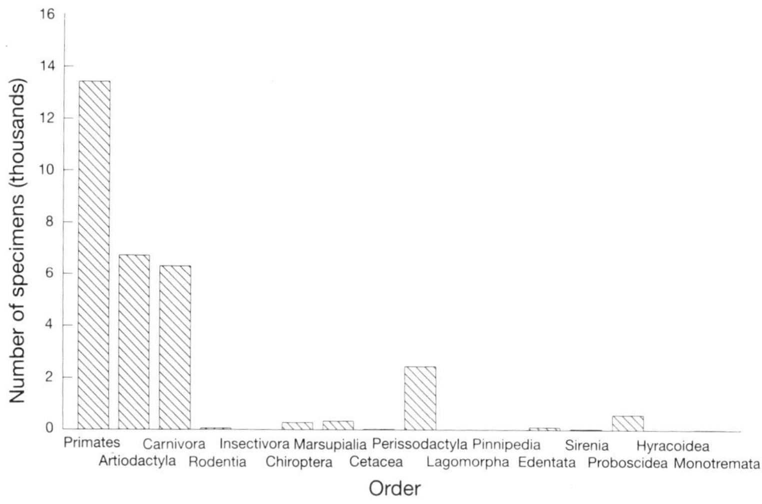

Figure 1.12 Number of specimens of IUCN threatened mammalian taxa held in captivity. Data obtained from the International Zoo Yearbook Census of Rare Animals in Captivity 1989/1990 (Olney and Ellis, 1991), and from International Species Inventory System records for 31 December 1991 (Flesness, pers. comm.). Where the two systems gave different numbers for particular taxa, the larger was used. These figures are nevertheless underestimates, since some captive breeding institutions do not report to either system.

devoted to threatened taxa, since the global zoological capacity for mammals is around 200 000 specimens (Olney and Ellis, 1991). The taxonomic distribution of specimens held is very uneven (Figure 1.12). The vast majority (87%) belong to just three orders: Primates, Artiodactyla and Carnivora. In contrast only four insectivore and four lagomorph specimens are registered.

Zoos hold *ex situ* populations of 32% of extant threatened mammalian taxa (i.e. species and subspecies, see Table 1.4). All threatened Proboscidea and Monotreme taxa, 87% of Perissodactyla, and 75% of Sirenia are held in captivity, while less than 60% of any other order is represented (Figure 1.13). Only around 50% of threatened Primates, Artiodactyla, and Carnivora are represented, which is surprising given the public appeal of these taxa and the large number of specimens held. Orders consisting mainly of smaller mammals such as Rodentia, Insectivora, Chiroptera and Lagomorpha are the most poorly represented. Although there are obvious difficulties associated with keeping some taxa, such as Cetacea, in captivity, those taxa which are not held are mostly small animals in which the public and scientists show little interest.

Zoos hold relatively more taxa and specimens of the taxa assigned to the highest risk categories – Endangered and Vulnerable. IUCN (1990a)

Table 1.4 Analysis of captive-held mammalian taxa by threat category[a]

IUCN threat category	*No. extant taxa*	*No. taxa in captivity*	*% extant taxa in captivity*	*No. of specimens in captivity*
Extinct* (Ex)	2	2	100	967
Endangered (E)	225	85	38	12 262
Vulnerable (V)	167	90	54	13 847
Rare (R)	69	15	22	1532
Indeterminate (I)	61	8	13	336
Insufficiently Known (K)	133	12	9	732
Threatened (T)	1	1	100	656
Total	658	213	32	30 332

[a]Numbers of extant threatened taxa (including species and subspecies) taken from IUCN (1990a). Numbers of taxa and specimens held in captivity, obtained from combining the IZY Census of Rare Animals in Captivity 1989/1990 (Olney and Ellis, 1991) with data from ISIS for 31 December 1991 (Flesness, pers. comm.): where both listed the number of specimens of a threatened taxon held or bred in captivity the higher figure was used. These figures are nevertheless underestimates since some captive breeding institutions do not report to either IZY or ISIS. *Populations of the North China Sika (*Cervus nippon mandarinus*) (listed as Ex) and Przewalski's Horse (*Equus przewalskii*) (listed as Ex?) are held in captivity.

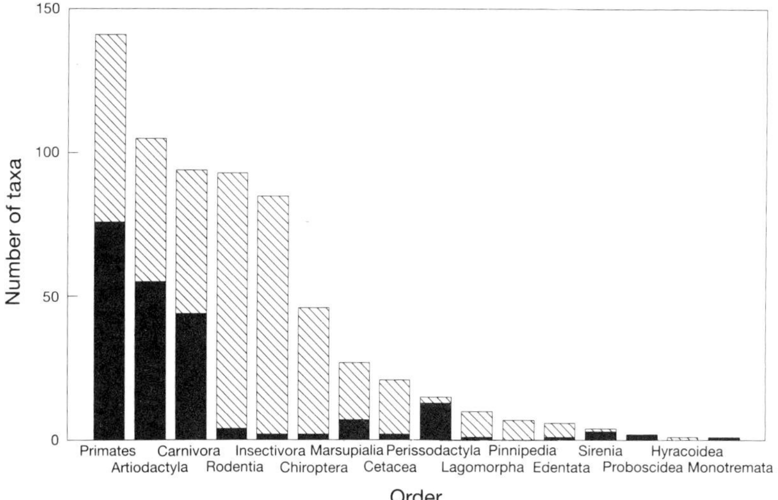

Figure 1.13 Numbers of IUCN threatened mammalian taxa held in captivity. Total number of extant threatened taxa (i.e. including both species and subspecies) taken from IUCN (1990a). Numbers of threatened taxa taken by combining the IZY Census of Rare Animals in Captivity 1989/1990 (Olney and Ellis, 1991) with ISIS records for 31 December 1991 (Flesness, pers. comm.).
Key: ■ Held in captivity ▧ Not held in captivity

categorizes 34% of threatened mammalian taxa as Endangered, and 25% as Vulnerable. In comparison, 40% of threatened taxa and specimens held in zoos are Endangered, while 42% of taxa and 46% of specimens are Vulnerable (Table 1.4).

The numbers of many threatened taxa held are too low to ensure their successful preservation in captivity (Table 1.5). Around 50% have populations of less than 50 specimens, while only 6.5% exceed 500 specimens. Many of the taxa with small captive populations are amongst those most severely threatened in the wild: 58% of Endangered taxa and 43% of Vulnerable taxa held have captive populations of fewer than 50 specimens.

Table 1.5 Captive population sizes of threatened mammalian taxa[a]

Size of known captive population	1	2–9	10–49	50–99	100–249	250–499	>500
No. of threatened taxa (Total = 213)	11	45	53	26	41	23	14

[a]Details as in Table 1.4.

In the short term an effective population size of 50 is the minimum necessary to prevent inbreeding depression for a population of large mammals with no immigration or introduction of unrelated stock (Franklin, 1980). Such a population will nevertheless become inbred over time, at a rate directly related to the generation interval. In the long term an effective population size of 500 (corresponding to an actual population several times larger) may be a more realistic 'genetic' minimum viable population (Franklin, 1980; Lande and Barrowclough, 1987). In a population of this size mutation is expected to renew genetic variation as quickly as it is lost by inbreeding and genetic drift. Whilst suitably detailed and supervised captive breeding programmes can help avoid the deleterious effects of inbreeding, it would still appear that the captive populations of most taxa are too small to ensure the long-term preservation of their genetic diversity.

In summary, although some 507 (12%) of the world's mammalian species are considered threatened, only 174 (34%) of these are represented in captivity. Around 15% of the mammalian places in zoos are devoted to threatened taxa, roughly what one would expect if zoos were placing no special emphasis on holding and breeding threatened mammals at all. Moreover, zoos included in the IZY and ISIS censuses currently maintain captive populations of no more than 90 threatened mammal species (17% of the total) and perhaps a further 15 threatened subspecies that in the absence of future genetic interchange with wild populations will be viable in the long term.

1.5 DISCUSSION

The world is currently facing an unprecedented extinction crisis. The list of documented extinctions, caused mainly by human activities, is increasing exponentially. Estimating the number of species on earth and predicting future rates of species extinction does nothing to address the causes of the current crisis, but is none the less crucial to convince a sceptical public and politicians of the scale of the problem. Meanwhile, each new edition of the Red List catalogues more, not fewer, taxa threatened with extinction, as human activities jeopardize the survival of increasing numbers of species, and scientific knowledge of the threats facing the world's fauna improves. Faced with this situation, *in situ* conservation efforts must be increasingly focused on localities of high species diversity and endemism, while captive breeding resources must be re-allocated to the *ex situ* preservation of threatened taxa.

The captive breeding community is a powerful and influential component of the world conservation scene, but its efforts are often criticized as a misallocation of scarce resources. The world's zoos and aquaria spend US$3 billion annually, while in comparison the management budget of the 4300 protected areas of the developing world is just US$500 million (Foose, pers. comm.). Critics suggest that the overall cause of conservation could be substantially enhanced if much of the funds available to the zoo community were used to support the *in situ* protection of habitats and wildlife.

Captive breeding is widely regarded as less cost-effective than *in situ* preservation. For example, Leader-Williams (1990) calculated that keeping African elephants and black rhinos in zoos was 50 times more expensive than protecting equivalent numbers in the wild. In addition, the maintenance of captive populations does not have the associated benefits of protecting an animal's habitat, and by extension many other species.

One way for zoos and aquaria to counter these criticisms is to expand their contribution to conservation. Captive breeding has so far saved only a handful of species from total extinction, but could potentially save many others for which *in situ* efforts alone may prove to be inadequate. A significant number of institutions have already increased their commitment to conservation via improved international cooperation; the clearer setting of priorities for breeding threatened species; and the devotion of larger fractions of their budgets to field conservation. Coordination of these efforts is carried out through a variety of interconnected mechanisms, including studbooks, the CBSG, ISIS, and a number of regional cooperative captive breeding programmes (de Boer, Chapter 15).

However, much remains to be done. Captive breeding of threatened animals has until now been most successful for the 'charismatic declining megafauna', which evoke widespread public interest. Institutions need to expand their programmes to include threatened taxa with less popular appeal, such as amphibians and small mammals (see also Seal, Foose and Ellis, Chapter 16). Captive breeding priorities could perhaps best be determined by reference to the Red List (IUCN, 1990a) and the distribution and causes of past extinctions. Red List taxa are categorized according to their perceived degree of threat. Other factors, such as probability of success and cost, being equal, taxa assigned to the most threatened category (Endangered) should take priority in captive breeding programmes over those assigned to less threatened categories such as Insufficiently Known or Rare. Prioritization within a particular threat category could be based on the current conservation status of taxa, coupled with analysis of past extinctions. The latter shows that taxa most at risk of extinction are likely to be single-country endemics confined to islands or aquatic habitats, threatened by introduced animals and habitat destruction.

In the public imagination zoos and aquaria are still all too often identified as animal prison camps. They need not only to promote and publicise their conservation-oriented activities, but to increase the allocation of resources to their breeding programmes for threatened taxa. Given the scale of the current extinction crisis, the magnitude of the threats facing many animal species, and the present mood of environmental awareness, this may be the prescription not only to safeguard the future of threatened animals, but that of zoos and aquaria themselves.

ACKNOWLEDGEMENTS

The authors would like to thank Robert Cubey, Crawford Allan and Neil Cox for assistance in the preparation of figures and data analysis, and Nathan Flesness for the generous contribution of unpublished ISIS data. Georgina Green, Robin Pellew, and Simon Stuart are thanked for their comments on an earlier draft of this paper.

REFERENCES

Alvarez, W., Kauffman, E.G., Sutlyk, F., Alvarez, L.W. and Asaro, F. (1984) Impact theory of mass extinctions and the invertebrate fossil record. *Science*, **223**, 1135–41.

Barnes, R.D. (1989) Diversity of organisms: how much do we know? *American Zoologist*, **29**, 1075–84.

Corbet, G.B. and Hill, J.E. (1991) *A World List of Mammalian Species*, 3rd edn, Oxford University Press, Oxford.

Ellis-Joseph, S., Hewston, N., and Green, A. (1992) *Global Waterfowl Conservation Assessment and Management Plan*, The Wildfowl and Wetlands Trust, Slimbridge, UK.

Erwin, D.H. (1990) The End-Permian mass extinction. *Annual Review of Ecology and Systematics*, **21**, 69–91.

Franklin, I.R. (1980) Evolutionary change in small populations, in *Conservation Biology: An Evolutionary–Ecological Perspective*, (eds M.E. Soulé and B.A. Wilcox), Sinauer Associates, Sunderland, Mass., pp. 135–49.

Frost, D.R. (1983) *Amphibian Species of the World, A Taxonomic and Geographical Reference*, Allen Press, Lawrence, USA.

Halliday, T., Adler, K. and O'Toole, C. (1986) *The Encyclopaedia of Reptiles and Insects*, Equinox, Oxford.

Heywood, V.H. and Stuart, S.N. (1992) Species extinctions in tropical forests, in *Tropical Deforestation and Species Extinction*, (eds T.C. Whitmore and J.A. Sayer), Chapman & Hall, London, pp. 91–117.

IUCN (1986) *1986 IUCN Red List of Threatened Animals*, IUCN, Gland, Switzerland, and Cambridge, UK.

IUCN (1988) *1988 IUCN Red List of Threatened Animals*, IUCN, Gland, Switzerland, and Cambridge, UK.

IUCN (1990a) *1990 IUCN Red List of Threatened Animals*, IUCN, Gland, Switzerland, and Cambridge, UK.

IUCN (1990b) *Species Survival Commission Membership Directory*, IUCN, Gland, Switzerland, and Cambridge, UK.

Lande, R. and Barrowclough, G.F. (1987) Effective population size, genetic

variation, and their use in population managem
Conservation (ed. M.E. Soulé), Cambridge Univer
pp. 87–124.

Leader-Williams N. (1990) Black rhinos and Afr
conservation funding. *Oryx*, **24** (1), 23–9.

Lewin, R. (1991) Too many insects, and not enough
1751, 31.

Lovejoy, T.E. (1980) A projection of species extinctions, in *The Globa... Report to the President*, Vol. 2, Council on Environmental Quality, Washington DC, pp. 328–31.

Mace, G.M. and Lande, R. (1991) Assessing extinction threats: toward a re-evaluation of IUCN threatened species categories. *Conservation Biology*, **5** (2), 148–57.

Mittermeier, R.A. and Werner, T.B. (1990) Wealth of plants and animals unites 'megadiversity' countries. *Tropicus*, **4** (1), 1,4–5.

Myers, N. (1979) *The Sinking Ark: A New Look at the Problem of Disappearing Species*, Pergamon Press, Oxford.

Myers, N. (1988) Threatened biotas: 'hot spots' in tropical forests. *The Environmentalist*, **8** (3), 187–208.

Myers, N. (1990) The biodiversity challenge: expanded hot-spots analysis. *The Enviromentalist*, **10**, 243–56.

Nelson, J.S. (1984) *Fishes of the World*, 2nd edn, John Wiley & Sons New York.

Olney, P.J.S. and Ellis, P. (eds) (1991) 1990 *International Zoo Yearbook Volume 30*, Zoological Society of London, London.

Raven, P.H. (1988) Our diminishing tropical forests, in *Biodiversity*, (eds E.O. Wilson and F.M. Peter), National Academy Press, Washington, DC, pp. 119–22.

Reid, W.V. (1992) How many species will there be?, in *Tropical Deforestation and Species Extinction*, (eds T.C. Whitmore and J.A. Sayer), Chapman & Hall, London, pp. 55–73.

Sibley, C.G. and Monroe, B.L. (1990) *Distribution and Taxonomy of Birds of the World*, Yale University Press, New Haven, USA.

WCMC (World Conservation Monitoring Centre) (1992) *Global Biodiversity: Status of the Earth's Living Resources*, Chapman & Hall, London.

Wheeler, Q. (1990) Insect diversity and cladistic constraints. *Annals of the Entomological Society of America*, **83**, 1031.

Wilson, E.O. (1989) Threats to biodiversity. *Scientific American*, **261** (3), 108–16.

2

The effective use of flagship species for conservation of biodiversity: the example of lion tamarins in Brazil

J.M. Dietz
University of Maryland, Maryland, USA
L.A. Dietz
World Wildlife Fund-US, Washington, DC, USA
and
E.Y. Nagagata
Michigan State University, Michigan, USA

2.1 INTRODUCTION

A major role of modern zoological parks is to maintain for future reintroduction some 2000 species of larger vertebrates that are likely to become extinct in nature as the result of large-scale habitat changes over the next few hundred years (Soulé *et al.*, 1986). One assumption necessary to fulfill this goal is that there will be a natural community waiting to receive the species to be reintroduced. This assumption may not be warranted for two reasons. First, factors which caused the extinction of large vertebrate species are also likely to have had a negative impact on the remainder of the ecosystem. For example, deforestation in the tropics, the major agent of extinction in this century (Erwin, 1988), ultimately results in the localized extinction of most resident plants and animals. Second, the removal of larger vertebrates, particularly those at the top of the food web, may cause additional waves of extinction in

Creative Conservation: Interactive management of wild and captive animals.
Edited by P.J.S. Olney, G.M. Mace and A.T.C. Feistner.
Published in 1994 by Chapman & Hall, London.
ISBN 0 412 49570 8

complex natural systems (Pimm, 1991). For example, the decline of elephants in South Africa has resulted in forestation of large areas of savannah habitat, a process which may lead to the localized extinction of several species of grazing herbivores (Owen-Smith, 1989). The resulting assemblage of plants and animals will hardly resemble the African savannah ecosystem and may be unsuitable for eventual reintroduction of elephants.

Some one million 'less-glamorous species', mostly plants and arthropods, are predicted to become extinct in each of the coming decades (Reid and Miller, 1988). These species comprise the ecological communities and habitats into which large vertebrates may one day be reintroduced. The only way to guarantee a future for many of these species may be to focus conservation efforts on the habitats in which they are, or were, found. Allowing for some simplistic assumptions, conservation efforts on behalf of each large vertebrate must also guarantee the survival, on average, of an additional 2500 species necessary for it to survive and evolve *in situ*. Quoting Devra Kleiman (Weinberg, 1992): 'If our goal is to protect, preserve and restore habitat, then the monkeys will naturally follow'. The question of interest then becomes one of how best to conserve intact habitats.

The term 'flagship species' has recently been used to describe an animal that stands for or promotes conservation in a general or regional sense (e.g. Mittermeier, 1986, 1988; Mallinson, 1991). For example, the giant panda used in the logo of the World Wide Fund for Nature may be considered a flagship species for nature conservation efforts worldwide, particularly for securing financial support. From a practical standpoint, the image helps to focus attention on a single species, rather than on the fuzzy concept of species richness, or fuzzier concept of genetic diversity. Flagship species, then, are those that when conserved *in situ* result in the conservation of a significant number of other species across a wide array of taxonomic groups, and in functioning natural systems. Mallinson (1991) writes: 'it is easier to generate interest and pride in a spectacular national animal, like a lion tamarin, than in attempting to communicate to the rank and file the complicated ecology of threatened rain forest habitat'.

In this chapter conservation efforts that focus on a strategically selected single species are demonstrated to contribute effectively to conservation of intact habitats and natural systems. Conservation programmes designed around three lion tamarins (Callitrichidae: *Leontopithecus* spp.), squirrel-sized primates found only in the Atlantic coastal forest of Brazil, are used as examples: golden-headed lion tamarins (*L. chrysomelas*) in southern Bahia State, golden lion tamarins (*L. rosalia*) in Rio de Janeiro State, and black lion tamarins (*L. chrysopygus*) in São Paulo State.

2.2 CRITERIA FOR SELECTION OF EFFECTIVE FLAGSHIP SPECIES

The costs of *in situ* conservation campaigns on behalf of endangered species are substantial. Kleiman *et al.* (1991) estimated the annual budget of the Golden Lion Tamarin Conservation Project at US$136 000, now extended over a decade. The cost of the recovery programme for the California condor (*Gymnogyps californianus*) has been estimated at US$25 million (Nash, 1992), and black-footed ferrets (*Mustela nigripes*) at US$10 million. In 1990, nearly US$30 million in US state and federal funds was spent on conservation measures for just four species: the northern spotted owl (*Strix occidentalis*), grizzly bear (*Ursus arctos horribilis*), least Bell's vireo (*Vireo bellii pusillus*) and red-cockaded woodpecker (*Picoides borealis*) (Gibbons, 1992). Clearly, time and money are insufficient to allow individual conservation projects for every vertebrate that becomes listed as 'endangered', 'threatened' or 'vulnerable'. How does one optimally select a flagship species for conservation of biodiversity? Three criteria are proposed here: geographical location, ecological role and the potential for attracting local support.

2.2.1 Geographical location

Roughly half the world's biodiversity is concentrated in approximately 7% of the land surface: the tropical forests (Wilson, 1988). Within the tropics, 'megadiversity regions' or 'hot spots' containing disproportionately large numbers of species and/or endemic species have been identified by several authors (review by McNeely *et al.*, 1990). In general, priority should be given to *in situ* conservation projects in one of these regions of high biological diversity/endemism.

The current rate of habitat degradation in the region should also be considered in planning an *in situ* conservation programme. Estimates of rates of deforestation and degradation in the tropics vary but may be of the order of 17 million hectares per year, which may result in the extinction of 10–30% of tropical forest species in the next 30 years (Reid, 1992). Given the expense and long time periods necessary for habitat restoration (Janzen, 1988; Uhl, 1988), we recommend selecting an area in which the natural ecosystems are relatively undisturbed.

Should one choose as a flagship a species that urgently needs conserving, for example, one with effective population sizes of less than 100? Selecting such a species has advantages and disadvantages. The sense of urgency related to saving the remaining few animals in a species may facilitate securing funding, legislative intervention, press attention and other necessary resources. In addition, critically endangered species such as the northern spotted owl have legal mandates for conservation in many countries. On the negative side, the high value (financial, conservation and genetic) of each surviving animal favours a conservation strategy designed around protecting individuals and not populations or communities. Priorities for allocating resources must be set by the survival of each individual. Rescue and triage may become a higher priority than conservation and expansion of intact habitats. These projects run the real risk of losing their flagship before achieving higher priority objectives such as conservation of habitat.

2.2.2 Ecological characteristics

Ecological keystone species are those upon which a number of others depend in one way or another (DeBach, 1974; Paine, 1966). For example, keystone species may provide food resources during a season of scarcity or maintain habitat conditions on which other species rely. By definition, no other species in the system performs the same critical role (Solbrig, 1991). Thus, extinction of a keystone species may result in loss of many other species that depend on it.

Keystone species may be grouped into four general types: predators, mutualists, those that provide critical resources during seasons of scarcity, and those with large ranges. Keystone predators include carnivores, herbivores and parasites that reduce the population size of a dominant species in a competitive interaction, thus preventing it from driving its less successful competitors to extinction (Brown and Heske 1990; Paine, 1966, 1980; for an alternative model see Hixon and Brostoff, 1983). For example, Terborgh (1988) describes how top carnivores such as jaguar (*Panthera onca*) and puma (*Felis concolor*) may hold populations of seed predators in check, thus preserving community stability.

Keystone mutualists are often part of coevolved systems containing many partner species (e.g. Gilbert, 1980). Examples include generalist pollinators and seed dispersers. For example, flying foxes (*Pteropus* spp.) have been suggested as keystone species on Pacific islands, because of their role in pollination and seed dispersal (Cox *et al.*, 1991).

Fig trees (*Ficus* spp.) and certain nectar-bearing vines are examples of species that provide keystone food resources for several frugivorous mammals during seasonal periods of food shortage in Neotropical rain forest (Terborgh, 1986; Terborgh and Stern, 1987). Howe (1977) identified a single tree species (*Casearia corymbosa*) that fruits during annual periods of food scarcity in regions of the American tropics, thus allowing the survival of at least three avian frugivores. However, even abundant ‘keystone’ resources such as figs may be unpredictable on a small spatial scale (Janzen, 1979).

In general, a species that occupies a large range will be sympatric with a larger proportion of the resident species than would a species with a smaller range (Eisenberg and Harris, 1989). Thus large-bodied animals, which generally have larger ranges than smaller species, may also be considered ecological keystone species. For example, Eisenberg (1980) suggested that the ranges of large carnivores may be used to define the minimum landscape area needed to preserve entire ecosystems in the tropics. East (1981) writes that preservation of viable populations of African hunting dogs (*Lycaon pictus*) and cheetah (*Acinonyx jubatus*), because of their large hunting ranges, should be sufficient for long-term conservation of large African faunal communities. However, conserving vertebrates with large ranges does not necessarily result in the conservation of invertebrates (Murphy and Wilcox, 1986).

Superficially it would seem a simple task to identify the ecological keystone species in a geographical region and conserve them as flagships, thereby conserving the biodiversity that is retained with them. Unfortunately, most of the research on the stability of food-web structure employed models and experiments in which species were removed and the resulting changes in species composition observed (e.g. Pimm, 1980, 1986). The questions often asked are, 'Which species will disappear if we remove from the food chain, e.g., a predator?' – this line of questioning is interesting, but it does not necessarily provide answers to our concern, 'Which species remain if a particular flagship species remains?'.

A second concern related to choosing keystone species as flagships is that it may take years of research to understand the interactions in a specific food web, particularly in, for example, a complex tropical forest. As we discuss below, flagships need to be identified at the onset of any *in situ* conservation programme. Also, the interactions among the elements of any natural system are complex and dynamic, and vary with location, season and environmental factors. A species with a keystone role in one ecological system may function in a different fashion in another (e.g. Paine, 1980). In order to conserve biodiversity at a diverse array of taxonomic levels, it is necessary to conserve the ecological processes which allow complex natural systems to function (Pickett, Parker and Fiedler, 1992; Rojas, 1993). Thus, it isn't necessary that ecological keystone species be selected as flagships, rather, the role of flagships may be to ensure the survival of one or more keystone species.

Finally, the current major threat to world biodiversity is rapid and widespread deforestation in the tropics: the virtual collapse of an entire biome. **Any** species that functions to reduce anthropogenic deforestation, or that facilitates habitat recovery, is effectively an ecological keystone species in these regions, and may be selected as a flagship species, regardless of whether it is a large specialist carnivore with a large home range or a small generalist herbivore with a small range. In our opinion, this generalization outweighs the other ecological considerations mentioned above.

2.2.3 Potential for building public support

Birds and mammals can serve as effective focal points to gain public support for habitat conservation, and thus conservation of biodiversity. Recent surveys conducted in the USA (Kellert, 1979; Kellert and Berry, 1980; Kellert and Westervelt, 1983; Westervelt and Llewellyn, 1985) and research by Dietz and Nagagata (1986) demonstrated that adults and children preferred animal species that they considered attractive. Children liked animals they considered beautiful and loveable significantly more than animals important to the balance of nature.

Factors which had a substantial negative influence on adult preferences were fear of the animal, responsibility for property damage, predatory or carnivorous behaviour, association with the wilderness, and cultural/historical antipathies. Factors which influenced preferences positively were large size, advanced intelligence, phylogenetic relatedness to human beings and complex social organization.

Additional criteria may include economic, cultural or religious values. For example, the 'jupará' or kinkajou (*Potos flavus*) is highly regarded in southern Bahia because it is thought to spread the seed of economically valuable cocoa in its faeces. A potential economic pitfall may be a high black-market price for individuals of the flagship species. The highest source of mortality in reintroduced captive-born golden lion tamarins is theft (Beck *et al.*, 1991).

A final criterion which should be considered is the practical aspect of visibility in the forest. A nocturnal species, for example, might be difficult for the public to see. A crucial role of a flagship species is to provide the local public with direct experience to which they can relate the need for conservation of habitat.

A high regard for the flagship species is generally insufficient to guarantee conservation of its habitat. It is also necessary to convert public interest to action on behalf of intact habitat. This is accomplished by increasing the understanding of the relationship between species, habitat and human well-being.

2.3 ELEMENTS OF A CONSERVATION PROGRAMME: THE EXAMPLE OF LION TAMARINS IN BRAZIL

The historical geographical distribution of all four forms of lion tamarins is the central portion of the Atlantic coastal forest of Brazil: a narrow strip extending from Bahia State to Paraná State (Hershkovitz, 1977; Lorini and Persson, 1990). Golden-headed lion tamarins (*Leontopithecus chrysomelas*) are found in southern Bahia State, golden lion tamarins (*L. rosalia*) in Rio de Janeiro State, black lion tamarins (*L. chrysopygus*) in south-west São Paulo State, and *L. caissara* in coastal São Paulo and Paraná States. (We will not consider the recently discovered *L. caissara*

Table 2.1 Native and captive population sizes, and remaining habitat for golden-headed lion tamarins (GHLTs), golden lion tamarins (GLTs) and black lion tamarins (BLTs) (after Seal, Ballou and Pádua, 1990)

	Number of tamarins			*Forest habitat (ha)*	
	In wild	*In protected areas*	*In captivity*	*Protected*	*Unprotected*
GHLTs	850–3100	550	285	6200	9000
GLTs	450	290	550	2900	6650
BLTs	450	88–450	64	29 000	2500

in our discussions since conservation efforts on its behalf have just begun.)

Deforestation, the major threat to *L. chrysomelas, rosalia* and *chrysopygus* has reduced and fragmented the wild populations. Overall, less than 2% of the original area of geographical distribution remains as forest (Seal, Ballou and Pádua, 1990). However, the number of individuals in captivity and in the wild, and the amount of forest remaining vary widely among the three forms (Table 2.1). All three have international management committees which make recommendations to participating zoos and the Brazilian government. The conservation strategy elaborated for each of the three primates is specific to its needs.

2.3.1 Golden-headed lion tamarin

Golden-headed lion tamarins (GHLTs) have a large captive population with a sufficient number of founders to preclude the need for additional animals from the wild. However, deforestation rates in southern Bahia are among the highest in Brazil. A relatively large amount of unprotected forest remains, but this is because cocoa farming requires an intact overstory (Alves, 1988, 1991). The great part of the protected forest is in Una Biological Reserve, which was heavily occupied by squatters. Thus the focus of conservation efforts for this species include: protection of Una Reserve by compensating squatters for their land and by purchasing a corridor joining disjunct portions of the reserve; more precisely defining the size of the wild population of tamarins; and conservation education aimed at landowners and school children.

2.3.2 Golden lion tamarin

The captive population of golden lion tamarins is adequate in size but the number of founders is insufficient to meet demographic goals (Seal

Ballou and Pádua, 1990). A multifaceted international programme has been working toward conservation of this primate in Brazil since 1983 (Kleiman *et al.*, 1986), including a successful reintroduction component (Beck *et al.*, 1991). Of major concern are the small number of individuals of this species in the wild, and the reduced amount of protected and unprotected area under forest. Thus, conservation efforts were focused on reforestation in Poço das Antas Reserve; on location of additional populations of tamarins and of adequate forest outside the reserve; and on community education promoting forest conservation.

2.3.3 Black lion tamarin

The major conservation problems facing black lion tamarins are the small number of individuals in the wild and the very small amount of protected forest. They are breeding well in captivity but the small number of founders (22 individuals) causes concern. Although the total area of protected habitat is the largest among the three species, the majority are in Morro do Diabo State Park, and are distantly separated from the few small habitat fragments known to contain individuals (Seal Ballou and Pádua, 1990). Conservation efforts were focused on exhaustive censuses to locate every tamarin in the wild and on conservation education aimed at landowners and visitors to Morro do Diabo State Park (Pádua, 1991).

2.4 AN ACTION PLAN FOR BUILDING PUBLIC SUPPORT FOR FLAGSHIP SPECIES

To build public support for habitat conservation efficiently it is necessary to develop systematically a strategic plan tailored to local needs and to take into account the resources available to project leaders. A systems model for generating public support for conservation is described below, illustrating the steps, with examples from the three lion tamarin projects. This model was adapted by Dietz and Nagagata as part of the Golden Lion Tamarin Conservation Project (GLT project) (Dietz and Nagagata, 1986). A model of this type is useful in focusing attention on priority problems, systematically developing appropriate solutions, and maximizing the effectiveness of solutions.

2.4.1 Step 1: defining priority problems

What are the most important conservation problems? Can they realistically be addressed with education alone or must some other type of solution, such as an economically viable alternative, be found as well?

Researchers studying the ecology of each primate species identified the factors limiting population expansion. In all three projects habitat destruction was seen as the principal problem to be addressed. Hunting and capture for the pet trade were perceived to be secondary threats. Project scientists formed teams with managers and/or educators to develop solutions, the first step of which was to set specific conservation objectives.

All three projects identified as primary objectives the effective protection of the existing park or reserve and all other lands containing habitat suitable for tamarins. All recognized that without the collaboration of the people in the region, other efforts would have limited success in conserving the tamarins and their habitats. Each programme adopted a lion tamarin as a symbol for forest conservation in the region.

2.4.2 Step 2: identifying and evaluating the population, resources, and setting

This stage may be equated with market research. To sell a product or idea one must understand the consumer. Who are the people who must be influenced by the programme, and what are their needs and interests? What resources and support are locally available? Given the realities of the local setting, what is feasible and what is not? Also, a systematic evaluation prior to initiating education activities is necessary to serve as a benchmark for measuring any subsequent changes in knowledge, attitudes and behaviours.

All three lion tamarin projects conducted surveys of local knowledge and attitudes regarding wildlife and local protected areas, the GLT project in 1984, and the other two projects a few years later. In the light of limited resources, and the lack of local telephones, mail delivery and passable roads in the region of the GLT project, direct contact was limited to the population of three municipalities immediately surrounding the reserve. The project's media effort was focussed on residents of the cities of Rio de Janeiro and São Paulo who were illegally purchasing wild animals; government bureaucrats and politicians in Rio and Brasilia; and the public at large. Each of the three projects noted a general public ignorance of the local wildlife and conservation issues. Of the surveyed adults living within the geographical distribution of the golden lion tamarin, 41% did not recognize the tamarin in a photo. Most of the adults interviewed by that project did not know that Poço das Antas Reserve existed.

This kind of information served as a basis for planning strategies and capitalizing on local interests that were compatible with project conservation objectives. The GLT project found no negative attitudes about the golden lion tamarin or the forest in the region. Of those who indicated they valued the tamarin or the forest, many indicated aesthetic as well as economic reasons. The project thus determined that the golden lion tamarin would be an appropriate vehicle, or flagship, to increase levels of knowledge concerning the relationships of local wildlife, habitat, humans and the well-being of all.

The GLT project was designed to increase awareness of the importance of each of the various elements of the forest, stressing the long-term and global consequences of local human actions on the environment. Local residents were encouraged to think about conservation issues using arguments such as: One forest patch might seem like a little, but cutting many small patches destroys a large total area. Where does the wildlife go when you cut the forest? Where does the water go? What happens to the soil? What would happen if we killed all the snakes? When local residents became aware of conservation problems, the project provided them with information on how they could contribute to solutions.

As in many rural areas of Brazil, central Rio de Janeiro State provides few economic and educational options for young people. When asked what they would show to a newcomer, many interviewees said, 'Nothing. All we have is forest here'. Creating pride in local natural resources became an important part of the project's conservation strategy.

Interviews done by the GLT project also showed that the mass media were a powerful means of reaching the general public in the area. Of the local residents interviewed, 80% watched television regularly, and 99% listened to the radio. However, the majority of newspaper readers were absentee landlords.

2.4.3 Step 3: building a positive relationship

Because human attitudes and behaviours change slowly, a conservation –education effort must be designed to function over the long term. To ensure this, the GLT project encouraged community involvement in the planning and implementation of its educational activities. From its inception in 1983, the GLT project worked to build a constructive relationship with the local community leaders (politicians, teachers, judges, police, businessmen, priests, ministers, etc.). Through informal conversations with these people, GLT project personnel learned about the community and its needs, and explained project objectives. As the GLT project began to be perceived as a local resource, its participants were invited to speak about conservation of the local environment at community meetings. These leaders also began to see the reserve and project activities as things which could draw positive public attention to their community. Nearly all of the 30+ project participants are now from the local community. They came to the project not only for jobs, but also out of a sincere desire to contribute to local conservation and because of the status with which the project was regarded in the local community. The other two lion tamarin projects are utilizing similar strategies involving interested local institutions and employing local participants.

2.4.4 Step 4: select and test methods

The GLT project selected methods which interested the local leaders and/or met local needs, and which seemed likely to produce the desired result of motivating people actively to participate in forest conservation. Project educational materials were multi-purpose, short, simple and low cost. Since no other information existed, content on local fauna and flora often came from the unpublished results of the ecological studies underway in the reserve.

Prototypes of all materials were tested and revised before final production. Something which was perfectly understandable in Rio de Janeiro, 60 km away, was not necessarily understandable in this area. Since project surveys indicated that 41% of local adults had no formal education, audio-visual and live presentations were used to address this segment of the population.

2.4.5 Step 5: implementing activities

GLT project personnel concentrated efforts on working with existing local groups and institutions to achieve mutual objectives. Direct contact with the public was accomplished by interns, volunteers and five young graduates of a local teacher-training high school. Activities were continually developed and/or changed as the need arose in the local communities. Work began in one municipality, and after gaining momentum and experience extended the programme into two other municipalities.

The other projects also work through existing organizations: school groups and teachers, a local university, local non-governmental organizations, farmers' organizations, and rural extension programmes, as well as with the government agencies responsible for the local park and reserve.

Materials produced by the GLT project included press releases, video copies of news and other programmes on conservation, public service messages for radio and TV, educational posters, pamphlets, school notebooks with an appropriate conservation-based story on the cover, a slide collection for use in educational activities, slidetape programmes, information packages for landowners, a tamarin logo designed for use on all Reserve vehicles and project materials, an electric question/answer board, a travelling exhibit used at local festivals, and T-shirts, stickers and badges sold or given in recognition of contributions to conservation.

Materials used in the other projects were similar. In some cases it was possible for the projects to share materials, thus saving time and expense. Indeed, the black lion tamarin (BLT) and golden-headed lion tamarin (GHLT) projects noted changes in local knowledge resulting from the considerable national media attention given to golden lion tamarins.

Activities conducted by all three projects were remarkably similar: classes for schools, farm workers, and other groups; training classes for guards, project personnel, local teachers; lectures for local groups; press events; a children's play; musical events; parades and festivals; encouragement of planting native trees for watershed protection; encouraging young people to join or establish wildlife clubs; and observing the primates themselves.

In the GLT project, educational field trips to the reserve were conducted for farmers, school groups and families. Free-ranging tamarins accustomed to humans were located with the aid of radio-telemetry. Guests were also accompanied on a guided trail which encouraged observation and discovery of details of the forest. More recently, personal visits have been made to landowners, encouraging them to register their remaining forest as a permanent reserve and thus qualifying to receive captive-born or translocated wild-born golden lion tamarins.

The BLT project concentrated on bringing local school children to Morro do Diabo State Park, where wild tamarins were occasionally seen along an educational trail. The GHLT project used a similar approach, showing visitors rehabilitated tamarins released into a small forest island. In all three projects, emphasis was placed on showing the tamarin in its natural environment rather than in a captive situation.

2.4.6 Step 6: evaluation

This is a crucial step in an effective education programme. The first evaluation for the GHLT project was initiated in 1992. In the BLT project changes in the behaviour of park visitors as a result of education programme activities were documented. The formal evaluation of the GLT project education programme involved comparison of the pre-questionnaires and interviews with similar questionnaires administered after two years of community activities. Results indicated significant changes in the knowledge and attitudes of local Brazilian adults residing in the municipality of Silva Jardim, location of Poço das Antas Reserve.

Interviewers from the local community, trained by Project leaders, showed a photo of a golden lion tamarin to local adult residents and

Table 2.2 Responses to some questions used in interviews of adults before and after two years of GLT project activities

Question	*Responses to questions*	
	Before project activities	*After project activities*
What is the name of this animal? (photograph of golden lion tamarin)	212 of 518 (59%) responded correctly	394 of 497 (79%) responded correctly
How does the tamarin live?	33 of 140 (24%) responded correctly	223 of 406 (55%) responded correctly
Is the golden lion tamarin important or beneficial?	397 of 515 (77%) responded, 'Don't know'	88 of 400 (22%) responded, 'Don't know'
	72 of 515 (14%) responded, 'Yes'	248 of 400 (62%) responded, 'Yes'
What would you do with a little bird you found in the woods?	281 of 512 (55%) responded, 'Raise it at home'	145 of 499 (29%) responded, 'Raise it at home'
	225 of 512 (44%) responded, 'Leave it alone'	344 of 499 (69%) responded, 'Leave it alone'
What would you do with a snake you found in the woods?	374 of 512 (73%) responded, 'Kill it'	274 of 499 (55%) responded, 'Kill it'
	128 of 512 (25%) responded, 'Leave it alone'	160 of 499 (32%) responded, 'Leave it alone'

asked then what animal it was. In two years of project activities the proportion of correct responses increased significantly (Table 2.2) a chi-square goodness of fit test (Sokal and Rohlf, 1989, p. 702, $P < 0.001$) was used to reject the null hypothesis of no significant difference in percentage of pre- and post-treatment responses.

In its educational activities, the GLT project tried to communicate some of the habits of tamarins, both to interest people and to emphasize the tamarins' relationship with the forest. This information was not available to the local public through other sources. To evaluate these activities interviewers asked questions such as 'How does the tamarin live?'. Respondents were given the choices of 'alone', 'in pairs with young' (the correct answer), or 'in large groups'. After two years of project activities, there was a significant increase in the proportion of interviewees who chose the correct answer and a decrease in the proportion who thought they lived in large groups as do some other monkeys in the region.

To measure values perceived by local residents, interviewers asked questions such as: 'Is the golden lion tamarin important or beneficial?'. The number of respondents who answered 'no' remained unchanged over two years, but there was a significant decrease in the percentage of respondents who chose 'don't know' and a significant increase in the percentage who responded 'yes, the tamarin is important'. Many responses in pre- and post-interviews stated reasons related to beauty or pleasure in seeing the animal. But others gave reasons such as 'the tamarin is a part of nature' or 'it has a right to live'. These answers suggest new moral values not recorded in the pre-interviews.

Interviewers could not directly ask people about activities such as hunting or keeping wild animals as pets, because these are illegal in Brazil. However, one question which indicates the respondents' intentions regarding certain activities was included. Respondents were asked what they would do if they found several animals in the woods. There were no post-treatment changes regarding 'a little monkey': most respondents wanted to take it home or leave it alone. But with 'a little bird', there was a significant decrease in the percent who planned to 'raise it at home' and a significant increase in those who would 'leave it alone'.

The answers concerning what local residents would do with a snake they found in the woods gave insight on the breadth of the project's impact. Although project activities were not aimed at conservation of snakes, there was a significant decrease in the percentage of the respondents who said they would kill the snake and a significant increase in the percentage who would 'leave it alone'. This attitude change is an indication that the golden lion tamarin played a flagship role in promoting the emergence of a broader conservation ethic.

To determine which methods reached the most adults, interviewers asked those who correctly identified the tamarin how they had seen or heard of it. The responses included places and media related and not related to project activities. The most-often mentioned project activities included television (14.9% of 1599 responses), radio (6.9%) and in a newspaper or magazine (5.0%). Television was cited almost as often as all the non-project activities combined. Clearly the news media were an effective way to transmit the message to the local community. Other project activities that scored highly for this question were project-produced materials such as T-shirts/stickers/buttons (9.0%), and a photo or poster (6.7%). Only 6.9% of the responses mentioned seeing tamarins in the forest.

Data collected opportunistically also indicated significant progress toward achieving project objectives. More than 20 illegally held golden lion tamarins and 25 maned sloths (*Bradypus torquatus*), another endangered species endemic to the region, have been voluntarily handed over to Reserve officials since project activities began. The Brazilian Institute for the Environment (IBAMA), which previously had barred visitors to Poço das Antas Reserve, constructed the first centre dedicated to community and visitor education in any Brazilian national biological reserve. Finally, in 1990 several hundred local residents turned out to help fight a large fire that burned out of control in the reserve.

2.4.7 Step 7: recycling

The GLT project is currently in the process of recycling. The results of the first formal evaluation of the programme made it possible to determine the cost-effectiveness of individual activities. With this information the project team can better decide which conservation–education activities to continue in this region and elsewhere.

In conclusion, the results from the three lion tamarin conservation projects suggest that these primates served as effective flagship species to facilitate changing local attitudes and behaviours about forest conservation. Long-term, collaborative efforts toward conserving these flagship species, including ecological research, species and habitat management, and building public support can increase the likelihood that forest will exist for these primates in the future.

ACKNOWLEDGEMENTS

We gratefully acknowledge the contribution of all those involved in the projects described in this paper. Without their perseverance, dedication, professionalism, vision, and hard work there would be no real experience from which to draw these conclusions. We also thank the following agencies and institutions for their financial and logistic support for the community education efforts of the GLT project: Instituto Brasileiro do Meio Ambiente e dos Recursos Naturais Renovaveis; Centro de Primatologia do Rio de Janeiro-FEEMA; Fundação Brasileira para a Conservação da Natureza; Fundação Roberto Marinho; Canadian Embassy in Brazil; Golden Cross-Brasil; World Wildlife Fund-U.S.; the Smithsonian Institution's National Zoological Park, International Environmental Sciences Programme and Education Outreach Programme; Frankfurt Zoological Society; Friends of the National Zoo; and Jersey Wildlife Preservation Trust.

REFERENCES

Alves, M.C. (1988) Tamarins and cocoa don't mix . . . or do they? *On the Edge*, **36**, 3–4 and 12.

Alves, M.C. (1991) *The role of cacao plantations in the conservation of the Atlantic Forest of southern Bahia, Brazil.* University of Florida, Gainesville, Master's thesis.

Beck, B.B., Kleiman, D.G., Dietz, J.M., Castro, I., Carvalho, C., Martins, A. and Rettberg-Beck, B. (1991) Losses and reproduction in reintroduced golden lion tamarins *Leontopithecus rosalia*. *Dodo*, **27**, 50–61.

Brown, J.H. and Heske, E.J. (1990) Control of a desert grassland transition by a keystone rodent guild. *Science*, **250**, 1705–7.

Cox, P.A., Elmqvist, T., Pierson, E.D. and Rainey, W.E. (1991) Flying foxes as strong interactors in South Pacific island ecosystems: a conservation hypothesis. *Conservation Biology*, **5** (4), 448–54.

DeBach, P. (1974) *Biological Control by Natural Enemies*, Cambridge University Press, Cambridge.

Dietz, L.A. and Nagagata, E.Y. (1986) Programa de educação comunitária para a conservação do mico leão dourado *L. rosalia* (Linnaeus, 1766): desenvolvimento e avaliação de educação como uma tecnologia para a conservação de uma espécie em extinção, in *A Primatologia no Brasil*, Vol. 2, (ed. M.T. de Mello), Sociedade Brasileira de Primatologia, Brasilia, D.F., pp. 249–56.

East, R. (1981) Species–area curves and populations of large mammals in African savanna reserves. *Biological Conservation*, **21**, 111–26.

Eisenberg, J.F. (1980) The density and biomass of tropical mammals, in *Conservation Biology: an Evolutionary–Ecological Perspective*, (eds M.E. Soulé and B.A. Wilcox), Sinauer Associates, Sunderland, Mass., pp. 35–55.

Eisenberg, J.F. and Harris, L.D. (1989) Conservation: a consideration of evolution, population, and life history, in *Conservation for the Twenty-first Century*, (eds D. Western and M. Pearl), Oxford University Press, New York, pp. 99–108.

Erwin, T. (1988) The tropical forest canopy, in *Biodiversity*, (ed. E.O. Wilson), National Academy Press, Washington, DC, pp. 123–9.

Gibbons, A. (1992) Mission impossible: saving an endangered species. *Science*, **256**, 1386.

Gilbert, L.E. (1980) Food web organization and the conservation of Neotropical diversity, in *Conservation Biology*, (eds M.E. Soulé and B.A. Wilcox), Sinauer, Sunderland, Mass., pp. 11–33.

Hershkovitz, P. (1977) *Living New World Monkeys* (Platyrrhini), University of Chicago Press, Chicago.

Hixon, M.A. and Brostoff, W.N. (1983) Damselfish as keystone species in reverse: intermediate disturbance and diversity of reef algae. *Science*, **220**, 511–13.

Howe, H.F. (1977) Bird activity and seed dispersal of a tropical wet forest tree. *Ecology*, **58**, 539–50.

Janzen, D.H. (1979) How to be a fig. *Annual Review of Ecology and Systematics*, **10**, 13–51.

Janzen, D.H. (1988) Tropical dry forests: the most endangered major tropical ecosystem, in *Biodiversity* (ed. E.O. Wilson), National Academy Press, Washington, DC, pp. 130–7.

Kellert, S.R. (1979) *Public Attitudes Toward Critical Wildlife and Natural Habitat Issues: Phase I*, U.S. Government Printing Office No. 024–010–00–623–4, Washington, DC.

Kellert, S.R. and Berry, J.K. (1980) *Knowledge, Affection and Basic Attitudes Toward Animals in American Society: Phase III.* U.S. Government Printing Office No. 024–010–00–625–1, Washington, DC.

Kellert, S.R. and Westervelt, M. (1983) *Children's Attitudes, Knowledge and Behaviors Toward Animals: Phase V.* U.S. Government Printing Office No. 024 –010–00 641–2, Washington, DC.

Kleiman, D.G., Beck, B.B., Dietz, J.M. and Dietz, L.A. (1991) Costs of a re-introduction and criteria for success: accounting and accountability in the Golden Lion Tamarin Conservation Program. *Symposia Zoological Society London*, **62**, 125–42.

Kleiman, D.G., Beck, B.B., Dietz, J.M., Dietz, L.A., Ballou, J.D. and Coimbra-Filho, A.F. (1986) Conservation program for the golden lion tamarin: captive research and management, ecological studies, educational strategies, and reintroduction, in *Primates – The Road to Self-Sustaining Populations*, (ed. E. Benirschke), Springer–Verlag, New York, pp. 959–79.

Lorini, M.L. and Persson, V.G. (1990) Nova espécie de *Leontopithecus* Lesson, 1840, do sul do Brasil (Primates, Callitrichidae). *Boletim do Museu Nacional*, Nova Serie, Zoologia, **338**, 1–14.

Mallinson, J.C. (1991) 'Flagship' species aiding the conservation of animals and associated habitat. Unpublished report delivered to 46th annual conference of International Union of Directors of Zoological Gardens, Singapore.

McNeely, J.A., Miller, K.R., Reid, W.V., Mittermeier, R.A. and Werner, T.B. (1990) *Conserving the World's Biodiversity*, IUCN, Gland; World Res. Inst, World Wildl. Fund-U.S., World Bank, Washington, DC.

Mittermeier, R.A. (1986) Primate conservation priorities in the Neotropical region, in *Primates – The Road to Self-Sustaining Populations*, (ed. E. Benirschke), Springer–Verlag, New York, pp. 221–40.

Mittermeier, R.A. (1988) Primate diversity and the tropical forest, in *Biodiversity* (ed. E.O. Wilson), National Academy Press, Washington, DC, pp. 145–54.

Murphy, D.D. and Wilcox, B.A. (1986) Butterfly diversity in natural habitat fragments: a test of the validity of vertebrate-based management, in *Wildlife 2000: Modeling Habitat Relationships of Terrestrial Vertebrates*, (eds J. Verner, M. Morrison and C. Ralph), University of Wisconsin Press, Madison, pp. 287–92.

Nash, J.M. (1992) The $25million bird. *Time Magazine*, 27 January, pp. 56–7.

Owen-Smith, N. (1989) Megafaunal extinctions: the conservation message from 11 000 years B.P. *Conservation Biology*, **3** (4), 405–12.

Pádua, S. (1991) *Conservation Awareness Through an Environmental Education School Program at Morro do Diabo State Park, S.P. Brazil.* University of Florida, Gainesville, Master's thesis.

Paine, R.T. (1966) Food web complexity and species diversity. *American Naturalist*, **100**, 65–75.

Paine, R.T. (1980) Food webs: linkage, interaction strength and community structure. *Journal of Animal Ecology*, **49**, 667–85.

Pickett, S.T.A., Parker, V.T. and Fiedler, P.L. (1992) The new paradigm in ecology: implications for conservation biology above the species level, in *Conservation Biology, The Theory and Practice of Nature Conservation, Preservation and Management*, (eds P. Fiedler and S. Jain), Chapman & Hall, New York, pp. 65–88.

Pimm, S.L. (1980) Species deletion and the design of food webs. *Oikos*, **35**, 139–49.

Pimm, S.L. (1986) Community structure and stability, in *Conservation Biology: The Science of Scarcity and Diversity*, (ed. M. Soulé), Sinauer, Sunderland, Mass., pp. 309–29.

Pimm, S.L. (1991) *The Balance of Nature?*, Chicago University Press, Chicago.
Reid, W.V. (1992) How many species will there be?, in *Tropical Deforestation and Species Extinction*, (eds T. Whitmore and J. Sayer), Chapman & Hall, London.
Reid, W. and Miller, K. (1988) *Keeping Options Alive: The Scientific Basis for Conserving Biodiversity*. World Resources Inst., Washington, DC.
Rojas, M. (1993) The species problem and conservation: what are we protecting? *Conservation Biology* **6** (2), 170–8.
Seal, U.S., Ballou, J.D. and Pádua, C.V. (1990) Leontopithecus *Population Viability Analysis Report*, Captive Breeding Specialist Group, Species Survival Commission Union, Apple Valley, MN, USA.
Sokal, R.R. and Rohlf, F.J. (1989) *Biometry*, 2nd edn, W.H. Freeman & Co., New York.
Solbrig, O.T. (1991) *From Genes to Ecosystems: A Research Agenda for Biodiversity*, International Union of Biological Sciences, Cambridge, Mass.
Soulé, M.E., Gilpin, M., Conway, W. and Foose, T. (1986) The millenium ark: how long a voyage, how many staterooms, how many passengers? *Zoo Biology* **5**, 101–13.
Terborgh, J. (1986) Keystone plant resources in the tropical forest, in *Conservation Biology: the Science of Scarcity and Diversity*, (ed. M. Soulé), Sinauer Associates, Sunderland, Mass., pp. 330–44.
Terborgh, J. (1988) The big things that run the world – a sequel to E.O. Wilson. *Conservation Biology*, **2** (4), 402–3.
Terborgh, J. and Stern, M. (1987) The serreptitious life of the saddle-backed tamarin. *American Scientist*, **75**, 260–9.
Uhl, C. (1988) Restoration of degraded lands in the Amazon Basin, in *Biodiversity*, (ed. E.O. Wilson), National Academy Press, Washington, DC., pp. 326–32.
Weinberg, S. (1992) Return of a native. *Pacific Discovery*, **45** (2), 8–14.
Westervelt, M.O. and Llewellyn, L.G. (1985) *Youth and Wildlife: The Beliefs and Behaviors of Fifth and Sixth Grade Students Regarding Non-Domestic Animals*, U.S. Government Printing Office, Washington, DC.
Wilson, E.O. (1988) The current state of biological diversity, in *Biodiversity*, (ed. E.O. Wilson), National Academy Press, Washington, DC., pp. 3–18.

3

Meta-populations: is management as flexible as nature?

J.L. Craig
University of Auckland, New Zealand.

3.1 INTRODUCTION

Conservation is a very recent science that has been built from a basis in ecology and evolutionary biology (Soulé, 1985; Simberloff, 1988). Early players in the formulation of the applied science of conservation were the geneticists who, seeing that they could assist conservation, offered the idea of the preservation of genetic diversity as a component of conservation. Unfortunately for conservation and genetics, these first ideas rapidly became enshrined as dogma by those unable to question the assumptions of the new application of genetics. Ecologists joined the geneticists, and offered viability analyses as a start to adding demographic information to that supplied by the geneticists. Both of these sources of information may help any particular conservation initiative but how they are applied and their relative importance can only be determined by how the assumptions of the different genetic, ecological and behavioural models match the organisms involved. It is important to ask whether we are trying to conserve real or theoretical populations.

Financial and public opinion statistics are further additions to the types of information needed to produce a realized conservation programme. Resources for conservation will rarely equate with the poten-

Creative Conservation: Interactive management of wild and captive animals.
Edited by P.J.S. Olney, G.M. Mace and A.T.C. Feistner.
Published in 1994 by Chapman & Hall, London.
ISBN 0 412 49570 8

tial for action. The widespread emphasis on biological criteria in management planning is important, but planning will benefit from an equal stress on innovative financial and advocacy management.

This paper addresses the biological and management issues in conservation and advocates a meta-population approach for the management of rare and endangered species. (Meta-population is used to describe the situation where the total population of a species is divided into a number of separate sub-populations.) Considering the flexibility of animals in nature, it is important to consider whether management of rare animals is as flexible as nature, and if not, why not.

3.2 MANAGEMENT GUIDELINES FROM THEORY

Theoretical models based on effective population size, viability analyses and similar calculations may provide useful recommendations for management. As Shaffer (1987) suggested, 'in the absence of scientifically validated predictive relationships, conservation **must** be guided in large part by theory and inference' (my emphasis).

From the theory of island biogeography have come recommendations that large reserves are typically more beneficial than small reserves (Dawson, 1984; IUCN, 1980; Simberloff, 1988 and references therein). Advantages of size are also advocated for other reasons. Genetics theory, based on laboratory results, led to recommendations that to avoid the consequences of inbreeding a minimum effective population size of 50 is needed, and that to avoid long-term loss of genetic variability a minimum effective population size of 500 is required – see Franklin (1980) and Simberloff (1988) for reviews. While these numbers are an order of magnitude only (M. Soulé, pers. comm.) and will not hold in all situations (e.g. Ewens *et al.*, 1987), they have appeared with remarkable frequency in both theoretical and management publications (see below). Similarly, genetical arguments are the basis of recommendations for migration in order to mimic panmixia and achieve random mating.

Unfortunately, guidelines used by managers in one situation have become recommendations in other less appropriate situations, as the word '**must**' in the quotation of Shaffer (above) implies. For example, in New Zealand the ideas that inbreeding is universally deleterious and that 500 represents the desired minimum number for every effective sub-population, rapidly became adopted in local management recommendations and recovery plans (Triggs, 1988; Crouchley, 1990; Rasch, 1988). Research has shown that high levels of inbreeding occur in many island species and hence models derived for continental northern hemisphere species may not apply in most island situations (Craig, 1991a and references therein).

These problems highlight the dangers of unquestioned adoption of models without regard to their underlying assumptions. For example, when investigating the effective population size, actual behaviour such as dispersal and reproductive variance are modelled to determine the number of individuals required to mirror an ideal population that will undergo random genetic loss of alleles (see Lande and Barrowclough, 1987 for more detail). Typically, census sizes are estimated to be upward of 2–4 times the effective population size. As Reed, Doerr and Walters (1988) and Shaffer and Samson (1985) have argued for red-cockaded woodpeckers (*Picoides borealis*) and grizzly bears (*Ursus arctos*), these numbers may not be useful for managers as they are left with the conclusion that they cannot reserve adequate habitat for long-term persistence. One option is to give up; another is to ask whether current genetic modelling is relevant to conservation management. Certainly the idea that all genotypes have equal fitness (selectively neutral) is being questioned (e.g. Lesica and Allendorf, 1992). A further option is to assume that individual sub-populations within a meta-population need to be modelled together. As Lacy (1987) has shown for a number of small sub-populations, each may lose variability but the probability is low that the loss will occur to the same extent in each population. Moreover, studies of wild and laboratory populations have shown that some populations with small numbers and high levels of inbreeding can be very successful (Craig, 1991a; L. Pray, pers. comm.; below) Perhaps our original models require extension to incorporate more realistic genetic characteristics (by, for example, moving from single locus models and from the assumption that all genotypes are equal or selectively neutral) or to increase the emphasis on demographic features, as Gilpin (1987) and Mace and Lande (1991) have.

Other researchers argue that genetic considerations are of minimal consequence as the threats associated with genetic issues are greatly overshadowed by the chance demographic events that threaten small populations (e.g. Shaffer, 1987). Others (e.g. Saunders, Hobbs and Margules, 1991) have suggested that theory has given minimal assistance to managers as, especially in the case of reserve size, they are faced with what exists rather than what is ideal. For example, many New Zealand birds are restricted to offshore islands where maximum population size may not even reach 100 let alone a theoretical effective size.

Theoretical modelling is undergoing considerable extension, and many models do not take genetic effects into consideration (e.g. Possingham *et al.*, 1992). Results may be very different from those that do consider genetics. For example, a recent model (Hoyle, 1993) using seven years' demographic data for a subdivided island population of a rare New Zealand wattlebird, the saddleback (*Philisturnus carunculatus*), has suggested that a population of five to eight pairs in an area of 2.4 – 3.6 hectares is all that is required for a viable population. Given that island reserves suitable for 5–50 pairs are common whereas islands capable of sustaining an effective population size of 500 are not, this model offers greater hope to managers.

Other theoretical ideas, such as 'habitat requirements', that are part of a major research agenda in worldwide conservation, similarly have assumptions that must be considered in every situation (Gray and Craig, 1991). In particular the assumptions that historical distributions and behavior are a reliable guide to the limits of management options is particularly questionable for rare and endangered species (Craig, 1992). Ideas of behavioural inflexibility and adaptation have more complex problems (Bateson, 1983; Gray, 1987). How these various models relate to natural systems requires urgent consideration.

3.3 CHARACTERISTICS OF REAL POPULATIONS

A consequence of habitat loss has been fragmentation. Consideration of this has been coloured by the implicit assumption that continuous habitat coincides with free social interchange, and hence with panmixia or random mating. Thus for many researchers (see Lande and Barrowclough, 1987; Saunders, Hobbs and Margules, 1991; Simberloff, 1988 and references therein), habitat loss and fragmentation have resulted in a forced change in social and reproductive strategies, which aggravates the long-term genetic consequences for rare species. While this is undoubtedly true for some species, the universal validity of this conclusion has rarely been challenged. This conclusion that social and reproductive patterns must have changed probably reflects the strong contribution of the early players in conservation (geneticists and demographers) and the minimal contribution of behavioural ecologists. Thus for behavioural ecologists, the assumption in meta-population modelling that territories are randomly or evenly spread through existing habitat (e.g. Lande, 1987) is often unrealistic and denies the importance of localized hot spots of favourable habitat (Goodman, 1987).

In contrast with the above conclusion, studies of dispersal backed by genetic studies have shown that many populations are naturally subdivided into small units, and that these may have a high degree of genetic isolation from neighbouring units (e.g. McCauley, 1991). For example, primates are often found in matrilineal groups that are genetically distinct from other neighbouring groups that they interact with socially (Oliver *et al.*, 1981). Similarly some birds fit the same pattern. Pukeko (*Porphyrio melanotus*) live in communal family groups that retain young (Craig and Jamieson, 1988), and these units are genetically distinct from each other (Craig *et al.*, in prep.) even though they interact daily. Similarly Erlich's studies of butterflies have shown a naturally occuring meta-population structure (Erlich, 1965; Erlich and Gilbert,

1973; Erlich *et al*. 1975). Genetic studies of plants suggest similar subdivision of populations (Schwaegerle and Schaal, 1979).

Studies of dispersal have shown that many birds are strongly philopatric and they settle near their natal territory (Greenwood and Harvey, 1982; Woolfenden and Fitzpatrick, 1984; Craig and Jenkins, in prep.), even to the extent of mating with close relatives (Craig, 1991a; Stewart, 1980; Triggs *et al.*, 1991).

The assumption that most populations are continuous and not subdivided has limited interpretations of genetic surveys of remaining units of rare species. For example, failure to show some predetermined level of isozyme variability is assumed to mean that the subdivision of the population has increased inbreeding and genetic drift, leading to the measured loss of genetic diversity (e.g. O'Brien *et al.*, 1983). Whether there were ever greater levels of genetic diversity are not considered, as the population is assumed to have been large, panmictic, continuous and genetically diverse (see Pimm, Gittleman and McCracken, 1989). Should there not be widespread empirical support for this assumption before conclusions on genetic loss are credited? Even then it is likely that results will differ for animals with different social systems. For some species, could natural populations be a variable mosaic of genetic diversity (Avise, 1989)?

Given this lack of verification is it always necessary immediately to assume that habitat fragmentation has caused special and highly deleterious effects? Especially in island populations like those found in New Zealand, it is highly likely that migration between islands has always been limited. This is especially so for invertebrates and non-flying vertebrates. New Zealand has in excess of 700 islands off the three main islands, and hence some species have been divided into numerous subpopulations for long periods of time. As Pimm, Jones and Diamond (1988), Shaffer (1987), Simberloff (1988) and Craig (1991b) have suggested, it is the demographic, problems of small size that are likely to be of far greater importance than any genetic considerations. Even demographic problems may be of limited importance above some low threshold of 8–20 pairs if the evidence of some wild populations is considered (Pimm, Jones and Diamond, 1988; Craig, 1991a).

3.3.1 Inbreeding

Inbreeding is often assumed to be a major problem with rare-species management, especially when captive breeding is involved (Ralls and Ballou, 1983, 1986a,b). Ralls and Ballou have collated considerable evidence on the survival differences between inbred and outbred young produced in zoo environments. Such effects are expected when inbreeding is imposed on a previously outbreeding stock (but see Templeton and Read, 1983) but may be less likely for animals or plants that have high levels of natural inbreeding (Craig, 1991a). Moreover, the results of inbreeding may be exacerbated by other issues of captive breeding. A lack of social stimulation typically associated with reproductive competition in the wild but severely controlled in captive conditions, the confounding influence of long-term social dominance, and the management of matings to ensure genetic contribution by all founders may confound problems currently attributed to inbreeding alone.

Numerous examples of breeders finding low levels of fertility or survivorship among inbred captive stock but a failure to observe similar effects among inbred wild stock help demonstrate the complexities of the problem. For example, populations of less than 20 individuals of the pink pigeon (*Columba mayeri*) have survived for over 150 years with close to 100% fertility of eggs in the wild but much less than this in captivity (C. Jones, pers. comm.). Other captive animal populations have been established from few founders and appear to suffer no negative effects of inbreeding (e.g. Przewalski's horse [*Equus caballus przewalskii*]; O. Ryder, pers. comm.).

Numerous wild populations have survived for 100 years or more as small populations: pink pigeon (see above), red-tailed hawk (*Buteo jamaicensis socorroensis*; Walter, 1990), birds on British Islands (Pimm, Jones and Diamond, 1988), some New Zealand birds (Craig, 1991a), Hawaiian crow (*Corvus tropicus*; T. Cade, pers. comm.). Moreover numerous animal populations have successfully recovered from very small numbers (e.g. Chatham Island black robin *Petroica traversi*; Merton, 1990) or have been established from the introduction of a small number of founders (Atkinson, 1990; Craig, 1990a; Craig and Reed, in prep; Griffith *et al.*, 1989). A comprehensive analysis of the relative successes and failures of populations established from small numbers is needed to lift the argument from either perspective above the anecdotal.

The social organization of some animals means that natural levels of inbreeding may be significantly higher than that expected from chance alone. For example, communally breeding pukeko in a number of areas in New Zealand have observed levels of inbreeding among close relatives (r = 0.5) that may reach as high as 70% (Craig and Jamieson, 1988). DNA fingerprinting has confirmed observations, and the degree of band sharing among individuals in the same group makes the assignment of parentage difficult, whereas the degree of band sharing between neighbouring groups is low (Craig *et al.*, in prep). In spite of such high levels of inbreeding, pukeko are extremely common, and are shot as game. Other, rarer, New Zealand birds, such as blue duck (*Hymenolaimus malacorhyncos*) have levels of close inbreeding of 50% or greater (Triggs *et al.*, 1991). Where information is available for other forest birds, levels of inbreeding among family members in excess of 9% is not unusual

(Craig 1991a). Such levels have been reported among communally breeding birds in other parts of the world (Craig, 1991c; Craig and Jamieson, 1988).

Inbreeding is but one extreme of a continuum of mating behaviour, with outbreeding at the other extreme. While inbreeding can be deleterious for normally outbred individuals, it is equally important to guard against the potential of outbreeding depression among naturally inbreeding individuals (Templeton *et al.*, 1986). While some argue that outbreeding depression is merely the sign that animals of different but unrecognized taxa have been mixed, examples of outbreeding depression are claimed (e.g. Templeton *et al.*, 1986; Greig, 1979)

3.3.2 Behavioural and habitat flexibility

Investigation of 'habitat requirements' is a common component of research on rare and endangered species (Gray and Craig, 1991). Results of habitat usage are used to determine likely habitat features that are important for the long-term survival of the species or for choosing new release sites. Unfortunately, underlying assumptions in the associated research programmes include the ideas that present habitats are optimal, that current ecological factors are more important than historical ones and that the species' requirements are fixed.

The large majority of rare species have had their geographical and habitat ranges greatly reduced during their immediate history, and hence the extent to which current habits and habitat are a useful guide for management should be considered carefully. A lively debate over two rare species, takahe and saddle back is in progress in New Zealand at the moment.

Takahe (*Porphyrio mantelli*) have been managed in the alpine grasslands where the remnant population was rediscovered 40 years ago. Their total population has fluctuated around 200 since discovery. Research on habitat use (Mills, Lee and Mark, 1978; Mills, Lavers and Lee, 1984) showed that they are selective feeders that take tussock of the highest nutrient value. Birds are three years old at least before breeding, and at most lay a single clutch per year. Adult and especially chick mortality is high. Historical distributions show that they were once found throughout the country and were found in a wide range of habitat types, including grasslands, forest and dune country. Recently some birds have been put on islands that lack alpine tussocks; they breed at an earlier age, mortality rates are lower, some pairs have laid as many as four replacement clutches, and they use both grass and forest habitats. Captive trials have shown that the birds prefer pasture grasses over their most preferred alpine tussocks. Moreover, the cost of management in the alpine region greatly exceeds that for islands. In spite of all this information, the idea of optimality of relic alpine habitat prevails, and no priority is being given to establishing a meta-population through island populations.

The case of the saddleback is equally informative. Once widespread, they declined rapidly in the latter part of last century, to survive on a single island. Habitat requirement studies suggested the need for forest, especially tall, older forest. Releases onto offshore islands meeting these requirements were successful, and the species, has now been reduced in its rarity-ranking in the Red Data Book. Tiritiri Island (see below) was considered inadequate as the amount of mature forest was minimal. Eventually birds were released onto the island, where they have shown that they will use any scrub vegetation, including regenerating forest planted by the public, and they have greatly increased their reproductive rate. On the original island, saddleback of two years of age or more laid a single clutch of two eggs. In the initial years on most islands, some birds bred at age one, some laid up to two clutches and a few laid a third egg in some clutches. On Tiritiri, the island where suitable habitat was considered lacking, birds consistently bred at age one, and all pairs laid more than one clutch, with some consistently rearing as many as four clutches of four eggs!

These examples demonstrate the difficulty of determining habitat needs, and how management based on an adherence to a belief in rigid habitat requirements and a fixed breeding rate may reduce the chances of rare and endangered species recovering. Planning for a meta-population approach, and using different populations to test the behavioural and habitat flexibility of rare species, may allow more effective conservation.

3.3.3 Advantages of meta-population management

From theory, subdivision of the total population of a species into a number of smaller independent or semi-independent populations (a meta-population) has a number of potential advantages. Genetic diversity can be maintained at higher levels. This is achieved by the preservation of different alleles in different populations, although the small population sizes of each of the separate populations may result in lower variability in each of these compared with what would be possible in a single large population. Thus genetic variability is maintained among the populations rather than within each one (e.g. Wright, 1931; Quinn and Hastings, 1987; Craig, 1991b and references therein).

Chance demographic changes will predispose each of the separate populations to a higher chance of extinction within any time period, but it is unlikely that all or even more than one population will go extinct at the same time. Hence in the rare event of an extinction of one population, the lost population can be readily re-established from the other remaining parts of the meta-populations. Other advantages from having a subdivision into isolated populations is that as long as these are spread geographically, there is increased insurance against catastrophes such as fire, volcanoes, earthquakes, cyclones and disease.

Recommendations for migration between the separate populations vary. In order to maintain higher levels of genetic variability within each of the sub-populations, migration of one to five successfully reproducing individuals per generation is recommended (e.g. Crow and Kimura, 1970; Ewens, 1979; Ewens *et al.*, 1987; Quinn and Hastings, 1987). Such migration will maintain higher genetic variability within each population but may reduce levels for the overall meta-population as it will tend to homogenize the populations and destroy the separate genetic characteristics of each individual population. Such migration has other disadvantages. It will increase the risk of disease in the total population of the species and it will tend to counter any selection for adaptation to local conditions (Simberloff and Cox, 1987).

In addition to theoretical considerations, a meta-population approach has practical management advantages as well. These are especially important when attempting to increase numbers of rare and endangered species. Captive populations can be managed as separate units or can be linked to selected wild populations. This will allow segregation of populations derived from captive stock from those derived from wild releases to reduce the potential for disease. Repairing or restoring habitat for re-establishing a species in part of its former range can be achieved on a piecemeal approach rather than having to restore a single large area for an undivided population. Establishing individual units of a meta-population allows managers to make use of habitat patches of varying sizes – a situation that most typically matches reality. In New Zealand, islands offer the only chance for many of our rare animals, as the mainland is totally compromised by the introduction of herbivorous and predatory placental mammals into an environment that naturally lacked such animals. Islands are of widely varying sizes, and demands for single large reserves to allow one large contiguous population would lead to compromises in the natural values of the few large islands that are suitable. Moreover, meta-populations allow managers to meet the seemingly contradictory demands of total protection and public involvement (see below).

3.4 CONSERVATION IS A BUSINESS

Conservation, like most other activities, cannot be carried out in isolation from the economic, political and social environment. While this has been recognized by some (Janzen, 1989), many programmes continue in the face of reality. For example, in New Zealand, species recovery plans do not include realistic costing of the various options. Research on social attitudes and the wishes of taxpayers who fund conservation is just beginning and has low priority. Moreover, rare-species management is excluded from public activities (Department of Conservation, 1990), and is planned nationally even though it has strong regional basis. The suggestion that priority should include achieving the most for the greatest number of species and for the greatest number of people is strongly rejected.

Conservation is variably defined, but predominantly includes the idea of management of human use (IUCN, 1980; Nature Conservation Council, 1981). Preserving the heritage of people is included by law (Fisher, 1991) and gains importance since national, regional and even personal identity often includes a close association with particular plants and animals (e.g. Nature Conservation Council, 1981; Janzen, 1989). Hence conservation is not just for scientists and other conservation professionals but is most likely to succeed when it involves and is 'owned' by local people (Janzen, 1989; Galbraith, 1990). Moreover, because resources are always limited, there is a need for public support and finance that is more likely to be forthcoming if people are involved at all levels.

Meta-populations allow special advantages for the involvement of people even in the most sensitive programmes with highly endangered species. Some populations can be maintained with no public access, while others can be set up for open public access (Craig, 1990b). The former can be considered primarily for protection whereas the latter will serve the functions of both advocacy and protection. Close public involvement in the restoration of habitat for reintroduction of a rare species to a part of its former range will not only allow low set-up costs but will likely provide financial support for the contributing breeding programme.

A good example of this approach is Tiritiri Matangi Island near Auckland, New Zealand. The island was farmed for more than 100 years and was burnt regularly. Remnant populations of some native birds, including bellbirds (*Anthornis melanura*) that had been extinct on the neighbouring mainland for at least 120 years, had survived. The last farm stock were removed in 1972, and in 1982 a plan was adopted to replant the island to provide habitat for rare birds. A nursery on the island was run with public help, and the public paid for the privilege of being able to replant the trees they had helped rear. Seven species of birds that are either locally or nationally rare have been reintroduced to the island. Two originated from captive-reared stock, and all have survived and bred better on this island than on any other (Craig, 1990a). Birds are using newly planted areas after as few as three years growth.

For each of the species. Tiritiri represents one of the populations in a much larger meta-population. For example, the saddleback is now on eight closed-access islands, two open-access islands, one controlled-access sanctuary and in two captive populations.

Bird releases are public events that are greatly oversubscribed in spite of charging a premium to attend. Concerned members of the public have initiated their own incorporated supporters group that is primarily concerned with raising money for the island restoration (Galbraith, 1990). Not only are they paying for public amenities, they arrange finance for bird transfers, and have even paid for setting up other wild and captive populations that have originated from birds earlier introduced to Tiritiri. The island is an 'open sanctuary' that allows unrestricted access, with the public constrained to formed trails but with the animals wild. The programme was a bold move by the government conservation authority that has succeeded. In a bizarre twist, the greatest constraint to progress is now the failure of the government authority to move with the increasing momentum of public interest.

The success of the restoration programme can be easily judged from the positive outcome of all introductions, the statistic that the public have replanted the island in eight years with 300 000 trees, and by the ever increasing visitation by the public. The island is near Auckland, a city of over a million people who live in an area that saw major extinctions of native animals over 100 years ago and where native animals are continuing to decline. By visiting Tiritiri the public can see more than twice the number of native birds and see three times the density of birds than they can see within 100 km on the mainland (Craig, unpublished).

A survey of public opinion (Craig *et al.*, in prep) showed that the public of New Zealand wanted greater access to rare native animals, and that they were prepared to pay for such access. People who had already been to Tiritiri and hence had had the experience of interacting with endangered animals in the wild were prepared to pay 50% more than those who were yet to have such an experience. The increased conservation opportunities offered by planning for public involvement and its associated revenue generation are enormous (see also Craig, 1990b).

3.4.1 Meta-populations in the conservation business

Most businesses evaluate the advantages and disadvantages of different strategies before committing limited resources. Conservation would benefit from the same approach. The wants and needs of clients, the offerings of competitors, and the costs of various alternatives need as close attention as the biological issues. An evaluation of Strengths, Weaknesses, Opportunities and Threats (SWOT) (Table 3.1) is one

Table 3.1 SWOT analysis for conservation management of a meta-population versus a single large population

	Meta-population	*Single large population*
Strengths	High genetic variability	High genetic variablity
	Independent sub-populations – isolated from disease – isolated from catastrophes	Low chance of demographic loss
Weaknesses	Some populations small – loss of genetic diversity – increased inbreeding – high chance of demographic loss	Single integration location – threat of fire, disease, catastrophes, predators
Opportunities	Can easily involve public Can innovate with different: – species assemblages – translocation methods	
Threats	Taxonomists may declare each population a different taxon	Threat that location will be reclassified for development

useful method for evaluating meta-populations versus single large populations in rare species management.

Strengths: Both a single large population and a meta-population offer the advantage of maintaining high levels of genetic variation. Similarly both a meta-population and a single large population have a very low probability of extinction from chance demographic events. An additional advantage for the meta-population is that, with the sub-populations independent of each other, there is a very low probability that catastrophes or disease could threaten the total population.

Weaknesses: One of the greatest weaknesses of the single large population is that the species is found in a single integrated location. This greatly increases the risk of catastrophes such as fire but also increases the risk of disease or a newly arrived habitat change such as invasive weeds or a predator threatening the species. Moreover, should the population be threatened, there is no backup source of individuals for re-establishment. In a meta-population, the individual sub-populations may each maintain lower levels of genetic variability and hence, because of their small size, suffer genetic or demographic problems. The importance of such issues declines exponentially with the number of sub-populations.

Opportunities: The opportunities of a single large population in a single locality are highly constrained. While it may be possible to involve the public in part of the programme, the need to constrain their activities will greatly restrict their feeling of belonging and the amount they are likely to contribute.

Meta-populations offer major opportunities for involving the public as some populations can be managed for public access and involvement (Figure 3.1). The involvement of the public may marginally increase the risks to that sub-population, but, as demonstrated above, this is greatly offset by the lower set-up and maintenance costs as the public are prepared to pay directly. They may even assist in paying to set up other sub-populations, as has happened from Tiritiri.

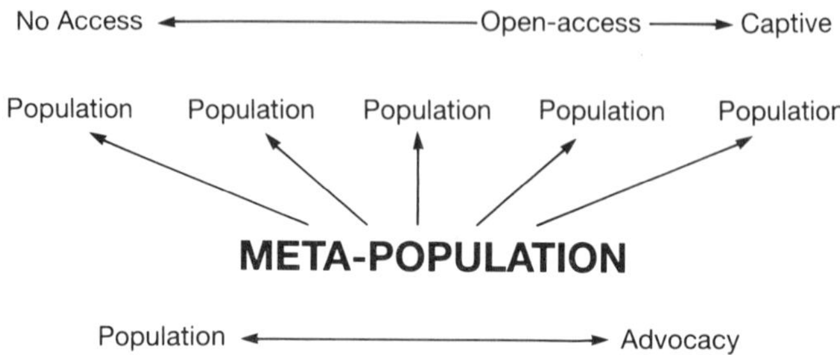

Figure 3.1 The different populations making up a meta-population provide opportunities for the involvement of people. Populations can be spread along a spectrum from outright protection with no public access, through to protection plus advocacy for wild and captive populations with open public access.

A further opportunity provided by meta-populations is that different sub-populations can be set up in different ways, both as a test of establishment and translocation techniques, and also as a way of restoring different types of ecosystems. For example, in New Zealand there has been considerable debate over whether it is better to restore to pre-human or to pre-European ecosystems. With a meta-population approach both are possible. The conservation opportunities and the revenue opportunities of the latter approach are enormous.

Threats: The single large population in one locality is always under the threat that short-term political planning may change priorities and allow commercial activity, such as agriculture, forestry or mining, that will put the species under renewed threat. The genetic differentiation of populations in the meta-population approach may occasionally be threatened by taxonomists: different populations may be declared different taxa!

3.5 SUMMARY AND CONCLUSIONS

Meta-populations offer special opportunities for the conservation of rare and endangered species. Not only are there theoretical advantages in terms of preservation of high levels of genetic diversity and low risk of chance extinction, they also offer ways of increasing public interaction and lowering costs. Moreover, meta-populations allow a way to structure small captive populations, such as those found in zoos, into meaningful units that are linked to selected sub-populations of wild animals. The dominance of theoretical models in conservation texts (e.g. Schonewald-Cox *et al.*, 1983; Soulé, 1986, 1987) that are derived from ideas of fixed species characteristics, as well as perceptions of population structures in large continental areas, is influencing conservation management, worldwide. The applicability of these models, especially in island situations, requires urgent evaluation.

Consideration of natural populations suggests animals and plants are extremely flexible in their behaviour and population structures. Meta-population management allows similar flexibility with the chance to try novel solutions to persistent problems. Moreover, conservation from an inflexible biological perspective will be unlikely to achieve as much as flexible management that considers biology, finances, politics and sociology together. Can we meet the challenge?

ACKNOWLEDGEMENTS

I wish to thank Craig Miller, Anne Stewart, Chris Craig and Bob Lacy for comments on the manuscript. My research is supported by the University of Auckland, the Department of Conservation and the New Zealand Lotteries Board.

REFERENCES

Atkinson, I.A.E. (1990) Ecological restoration on islands: prerequisites for success, in *Ecological Restoration of New Zealand Islands* (eds D.R. Towns, C.H. Daugherty and I.A.E. Atkinson), Dept of Conservation, Wellington, pp. 73–90.

Avise, J.C. (1989) A role for molecular genetics in the recognition and conservation of endangered species. *Trends in Ecology and Evolution*, **4**, 279–81.

Bateson, P. (1983) Genes, environment and the development of behaviour, in *Genes, Development and Learning* (eds T.R. Halliday and P.J.B. Slater), Blackwell, Oxford, pp. 52–81.

Craig, J.L. (1990a) Islands: refuges for threatened species. *Forest and Bird*, **21**, 28–9

Craig, J.L. (1990b) Potential for ecological restoration of islands for indigenous fauna and flora, in *Ecological Restoration of New Zealand Islands* (eds D.R. Towns, C.H. Daugherty and I.A.E. Atkinson), Dept of Conservation, Wellington, pp. 156–65

Craig, J.L. (1991a) Are small populations viable? *Acta XX Congressus Internationalis Ornithologici*, **4**, 2546–51.

Craig, J.L. (1991b) Genetic roulette. *NZ Science Monthly*, **8**, 8–10.

Craig, J.L. (1991c) Communal breeding along the changing face of theory. *Acta XX Congressus Internationalis Ornithologici*, **1**, 233–46.

Craig, J.L. (1992) Kokako recovery: for whom and at what cost? *Forest and Bird*, **23**, 30.

Craig, J.L. and Jamieson, I.G. (1988) Incestuous mating in a communal bird: a family affair. *American Naturalist*, **131**, 58–70.

Craig, J.L. and Reed, C. (in prep) Translocations and island conservation: is the number of founders important?

Craig, J.L. and Jenkins, P.F. (in prep) Song dialects, dispersal and inbreeding: an experimental test with a newly established population.

Craig, J.L. Lambert, D.A., Jack, K. and Anderson, S.A. (in prep). Relatedness and band sharing among and between territories of a communal bird.

Crouchley, D. (1990) *Takahe Recovery Plan*, Department of Conservation, Wellington.

Crow, J.F. and Kimura, M. (1970) *An Introduction to Population Genetic Theory*, Harper & Row, New York.

Dawson, D.G. (1984) Principles of ecological biogeography and criteria for reserve design. *J. Roy. Soc. NZ*, **14**, 11–15.

Department of Conservation (1990) *Corporate Plan 1990/91*, Department of Conservation, Wellington.

Erlich, P.R. (1965) The population biology of the butterfly, *Euphydras editha*. II The structure of the Jasper Ridge colony. *Evolution*, **19**, 327–36.

Erlich, P.R. and Gilbert, L.E. (1973) Population structure and dynamics of the tropical butterfly *Heliconius ethilla*. *Biotropica*, **5**, 69–82.

Erlich, P.R., White, P.R., Singer, M.C. *et al.* (1975) Checkerspot butterflies: a historical perspective. *Science*, **188**, 221–8.

Ewens, W.J. (1979) *Mathematical Population Genetics*, Springer–Verlag, Berlin.

Ewens, W.J., Brockwell, P.J., Gani, J.M. and Resnick, S.I. (1987) Minimum viable populations in the presence of catastrophes, in *Viable Populations for Conservation* (ed. M.E. Soulé), Cambridge University Press, Cambridge, pp. 59–68.

Fisher, D.E. (1991) The resource management legislation of 1991: a judicial analysis of its objectives, in *Resource Management*, Brooker & Friend, Wellington, pp. 1–30.

Franklin, I.A. (1980) Evolutionary changes in small populations, in *Conservation Biology. An Evolutionary–Ecological Perspective* (eds M.E. Soulé and B.A. Wilcox) Sinauer Associates, Sunderland, Mass, pp. 135–49.

Galbraith, M.P. (1990) Volunteer's view of the ecological restoration of an offshore island, in *Ecological Restoration of New Zealand Islands* (eds D.R. Towns, C.H. Daugherty and I.A.E. Atkinson), Dept of Conservation, Wellington, pp. 170–4.

Gilpin, M.E. (1987) Spatial structure and population vulnerability, in *Viable Populations for Conservation* (ed. M.E. Soulé), University of Cambridge Press, Cambridge, pp. 125–39.

Goodman, D. (1987) How do any species persist? Lessons for conservation biology. *Conservation Biology*, **1**, 59–62.

Gray, R.D. (1987) Beyond labels and binary oppositions: what can be learnt from the nature/nurture dispute? *Biology Forum*, **80**, 192–6.

Gray, R.D. and Craig, J.L. (1991) Theory really matters: hidden assumptions in the concept of 'habitat requirements'. *Acta XX Congrssus Internationalis Ornithologici*, **4**, 2553–60.

Greenwood, P.J. and Harvey, P.H. (1982) The natal and breeding dispersal of birds. *Annual Review Ecology and Systematics*, **13**, 1–21.
Greig, J.C. (1979) Principles of genetic conservation in relation to wildlife management in southern Africa. *South African Journal of Wildlife Research*, **9**, 57–78.
Griffith, B., Scott, J.M., Carpenter, J.W. and Reed, C. (1989) Translocation as a species conservation tool: status and strategy. *Science*, **244**, 477–80.
Hoyle, S.D. (1993) North Island saddlebacks on Tiritiri Matangi Island: a computer model of population dynamics. University of Auckland, NZ, MSc dissertation.
IUCN (1980) *World Conservation Strategy*, IUCN, Gland, Switzerland.
Janzen, D.H. (1989) The evolutionary biology of national parks. *Conservation Biology*, **3**, 109–10.
Lacy, R.C. (1987) Loss of genetic diversity from managed populations: interacting effects of drift, mutation, immigration, selection, and population subdivision. *Conservation Biology*, **1**, 143–58.
Lande, R. (1987) Extinction thresholds in demographic models of territorial populations. *American Naturalist*, **130**, 624–35.
Lande, R. and Barrowclough, G.F. (1987) Effective population size, genetic variation, and their use in population management, in *Viable Populations for Conservation*. (ed. M.E. Soulé), Cambridge University Press, Cambridge, pp. 87–124.
Lesica, P. and Allendorf, F.W. (1992) Are small populations of plants worth preserving? *Conservation Biology*, **6**, 135–9.
McCauley, D.E. (1991) Genetic consequences of local population extinction and recolonization. *Trends in Ecology and Evolution*, **6**, 5–8.
Mace, G.M. and Lande, R. (1991) Assessing extinction threats: toward a re-evaluation of IUCN threatened species categories. *Conservation Biology*, **5**, 148–57.
Merton, D. (1990) The Chatham Island black robin. *Forest and Bird*, **21** (3), 14–19.
Mills, J.A., Lee, W.G. and Mark, A.F. (1978) Takahe feeding studies – preferences and requirements, in *Proceedings of a seminar on the takahe and its habitat*, (ed. Anon) Fiordland National Park Board, Invercargill, NZ, pp. 74–94.
Mills, J.A., Lavers, R.B. and Lee, W.G. (1984) The takahe – a relict of the Pleistocene grassland avifauna of New Zealand. *NZ J. Ecology*, **7**, 57–70.
Nature Conservation Council (1981) *New Zealand Conservation Strategy: Integrating Conservation and Development*, Nature Conservation Council, Wellington.
O'Brien, S.J., Wildt, D.E., Goldman, D. *et al.* (1983) The cheetah is depauperate in genetic variation. *Science*, **221**, 459–62.
Oliver, T.J., Ober, C., Buettner-Janusch, J. and Sade, D.S. (1981) Genetic differentiation among matrilines in social groups of rhesus monkeys. *Behavioral Ecology and Sociobiology*, **8**, 279–85.
Pimm, S.L., Jones, H.L. and Diamond, J. (1988) On the risk of extinction. *American Naturalist*, **132**, 757–85.
Pimm, S.L., Gittlemen, J.L. and McCracken, G.F. (1989) Plausible alternatives to bottlenecks to explain reduced genetic diversity. *Trends in Ecology and Evolution*, **4**, 176–7.
Possingham, H.P., Davies, I., Noble, I.R. and Norton, T.W. (1992) A metapopulation simulation model for assessing the likelihood of plant and animal extinctions. *Mathematics and Computers in Simulation*, **33**, 367–72.
Quinn, J.F. and Hastings, A. (1987) Extinction in subdivided habitats. *Conservation Biology*, **1**, 198–208.
Ralls, K. and Ballou, J.D. (1983) Extinction: lessons from zoos, in *Genetics and Conservation: A Reference for Managing Wild Animal and Plant Populations* (eds

C.M. Schoenwald-Cox, S.M. Chambers, B. MacBride and L. Thomas, Benjamin/Cummings, London, pp. 164–84.

Ralls, K. and Ballou, J.D. (1986a) Preface to the preceedings of the workshop on genetic management of captive populations. *Zoo Biology*, **5**, 81–6

Ralls, K. and Ballou, J.D. (1986b) Captive breeding programs for populations with a small number of founders. *Trends in Ecology and Evolution*, **1**, 19–22.

Rasch, G. (1988) *Kokako Recovery Plan*, Department of Conservation, Wellington.

Reed, J.M., Doerr, P.D. and Walters, J.R. (1988) Minimum viable population size of the red-cockaded woodpeckers. *Journal of Wildlife Management*, **52**, 385–91.

Saunders, D.A., Hobbs, R.J. and Margules, C.R. (1991) Biological consequences of ecosystem fragmentation: a review. *Conservation Biology*, **5**, 18–32.

Schonewald-Cox, C.M., Chambers, S.M. MacBride, B. and Thomas, W.L. (eds) (1983) *Genetics and Conservation: A Reference for Managing Wild Animal and Plant Populations*, Benjamin/Cummings, Menlo Park, California.

Schwaegerle, K.E. and Schaal, B.A. (1979) Genetic variability and founder effect in the pitcher plant *Sarracenia purpurea* L. *Evolution*, **33**, 1210-18.

Shaffer, M. (1987) Minimum viable populations: coping with uncertainty, in *Viable Populations for Conservation* (ed. M.E. Soulé), Cambridge University Press, Cambridge, pp. 69–86.

Shaffer, M.L. and Samson, F.B. (1985) Population sizes and extinction: a note on determining critical sizes. *American Naturalist*, **125**, 144–52.

Simberloff, D. (1988) The contribution of population and community biology to conservation science. *Annual Review of Ecology and Systematics*, **19**, 473–511.

Simberloff, D. and Cox, J. (1987) Consequences and costs of conservation corridors. *Conservation Biology*, **1**, 63–71.

Soulé, M.E. (1985) What is conservation biology? *Bioscience*, **35**, 727–34.

Soulé, M.E. (1986) *Conservation Biology: Science of Scarcity and Diversity*, Sineauer Associates, Sunderland, Mass.

Soulé, M.E. (ed.) (1987) *Viable Populations for Conservation*, Cambridge University Press, Cambridge.

Stewart, A.M. (1980) The social organization and foraging ecology of the tui *Prosthemadera novaeseelandiae*. University of Auckland, New Zealand, MSc thesis.

Templeton, A.R., Hemmer, H., Mace, G. *et al.* (1986) Local adaptation, coa-adaptation, and population boundaries. *Zoo Biology*, **5**, 115–25.

Templeton, A.R. and Read, B. (1983) The elimination of inbreeding depression in a captive herd of Speke's gazelle, in *Genetics and Conservation: A Reference for Managing Wild Animal and Plant Populations* (eds C.M. Schoenwald-Cox, S.M. Chambers, B. MacBride, and L. Thomas). Benjamin/Cummings, London, pp. 241–61

Triggs, S.J. (1988) *Conservation genetics in New Zealand: a brief overview of principles and applications*. Report No. 22, Department of Conservation, Wellington.

Triggs, S., Williams, M.J., Marshall, S. and Chambers, G. (1991) Genetic relationships within a population of blue duck (*Hymenolaimus malacorhynchos*). *Wildfowl*, **42**, 87–93

Walter, H.S. (1990) Small viable population: the red-tailed hawk of Sorroco Island. *Conservation Biology*, **4**, 441–3.

Woolfenden, G.E. and Fitzpatrick, J.W. (1984) *The Florida Scrub Jay*, Princeton University Press, Princeton.

Wright, S. (1931) Evolution in Mendelian populations. *Genetics*, **6**, 114–38.

4

Species differences and population structure in population viability analysis

S.M. Durant and G.M. Mace
Zoological Society of London, UK.

4.1 INTRODUCTION

As the number of threatened species increases, and as we understand more about the extent of the problem, the need for effective and efficient means of identifying endangered species and populations becomes more pressing. However, recognizing that a species has a problem is only a first step – a second crucial stage is to be able to assess accurately the situation and to devise an appropriate management plan to aid recovery. Population viability analysis (PVA) is the methodology invoked to address these problems, and can be used in a variety of contexts (e.g. see Gilpin and Soulé, 1986). In this paper we show how different kinds of analysis are appropriate at different stages of devising a conservation management plan, and for different kinds of problem. If PVA is really to play a valuable role in the recovery of many species, close attention must be paid to the assumptions underlying different methods and to the biology of the species being considered (Caro and Durant, in press).

4.2 POPULATION VIABILITY METHODS

Broadly speaking there are four kinds of population viability analysis. Those using subjective assessment, rules of thumb, analytical popula-

Creative Conservation: Interactive management of wild and captive animals.
Edited by P.J.S. Olney, G.M. Mace and A.T.C. Feistner.
Published in 1994 by Chapman & Hall, London.
ISBN 0 412 49570 8

tion models or computer simulation models. Generally these methodologies become more time consuming as one moves down the list. Models, particularly those using computer simulation, can also be used to devise or test the first two methodologies.

Subjective assessments have been used for a long time, and form the basis of many species listings, such as those in the IUCN Red Data Books. Here basic biological information on the species is used by the compiler to make a simple judgement about the likelihood of extinction in very general terms (Endangered, Vulnerable, Rare, etc.)

In an attempt to increase the objectivity of this process, various qualitative rules of thumb have been proposed more recently. These have used results from demographic and genetic models to establish particular values below which a population is said to be in danger or non-viable. The most widely applied of these has been the 50/500 rule (Franklin, 1980; Soulé, 1980). This rule advocates a short-term minimum effective population size of 50 to limit the deleterious consequences of inbreeding, and a longer-term minimum effective population size of 500 to allow the maintenance of genetic variation for adaptive evolution. These values were based largely upon population-genetic considerations, and with due allowance for the relationship between census and effective population size (see Harris and Allendorf, 1989), will generally convert to census population sizes of about 100–200 and 1000–2000, respectively (Groves and Clark, 1986; Reed, Doerr and Walters, 1988; Harris and Allendorf, 1989). These across-the-board values have been criticized on the grounds that they pay undue attention to genetics and do not incorporate environmental and demographic extinction factors (Shaffer, 1981; Pimm, Jones and Diamond, 1988; Lande, 1988). Additionally, many authors have failed to take account of the difference between census and effective population size, and so grossly underestimated the population size required when applying this rule to real situations.

A more recent set of criteria proposed by Mace and Lande (1991) may also be considered as rules of thumb. However, here an attempt was made to allow for more than genetic factors, and the criteria are based on a number of population parameters in addition to the total population size, including the degree of population subdivision and historical and projected changes in population size or habitat area. The rigorous application of both kinds of rule of thumb is likely to increase the number of species listed as threatened over the application of the subjective criteria.

However, whilst subjective criteria are unreliable, rules of thumb make widespread generalizations which may not apply to many species. Population models can be used for a species specific PVA where good, quantitative and/or qualitative information is available on the demography of a particular species. Two basic kinds of approach have been taken here: analytical methods where population processes are reduced to a series of analytical formulae (e.g. Goodman, 1987; Reed, Doerr and Walters, 1988) and simulation approaches based upon direct population modelling by simulation (e.g. Shaffer, 1983; Durant and Harwood, 1992; Durant and Harwood, in press). Each has its own merits. Analytical models are particularly appropriate when only quantitative information is available. For example, a series of population sizes are suitable for analytical treatment. However these models tend to be limited in the range of population and environmental data that they can incorporate (see Thompson, 1991) and may be based on population parameters that are very difficult to measure accurately, e.g. r and variance in r (Goodman, 1987) or reproductive value (Lande and Orzack, 1988). None the less, they do have the advantage of being based on relatively few parameters and being simple to use in many cases.

Simulation approaches, by contrast, can use a wide variety of information about a species' biology and life history. In particular, provided adequate sensitivity analyses are made, they can make use of purely qualitative data. For example, they can be used to investigate the effect a polygynous breeding structure has upon population dynamics, or the effect of reproductive suppression. To some extent, therefore, simulation models can avoid the problem of data availability, provided there is careful examination of the sensitivity of the results to changes in the parameters of the model (e.g. Shaffer, 1983). However, such detailed sensitivity analyses are extremely time consuming. If this stage is to be bypassed, simulation approaches require accurate data on many different parameters, and, as with analytical approaches, some of these may be difficult to estimate accurately. As a consequence results are often presented from data which are based on 'guesstimates'. In addition, for many models the process and sequence of events involved in the simulation is undocumented, and known only to the program's author. This makes them difficult to apply to some real situations and to compare results from different programs. Furthermore, results often have high variances and error margins, which have not been examined routinely in the presentation of results. Simulation models are also often extremely species specific and may not be appropriate for making broad generalizations. Nevertheless, despite these drawbacks simulation models are powerful tools as they allow exploratory analyses that are of great value to managers, and in the development of species recovery plans. In addition they can clearly incorporate all extinction factors which are significant for small populations, which are often not well represented in analytical methods.

4.3 SPECIES DIFFERENCES AND PVA

Population models are invaluable for investigating the implications of differences between species in population viability. Population models may also be used to refine rules of thumb and subjective assessment, since they can be used to test or devise new criteria. Indeed all the rules of thumb proposed to date are based on results from population models. Whether or not to use analytical or simulation models depends on the type of data available and the time available for analysis. Until recently population models, and hence PVAs, with the single exception of subjective assessments, have neglected species differences, apart from those differences in standard demographic rates. None the less it is likely that these differences will influence every aspect of PVAs differences between species in social organization, ecology and levels of inbreeding will affect population-genetic and demographic projections. Recent work using simulations has shown that certain population characteristics that frequently precede extinction can be identified and used with a monitoring and management programme (Durant, 1991). Species' differences may affect those characteristics which are suitable for this purpose as well as the success of any management interventions. In the following sections we examine the role in the application of population viability analyses of genetic and demographic extinction factors, spatial structure and dispersal patterns, the relevance of analytical models, indicators of extinction, and effective management strategies.

4.3.1 The role of genetics and demography

Several recent studies using simulation models that incorporate demographic stochasticity as well as genetic effects (usually the effects of inbreeding depression) indicate that for several real populations that can be studied, demographic extinction factors are more significant over the short term than genetic factors. This may result from deterministically declining population growth rates, or from the effects of demographic stochasticity on very small population sizes (Dobson *et al.*, 1992). Generally, as expected, it is only over the longer term that genetic considerations become important (Lande, 1988). However this should not be taken to signify that genetic concerns can be dismissed, as those populations that do survive may then be exposed to genetic risk factors (Soulé and Mills, 1992) and there are interacting and synergistic relationships between genetic and demographic factors (Gilpin and Soulé, 1986). In general it is likely that demographic extinction factors will be important for many wild populations. A population sufficiently large to withstand random fluctuations in its environment and demography is not likely to suffer any short-term genetic effects. However, species differ in their sensitivities to inbreeding depression (Ralls, Ballou and Templeton, 1988) and this should always be taken into account when considering the relative costs of genetic and demographic effects.

4.3.2 Spatial structure and dispersal

In most cases populations are not panmictic, but instead comprise sets of interacting sub-populations. This is particularly the case with social species. In this section the consequences of population subdivision on the extinction rates of three species with broadly similar population sizes and structures but with different patterns of social behaviour are contrasted. The model used is POPGEN, a population simulation model (Durant, 1991; Durant, Harwood and Beudels, 1992). We compare the results from the Mediterranean monk seal (*Monachus monachus*), the mountain gorilla (*Gorilla g. berengei*) and the roan antelope (*Hippotragus equinus*). Results are presented from 1000 simulations.

The monk seal population in the Mediterranean is highly subdivided since it is scattered around numerous islands and rocky coastlines (Marchessaux, 1989). There are little data on individual movements in this species, though radio-tracking has revealed that they are capable of moving over quite large distances (Reijnders and Ries, 1989). It was assumed that each individual had an equal probability of migrating in any particular year, and this value was termed the migration probability. The destination of any migrating animal was chosen randomly with an equal probability from all sub-populations, including those with no resident individuals. An individual was only allowed to migrate more than once in any particular year if the first destination sub-population was at its ceiling, when it was assumed to have insufficient space for an additional individual. In this case a new destination was chosen. In the rare event that all sub-populations were at their ceilings, an individual was allowed to stay in a full sub-population after 50 attempts at finding one with space had failed. The model simulated the eastern population of Mediterranean monk seals with a total of 300 individuals, initially divided into 30, 15, 10, 5 and 1 sub-populations. Each sub-population had a ceiling of 40, 80, 120, 240 and 1200 individuals, respectively (set so the overall meta-population ceiling was constant at 1200 individuals). The migration probability was set at 0.1.

The mountain gorilla is organized into social groups containing adult males and females and their offspring (Schaller, 1963; Weber and Vedder, 1983). Some adult males are solitary. Movement of individuals between groups is known to be relatively common, but only adults and sub-adults migrate, and females will only move to join bachelor adult males or a breeding group, but are never solitary (Harcourt, Fossey and Sabater-Pi, 1981). It was assumed in the model that adult males will not migrate from breeding groups in which they are the only male, and adult males that cannot find space in breeding groups will not breed. Migration is therefore not random in this species. For the purposes of comparison, and to examine the consequences of subdivision, populations were initially set at a total of 300 individuals and were subdivided into 30, 15, 10, and 1 sub-populations as with the monk seal.

Roan antelope are also separated into breeding groups, and individuals do not migrate randomly. However roan differ from gorillas in that females do not migrate between groups. Instead, all females tend to remain within their natal group for life, forming a cohesive unit, and each group of females (a herd) is held by a single adult male who drives out male offspring before they reach sexual maturity (Joubert, 1974). Herds split after they reach a certain size and are taken over by another male (Beudels, Durant and Harwood, 1992). A population ceiling was fixed in terms of the number of adult females in a herd rather than the number of individuals. As with monk seals and gorillas a starting population of 300 was investigated and the ceiling was fixed at 420 adult females, a value which was equivalent to the total ceiling in the monk seal and gorilla models. Populations made up of 30, 15, 10 and 5 herds were investigated.

Persistence of meta-populations changed radically depending on the species investigated. In monk seals persistence decreased with increasing levels of subdivision (Figure 4.1). This was a consequence of the random nature of migration in the monk seal model. As subdivision was increased there was a greater likelihood of the animal migrating to a locality with no resident individuals of the opposite sex. In effect this acted as an additional mortality. The pattern was very different with

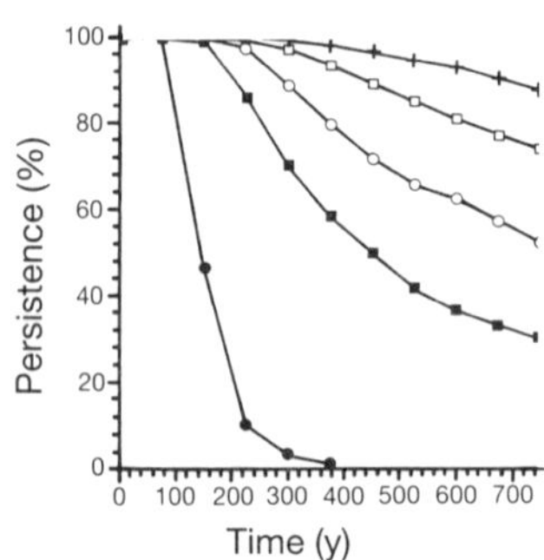

Figure 4.1 Persistence over time for meta-populations of Mediterranean monk seals under different levels of subdivision. The annual migration probability was set at 0.1 and populations were subjected only to demographic stochasticity. Initial population size was 300 individuals divided equally across sub-populations, and the overall meta-population ceiling was 1200, also divided equally across sub-populations.

Key: —+— panmictic population; —□— meta-population of five sub-populations; —○— meta-population of 10 sub-populations; —■— meta-population of 15 sub-populations; —●— meta-population of 30 sub-populations.

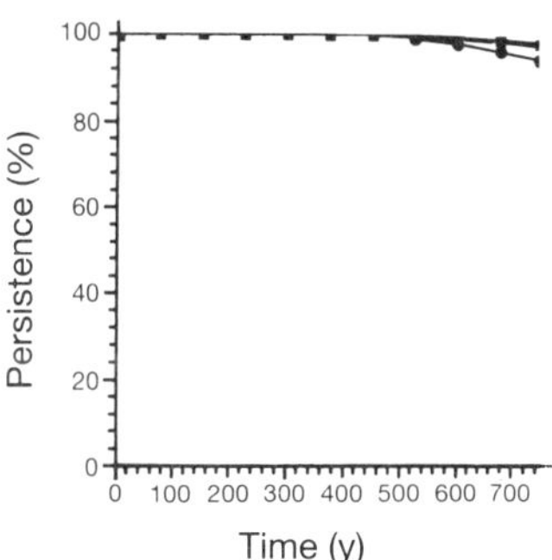

Figure 4.2 Persistence over time for meta-populations of mountain gorillas under different levels of subdivision. The annual migration probability was set at 0.1 and populations were subjected only to demographic stochasticity. Initial population size was 300 individuals divided equally across groups, and the overall meta-population ceiling was 1200, also divided equally across groups. Key as in Figure 4.1.

mountain gorillas and with roan antelope. Persistence of panmictic populations of the mountain gorilla were generally higher than those of the monk seal (Figure 4.2). This was largely a consequence of longer generation times and lower variances in the growth rate, which were a direct result of the low breeding rate and low mortality rate found in this species. Persistence only changed marginally with subdivision. Roan antelope had lower persistence rates than gorillas; however persistence of this species also showed very little change with subdivision (Figure 4.3). In the gorilla and roan models the breeding capability of the population was not reduced by migration as migrating females only join groups where they can reproduce. Unlike the monk seal example above, therefore, migration did not act as an extra mortality factor.

An interesting effect arises in the genetic diversity recorded in these simulated populations. The genetic diversity of monk seals appears to increase with subdivision (Figure 4.4); however this was largely a consequence of the large number of populations going extinct during these simulations. Small populations with low levels of genetic diversity were more likely to become extinct, and extinction was more likely as subdivision increased resulting in an apparent increase in genetic diversity. Overall genetic diversity in gorilla populations was higher than in monk seal populations (Figure 4.5). This was again a consequence of the longer generation times and lower variance in population growth, as was the case with the higher levels of persistence observed in this species. Genetic diversity did not change much with subdivision of gorilla populations. A population subdivided into 15 had significantly higher levels of allelic diversity than any other, however the difference

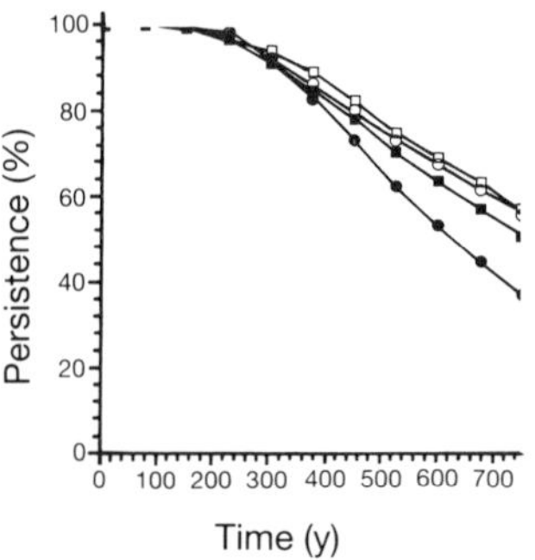

Figure 4.3 Persistence over time for meta-populations of roan antelope under different levels of subdivision. Populations were subjected only to demographic stochasticity. The initial population contained 150 females divided equally across herds. The overall meta-population ceiling was 420 adult females also divided equally across herds. Key as in Figure 4.1.

was not large. By far the strongest effect of subdivision on genetic diversity was observed in roan antelope (Figure 4.6). In this species the rate of decline in genetic diversity increased markedly as the number of herds decreased. Genetic diversity declined at a much faster rate than in the gorilla and monk seal models. These results are not surprising, since here the number of herds is an exact reflection of the number of breeding males in the population.

These examples illustrate the importance of social behaviour in population dynamics and therefore in meta-population models. If social patterns in the gorilla and roan had been neglected in this model, inappropriate management strategies could result. For example, the

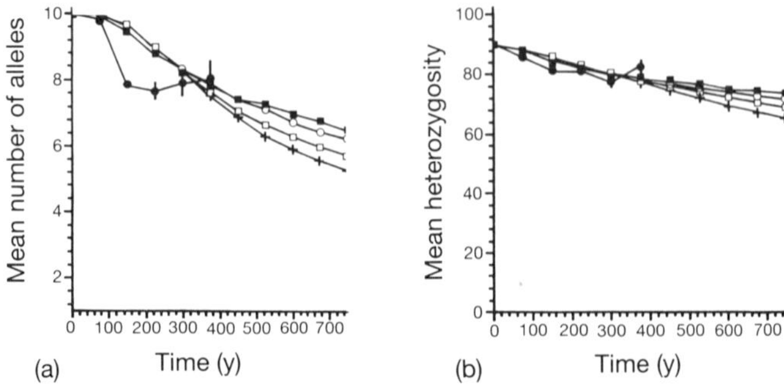

Figure 4.4 Mean number of alleles (a) and mean percentage heterozygosity (b) measured during simulations in Figure 4.1. Vertical lines indicate the standard error; key as in Figure 4.1.

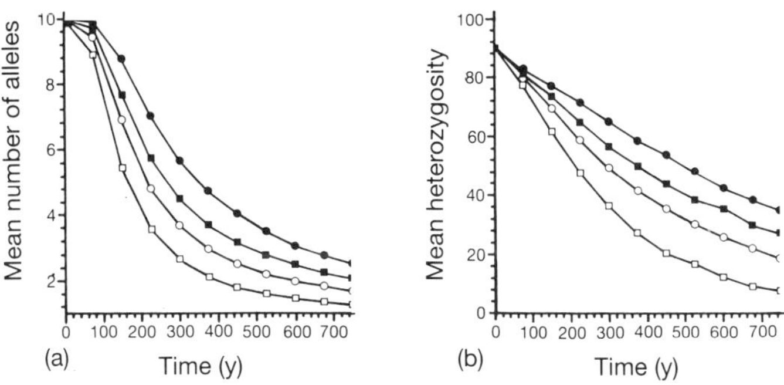

Figure 4.5 Mean number of alleles (a) and mean percentage heterozygosity (b) measured during simulations depicted in Figure 4.2. Vertical lines indicate the standard error; key as in Figure 4.1.

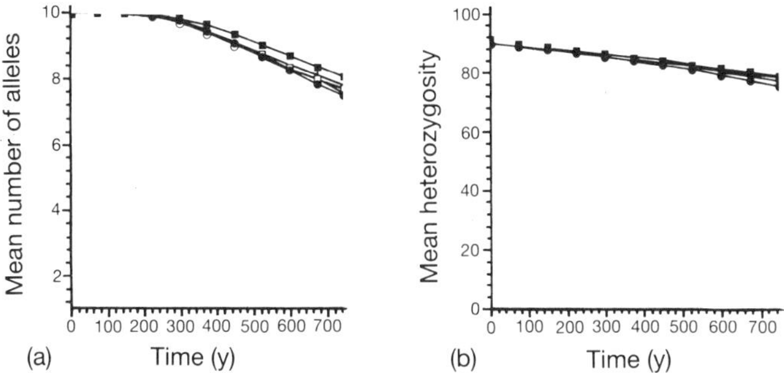

Figure 4.6 Mean number of alleles (a) and mean percentage heterozygosity (b) measured during simulations depicted in Figure 4.3. Vertical lines indicate the standard error; key as in Figure 4.1.

assumption of a random migration pattern for gorillas could have resulted in a conclusion that in order to minimize extinction risk subdivision of the populations should not be encouraged. In fact, this would have been quite inappropriate for this species.

4.3.3 The accuracy of analytical models

There are two main analytical models which are used widely within conservation biology. The first of these is based on a modified birth–death process (Goodman, 1987). The following equation (1) gives the mean time to extinction in terms of the mean and variance in the intrinsic growth rate for a population. The more accurate version of this equation – which we can call 'Model A' – uses population specific growth rates:

$$T(N) = \sum_{x=1}^{N} \sum_{y=x}^{N_c} \frac{2}{y[y\,v(y) - \mathrm{r}(y)]} \prod_{z=x}^{y-1} \frac{z\,v(z) + r(z)}{z\,v(z) - r(z)} \qquad \text{(Equation 1)}$$

where $r(z)$ and $t(z)$ are the mean and variance in the logarithm of the population growth rate for a population of z individuals. $T(N)$ is the expected time to extinction for a population of size N, and N_c is the population ceiling. An approximation – which we can call 'Model B' – uses the overall growth rate mean and variance:

$$T(N) = \sum_{x=1}^{N} \sum_{y=x}^{N_c} \frac{2}{y[y\,v - r]} \prod_{z=x}^{y-1} \frac{z\,v + r}{z\,v - r} \qquad \text{(Equation 2)}$$

The second main analytical model – which we can call the 'diffusion model' – is based on a diffusion process and uses the mean multiplicative growth rate and the initial reproductive value (Lande and Orzack, 1988):

$$T(N) = \ln V_0 / |r| \qquad \text{(Equation 3)}$$

where r is the mean logarithmic growth rate and V_o is the reproductive value of the initial population.

These models are used widely yet little is known about their accuracy for real populations. Ideally they should be tested against real populations; however the time scale for such an experiment would be unfeasible. As an alternative, the equations can be tested against simulation models. If there are clear departures against some or all of the assumptions of the models, then this gives an indication of how applicable the equations are to real populations. This strategy was used by Harris and Allendorf (1989) for a test of various estimates of genetically effective population size.

POPGEN was used to generate mean extinction times under 1000 simulations and a record was kept of population size and growth rate for each year of the simulation. Under these circumstances the diffusion model gave extinction times which were mostly within 10% of the mean extinction time generated by simulation as the deterministic growth rate was changed and the population subjected to increasing levels of environmental stochasticity (Figures 4.7 and 4.8). The birth–death process performed less well, and gave extinction time estimates as much as 70% below the simulated value. Interestingly, in none of the three species investigated did the models become markedly less accurate as populations were subdivided (Table 4.1). This suggests that the extinction dynamics of these meta-populations is reflected entirely in the growth rate mean and variance. Therefore, if the way in which subdivision might affect growth rate mean and variance can be predicted it,

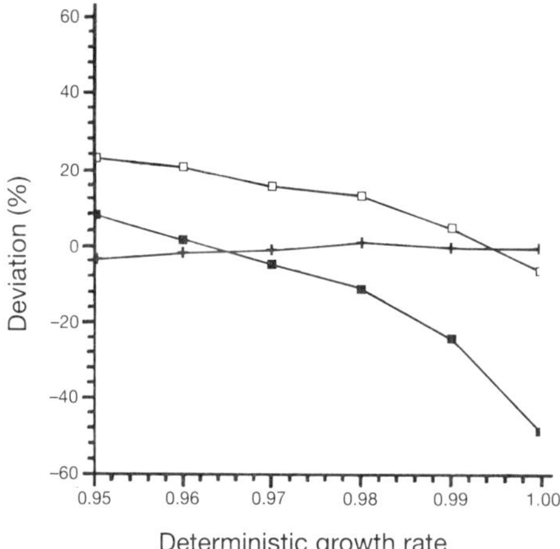

Figure 4.7 Deviation of expected time to extinction, calculated using published equations from realized values generated by simulation for panmictic populations of Mediterranean monk seals with a varying deterministic growth rate. Populations were subjected only to demographic stochasticity, and initially numbered 30 individuals with a ceiling of 100.
Key: —■— Birth–death process Model A (Equation 1); —□— Birth–death process Model B (Equation 2); —+— Diffusion model (Equation 3).

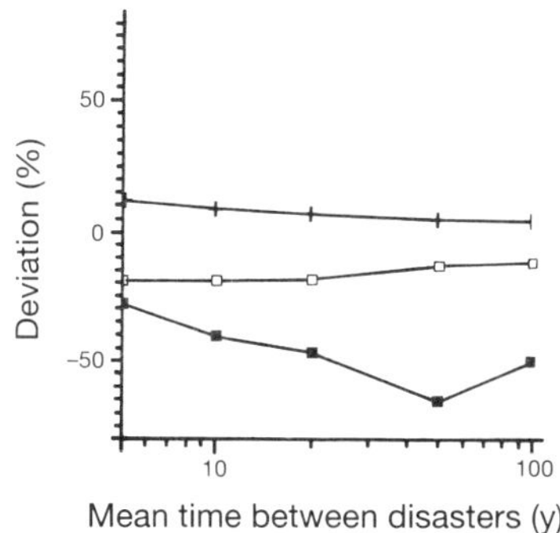

Figure 4.8 Deviation of expected time to extinction, calculated using published equations from realized values generated by simulation for panmictic populations of Mediterranean monk seals subjected to disasters of varying frequency. Disasters acted on both juvenile and adult survival and reduced survival by one half in each year they occurred. Populations were subjected only to demographic stochasticity, and initially numbered 30 individuals with a ceiling of 100.
Key: —■— Birth–death process Model A (Equation 1); —□— Birth–death process Model B (Equation 2); —+— Diffusion model (Equation 3).

Table 4.1 Deviation of expected time to extinction, calculated using published equations (see text) from realized values generated by simulation for meta-populations of three species under various levels of subdivision; populations initially numbered 100 individuals, and the overall ceiling was 400 individuals

Species	*Number of sub-populations*	*Mean intrinsic growth rate*	*Variance of intrinsic growth rate*	*Deviation (%) from mean extinction time generated by simulation*		
				Birth–death Model A (Equation 1)	*Birth–death Model B (Equation 2)*	*Diffusion model (Equation 3)*
Mediterranean monk seal	1	−0.00363	0.00430	−55	−17	−3
	5	−0.00789	0.00485	−42	21	−1
	10	−0.02439	0.00974	−21	10	1
Mountain gorilla	1	−0.00160	0.00164	−53	−12	−15
	5	−0.00197	0.00177	−48	−9	−12
	10	−0.00307	0.00237	−39	−4	−11
Roan antelope	1	−0.00385	0.00916	−66	−44	−1
	5	−0.00631	0.01290	−57	−40	−1
	10	−0.00804	0.01434	−52	−36	3

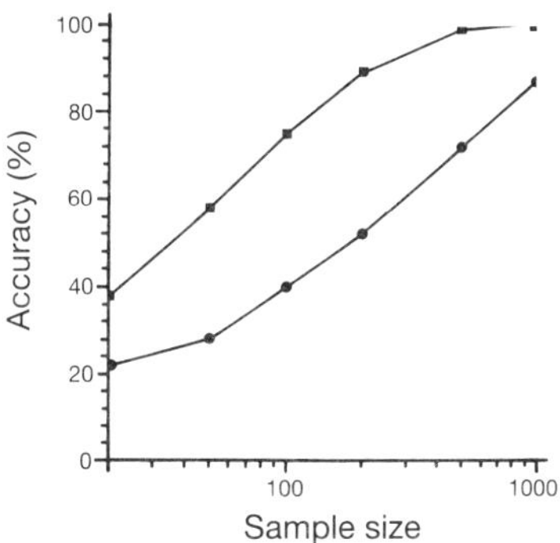

Figure 4.9 Sensitivity of estimates of expected time to extinction to varying growth-data sample sizes. Estimates of expected time to extinction were calculated using published equations. Accuracy was calculated as the percentage chance that an estimate was within 20% of the realized value generated by simulation. Data were taken from simulations of panmictic populations of Mediterranean monk seals subjected only to demographic stochasticity. Populations initially numbered 30 individuals and the ceiling was 100.
Key: —●— Birth–death process Model B (Equation 2); —■— Diffusion model (Equation 3).

should be possible to make estimates of extinction time using purely analytical models that are no less accurate than estimates for panmictic populations.

Unfortunately the accuracy of estimates declined rapidly with sampling. If only incomplete data were available, such as only a few years of consecutive data on one population, estimates were highly inaccurate (Figure 4.9). Since short sequences of growth-rate data are likely to be the norm for wild populations this has rather serious consequences for these models. It may well be the case that in the majority of real situations where there is a need to estimate viability, and only limited wild data are available, these analytical models will be highly inaccurate.

4.3.4 Indicators of extinction

The conclusions from the above section have serious consequences, since they suggest that analytical models require impossibly good growth rate data if they are to produce meaningful predictions of mean extinction time. If it were possible to test them, the same is likely to be true of simulation models. This suggests that a different approach is needed. Rather than concentrating on obtaining accurate estimates of difficult parameters, models can be used to suggest new population parameters which are sufficiently robust to be useful in field situations and which are useful for assessing extinction risk. This leads into a quest for 'indicators of extinction', easily measurable population characteristics which, when exhibited in a population, provide good evidence that the population is at some level of risk of extinction.

If such population characteristics could be identified, they would be appropriate for wildlife managers to monitor with a view to appropriate interventions being put in place as and when necessary. The metapopulation models described above were used to compare the effectiveness of a number of candidate indicators of extinction. Suitable indicators will be based on characteristics that are easily measured in natural populations, that give adequate warning of extinction (to allow time for management interventions), that have a high predictive power (frequently occur prior to extinction) and occur regularly in the course of most population histories.

Twelve different sub-population indicators based on a variety of demographic characteristics were found to be useful as indicators of extinction (Durant, 1991). These indicators included population sex

Table 4.2 Indicators of extinction used in analysis

Indicator number	*Description*
1	No offspring born in one year, population less than 20 individuals
2	No offspring born in one year, population less than 15 individuals
3	No offspring born in two successive years, population falls to less than 15 within 10 years
4	No offspring born in three years out of five
5	Population less than 20% or more than 80% female
6	Population less than 23% or more than 80% female
7	Population less than 23% or more than 80% female and population size less than 20
8	Adults less than 26% or more than 80% female and population size less than 20
9	Juveniles less than 26% or more than 90% female and population size less than 20
10	Juveniles less than 23% or more than 90% female and population size less than 20
11	Juveniles less than 26% or more than 90% female and population size less than 20 within 10 years
12	Juveniles less than 26% or more than 90% female and population size less than 15 within 10 years

ratios, absolute population size and frequency of years of zero recruitment as population characteristics (Table 4.2). The performance of these indicators was compared between species, and each indicator was scored for its effectiveness in 'warning', 'power' and 'occurrence'. The 'warning' was defined as the percentage chance that the species being simulated persisted for 10 years after the indicator was exhibited. The 'power' was defined as the percentage chance of extinction within 60 years from an indicator. Finally, the 'occurrence' was defined as the percentage chance that an indicator occurred during a simulation. The results (Table 4.3) show that the indicators were most powerful at predicting extinction in the monk seal and roan antelope simulations. On the whole, indicators with high power tended to give poor 'warning' and low 'occurrence' (e.g. indicators 3, 4, 5, 6 and 7), whereas indicators which gave good 'warning' and had a high 'occurrence' tended to have low 'power' (e.g. indicators 1, 2, 8, 9, 10, 11 and 12).

Presence/absence indicators, which measured the number of breeding groups still extant, were also found to be useful indicators of extinction within meta-population structures (Durant, 1991). The effectiveness of this measure was also compared between the three species.

These indicators tended to perform well on 'warning' and 'occurrence', but less well on 'power' (Table 4.4). As with the sub-population indicators, the presence/absence indicators were most powerful at predicting extinction for the monk seal and roan antelope models, and were much less powerful for gorillas (Tables 4.3 and 4.4). None the less, most indicators were nearly always exhibited before extinction, and so a population would be unlikely to go extinct without warning. In addition there was generally a period of at least 15 years before extinction occurred (warning), allowing time for management action.

4.3.5 Management at indicators

If indicators of extinction are to be useful for management then the effectiveness of various kinds of management intervention put into effect when indicators are observed needs to be examined. Management must be aimed at one or more of the demographic rates in a population. Age of first reproduction and longevity are unlikely to be amenable to manipulation, but survival and reproduction could be altered with varying degrees of difficulty depending on the species concerned. The consequences of management actions directed at increasing juvenile survival, such as might be achieved by captive rearing, were examined. Juvenile survival rates were increased to 1.0 when an indicator was exhibited in a population and were maintained at that level until the indicator had not been exhibited for at least 10 years. This management strategy was effective at reducing the risk of extinction both when

Table 4.3 Performance of indicators of extinction for meta-populations of monk seals, gorillas and roan antelope subjected to only demographic stochasticity[a]

	Species	*Indicator*[b]											
		1	*2*	*3*	*4*	*5*	*6*	*7*	*8*	*9*	*10*	*11*	*12*
Warning (%)	Monk seal	92	86	71	64	62	67	65	80	90	98	93	88
	Gorilla	100	99	99	100	81	86	83	95	99	99	100	99
	Roan antelope	94	93	73	67	–	–	–	–	99	99	99	98
Power (%)	Monk seal	76	82	88	91	89	89	91	82	77	79	73	81
	Gorilla	21	36	36	5	66	61	67	42	27	30	25	39
	Roan antelope	57	64	79	83	–	–	–	–	46	48	44	53
Chance of occurrence (%)	Monk seal	100	100	96	93	54	70	68	91	99	99	99	99
	Gorilla	100	100	100	100	69	85	83	99	100	100	100	100
	Roan antelope	100	100	91	87	–	–	–	–	100	100	100	100

[a]Populations initially numbered 100 individuals divided across 10 sub-populations each with a ceiling of 40 (or equivalent number of females in the case of roan). The migration probability was set at 0.01.
[b]Indicators were classified according to Table 4.2. Because the roan model does not give an accurate reflection of the number of males in the population indicators 5 to 9 were not appropriate for this species.

Table 4.4 Performance of presence/absence indicators of extinction for simulated meta-populations of monk seals, gorillas and roan antelope subjected to only demographic stochasticity[a]

Species	*Monk seal*				*Gorilla*				*Roan antelope*			
Number of breeding groups	*1*	*2*	*3*	*4*	*1*	*2*	*3*	*4*	*1*	*2*	*3*	*4*
Warning (%)	97	100	100	100	99	100	100	100	87	98	100	100
Power (%)	52	30	17	10	11	1	0	0	61	41	24	15
Chance of occurrence (%)	100	100	100	100	100	100	100	100	99	100	100	100

[a]Indicators were registered when meta-populations were reduced to 1, 2, 3 and 4 breeding groups. Populations initially numbered 100 individuals divided across 10 sub-populations each with a ceiling of 40 (or equivalent number of females for roan).

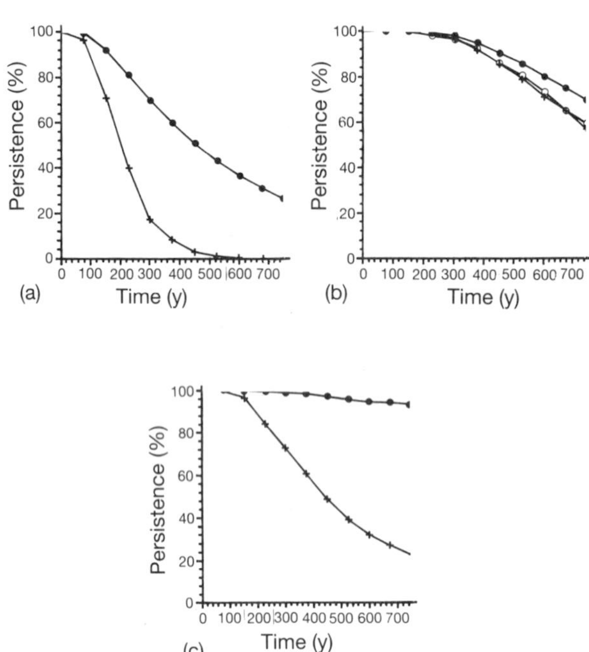

Figure 4.10 The effect of management on persistence over time for populations of Mediterranean monk seal (a), mountain gorilla (b) and roan antelope (c). Meta-populations were made up of 10 equal sized subunits with a migration probability of 0.01, and were subjected only to demographic stochasticity. Each sub-unit had an initial size of 10 individuals (five females for roan antelope) and had a ceiling of 40 (14 for roan antelope). Indicators were measured across the entire meta-population. Management, which increased juvenile survival to 1.0, was initiated whenever an indicator occurred, and was maintained until 10 years had passed without the indicator being exhibited.
Key: —x— no management; —●— management using indicators 1, 2, 8, 9, 10, 11 and 12 (see Table 4.2).

demographic indicator number 10 (see Table 4.2) was used (Figure 4.10), and when presence/absence indicators were used (Figure 4.11). Management was most intensive in the case of the monk seal, where 17.5% of the time was engaged in active management (Table 4.5). Interestingly, when presence/absence indicators were used, management did not necessarily increase in intensity if management was initiated at weaker indicators, such as when the meta-population had dropped to six or eight sub-populations (Table 4.6). However, if allowance is made for the fact that populations are larger at weaker indicators, and may thus use more resources at each intervention, then there is a consistent increase in time spent managing as indicators decrease in strength.

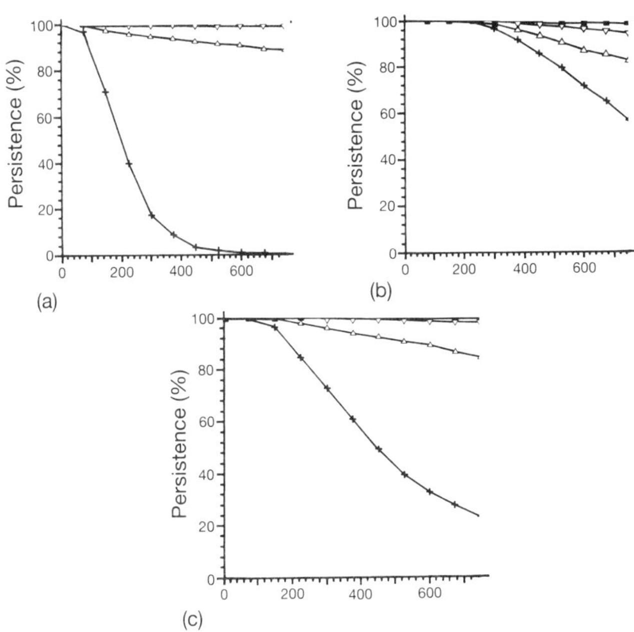

Figure 4.11 The effect of management at presence/absence indicators on persistence over time for meta-populations of Mediterranean monk seal (a), mountain gorilla (b) and roan antelope (c). Meta-populations were subjected only to demographic stochasticity, and were made up of 10 equal sized sub-units as in Figure 4.10. Management was initiated whenever the number of subunits dropped to the numbers indicated.
Key: —+— no management; —△— management at one breeding group; —◇— management at two breeding groups; —□— management at four breeding groups; —○— management at six breeding groups; —●— management at eight breeding groups.

4.4 METHODS IN CONTEXT

The previous section shows that there are three different though related processes that need to be developed in the context of species conservation. The first is to signal that a species is in trouble and to alert observers that something needs to be done (warning). The second is to examine the status of the species or population in question and to make some accurate prediction of the probability of extinction if current circumstances prevail. The third, where there does seem to be a serious problem, is to develop a coherent management plan to reduce extinction risk to acceptable levels.

Table 4.5 Frequency and length of management interventions at indicator 10 during 750 years of simulation for meta-populations of monk seals, gorillas and roan antelope[a]

Species	*Monk seal*	*Gorilla*	*Roan antelope*
Mean number of management interventions every 100 years	1.25	0.37	0.42
Mean number of years spent in management out of every 100 years	17.5	7.6	5.6
Mean duration of each management intervention	14.2	20.1	13.4

[a]Subjected to demographic stochasticity only with a migration probability of 0.01 where appropriate. Populations initially numbered 100 individuals, and were subdivided into 10 sub-populations, each with a ceiling of 40 (or an equivalent number of females in the case of roan).

Each of these three stages has different properties. In the warning phase, many species will be involved and the method needs to be relatively simple and of general applicability. The method should also be conservative since the function of this process is to take the species into the prediction and recovery phase; warning on too many species has less severe consequences than warning on too few. Second, at the prediction phase, the highest priority is for methods to be reliable, since valuable resources should not be wasted on species which would survive without intervention. At the last stage – management planning – both reliability and flexibility are important. Table 4.7 indicates how appropriate each method is in these various stages. A progression from simple rules of thumb at the assessment phase to simulation models at the management phase emerges.

Genetic and population dynamic models have a strong part to play in providing warnings when a population is in danger of extinction. It is then essential to consider every aspect of a species' biology to assess how critical the position is. For example, where a species can be considered to be showing an indicator of extinction such as those examined here, it should be immediately brought into the management phase. This phase by necessity has to be the most species specific of all. Management options are limited by the practical possibilities for the species. In the case of the monk seal, Hawaiian monk seals have been successfully reared as pups and re-released, dramatically raising the juvenile survival for this species (Gerodette and Gilmartin, 1990). This has been a very effective strategy for this species. Such a strategy could also be feasible for roan antelope, but not for gorillas. It would be virtually impossible to reintroduce captive reared animals into existing

Table 4.6 Frequency and length of management interventions at prescence/absence indicators[a]

Number of breeding groups at management intervention	*1*	*2*	*4*	*6*	*8*
Mediterranean monk seal					
Mean number of management interventions every 100 years	1.18	1.32	1.17	1.04	1.20
Mean number of years spent in management out of every 100 years	40.3	30.2	19.1	15.4	17.1
Number of management sup-population years	40.3	60.4	76.4	92.4	136.0
Mean duration of each intervention	35.1	22.9	16.3	14.8	14.3
Mountain gorilla					
Mean number of management interventions per 100 years	0.45	0.63	1.04	1.23	1.16
Mean number of years spent in management out of every 100 years	16.9	25.2	36.1	44.3	56.4
Number of management sub-population years	16.9	50.4	144.4	265.8	451.2
Mean duration of each intervention	37.2	39.7	34.5	35.9	48.6
Roan antelope					
Mean number of management interventions per 100 years	0.30	0.31	0.32	0.39	1.00
Mean number of years spent in management out of every 100 years	5.0	5.1	4.7	5.3	10.8
Number of management sub-population years	5.0	10.2	18.8	31.8	86.4
Mean duration of each intervention	17.3	16.7	14.5	13.4	10.8

[a]A migration probability of 0.01 was set over 750 years of simulation for meta-populations of three species subjected only to demographic stochasticity. Management was initiated when the meta-population dropped below the stated number of breeding groups. Populations initially numbered 100 individuals, and were subdivided into 10 subunits, each with a ceiling of 40 individuals. Mean number of management sub-population years was calculated from the mean number of years spent in management multiplied by the number of breeding groups at intervention.

Table 4.7 The applicability of different kinds of population analysis for different kinds of conservation actions (Yes = suitable; No = unsuitable; other comments specify problems of or limits to use)

	Conservation actions		
Kinds of population analysis	*Warning*	*Prediction*	*Management guidance*
Subjective assessment	Yes	No	No
Rules of thumb	Yes	Very unreliable	No
Analytical population models	Impractical	Only large populations	No
	Low reliability	Good data, low env. variation	
Stochastic population models	Impractical	Yes	Yes

gorilla groups. Even if it were possible, captive rearing is less successful than wild rearing in western lowland gorillas (*Gorilla g. gorilla*) (Mace, 1988) though it is improving all the time, and there is little experience with mountain gorillas. For this species a more cautionary scheme would be advisable, perhaps where veterinary care could be provided at a constant level, or just when numbers dropped to a lower but still fairly high number of groups, such as six out of a maximum of 10. Then it would still be possible for a slightly raised survival rate to keep the population from dropping to a critical level.

4.5 CONCLUSIONS

Population models have different roles to play at different stages in identifying endangered species and in planning for their recovery. Stochastic models need to be developed with thought and due regard to the species they are being applied to, but can provide useful information on the factors influencing extinction risk, as well as on appropriate monitoring and management strategies. Here we have examined only a very limited number of species. However we have shown that models can aid in the formulation of monitoring and management strategies. Furthermore, models can be used to identify those strategies which are likely to be least intensive in their use of resources and time.

By their very nature, indicators of extinction are not sensitive to the population ceiling and initial conditions. They therefore form a basis for a pragmatic approach to conservation which is particularly suited to situations where little information is available. A skewed sex ratio was found to be a particularly useful characteristic for indicating extinction. As a population becomes smaller, demographic stochasticity increased, and the sex ratio was more likely to become skewed. However other factors could also be useful in this context. In particular, frequencies of alleles at a locus may also become skewed as population size decreases and the effects of genetic drift come to predominate. By choosing an appropriate number of loci with sufficient variation at each it should be possible to monitor changes in population size. Current advances in DNA technology enable single loci to be monitored using small samples of skin, hair and even faeces. Therefore it is possible to envisage a strategy where a population could be monitored demographically using genetic analyses. Management based upon intervention when extinction indicators are observed can maintain populations which are at low numbers, and was effective for all three species simulated here.

ACKNOWLEDGEMENTS

S.M.D. was supported by a National Environment Research Council Studentship and thanks John Harwood for his support and encouragement, and the Department of Zoology, Cambridge University for providing bench space. G.M.M. was supported by the Pew Scholars Program in Conservation and the Environment.

REFERENCES

Beudels, R., Durant, S.M. and Harwood, J. (1992) Assessing the risks of extinction for local populations of roan antelope (*Hippotragus equinus*). *Biological Conservation*, **61**, 107–16.

Caro, T.M. and Durant, S.M. (in press) The importance of behavioural ecology for conservation biology: examples from studies of Serengeti carnivores, in *Serengeti II: Research, Management and Conservation of an Ecosystem* (eds A.R.E. Sinclair and P. Arcese), University of Chicago Press, Chicago.

Dobson, A.P., Mace, G.M., Poole, J. and Brett, R.A. (1992) Conservation biology: the ecology and genetics of endangered species, in *Genes in Ecology* (eds R.J. Berry, T.J. Crawford and G.M. Hewitt), Blackwell Scientific Publications, Oxford, pp. 405–30.

Durant, S.M. (1991) Individual variation and dynamics of small populations: implications for conservation and management. Ph.D. Thesis, Cambridge University.

Durant, S.M. and Harwood, J. (1992) Assessment of monitoring and management strategies for local populations of *Monachus monachus*, the Mediterranean monk seal. *Biological Conservation*, **61**, 81–92.

Durant, S.M. and Harwood, J. (in press) The effects of social organisation on the genetics and extinction dynamics of small populations of Mediterranean monk seal (*Monachus monachus*) and mountain gorilla (*Gorilla gorilla berengei*). *Conservation Biology*.

Durant, S.M., Harwood, J. and Beudels, R. (1992) Monitoring and management strategies for endangered populations of marine mammals and ungulates, in *Wildlife 2001* (eds D.R. McCullough and R.H. Barrett), Elsevier Applied Science, New York, pp. 252–61.

Franklin, I.R. (1980) Evolutionary change in small populations, in *Conservation Biology: An Evolutionary–Ecological Perspective* (eds M.E. Soulé and B.A. Wilcox), Sinauer Associates, Sunderland, Mass, pp. 135–49.

Gerodette, T. and Gilmartin, W.G. (1990) Demographic consequences of changed pupping and hauling sites of the Hawaiian monk seal. *Conservation Biology*, **4**(4), 423–30.

Gilpin, M.E. and Soulé, M.E. (1986) Minimum viable populations: processes of species extinctions, in *Conservation Biology – the Science of Scarcity and Diversity* (ed. M.E. Soulé), Sinauer Associates, Michigan pp. 19–34.

Goodman, D. (1987) The demography of chance extinction, in *Viable Populations for Conservation* (ed. M.E. Soulé), Cambridge University Press, Cambridge, UK., pp. 11–34.

Groves, C.R. and Clark, T.W. (1986) Determining minimum population size for recovery of the Black-footed ferret. *Great Basin Naturalist Memoirs*, **8**, 150–9.

Harcourt, A.H., Fossey, D. and Sabater-Pi, J. (1981) Demography of *Gorilla gorilla*. *Journal of Zoology, London*, **195**, 215–33.

Harris, R.B. and Allendorf, F.W. (1989) Genetically effective population size of large mammals: an assessment of estimators. *Conservation Biology*, **3**(2), 181–91.

Joubert, S.C.T. (1974) The social organisation of the roan antelope (*Hippotragus equinus*) and its influence on the spatial distribution of herds in the Kruger National Park, in *The Behaviour of Ungulates and its Relation to Management* (eds V. Geist and F. Walther), IUCN, Morges pp. 661–75.

Lande, R. (1988) Genetics and demography in biological conservation. *Science*, **241**, 1455–60.

Lande, R. and Orzack, S.H. (1988) Extinction dynamics of age-structured populations in a fluctuating environment. *Proceedings of the National Academy of Sciences, USA*, **85**, 7418–21.

Mace, G.M. (1988) The genetic and demographic status of the western lowland gorilla (*Gorilla g. gorilla*) in captivity. *Journal of Zoology, London*, **216**, 629–54.

Mace, G.M. and Lande, R. (1991) Assessing extinction threats: toward a reevaluation of IUCN threatened species categories. *Conservation Biology*, **5**(2), 148–57.

Marchessaux, D. (1989) The biology, status and conservation of the monk seal (*Monachus monachus*). Report to the Council of Europe. Nature and Environment series No. 41, Palais de l'Europe, Strasbourg.

Pimm, S.L., Jones, H.L. and Diamond, J.M. (1988) On the risk of extinction. *American Naturalist*, **132**, 757–85.

Ralls, K., Ballou, J.D. and Templeton, A. (1988) Estimates of lethal equivalents and the cost of inbreeding in mammals. *Conservation Biology*, **2**, 185–93.

Reed, J.M., Doerr, P.D. and Walters, J.R. (1988) Minimum viable population size of the red-cockaded woodpecker. *Journal of Wildlife Management*, **52**(3), 385–91.

Reijnders, P.J.H. and Ries, E.H. (1989) *Release and radio-tracking of two rehabilitated monk seals in the marine park 'Northern Sporades', Greece*, Research Institute for Nature Management, The Netherlands.

Schaller, G.B. (1963) *The Mountain Gorilla, Ecology and Behaviour*, University of Chicago Press, Chicago

Shaffer, M.L. (1981) Minimum population sizes for species conservation. *BioScience*, **31**, 131–4.

Shaffer, M.L. (1983) Determining minimum viable population sizes for the grizzly bear. *International Conference on Bear Research and Management*, **5**, 133–9.

Soulé, M.E. (1980) Thresholds for survival: maintaining fitness and evolutionary potential, in *Conservation Biology: An Evolutionary–Ecological Perspective*, (eds M.E. Soulé and B.A. Wilcox), Sinauer Associates, Sunderland, Mass pp. 151–69.

Soulé, M.E. and Mills, L.S. (1992) Conservation genetics and conservation biology: a troubled marriage, in *Conservation of Biodiversity for Sustainable Development* (eds O.T. Sandlund, K. Hindar and A.H.D. Brown), Scandinavian University Press, Oslo, pp. 55–69.

Thompson, G.G. (1991) Determining minimum viable populations under the Endangered Species Act. NOAA Technical Memorandum, F/NWC-198, US Department of Commerce, Springfield, Virginia.

Weber, A.W. and Vedder, A. (1983) Population dynamics of the Virunga gorillas: 1959–1978. *Biological Conservation*, **26**, 341–66.

5

Molecular genetics of endangered species

R.K. Wayne, M.W. Bruford, D. Girman[a], W.E.R. Rebholz, P. Sunnucks and A.C. Taylor[b]*
Zoological Society of London, UK.

5.1 INTRODUCTION

Management of rare and endangered species in captivity and in the wild requires an understanding of their relationships to other species and of the genetic units each species contains (Avise, 1989, 1992; May, 1990; Vane-Wright, Humphries and Williams, 1991; Dizon *et al.*, 1992). This paper includes a brief discussion of definitions of species and their population subdivisions, the genetic costs and risks associated with conservation programmes focused on systematic divisions below the species level, and the use of molecular techniques to reveal taxonomic units. We also provide some examples of the application of molecular techniques to current problems in the genetic management of rare and endangered species.

5.2 THE DEFINITIONS OF SPECIES AND SUBSPECIES

A vigorous debate has developed over the definition of a species (Templeton, 1989; O'Brien and Mayr, 1991; Mayr and Ashlock, 1991; Amato, 1991; Baum, 1992; Ereshefsky, 1992). Populations which show some measure of reproductive isolation in nature are commonly treated

[a] Department of Biology, UCLA, Los Angeles.
[b] School of Biological Science, University of New South Wales, Australia.
* Corresponding author.

Creative Conservation: Interactive management of wild and captive animals.
Edited by P.J.S. Olney, G.M. Mace and A.T.C. Feistner.
Published in 1994 by Chapman & Hall, London.
ISBN 0 412 49570 8

as species. The formalization of this concept, termed the 'biological species concept', maintains that species are freely interbreeding populations reproductively isolated from other populations (Mayr, 1963). As such, species are gene pools that evolve independently so as to define a unique phylogenetic heritage (Simpson, 1961; Mayr and Ashlock, 1991). They are the raw material of species selection and phylogenetic trends (Mayr, 1963; Stanley, 1979). The biological species concept maintains that subspecies are phylogenetically distinct populations that are normally allopatric but may naturally interbreed if barriers to gene flow are removed.

Although conceptually satisfying, in practice the biological species concept may be difficult to apply, and has been severely criticized on several grounds (e.g. Cracraft, 1983; McKitrick and Zink, 1988). The common occurrence of hybridization between species that are otherwise phenotypically and behaviourally distinct is a common criticism. More troubling is the difficulty of testing the biological species concept, because natural tests of reproductive isolation are rarely evident. Moreover, reproductive compatibility may be a phylogenetically primitive character, and taxa which have been separated for long periods of time may not necessarily have become reproductively isolated. Various alternatives have been developed (Templeton, 1989). One opposing viewpoint is embodied by the 'phylogenetic species concept' which maintains that any population of individuals that can be united by one or more unique traits may be considered a species (McKitrick and Zink, 1988). Reproductive incompatibility, because it may be a primitively common character, is not considered sufficient to define a species. Under the phylogenetic species concept many subspecies are likely to be elevated to the rank of species. Although the phylogenetic species concept can be applied objectively, it ignores apparent reproductive divisions among populations. Reproductive barriers, regardless of their phylogenetic polarity, constrain sympatric populations to independent phylogenetic trajectories, whereas populations with incomplete barriers may merge genetically. These dramatically different evolutionary consequences suggest a fundamental role for reproductive isolation which should not be overlooked in a species concept (O'Brien and Mayr, 1991).

A position that combines some of the strengths of both concepts was recently advocated by Avise and Ball (1990). These authors suggest that taxonomic units should be defined as in the phylogenetic species concept by the possession of uniquely derived traits (Wiley, 1981). Populations defined by such characters would be considered subspecies if they were primarily allopatric but interbred freely in the zones of overlap, or if barriers to gene flow were removed. In contrast, populations defined by unique traits but which were reproductively isolated under natural conditions would be considered species. An important additional element here is that taxa should be defined by a suite of phylogenetically concordant characteristics, i.e. multiple morphological and molecular genetic approaches should reveal similar phylogenetic units. The presence of such units is considered to reflect unique and independent evolutionary histories which should be separately conserved.

5.3 THE PROBLEM: TO PRESERVE OR INTERBREED?

In a world with unlimited resources, arguably any endangered population with a unique set of characteristics and a distinct evolutionary history should be separately conserved. However, considering the thousands of endangered species that are likely to exist, and the numerous taxonomic units they contain, it would be impossible with the limited space in protected areas and zoological parks to save them all (Soulé *et al.*, 1986). One concept that has been proposed to assist in determining a hierarchical ranking of species and populations involves an assessment of the relative phylogenetic distinction of a taxon (May, 1990; Vane-Wright, Humphries and Williams, 1991; Erwin, 1991; Dizon *et al.*, 1992). Taxa may be ranked according to the number of closely related forms that exist and the phylogenetic distance among them (Vane-Wright, Humphries and Williams, 1991). For example, the giant panda is the most phylogenetically distinct species of bear and has no close living relative (O'Brien *et al.*, 1985). Consequently, its genome is the only living record of a unique and long history of change and adaptation to a specific environment which could not readily be filled in a similar way by another species. The same arguments could be applied to other taxa ranging from beetles to bushbabies and can be applied to distinct species as well as the populations contained within them (e.g. Avise, 1989, 1992; Avise and Ball, 1990; Daugherty *et al.*, 1990).

However, the separate conservation of taxonomically distinct units within species may significantly increase the likelihood that the species and its subunits may go extinct. Phylogenetically distinct populations may be so small that, due to inbreeding effects or demographic factors associated with small population size, the probability is much higher that the species will go extinct if conserved as small independent phylogenetic units rather than as a single or several larger interbreeding populations. Moreover, depending on the specific sizes and distributions of populations, the loss of genetic variability in the entire species may be accelerated by division into small populations. Thus, the potential benefits of conservation schemes that preserve each phylogenetically distinct population need to be weighed against an increased risk of the loss of the species or larger population grouping (Ryder, 1986).

Apart from the loss of phylogenetic distinction, there are other potential hazards of interbreeding dissimilar populations (Templeton *et al.*, 1986; Templeton, 1986). For example, the loss through hybridization of specific adaptations to a particular environment. Often such adaptations are difficult to assess prior to interbreeding, but potential candidates are adaptations to differences in day-length, seasonality and food resources. More subtle is the suggestion that outbreeding depression may result from the dissolution of co-adapted gene complexes (Templeton, 1986). Distinct chromosomal races may indeed suffer some outbreeding depression, but the importance of maintaining the integrity of co-adapted gene complexes has not been well established. Interspecific hybrids often have reduced fertility and viability, but within species, hybrid vigour may occur resulting in increased fitness of hybrids. Theoretical analysis suggests that after an initial increase due to hybrid vigour, fitness of hybrids should decrease with increasing phylogenetic distance (Lynch, 1991).

Molecular genetic techniques may provide direct evidence of the potential for successful hybridization through documentation of gene flow between distinct populations. Otherwise, an indirect measure is provided by the relative genetic distance between populations. For example, the absence of appreciable genetic differentiation between East and South African populations of black rhino has been used to argue that they need not be conserved as separate conservation entities (Ashley, Melnick and Western, 1990). However, extreme caution must be used in this argument because the empirical relationship between genetic distance and fitness is likely to be species-specific and is unlikely to be linear (Lynch, 1991). Moreover, the extent of genetic divergence may strongly be affected by past demographic fluctuations and habitat fragmentation (e.g. Wayne *et al.*, 1992).

In consideration of these issues, rigid decision rules for separate population conservation as opposed to intermixing are inappropriate. Phylogenetic analysis of molecular data can provide relationship trees of populations whose divisions may allow a relative assessment of the taxonomic distinction of populations. Those monophyletic population groupings which represent the deepest divergence in the tree have the greatest distinction, and those which occupy the terminal branches are least distinct. The distinction of a taxon is diluted if it has many close relatives (see May, 1990; Vane-Wright, Humphries and Williams, 1991; Erwin, 1991; Dizon *et al.*, 1992). It is important that population groupings be supported by analysis of multiple independent data sets and that the genetic diversity of the species is well represented by the sampled populations. If these conditions are fulfilled, then populations that show relatively the greatest taxonomic distinction should perhaps receive highest priority for separate conservation so long as the persistence of the entire species is not severely threatened. If molecular and morphological analyses do not reveal population differences, or if evidence exists for high rates of genetic exchange between populations (e.g. Wayne *et al.*, 1992), intermixing may be advised so long as fertility barriers (e.g. chromosomal differences) or subtle adaptive differences (e.g. populations occupy distinct habitats) are unlikely to exist.

5.4 THE APPLICATION OF MOLECULAR GENETIC TECHNIQUES

Molecular genetic techniques can be used to identify phylogenetically distinct units within species, to assess the loss of genetic variability caused by inbreeding and to determine the relatedness of individuals within captive populations. Such knowledge is a prerequisite to sound genetic management, because by interbreeding genetically dissimilar individuals within a population the loss of genetic variability can be minimized (Mace, 1986; Lacy, 1989; Lacy, Petric and Warneke, in press). Questions concerning the effects of outbreeding can be assessed only indirectly, through analysis of the extent of past genetic exchange between populations, or the degree of genetic divergence between populations.

5.4.1 Karyological analysis

Species and even populations or individuals within species often differ in diploid number and chromosome morphology. Such differences may be apparent in standard chromosome preparations or may require more elaborate banding techniques such as G-, C- or R-banding (Sumner, 1990). Although field preparations are possible, chromosome preparations in mammals and birds commonly involve lymphocyte culture or the establishment of fibroblast cell lines. The former is the least time-consuming technique while the latter is useful in that an immortal and renewable source of tissue from individuals is established. The value of chromosome data is that they may potentially be used to distinguish between taxonomically distinct units, to delineate hybrid zones or to determine the suitability of various individuals for breeding (Benirschke and Kumamoto, 1991). However, informative differences generally become less discernible as closer taxonomic units are compared, and the phylogenetic analysis of chromosome data is clouded by the arbitrary nature of emphasis placed on different chromosome alterations (e.g. Modi, 1987). For conservation purposes, phylogenetic questions at the specific/subspecific level, and the identification of individuals with rare chromosomal abnormalities, are appropriately addressed by karyological studies. Fixed chromosome differences observed between populations within species may indicate that individuals should not be interbred because of potential reproductive problems (e.g. Rebholz, Williamson and Rietkerk, 1991).

5.4.2 Allozyme electrophoresis

The application to systematic and population genetic analyses of methods for separating the protein alleles of a single gene have dramatically changed ideas about the abundance and significance of genetic variation. The hundreds of vertebrate species that have been surveyed in the past 25 years provide a considerable comparative database (Smith *et al.*, 1982) and due to advances in technology and standardization of protocols, allozyme electrophoresis is now a relatively easy and inexpensive technique (Hillis and Moritz, 1990). Thus, although newer DNA-based methods are available, protein electrophoresis remains an important tool for studying the relationships and genetic variability of populations.

Essentially the technique identifies protein alleles from vertebrate tissues by separation of protein variants on starch, acrylamide, or cellulose acetate gels in an electric field, followed by histochemical staining for protein products of a single gene. Commonly, 20 to 30 gene products are assayed. However, surveys including more loci are often desirable as genetic heterozygosity measures have a large variance, especially if only a few loci are polymorphic. The principal problems with allozyme electrophoresis, as applied to identification of taxonomic units, is that closely related populations and species may differ primarily in allele frequency rather than the presence or absence of certain alleles, making phylogenetic analysis difficult (Buth, 1984; Swofford and Berlocher, 1987). Moreover, genotype frequencies may be influenced by natural selection among alleles (e.g. Allendorf, Knudson and Leary, 1983; Koehn, Diehl and Scott, 1988; Pemberton *et al.*, 1988; Mopper *et al.*, 1991). However, an important benefit of the technique is that existing allozyme protocols can be applied to a wide variety of taxa to reveal variation at homologous loci. Allozyme electrophoresis can be used to provide an estimate of the genetic distance between closely related species and the populations they contain, and to characterize levels of variation within populations (Nei, 1987). Analysis of allozyme variation provides an estimate of the loss of genetic variability in small captive populations relative to those in the wild and may be used to detect the genetic mixing of distinct populations (e.g. O'Brien *et al.*, 1987). Also, using allozyme electrophoresis, paternity within closed populations may potentially be deduced (Lacy *et al.*, 1988).

5.4.3 Mitochondrial DNA analysis

The mitochondrion, a cytoplasmic organelle, contains a small, circular, self-replicating genome of double-stranded DNA. The mitochondial DNA (mtDNA) sequence appears to evolve much faster than the average expressed nuclear gene in most vertebrates (Brown, 1986;

Shields and Wilson, 1987). Thus it is possible to discriminate closely related species and populations using mtDNA analysis often more effectively than by using methods such as allozyme electrophoresis. Generally, mtDNA is maternally inherited in a clonal fashion such that, barring mutation, all future progeny from a given cross will have a genome identical to that of their mother.

mtDNA sequence differences among individuals or populations can be assayed indirectly through restriction site analysis or directly through DNA sequencing. The former is often less expensive and allows rapid screening of large samples; the latter may be more laborious but provides data in their most elemental form. In restriction site analysis, isolated mtDNA is digested with a restriction enzyme and the resulting fragments separated by electrophoresis. Genotypes are defined as their composite restriction fragment pattern for a panel of 10–20 restriction enzymes (e.g. Lansman *et al.*, 1983; Avise *et al.*, 1987). Many individuals may share a common genotype. The pattern of restriction site loss or gain among genotypes for a panel of restriction enzymes can be used to construct phylogenetic trees of populations or species, and thus be used to estimate their taxonomic distinction and evolutionary history (Wilson *et al.*, 1985; Avise *et al.*, 1987; Avise, 1989, 1992; Dizon *et al.*, 1992). Genotype data can also be used to estimate components of within- and among-population variation (Lynch and Crease, 1990).

The direct sequencing of genes from many individuals has recently become more feasible with the development of the polymerase chain reaction (PCR) and the identification of universal primers that allow amplification of sequences as much as several thousand base pairs in length (Kocher *et al.*, 1989). Essentially, PCR involves the enzymatic amplification, through repeated temperature cycles, of a segment of DNA bounded by two primer sequences 10–30 base pairs long. Using PCR, a DNA sequence can be obtained from very small quantities of tissue, even from bone, skin, feathers, hair and faecal samples or from specimens in museum and archaeological collections (Pääbo, 1989; Wayne and Jenks, 1991; Hagelberg, Gray and Jeffreys, 1991; Ellegren, 1991; Taberlet and Bouvet, 1992; Höss *et al.*, 1992). Because different genes in the mtDNA genome evolve at different rates, rapidly evolving genes can be analysed to determine the relationships of recently diverged populations, whereas slowly evolving genes can be used for systematic questions involving distantly related species.

If mtDNA genotype data are analysed using phylogenetic methods, the structure of genotype trees often reflects geographical distance between populations or the presence of geographical barriers. This coincidence of phylogeny with the geographical distribution of populations (phylogeographical partitioning) may record the patterns of colonization and gene flow among populations (Avise *et al.*, 1987). Therefore, the primary application of mtDNA analysis to species conservation concerns the identification and assessment of genetic similarity of phylogeographical units within species (e.g. Lansman *et al.*, 1983; Avise, 1989, 1992; Avise and Nelson, 1989; Avise and Ball, 1990; Ashley, Melnick and Western, 1990; Wayne *et al.*, 1992), the degree of hybridization between phylogenetic units (e.g., Ferris *et al.*, 1983; Avise and Saunders, 1984; Carr *et al.*, 1986; Telgelström, 1987; O'Brien *et al.*, 1990; Lehman *et al.*, 1991), and the loss of variability in captive and small, isolated wild populations (Wayne *et al.*, 1991). In some cases, mtDNA analysis may be used to identify maternal relationships of individuals, so as to deduce the population or region of origin of animals or to assess the genetic purity of captive stocks.

5.4.4 Analysis of variable number of tandem repeat DNA

In the nuclear genome of many eukaryotes, arrays of repeated sequences are common. The number of repeats in an array may vary dramatically among alleles at the same or different loci. The length variation of these repeats can be assayed by using restriction enzymes having recognition sites outside the arrays, thus excising fragments that differ in length. The mixture of fragments can be separated by size on agarose gels in the presence of an electric field, and visualized by hybridization with a radioactively labelled probe based on the core sequence of the repeat. The first applications of this approach by Jeffreys, Wilson and Thein (1985a), utilized variation in tandem arrays of a core sequence of approximately 30 base pairs, which yielded a gradation of fragments ranging in size from two to over 20 000 base pairs. The results which resemble a supermarket bar code, have become widely known as a DNA fingerprint. The presence of diagnostic fragments in parents and offspring has allowed accurate paternity assessment (Burke and Bruford, 1987; Burke, 1989), and the overall number of shared fragments in a population has been used to deduce close and unrelated classes of individuals as well as overall variability (Gilbert *et al.*, 1990; Reeve *et al.*, 1990; Packer *et al.*, 1991; Wayne *et al.*, 1991; Lehman *et al.*, 1992). An apparent problem with DNA fingerprinting as commonly applied is that it is a multilocus approach which reveals fragments not all of which segregate independently of each other. Moreover, bands that are considered identical by inspection may actually be slightly different and not homologous, leading to inflated bandsharing coefficients between non-relatives. Also, non-random breeding or past population contractions may decrease the confidence of parentage estimates (Lynch, 1988, 1990).

Since Jeffreys's discovery, many hypervariable tandem repeat sequences have been described (Nakamura *et al.*, 1987; Vassart *et al.*, 1987).

Some of these are much smaller than the Jeffreys's core sequence and may even consist of two base pair (dinucleotide) repeats (see Tautz, 1989; Litt and Luty, 1989; Rassmann, Schlotterer and Tautz, 1991; Moore *et al.*, 1991). Thus, there are microsatellite and minisatellite hypervariable DNA classes, the former consisting of repeats of no more than a few base pairs, the latter generally having more than a dozen base pairs in the core sequence. An advantage of microsatellite repeats is that the total size of the arrays is much smaller, generally less than 300 base pairs, and single loci can be amplified using PCR (Rassmann, Schlotterer and Tautz, 1991). Thus, there is the potential for studying allele frequencies at single hypervariable loci in recent and historic populations, and for rapid forensic identification of individuals and species from bone and hair samples (Hagelberg, Gray and Jeffreys, 1991; Schlötterer, Amos and Tautz, 1991; Ellegren, 1991). Moreover, many of the interpretive problems associated with multilocus DNA fingerprinting may be alleviated by the development of single locus probes and locus-specific PCR primers.

At present the application of fingerprint analysis to conservation of endangered species focuses on breeding management of captive populations. A genetic fingerprinting analysis may identify the successful breeders in populations, and assist in better estimation of effective population size and estimates of long-term losses in genetic variability. Moreover, individuals within captive populations may be grouped into categories of genetic relatedness (Packer *et al.*, 1991; Lehman *et al.*, 1992) such that losses of genetic variability may be reduced by managed breeding among genetically dissimilar animals (see below). Future applications, especially as single locus methods are developed, may include the characterization of genetic variability in populations and the identification of population-specific markers (Gilbert *et al.*, 1990; Wayne *et al.*, 1991; Schlötterer, Amos and Tautz, 1991).

5.5 EXAMPLES OF THE APPLICATION OF MOLECULAR GENETIC TECHNIQUES

5.5.1 Molecular phylogenetics of antelopes

The subfamily Antilopinae is subdivided into two tribes, Neotragini (dwarf antelope) and Antilopini (gazelle), and contains species found in Africa, Arabia, Asia, and India. Most gazelle species are endangered or threatened (Ryder, 1987; IUCN, 1990), due to habitat loss, hunting and competition with domestic livestock (East, 1990). The taxonomy of the Antilopinae has long been a source of controversy, and as a result, different classification schemes have been used (Groves, 1969, 1985,

1988; Hallenorth and Diller, 1980; Honacki, Kinman and Koeppl, 1982; O'Regan, 1984; Corbet and Hill, 1991). Descriptions of pelage coloration and measurements of skulls and bones are common features used as a basis for classification (Groves, 1969). However, in the genus *Gazella* these features often overlap, and slight differences between species are obscured by considerable geographical variation within species.

Molecular genetic research has the potential to define better the systematic relationships of gazelle species and subspecies (Benirschke, 1977; Furley, Tichy and Uerpmann, 1988), which is of importance for conservation management strategies (Ryder, 1986). Chromosome banding analysis and DNA sequence analysis of mitochondrial genes are being applied to solve systematic problems of the Antilopinae. Chromosome research on the Saudi gazelle (*Gazella saudiya*) (Figure 5.1; Rebholz, Williamson and Rietkerk, 1991) showed that it had a unique karyotype, suggesting it was not a subspecies of dorcas gazelle (*G. dorcas*) as previously implied (Groves and Harrison, 1967; Groves, 1969, 1985). More samples need to be analysed, but the potential significance of this result is that the Saudi gazelle should be managed as a distinct unit, separate from dorcas gazelle, so avoiding hybridization and potential outbreeding depression (Benirschke and Kumamoto, 1991).

5.5.2 The African hunting dog (*Lycaon pictus*)

The African hunting dog has historically inhabited savannah-like habitat south of the Sahara desert. Wild dog populations have declined precipitously during the last 20 years and perhaps no more than several thousand now exist (Ginsberg and Macdonald, 1990). The east and west African populations are the most endangered, having recently suffered severe population declines. However, the representation of wild dogs in zoological collections is thought to be southern African in origin. Thus, a concern is that the east African wild dogs are a genetically distinct population that should be represented in a separate captive breeding programme. Moreover, the potential of southern African dogs as a source of individuals for vanishing or extinct east African populations may depend on the measure of phylogenetic distinction that exists between them. The probability of extinction in the east African population has to be weighed against the phylogenetic uniqueness that may be lost through hybridization with introduced southern African wild dogs.

Levels of phylogenetic distinctiveness were examined among populations of African wild dogs using mtDNA restriction site and cytochrome *b* sequence analyses. Results indicated that southern and eastern populations of wild dogs show about 1% divergence in mtDNA sequence and that they form two monophyletic clades containing three mtDNA

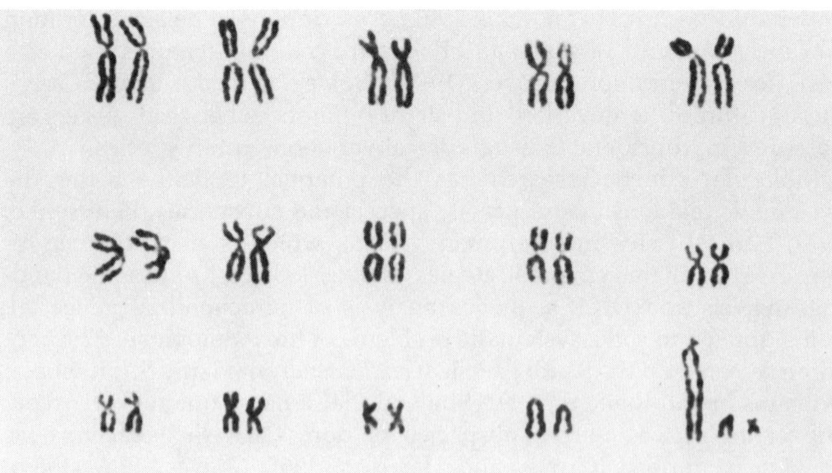

Dorcas gazelle (*Gazella dorcas*), male

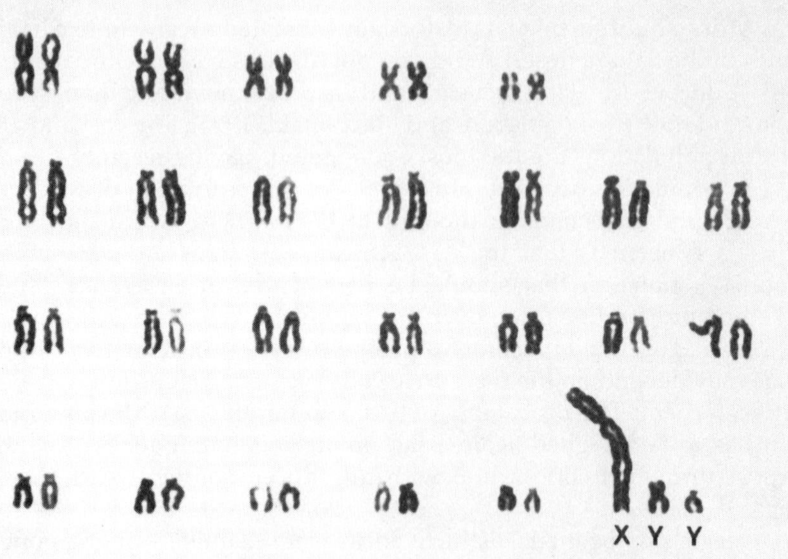

Saudi gazelle (*Gazella saudiya*), male

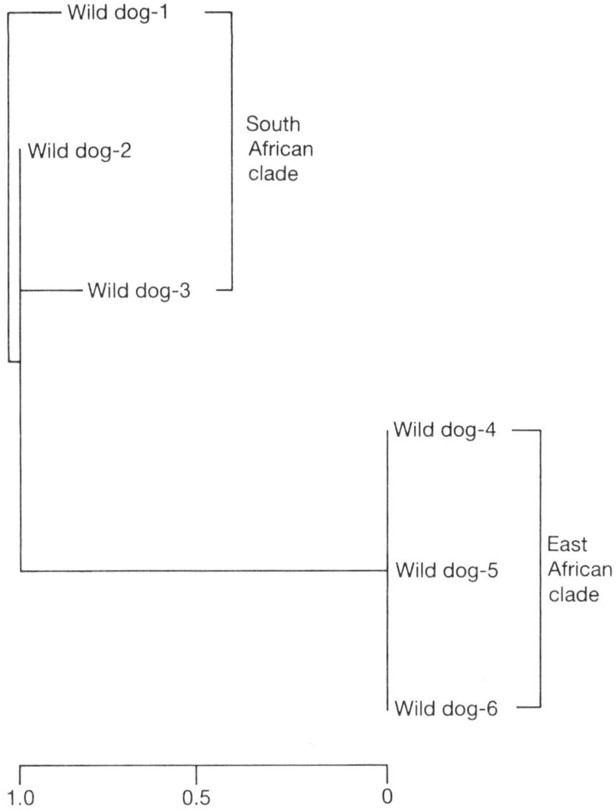

Figure 5.2 The single most parsimonious tree of mtDNA genotypes found in 87 east and south African wild dogs and generated from phylogenetic analysis of 736 bp of cytochrome *b* sequence (Girman *et al.*, in press). Percent sequence divergence indicated on scale. Tree length = 8, no homoplasy is evident.

genotypes each (Figure 5.2). The degree of sequence divergence between the clades is substantial relative to sequence divergence observed within other large canid species (Girman *et al.*, in press). For example, restriction site data on the two most divergent grey wolf (*Canis lupus*) genotypes known, one from China and the other from Mexico, show that they differ only in 0.5% of their mtDNA sequence. Therefore,

---

Figure 5.1 Top: Karyotype of a male dorcas gazelle (with XY_1Y_2 sex chromosomes, 2n = 31), which is dominated by metacentric chromosomes. Bottom: Karyotype of a Saudi gazelle (with XY_1Y_2 sex chromosomes, 2n = 51), which consists of five metacentric autosome pairs. This karyotype differs greatly from the usual dorcas gazelle karyotype. Reproduced from W.E.R. Rebholz, D. Williamson and F. Rietkerk, Saudi Gazelle (*Gazella saudiya*) is not a subspecies of dorcas gazelle, published by *Zoo Biology*, 1991.

African wild dogs from southern and eastern Africa are distinct subspecies under the definition suggested by Avise and Ball (1990). In the past, as many as seven African wild dog subspecies have been proposed (Allen, 1939). Our results suggest at least one genetic entity exists in the southern region of Africa and a second resides in eastern Africa in the Masai Mara–Serengeti ecosystem.

The conservation implications of these results are that reintroduction of southern African wild dogs into areas of eastern Africa may not be prudent, because it would unnecessarily mix distinct subspecies. In addition, interbreeding between the two forms may result in outbreeding depression in the hybrids due to dissolution of co-adapted gene complexes and loss of local adaptations (Templeton, 1986). Because the zoo population of African wild dogs appears primarily to be derived from southern African founders we recommended that captive breeding of eastern African wild dogs be initiated. Captive breeding of eastern African wild dogs will potentially allow for a reintroduction or augmentation programme that would avoid problems of mixing and possibly eradicating, through hybridization, the eastern African subspecies.

5.5.3 Northern hairy-nosed wombats (*Lasiorhinus krefftii*)

Many unique Australian species have suffered the consequences of recent human expansion. The northern hairy-nosed wombat is a dramatic example, now existing as only a single colony of about 70 individuals at Epping Forest National Park, near Clermont, Queensland (Figure 5.3). At the time of European settlement, about 120 years ago, the geographical range of the species spread several thousand kilometres through Queensland and New South Wales. The combined effects of drought and cattle grazing are thought to have been in part responsible for the decline (Crossman, Johnson and Horsup, 1993). Recent monitoring of the Epping Forest northern hairy-nosed wombat found that most individuals restrict their movement to burrow clusters from which females occasionally emigrate (Crossman, Johnson and Horsup, 1993; Figure 5.3). Two obvious concerns are the loss of genetic variation that may have resulted from the population decline, and the degree of inbreeding occurring within wombat burrow clusters.

Our preliminary allozyme, mtDNA and fingerprint studies of northern hairy-nosed wombats revealed low and uninformative levels of variation. Consequently, variation in 16 hypervariable microsatellite loci was assessed through polymerase chain reaction amplification of DNA followed by allele separation on acrylamide gels (Taylor, Sherman and Wayne, 1993). The results showed a marked decline in heterozygosity and allelic diversity of northern hairy-nosed wombats, relative to a

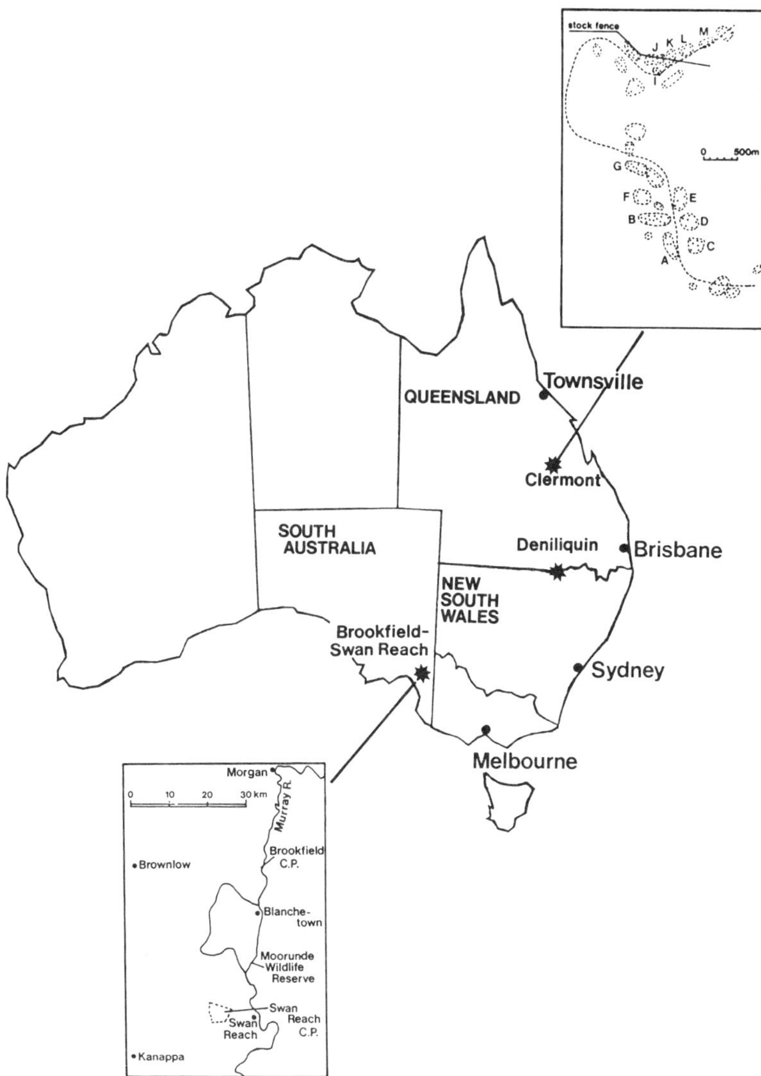

Figure 5.3 Map of sampling localities indicating extant (Clermont) and extinct (Deniliquin) northern hairy-nosed wombat populations and the southern hairy-nosed wombat population at Brookfield–Swan Reach. Burrows at Clermont, Epping Forest are indicated in the insert by dots and burrow clusters are symbolized by broken lines encircling dots. Reproduced from C.N. Johnson and D.G. Crossman, Dispersal and social organization of the northern hairy-nosed wombat *Lasiorhinus krefftii*, published by *Journal of Zoology*, 1991.

Table 5.1 Number of loci, mean number of alleles per locus (A), and Hardy–Weinberg (H_E) heterozygosity for northern hairy-nosed (NHN) and southern hairy-nosed (SHN) wombat populations

Number of loci	*Species/population*	*A*	H_E
16 loci	NHN wombat		
	Epping Forest	1.81 (0.21)[a]	0.27 (0.07)
	SHN wombat		
	Brookfield	4.38 (0.44)	0.65 (0.05)
	Swan Reach	3.94 (0.48)	0.67 (0.06)
	Combined	5.00 (0.58)	0.66 (0.05)
8 loci[b]	NHN wombat		
	Epping Forest	2.00 (0.3)	0.32 (0.10)
	Deniliquin	2.33 (0.40)	0.43 (0.13)
	SHN wombat		
	Brookfield	4.40 (0.70)	0.62 (0.10)
	Swan Reach	4.60 (0.80)	0.70 (0.11)
	Combined	5.30 (1.00)	0.65 (0.10)

[a]Standard errors in parenthesis.
[b]The eight loci used in comparisons involving Deniliquin are a subset of the 16 used for the other comparisons.

widely abundant close relative, the southern hairy-nosed wombat (*Lasiorhinus latifrons*). Northern hairy-nosed wombats had 41% of the level of heterozygosity and 36% the value of allelic diversity of the southern species (Table 5.1). An effective population size of less than 20 individuals over the 120 year decline is consistent with this observed loss of genetic variability. Because the polymerase chain reaction can be used to amplify loci from even small amounts of degraded DNA, we were also able to amplify loci from five museum skins of a now extinct northern hairy-nosed wombat population at Deniliquin. The Epping Forest wombats had slightly lower levels than the Deniliquin population, which was likely to be in decline when the museum samples were obtained in 1884. Finally, we found the relatedness of individuals within burrow clusters was not significantly greater than that of individuals from different burrow clusters. Consequently, our analysis suggests artificial movements of individuals is not required to maximize outbreeding; natural immigration appears to be doing an adequate job. In conclusion, the use of hypervariable microsatellite loci has allowed us to estimate the loss of variation and relatedness in a bottlenecked species for which other molecular methods have shown uninformative levels of variation.

5.5.4 The pink pigeon (*Nesoenas mayeri*)

An important aim of the genetic management of captive populations is the avoidance of inbreeding and the consequent losses of genetic variability through breeding individuals of low relatedness (Ralls, Ballou and Templeton, 1988). This approach depends on complete studbook information and often assumes founder individuals to be unrelated. However, rarely is such information available and substructuring in small endangered populations can lead to inbreeding among wild caught founders (Mace, 1986). Individuals can potentially be grouped into relatedness classes by using DNA fingerprinting and the most dissimilar individuals used as breeding pairs (Jeffreys, Wilson and Thein, 1985a, 1985b; Burke and Bruford, 1987; Packer *et al.*, 1991; Wayne *et al.*, 1991; Lehman *et al.*, 1992).

The Mauritius pink pigeon (*Nesoenas mayeri*) is an example of an endangered species with a complete studbook but where the relationships among the six founder birds are not known. Inbreeding depression in captive pink pigeons may have occurred (Jones, Todd and Mungroo, 1985), and therefore a clear need is apparent for genetic management. The relationship between fingerprint similarity (band-sharing) and genetic relatedness in captive pink pigeons, including four of the six founders, has been studied (for methods see Burke and Bruford, 1987; Bruford *et al.*, 1992). Mean probability of band-sharing x, was estimated by comparing individuals' DNA fingerprints in adjacent lanes (Figure 5.4) and equals the proportion of bands that match between two individuals divided by half the total number of bands (Lynch, 1988). The mean probability of two individuals sharing a band was 0.500 ± 0.014 for combined data from two fingerprint probes (77 comparisons). Thus two individuals selected at random from the population would be expected to share about 50% of their fingerprint bands.

To establish the efficacy of the technique in distinguishing among levels of relatedness in the population, pair-wise comparisons were grouped into three classes; first degree relatives, second degree relatives and individuals thought to be unrelated. The band-sharing coefficients of all classes are significantly different using Mann–Whitney ranked sample tests (Figure 5.5). As expected, some overlap exists between the classes. However the distribution indicates that birds with x = 0.25–0.45 are unrelated, x = 0.46–0.58 are second degree relatives and x = 0.59–0.75 are first degree relatives. Mean band-sharing coefficients of the founder individuals sampled ranged from 0.31 to 0.45 and are within the unrelated range, with the exception of one comparison whose coefficient (0.5) is in the second degree relationship range.

In summary, we have shown that DNA fingerprinting can be used to discriminate among individuals of different levels of relatedness within

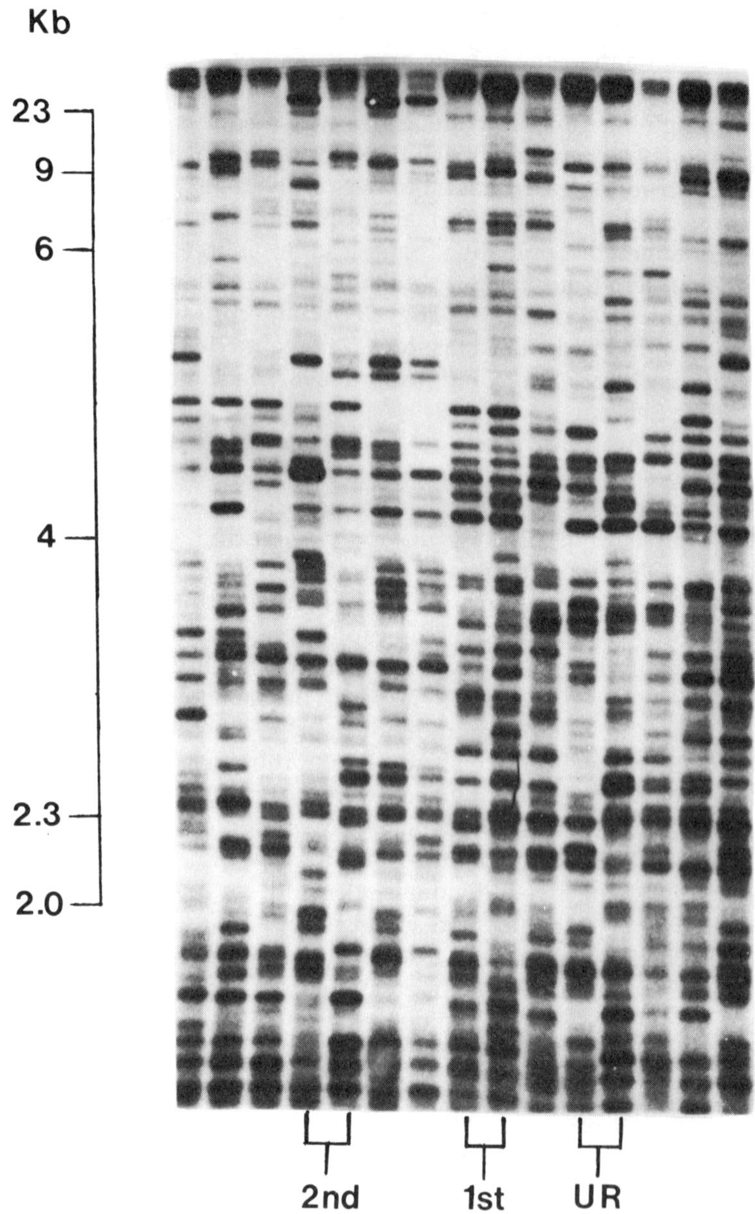

Figure 5.4 DNA fingerprint of pink pigeon individuals from the captive population at Jersey Zoo. Band sharing coefficients are compared for individuals of different levels of relatedness. 1st = first degree relative, 2nd = second degree relative, UR = putatively unrelated.

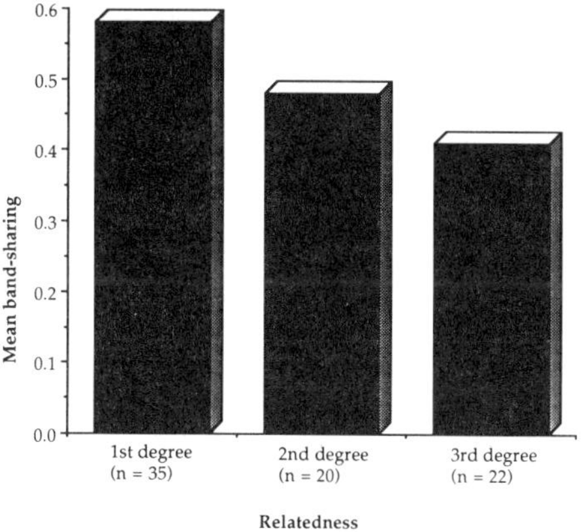

Figure 5.5 Average band-sharing (multiply by 100 for percentages) between first and second degree relatives and between unrelated (unrel) individual pink pigeons for Jeffreys's probes 33.15 and 33.6 (Jeffreys, Wilson and Thein, 1985a).

a captive population of pink pigeons. Some studbook information is required to first determine relatedness classes but using this approach practical management decisions can now be made about the breeding of individuals whose relationships are not known. Moreover, the analysis suggests most of the founding individuals appear unrelated, and hence should be treated as such in future breeding plans.

5.5.5 The Arabian oryx (*Oryx leucoryx*)

Oryx leucoryx disappeared from the wild in 1972 due to overhunting and habitat degradation (Abuzinada, Habibi and Seitre, 1988). A captive-bred 'World Herd' was established in the 1960s, and managed by various zoological institutions. Private collections were also maintained in the Middle East. In 1986, as part of a Saudi Arabian conservation programme, 56 Arabian oryx, derived from wild caught animals owned by King Khaled, Riyadh Zoo, other Middle Eastern collections, and the World Herd, were translocated to The National Wildlife Research Centre (NWRC) at Taif, Saudi Arabia. Subsequently, the number was reduced to 35 due to an outbreak of tuberculosis. The challenge was to maximize retention of genetic variability in the remaining stock given the absence of information about the pedigree of individuals. Use of molecular genetics to establish relationships was not practical at that time, and genetic management proceeded by identifying founders and minimizing future inbreeding by husbandry (Greth *et al.*, in press).

Although there are several polymorphic allozyme systems known in the Arabian oryx, allozyme analysis was not sufficient to determine relatedness between individuals in the Taif herd (Vassart, Granjo and Greth, 1991). Thus, multilocus minisatellite fingerprinting (Jeffreys, Wilson and Thein, 1985a,b) was chosen as the method to determine relatedness. In multilocus fingerprinting analysis, offspring may be matched to parents by possession of unique bands, or fingerprint similarity between pairs of individuals can be compared to a calibration curve derived from individuals of known relatedness (see above, Kuhnlein *et al.*, 1990; Packer *et al.*, 1991; Lehman *et al.*, 1992). The first approach can be used to match offspring to sires, and the second to determine relatedness of the founders.

A calibration curve was established relating band-sharing between pairs of individuals of known degrees of relatedness. The first degree relative band-sharing coefficient was 0.77 ± 0.044 (± 1 SD), second degree 0.64 ± 0.056 and unrelated 0.45 ± 0.064. These values are significantly different (Mann–Whitney Z test, <0.007; Greth *et al.*, in press). When all relevant data are considered, average band-sharing between unrelated Arabian oryx is 0.55 ± 0.075 and is high compared to most mammals: estimates of band-sharing among unrelated individuals determined with the 33.15 fingerprint probe ranges from 0.21 in humans to 0.46 in dogs and cats (Ely, Alfrod and Ferrell, 1991). Although band-sharing values obtained from different laboratories may be difficult to compare, the high fingerprint similarity between unrelated Arabian oryx probably reflects the recent genetic bottleneck. Bottlenecks have been known to reduce fingerprint variability in populations of the koalas (*Phascolarctos cinereus*) (Taylor *et al.*, 1991) and island foxes (*Urocyon littoralis*) (Gilbert *et al.*, 1990). Moreover, the band-sharing was significantly higher between NWRC unrelated animals than between these oryx and ones from a herd in Abu Dhabi. This suggests that it may be possible to use multilocus band-sharing to identify herds between which genetic exchange would increase levels of genetic variability.

In summary, fingerprint data will be used to improve the likelihood that genetic variability will be retained in the captive herd by supplying information on parentage and relatedness. Identification of parents allows for equalization of founder contributions and maximal retention of genetic variability whilst assessment of relatedness among potential breeders permits the mating of genetically dissimilar individuals and the avoidance of inbreeding depression.

5.6 CONCLUSION

Molecular genetic techniques can be used to address conservation questions at multiple levels of evolutionary divergence. We have discussed their application to the systematics of closely related species and the populations they contain. The information may be used to assess the level of phylogenetic distinction of species and populations, assess their relative levels of genetic variability and identify source populations for reintroduction or augmentation. New DNA-based techniques can contribute increased resolution among closely related populations and among individuals within the same population, and can potentially contribute to better management of captive populations through assessments of parentage and genetic similarity. Moreover, PCR-based techniques can be utilized on extremely degraded remains, and thus provide insights into past levels of genetic variability and have important forensic applications. Examples are provided of the application of molecular genetic techniques to several of these questions. Potentially, once the necessary molecular tools are developed by a specialist they can be routinely used with a minimal amount of expertise or expense. Thus, the challenge in the future is to build effective collaborations between field biologists, population managers and theoretical and molecular population geneticists such that techniques of the appropriate resolution and cost are matched to significant conservation management questions.

ACKNOWLEDGEMENTS

The authors would like to thank the following people who have contributed to each of our conservation genetics studies. Molecular studies of antelopes: this project is a collaboration with Douglas Williamson, Frank Rietkerk, Nick Lindsay, and Charlie Kichenside at King Khalid Wildlife Research Center, Thumamah, Saudi Arabia, and we would like to thank the National Commission of Wildlife Conservation and Development, Riyadh, Saudi Arabia for their support. African hunting dog: we would like to thank the generosity of the many field workers who provided blood samples for this study. Partial support was provided by NSF. Northern hairy-nosed wombat: this research was supported by the Institute of Zoology and collaborative grants from the Australian Research Council to the University of NSW, University of QLD and QLD Department of the Environment and Heritage; the project was made possible with the help of Deryn Alpers, Chris Johnson, Joan Dixon, Joe Stelman, Stephanie Williams, Jenny Graves and Neil Murray. Pink pigeons: this is a collaborative project involving MWB and Susan Haines (IOZ), Terry Burke and Andrew Krupa (University of Leicester), Carl Jones (Mauritius Wildlife Preservation Fund), John Hartley, Reece Lind and David Jeggo (Jersey Wildlife Preservation Trust) and was in part funded by JWPT. Arabian oryx: we would particularly like to thank Arnaud Greth at the National Wildlife Research Centre, Taif, Saudi Arabia.

REFERENCES

Abuzinada, A., Habibi, K. and Seitre, R. (1988) The Arabian Oryx programme in Saudi Arabia, in *Conservation and Biology of Desert Antelopes* (eds A. Dixon and D. Jones), Christopher Helm, London, pp. 41–6.

Allen, G.M. (1939) A checklist of African Mammals. *Bull. Mus Comp. Zool. Harvard*, **83**, 188–91

Allendorf, F.W., Knudson, K.L. and Leary, R.F. (1983) Adaptive significance of the differences in the tissue-specific expression of a phosphoglucomutase gene in rainbow trout. *Proc. Natl. Acad. Sci. USA*, **800**, 1397–400.

Amato, G.D. (1991) Species hybridization and protection of endangered animals. *Science*, **253**, 250.

Ashley, M.V., Melnick, D.I. and Western, D. (1990) Conservation genetics of the black rhinoceros (*Diceros bicornis*), 1. evidence from the mitochondrial DNA of three populations. *Conservation Biology*, **4**, 71–7.

Avise, J.C. (1989) A role for molecular genetics in the recognition and conservation of endangered species. *Trends in Ecology and Evolution*, **9**, 279–81.

Avise, J.C. (1992) Molecular population structure and the biogeographic history of a regional fauna: a case history with lessons for conservation biology. *OIKOS*, **63**, 62–76.

Avise, J.C. and Saunders, N.A. (1984) Hybridization and introgression among species of sunfish (*Lepomis*): analysis by mitochondrial DNA and allozyme markers. *Genetics*, **108**, 237–55.

Avise, J.C. and Nelson, W.S (1989) Molecular genetic relationships of the extinct Dusky seaside sparrow. *Science*, **243**, 646–8.

Avise, J.C. and Ball, R.M. (1990) Principles of genealogical concordance in species concepts and biological taxonomy, in *Oxford Surveys in Evolutionary Biology*, Vol. 7 (eds D. Futuyma and J. Antonovics), Oxford University Press, Oxford, pp. 45–67.

Avise, J.C., Arnold, J., Ball, R.M. *et al.* (1987) Intraspecific phylogeography: the mitochondrial DNA bridge between population genetics and systematics. *Ann. Rev. Ecol. Syst.*, **18**, 489–522.

Baum, D. (1992) Phylogenetic species concepts. *Trends in Ecology and Evolution*, **7**, 1–2.

Benirschke, K. (1977) Genetic management. *International Zoo Yearbook*, **17**, 50–60.

Benirschke, K. and Kumamoto, A.T. (1991) Mammalian cytogenetics and conservation of species. *J. Heredity*, **82**, 187–91.

Brown, W.M. (1986) The mitochondrial genome of animals, in *Molecular Evolutionary Biology* (ed. R. MacIntrye), Cornell University Press, New York. pp. 95–128.

Bruford, M.W., Hanott, O., Brookfield, J.F.Y. and Burke, T. (1992) Single-locus and multilocus DNA fingerprinting, in *Molecular Genetic Analysis of Populations: a Practical Approach* (eds T. Burke, G. Dolf, A.J. Jeffreys and R. Wolff), IRL Press, Oxford, pp. 225–69.

Burke, T. (1989) DNA fingerprinting and other methods for the study of mating success. *Trends in Ecology and Evolution*, **4**, 139–44.

Burke, T. and Bruford, M.W. (1987) DNA fingerprinting in birds. *Nature*, **327**, 139–44.

Buth, D.G. (1984) The application of electrophoretic data in systematic studies. *Ann. Rev. Ecol. Syst.*, **15**, 501–22.

Carr, S.M., Ballinger, S.W., Derr, J.N. *et al.* (1986) Mitochondrial DNA analysis of hybridization between sympatric white-tailed deer and mule deer in west Texas. *Proc. Natl. Acad. Sci. USA.*, **83**, 9576–80.

Corbet, G.B. and Hill, J.E., (1991) *A World List of Mammalian Species*, 3rd edn, Oxford University Press, Oxford.

Cracraft, J. (1983) Species concepts and speciation analysis, in *Current Ornithology* (ed. R.F. Johnson), Plenum Press, New York, pp. 150–87

Crossman, D., Johnson, C. and Horsup, A. (1993) Trends in the population of the northern hairy-nosed wombat in Epping Forest National Park. *Pacific Conservation Biology*.

Daugherty, C.H., Cree, A., Hay, J.M. and Thompson, M.B. (1990) Neglected taxonomy and continuing extinctions of tuatara (*Spenodon*). *Nature*, **347**, 177–9.

Dizon, A.E., Lockyer, C., Perrin, W.F. *et al.* (1992) Rethinking the stock concept: a phylogenetic approach. *Conservation Biology*, **6**, 24–36.

East, R. (1990) *Antelopes, global survey and regional action plans. Part 3. West and Central Africa* (compiled by R. East and the IUCN/SSC Antelope Specialist Group), IUCN, Gland, Switzerland.

Ereshefsky, M. (1992) Species, higher taxa, and the units of evolution in *The Units of Evolution* (ed. M. Ereshefsky), MIT Press, Cambridge, Mass, pp. 381–98.

Ellegren, H. (1991) DNA typing of museum birds. *Nature*, **354**, 113.

Ely, J., Alford, P. and Ferrell, R.E. (1991). DNA 'fingerprinting' and the genetic management of a captive chimpanzee population (*Pan troglodytes*). *Am. J. Primatol.*, **24**, 39–54.

Erwin, T.L. (1991) An evolutionary basis for conversation strategies. *Science*, **253**, 750–2.

Ferris, S.D., Sage, R.D., Huang, C.M. *et al.* (1983) Flow of mitochondrial DNA across a species boundary. *Proc. Natl. Acad. Sci. USA*, **79**, 2290–4.

Furley, C.W., Tichy, H. and Uerpmann, H.-P. (1988) Systematics and chromosomes of the Indian gazelle, *Gazella bennetti* (Sykes, 1831). *Zeit. fur Saugetierkunde*, **53**, 48–54.

Gilbert, D.A., Lehman, N., O'Brien, S.J. and Wayne, R.K. (1990) Genetic fingerprinting reflects population differentiation in the Channel Island Fox. *Nature*, **344**, 764–7.

Ginsberg, J.R. and Macdonald, D.W. (1990) *Foxes, Wolves, Jackals and Dogs. An Action Plan for the Conservation of Canids*. IUCN, Gland Switzerland.

Girman, D.J., Kat, P.W., Mills, G. *et al.* (in press) A genetic and morphological analysis of the African Wild Dog (*Lycaon pictus*). *J. Heredity*.

Greth, A., Sunnucks, P., Vassart, M. and Stanley, H. (in press) Genetic management of an Arabian oryx (*Oryx leucoryx*) population without known pedigree. Proceedings of '*Ungulates 91*' September, 1991, Toulouse, France.

Groves, C.P. (1969) On the smaller gazelles of the genus *Gazella* de Blainville, 1816. *Zeitschrift fur Saugetierkunde*, **34**, 38–60.

Groves, C.P. (1985) An introduction to the gazelles. *Chinkara*, **1**, 4–16.

Groves, C.P. (1988) A catalogue of the genus Gazella, in *Conservation and Biology of Desert Antelopes* (eds. A. Dixon and D. Jones), Christopher Helm, London, pp. 193–8.

Groves, C.P. and Harrison, D.L. (1967) The taxonomy of the gazelles (genus *Gazella*) of Arabia. *J. Zoology*, **152**, 381–7.

Hagelberg, E., Gray I.C. and Jeffreys, A.J. (1991) Identification of the skeletal remains of a murder victim by DNA analysis. *Nature*, **352**, 427–9.

Hallenorth, T. and Diller, H. (1980) *A Field Guide to the Mammals of Africa Including Madagascar*. William Collins Sons & Co, London.

Hillis, D.M. and Moritz, C. (1990) *Molecular Systematics*, Sinauer Press, Sunderland, Mass.

Honacki, J.H., Kinman, K.E. and Koeppl, J.W. (1982) *Mammal Species of the*

World. A Taxonomic and Geographic Reference. The Association of Systematics Collections, Lawrence, Kansas.

Höss, M., Kohn, M., Pääbo, S. *et al.* (1992) Excremental analysis by PCR. *Nature,* **359**, 199.

IUCN, (1990) *Red List of Threatened Animals,* IUCN, Gland, Switzerland, and Cambridge, UK.

Jeffreys, A.J., Wilson, V. and Thein, S.L. (1985a) Hypervariable minisatellite regions in human DNA. *Nature,* **316**, 76–9.

Jeffreys, A.J., Wilson, V. and Thein, S.L. (1985b) Individual-specific DNA 'fingerprints' of human DNA. *Nature,* **316**, 76–9.

Johnson, C.N. and Crossman, D.G. (1991) Dispersal and social organization of the northern hairy-nosed wombat *Lasiorhinus krefftii. J. Zoology,* **225**, 605–13.

Jones, C.G., Todd, D.M. and Mungroo, Y. (1985) Mortality, morbidity and breeding success of the pink pigeon *Columba (Nesoenas) mayeri. Proceedings of the Symposium on Disease and Management of Threatened Bird Populations.* ICBP, Ontario.

Kocher T.D., Thomas, W.K., Meyer A. *et al.* (1989) Dynamics of mitochondrial DNA evolution in animals: amplification and sequencing with conserved primers. *Proc. Natl. Acad. Sci. USA,* **86**, 6196–200.

Koehn, R.K., Diehl, W.J. and Scott, T.M. (1988) The differential contribution by individual enzymes of glycolysis and protein catabolism to the relationship between heterozygosity and growth rate in the coot clam, *Mulinia lateralis. Genetics,* **118**, 121–30.

Kuhnlein, U., Zadworny, D., Dawe, Y. *et al.* (1990) Assessment of inbreeding by DNA fingerprinting: development of a calibration curve using defined strains of chickens. *Genetics,* **125**, 161–5.

Lacy, R.C., Foster, M.L. *et al.* (1988) Determination of pedigrees and taxa of primates by protein electrophoresis. *Int. Zoo Yb.,* **27**, 159–68.

Lacy, R.C. (1989) Analysis of founder representation in pedigrees: founder equivalents and founder genome equivalents. *Zoo Biology,* **8**, 111–23.

Lacy, R.C., Petric, A. and Warneke, M. (in press). Inbreeding and outbreeding in captive populations of wild animal species in *The Natural History of Inbreeding and Outbreeding* (ed. N.W. Thronhill), University of Chicago Press, Chicago.

Lansman, R.A., Avise, J.C., Aquadro, C.F. *et al.* (1983) Extensive genetic variation in mitochondrial DNAs among geographic populations of the deer mouse, *Peromyscus maniculatus. Evolution,* **37**, 1–16.

Lehman, N., Eisenhawer, A., Hansen, K. *et al.* (1991) Introgression of coyote mitochondrial DNA into sympatric North American gray wolf populations. *Evolution,* **45**, 104–19.

Lehman, N., Clarkson, P. Mech, L.D. *et al.* (1992) The use of DNA fingerprinting and mitochondrial DNA to study genetic relationships within and among wolf packs. *Behavioural Ecology and Sociobiology,* **30**, 83–94.

Litt, M. and Luty, J.A. (1989) A hypervariable microsatellite revealed by *in vitro* amplification of dinucleotide repeat within the cardiac muscle actin gene. *Am. J. Genet.,* **44**, 397–401.

Lynch, M. (1988) Estimation of relatedness by DNA fingerprinting. *Mol. Biol. Evol.,* **5**, 584–99.

Lynch, M. (1990) The similarity index and DNA fingerprinting. *Mol. Biol. Evol.,* **7**, 468–84.

Lynch, M. and Crease, T.J. (1990) The analysis of population survey data based on DNA sequence variation. *Mol. Biol. Evol.,* **7**, 377–94.

Lynch, M. (1991) The genetic interpretation of inbreeding and outbreeding depression. *Evolution*, **45**, 622–9.

Mace, G.M. (1986) Genetic management of small populations. *Int. Zoo Yb.*, **24/25**, 167–74.

May, R. (1990) Taxonomy as destiny. *Nature*, **347**, 129–30.

Mayr, E. (1963) *Populations, Species and Evolution*, Belknap Press, Cambridge, Mass.

Mayr, E. and Ashlock, P.D. (1991) *Principles of Systematic Zoology*. McGraw-Hill, New York.

McKitrick, M.C. and Zink, R.M. (1988) Species concepts in ornithology. *The Condor*, **90**, 1–14.

Modi, W.S. (1987) Phylogenetic analysis of chromosomal banding patterns among the nearctic Arvicolidae (Mammalia, Rodentia). *Systematic Zoology*, **36**, 109–36.

Mopper, S., Mitton, J.B., Whitham, T.G. *et al.* (1991) Genetic differentiation and heterozygosity in pinyon pine is associated with resistence to herbivory and environmental stress. *Evolution* **45**, 989–1000.

Moore, S.S., Sargeant, L.L., King, T.J. *et al.* (1991) The conservation of dinucleotide microsatellites among mammalian genomes allows the use of heterologous PCR primer pairs in closely related species. *Genomics*, **10**, 654–60.

Nakamura, Y., Lepper, M., O'Connell, P. *et al.* (1987) Variable number of tandem repeat (VNTR) markers for human gene mapping. *Science*, **235**, 1616–622.

Nei, M. (1987) *Molecular Evolutionary Genetics*. Columbia University Press, New York.

O'Brien, S.J., Nash, W.G., Wildt, D.E. *et al.* (1985) A molecular solution to the riddle of the giant panda's phylogeny. *Nature*, **317**, 140–4.

O'Brien, S.J., Joslin, P., Smith, G.L. *et al.* (1987) Evidence for African origin of founders of the Asiatic lion species survival plan. *Zoo Biology*, **6**, 99–116.

O'Brien, S.J., Rolke, M.E., Yuhki, N. *et al.* (1990) Genetic introgression within the Florida panther *Felis concolor coryi*. *Natl. Geo. Res.*, **6**, 485–94.

O'Brien, S.J. and Mayr, E. (1991) Bureaucratic mischief: recognizing endangered species and subspecies. *Science*, **251**, 1187–8.

O'Regan, B.P. (1984) Gazelles and dwarf antelopes in *Hoofed Mammals* (ed. D. MacDonald), Torstar Books, New York, pp. 134–41.

Pääbo, S. (1989) Ancient DNA: extraction, characterization, molecular cloning, and enzymatic amplification. *Proc. Natl. Acad. Sci. USA*, **86**, 1939–43.

Packer, C., Gilbert, D.A., Pussey, A.E. and O'Brien, S.J. (1991) A molecular genetic analysis of kinship and cooperation in African lions. *Nature*, **351**, 562–5.

Pemberton, J.M., Albon, S.D., Guinness, F.E. *et al.* (1988) Genetic variation and juvenile survival in red deer. *Evolution*, **42**, 921–34.

Ralls, K., Ballou, J.D. and Templeton, A.R. (1988). Estimates of lethal equivalents and the costs of inbreeding in mammals. *Conservation Biology*, **2**, 185–93.

Rassmann, K., Schlotterer, C. and Tautz, D. (1991) Isolation of simple-sequence loci for use in polymerase chain reaction-based DNA fingerprinting. *Electrophoresis*, **12**, 113–18.

Rebholz, W.E.R., Williamson, D. and Rietkerk, F. (1991) Saudi gazelle (*Gazella saudiya*) is not a subspecies of dorcas gazelle. *Zoo Biology*, **10**, 485–9.

Reeve, H.K., Westneat, D.F., Noon, W.A. *et al.* (1990) DNA fingerprinting reveals high levels of inbreeding in colonies of the eusocial naked mole rat. *Proc. Natl. Acad. Sci. USA.*, **87**, 2496–500.

Ryder, O.A. (1986) Species conservation and systematics: the dilemma for subspecies. *Trends in Ecology and Evolution*, **1**, 9–10.

Ryder, O.A. (1987) Conservation action for gazelles: an urgent need. *Trends in Ecology and Evolution*, **6**, 143–4.

Schlötterer, C., Amos, B. and Tautz, D. (1991) Conservation of polymorphic simple sequence loci in cetacean species. *Nature*, **354**, 63–5.

Shields, G.F. and Wilson, A.C. (1987) Calibration of mitochondrial DNA evolution in geese. *J. Mol. Evol.*, **24**, 212–17.

Simpson, G.G. (1961) *Principles of Animal Taxonomy*. New York, Columbia University Press.

Smith, M.W., Aquadro, C.F., Smith, M.H. *et al.* (1982) *Bibliography of Electrophoretic Studies of Biochemical Variation in Natural Vertebrate Populations*, Texas Tech. Press, Lubbock, Texas.

Soulé, M., Gilpin, M., Conway, W. and Foose, T. (1986) The Millenium Ark: How long a voyage, how many staterooms, how many passengers. *Zoo Biology*, **5**, 101–13.

Sumner, A.T. (1990) *Chromosome Banding*, Unwin Hyman, London.

Stanley, S.M. (1979) *Macroevolution. Pattern and Process*, W.H. Freeman, San Francisco.

Swofford, D.L. and Berlocher, S.H. (1987) Inferring evolutionary trees from gene frequency data under the principle of maximum parismony. *Syst. Zool.*, **36**, 292–325.

Taberlet, P. and Bouvet, J. (1992) Bear conservation genetics. *Nature*, **358**, 197.

Tautz, D. (1989) Hypervariability of simple sequences as a general source for polymorphic DNA markers. *Nucleic Acids Research*, **17**, 6463–71.

Taylor, A.C., Marshall Graves, J.A. *et al.* (1991) Conservation genetics of the Koala (*Phacolarctos cinereus*) II. Limited variability in minisatellite DNA sequences. *Biochemical Genetics*, **29**, (7/8), 355–63.

Taylor, A.C., Sherwin, W.B. and Wayne, R.K. (1993) The use of simple sequence loci to measure genetic variation in bottlenecked species: the decline of the northern hairy-nosed wormbat (*Lasiorhinus krefftii*). *Molecular Ecology*.

Telgelström, H. (1987) Transfer of mitochondrial DNA from the northern red-backed vole (*Clethrionomys rutilus*) to the bank vole (*Clethrionomys glareolus*). *J. Mol. Evol.*, **24**, 218–27.

Templeton, A.R. (1986) Coadaptation and outbreeding depression, in *Conservation Biology: The Science of Scarcity and Diversity* (ed. M.E. Soulé), Sinauer Press, Sunderland, Mass. pp. 105–16.

Templeton, A.R., Helmut Hemmer, H., Mace, G. *et al.* (1986) Local adaptation, coadaptation, and population boundaries. *Zoo Biology*, **5**, 115–25.

Templeton, A.R. (1989) The meaning of species and speciation: A genetic prespective, in *Speciation and its Consequences* (eds. D. Otte and J.A. Endler), Sinauer Associates, Sunderland, Mass, pp. 3–27.

Vane-Wright, R.I., Humphries, C.J. and Williams, P.H. (1991) What to protect – Systematics and the agony of choice. *Biological Conservation*, **55**, 235–54.

Vassart, G., Georges, M., Monsieur, R. *et al.* (1987) A sequence in M13 phage detects hypervariable minisatellites in humans and animal DNA. *Science*, **235**, 683–4.

Vassart, M., Granjo, L. and Greth, A. (1991) Genetic variability in the Arabian Oryx (*Oryx leucoryx*). *Zoo Biology*, **10**, 399–408.

Wayne, R.K. and Jenks, S.M. (1991). Mitochondrial DNA analysis supports extensive hybridization of the endangered red wolf (*Canis rufus*). *Nature*, **351**, 565–8.

Wayne, R.K., Gilbert, D., Lehman, N. *et al.* (1991) Conservation genetics of the endangered Isle Royale gray wolf. *Conservation Biology*, **5**, 41–51.

Wayne, R.K., Lehman, N., Allard, M.W. and Honeycutt, R.L. (1992) Mitochondrial DNA variability of the gray wolf: genetic consequences of population decline and habitat fragmentation. *Conservation Biology*, **6**, 559–69.

Wiley, E.O. (1981) *Phylogenetics: The Theory and Practice of Phylogenetic Systematics*, Wiley Interscience, New York.

Wilson, A.C., Cann, R.L., Carr, S.M. *et al.* (1985) Mitochondrial DNA and two perspectives on evolutionary genetics. *Biol. J. Linn. Soc.*, **26**, 375–400.

6

Evolutionary biology, genetics and the management of endangered primate species

K. Vàsàrhelyi and R.D. Martin
University of Zürich, Switzerland.

6.1 INTRODUCTION

There is no longer any sharp distinction between wild and captive animal populations. The effects of human activities on natural environments have been so far-reaching that there is now a continuous spectrum extending from limited groups of animals maintained in small enclosures through a range of intermediate conditions to free-living populations in relatively undisturbed environments. In all cases, at least some degree of management is now necessary and the basic issues with respect to endangered species are essentially the same across the spectrum. Effective management of endangered species requires a thorough grounding in conservation biology, a newly-emergent synthetic discipline within universities. Among other things, this discipline should provide training for application of developments in reproductive biology, conservation genetics, evolutionary biology, wildlife ecology, wildlife health and disease, restoration ecology, reintroduction management and environmental economics. Such applications in turn benefit from close cooperation between university-based researchers and managers of endangered species in zoos and reserves. Only in exceptional cases do zoos or reserves have direct access to specialized facilities on site.

Creative Conservation: Interactive management of wild and captive animals.
Edited by P.J.S. Olney, G.M. Mace and A.T.C. Feistner.
Published in 1994 by Chapman & Hall, London.
ISBN 0 412 49570 8

Staff in universities and similar institutions commonly have access to the necessary techniques and expensive equipment that are often beyond the reach of managers of endangered species, and they also have the capacity to conduct fundamental research leading to new applications. Within the overall framework of such work, the theory and practice of monitoring and managing endangered species should occupy a prominent place.

In the following text the value of university-based research for the management of endangered primate species will be illustrated with examples drawn from extremes of the continuum linking evolutionary biology and genetics. These examples illustrate two different yet complementary approaches to the question of population viability. Evolutionary considerations yield information concerning the genetic status and prospects of wild populations of endangered species and provide models for fragmented captive populations. These results are applicable to the formulation of management goals, which in turn depend on a thorough understanding of the genetic composition of the population of interest. Consequently, information from both evolutionary biology and genetics should optimally be incorporated into a comprehensive approach to conservation.

The example taken here is the relationship between species number and habitat area (the species–area curve), which can be examined and applied using techniques developed for allometric scaling analyses within the context of comparative biology. Among other things, this species–area relationship suggests that there may be particular cause for concern about progressive genetic impoverishment of residual populations of endangered primate species in habitat islands (e.g. reserves) over the long term. For this and other reasons, genetic monitoring is clearly indispensable for effective long-term management of both captive and wild populations, and should be given high priority.

6.2 SPECIES–AREA CURVES AND PRIMATE POPULATIONS

6.2.1 Introduction

One of the main preoccupations of conservation biologists is the long-term future for reserve areas of natural habitat. This matter cannot be addressed properly with existing information derived from short-term experiments or calculations based on laboratory populations, and this is an area where information derived from evolutionary biology can provide useful general guidelines.

Inevitably, reserve areas are far smaller than the geographical ranges originally occupied by the species they are designed to protect. Even where several reserves exist within the original range of a given species, their total area will still be very small in relation to that original range, and they will commonly be separated from one another by regions in which habitat destruction has occurred. Hence, reserves can to some extent be regarded as 'islands' and it is in principle possible to apply models and methods developed in the context of island biogeography. Following pioneering studies by Darlington (1957), Preston (1962) and MacArthur and Wilson (1963, 1967), it is now widely accepted that – at least for real islands separated by significant expanses of water – there is a broad, reasonably consistent relationship between the size of a given island and the number of species of a particular taxonomic group (e.g. mammals) that that island can support over the longer term.

Using techniques originally developed for allometric scaling analyses in comparative morphology, it is possible to investigate this species–area relationship and to extract general guidelines with potential applications in conservation biology. It has been proposed that, given long enough to reach equilibrium, the number of species on a given island will represent a balance between the arrival or origin of new species and the extinction of existing species. Further, the number of species present at equilibrium will be related (among other things) to the size of any given island (MacArthur and Wilson, 1963). As an extrapolation of this, it follows that subdivision of a formerly contiguous landmass into a number of islands (e.g. as a result of rising sea-level) should lead to increased rates of species extinction on those islands, such that each will ultimately carry a reduced number of species commensurate with its area. This provides an approximate parallel to the fragmentation of natural mainland habitats into 'islands' through human activities and can provide a basis for identifying broad principles that are likely to govern the long-term survival of species in residual reserve areas. The applicability of this concept of the species–area relationship has been explored in a number of previous studies (e.g. Simberloff and Abele, 1976; Terborgh, 1976; Soulé, Wilcox and Holtby, 1979; Soulé and Simberloff, 1986; Nilsson, Bengtsson and Ås, 1988; Soulé *et al.*, 1988; Wilcox, 1980; Shafer, 1990).

The Sunda Shelf in South-East Asia is of special interest with respect to the species–area relationship, as the most recent stage in its geological history provides a particularly instructive model of the eventual loss of species over a long period following fragmentation of a previously cohesive landmass into islands (see Ashton and Ashton, 1972; Whitmore, 1975). As the last glaciation drew to a close at the end of the Pleistocene, melting ice led to a major rise in sea level by some 100 m over the period between 18 000 and 6500 years ago, such that the major islands that now make up the Sunda archipelago achieved their present pattern of separation approximately 8000–10 000 years ago. Undoubtedly, a large proportion of the Sunda Shelf region was originally inhabited by a relatively diverse rain-forest fauna and flora, probably richer than that in the present mainland region of the Malay Peninsula; but species were presumably lost from the individual islands that remained following the rise in sea level, until an eventual balance was reached between species number and available habitat area on each island. This process in effect represents a natural experiment demonstrating the way in which species may be lost over a period of approximately 10 000 years following substantial habitat fragmentation.

6.2.2 Methods

Using the notation employed by MacArthur and Wilson (1967), the relationship between the size of a given island and the number of species of a particular taxonomic group can be expressed as the power function $S = c \cdot A^z$ (in which S represents the number of species, A is island area, z is the exponent value and c is a constant). This relationship can be converted to logarithmic form:

$$\log S = z \cdot \log A + \log c$$

The values of z and c can be established empirically by determining an appropriate best-fit line for the logarithmically converted data set.

There are problems with respect to the determination of best-fit lines for biological data sets. Alternative line-fitting procedures may be used for determining the trend in logarithmically converted data, and they can lead to differing conclusions. At present, the most commonly used line-fitting procedure is the least squares regression, although the major axis and the reduced major axis are being increasingly used in allometric scaling studies (Aiello, 1992; Harvey and Mace, 1982; Martin and Barbour, 1989). The least squares regression, in contrast to the major axis and the reduced major axis, is an asymmetrical approach. It is assumed that the y-variable is dependent upon the x-variable and that the latter is both independent and free of error. In allometric scaling studies, these two conditions usually do not apply when two biological variables (e.g. tooth size and body size) are plotted against one another. In analyses of species–area relationships, however, it could be argued that species number (S) must depend upon island area (A) and that the latter is an independent variable which can be measured with negligible error. Nevertheless, there are additional arguments in favour of using a symmetrical line-fitting approach (e.g. major axis or reduced major axis) in scaling studies of two biological variables, notably in interspecific studies (Aiello, 1992; Martin and Barbour, 1989). In particular, a symmetrical approach should be used when the underlying **relationship** between two variables is at issue.

A least squares regression generally yields a lower exponent value (i.e. slope of the best-fit line) than either the major axis or the reduced major axis, and this necessarily affects the conclusions. In practice, the alternative line-fitting techniques yield only marginally different results when correlation coefficient values are high ($r \geqslant 0.97$); but in cases where correlations are not very strong (as is typically the case with studies of species–area relationships) markedly different results may emerge. In view of the continuing controversy over the correct choice of a line-fitting method, the best procedure at present is to apply several different approaches and to comment on any differences between the results obtained. Accordingly, in the analyses presented below all three methods have been compared, although this was not possible in cases where results were taken from authors who used only a single line-fitting technique.

6.2.3 Results

The inferred process of extinction of species in the Sunda Shelf has been analysed with respect to its mammalian fauna. In an initial study restricted to the large mammals of the region, Soulé, Wilcox and Holtby (1979) obtained the following equation relating the number of species (S_{lm}) to island area (in km^2):

$$\text{regression: } \log_{10} S_{lm} = 0.21 \, . \, \log_{10} A + 0.18$$
$$(r = 0.65;\ r^2 = 0.42)$$

In a subsequent study, Wilcox (1980) reported on an analysis of the species–area relationship for land mammals generally (excluding bats), which yields the following empirical formula relating species number (S_m) to island area:

$$\text{regression: } \log_{10} S_m = 0.30 \, . \, \log_{10} A + 0.26$$

More recently, a species–area analysis was undertaken for the primate fauna of the Sunda Shelf region, considering together data for the Sunda archipelago, the Philippines and Sulawesi (Tardent, 1988; Figure 6.1). That study identified a total of 28 primate species now inhabiting the mainland and/or various islands of the Sunda Shelf. Of these, 13 occur exclusively on the mainland and nine exclusively on islands of the Sunda archipelago, with six species occurring both on the mainland and on islands. Analysis of the data for island-living primate species leads to the following formulae for the relationship between number of primate species (S_p) and island area:

$$\text{regression: } \log_{10} S_p = 0.26 \, . \, \log_{10} A - 0.65$$
$$\text{major axis: } \log_{10} S_p = 0.28 \, . \, \log_{10} A - 0.71$$
$$\text{reduced major axis: } \log_{10} S_p = 0.34 \, . \, \log_{10} A - 1.00$$
$$(r = 0.77;\ r^2 = 0.60)$$

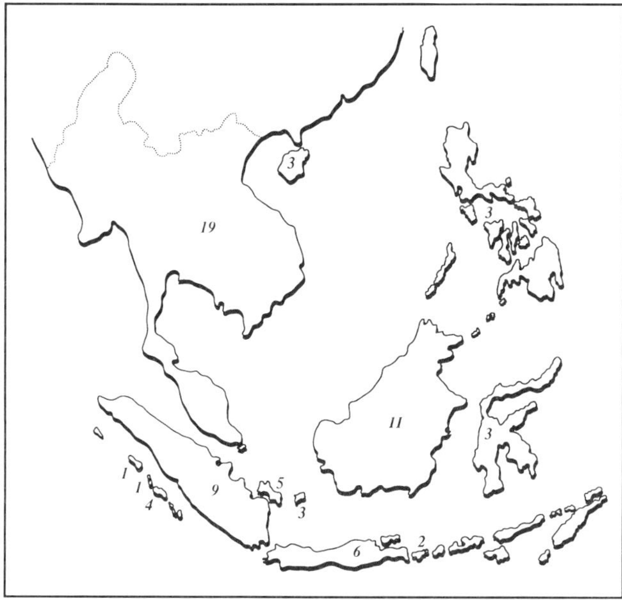

Figure 6.1 Outline of map of South-East Asia, showing the islands of the Sunda Shelf together with Hainan, the Philippines and Sulawesi. Figures indicate the numbers of primate species present on the mainland and on the larger islands of the region (data from Tardent, 1988).

When the results of the various analyses of Sunda mammals are combined for comparison (Figure 6.2), it can be seen that the three categories (land mammals excluding bats; large mammals; primates) show similar scaling patterns, with only relatively minor differences in the slopes of the lines (representing an approximately constant scaling exponent z). There are, however, major differences in the intercept values (reflecting different values of the coefficient c). It is noteworthy that the line for primates lies well down in the graph, reflecting a very low value of the coefficient c, and indicating that maintenance of a varied primate fauna requires a relatively large habitat area. In the terminology of allometric scaling studies, primates represent a distinct grade with a low value of the coefficient c. This indicates that primates may be valuable indicator species for determining the size of reserve areas and for following the processes of change induced by habitat fragmentation.

From the equations given above for primates, the following predictions can be made with respect to the island area required to maintain a single primate species in equilibrium on the Sunda Shelf over the long term: 316 km^2 (regression); 343 km^2 (major axis); 873 km^2 (reduced major axis). Similarly, the following predictions can be made with respect to

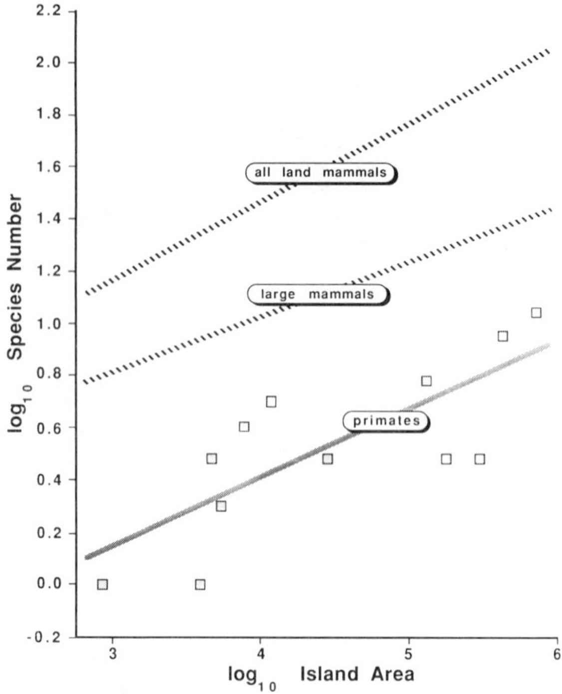

Figure 6.2 Scaling relationships between species number and island area (in km^2) on the Sunda Shelf for land mammlas excluding bats (data from Wilcox, 1980), for large land mammals (data from Soulé, Wilcox and Holtby, 1979) and for primates (data from Tardent, 1988). Individual points are shown only for the primates. For purposes of comparability, the best-fit lines are least squares regressions in all cases.

the island area required to hold a moderately diverse fauna containing six primate species in equilibrium over the long term: 311 092 km^2 (regression); 206 455 km^2 (major axis); 169 768 km^2 (reduced major axis). It should be noted, incidentally, that the reduced major axis gives markedly different predictions in comparison with the regression. Taking the reduced major axis formula, the predicted area required for a single primate species is almost three times bigger than that inferred from the regression formula. By contrast, the area predicted from the reduced major axis for maintenance of six primate species is approximately half of that predicted by the regression. In all cases, however, the predicted areas are substantial and indicate that primate populations inhabiting rain-forest areas are likely to encounter severe problems following habitat fragmentation.

The species–area relationship for primates hence suggests that a forested island area of at least 300 km^2 and possibly as much as 900 km^2 might be required to maintain a typical single primate species in equilibrium over the long term. The crab-eating macaque (*Macaca fascicularis*), which is a medium-sized primate inhabiting the Sunda region, can be taken as a representative species here. A survey of crab-eating macaques on the island of Sumatra (Crockett and Wilson, 1980) yielded an average figure of 55.4 individuals/km^2 for the population density of this species. This would correspond to about 16 600 macaques if extrapolated to an island with a total forested area of 300 km^2 and to almost 50 000 macaques if extrapolated to a total forested area of 900 km^2.

This provides a different perspective on the concept of 'minimum viable population size', which is one of the key concepts in conservation biology (Shaffer, 1981; Gilpin and Soulé, 1986; Lacy, 1992). Most approaches to the inference of a minimum viable population for conservation of a given species are, for understandable reasons, based on genetic considerations (e.g. minimal conditions required for survival of a population for 100 years) and have generated estimates of 50 or 500 individuals (Durant and Mace, Chapter 4).

The reasons for loss of species from an island area following isolation have yet to be established, but they are undoubtedly multiple. It seems quite likely that genetic processes play at least a contributory part in such extinction, but the magnitude of that part has yet to be established. We need to determine whether genetic variability exerts a major influence on the long-term survival of populations much larger than those usually considered in discussions of 'minimum viable population size'. Some indirect information relating to this problem can be obtained by examining genetic variability in primate populations on the islands of the Sunda Shelf. Once again, *Macaca fascicularis* presents itself as a model species, not only because of its middle-of-the-range body size but also because it has been subjected to extensive genetic investigation in a series of studies (Kawamoto, Nozawa and Ischak, 1981; Kawamoto, Ischak and Supriatna, 1984; Kawamoto *et al.*, 1988; Kondo *et al.*, 1993). Representative samples of crab-eating macaques from islands of the Sunda Shelf and from introduced populations on the island of Angaur and Mauritius have yielded data on average heterozygosity and proportions of polymorphic loci. When these data are plotted against island area on logarithmic coordinates (Figure 6.3), it emerges that there is a fairly consistent relationship ($r = 0.78$; $r^2 = 0.61$) between genetic variability and land area for seven islands of the Sunda Shelf, with the proportion of polymorphic loci (for example) increasing progressively with island area.

The data presented in Figure 6.3 indicate that genetic variability shows a significant association with island area (and hence population size of the primate species concerned) even with very large natural populations. Hence, it can be concluded that, even at the scale represented by the islands of the Sunda Shelf, habitat fragmentation is likely to be associated with a decline in variability in isolated populations.

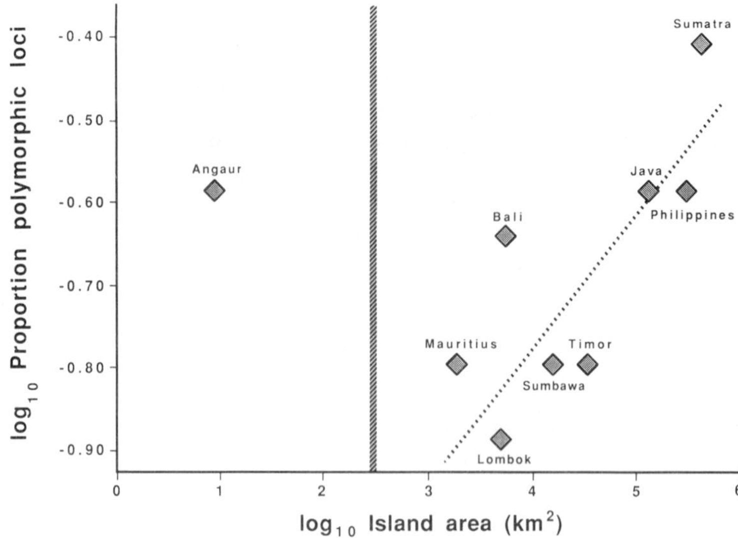

Figure 6.3 Relationship between the proportion of polymorphic loci and island area (in km^2) for seven natural populations of crab-eating macaques (*Macaca fascicularis*) on islands of the Sunda Shelf, and for two introduced populations on the islands of Mauritius and Angaur (data from Kondo *et al.*, 1993). The best-fit line for the seven natural populations is the major axis. The vertical hatched line indicates the approximate lower limit of island area for long-term survival of a single species, as inferred from the species–area curve for primates inhabiting islands of the Sunda Shelf (see Figure 6.2). The point for Angaur is clearly aberrant (see discussion in text).

Whether such reduction in genetic variability is directly associated with extinction of species on isolated islands is a moot point, but there is at least an indication here that genetic factors may play a role in long-term viability for populations much larger than those commonly considered in conservation biology.

The island of Mauritius, which now has a population of approximately 30 000 crab-eating macaques derived from an unknown number of introduced founders, fits the best-fit line for the natural populations of the Sunda Shelf islands quite closely. On the other hand, the island of Angaur shows a major departure from the trend. This tiny island, with a total land area of about 8.3 km^2, has a population of about 500 crab-eating macaques that were purportedly derived from a single pair of animals introduced at the beginning of this century. A study of genetic variation in this population (Kawamoto *et al.*, 1988) revealed an unexpectedly high level of heterozygosity, and also yielded evidence that ruled out derivation of the population from a single pair of founders. At present, there is no explanation for the high degree of genetic variability currently found among the crab-eating macaques of Angaur, but clarification of this issue could yield extremely valuable information for conservation biology.

6.2.4 Discussion

Soulé, Wilcox and Holtby (1979) explicitly used the species–area relationship as a model for predicting future developments in reserves. Taking the relationship established for the large mammal species inhabiting the islands of the Sunda shelf, they predicted that a substantial number of large mammal species will become extinct in 19 East African game reserves, with a typical reserve losing almost 50% of its large mammal species over 500 years. Other authors have argued that the species–area relationship implies that, for a given habitat, it is always preferable to maintain a single large reserve than a number of small reserves because larger areas contain more species. Such direct applications of the species–area relationship have attracted criticism, and it is important to recognize the limitations of island biogeography theory in this context (see also Magin *et al.*, Chapter 1). Several authors have criticized interpretations of island–area relationships, notably with respect to conservation issues. Such criticisms have concerned (i) the inference that these relationships represent equilibrium states, and (ii) the 'SLOSS' debate (single large or several small) about optimal reserve size. These issues have been effectively reviewed by Shafer (1990) and it should be emphasized that they have no direct bearing on the essentially empirical approach adopted here.

Caution is obviously required in the interpretation of species–area relationships for a number of reasons. In the first place, the correlations found are relatively low (indicating that factors other than land area affect species numbers to an appreciable extent), and uncertainty about the correct line-fitting method hence has major implications. There are also numerous uncertainties concerning the data used. For the analyses of the Sunda Shelf islands presented here, total island area has been taken for the sake of simplicity. It is quite likely, however, that the islands concerned were not completely covered by forest over most of the 10 000-year period concerned. Because of environmental heterogeneity, it is of course a crude approximation to equate island area with area of available habitat for any given species, and such approximation will lead to over-estimation of the habitat area required to support a given number of species. There may even be uncertainty about the number of species that should be recognized. For instance, the number of primate species indicated for Sulawesi in Figures 6.1 and 6.2 represents a very conservative estimate, recognizing only the three species *Macaca nigra, Tarsius pumilus* and *Tarsius spectrum*. Some authors divide *Macaca nigra* into as many as seven different species, and a new Sulawesi tarsier species (*Tarsius dianae*) was reported after the analysis illustrated had been conducted. Hence, the number of primate species on Sulawesi could be as many as 10. Taxonomic issues of this kind are likely to arise with any study of species–area relationships and introduce an additional element of doubt about the reliability of any results. Furthermore, any study of species–area relationships is typically restricted to a single taxonomic group, and ecological interactions at a higher taxonomic level are hence ignored. For these and several other reasons, any extrapolation of findings from species–area studies to conservation biology should be recognized as providing general guidelines rather than precise estimates.

At the same time, it should be recognized that investigation of general evolutionary relationships can make an invaluable contribution to conservation biology. The special value of the evolutionary approach is that it is based on processes taking place over the long term. Whatever their limitations, investigations of evolutionary processes and their outcome provide the only source of information about long-term processes in natural populations. Such investigations can provide general pointers to the changes associated with extinction and can hence provide vital information for conservation measures. The example taken here – the relationship between habitat area and species number – is of interest for a number of reasons. In addition to providing general guidelines concerning long-term expectations, this relationship can provide an explicit illustration of extinction of species following habitat fragmentation. Although the model has its limitations, it provides one of the few available clear examples of an inferred reduction in species number following a reduction in available habitat area.

For the time being, it seems justifiable to conclude that the long-term 'minimum viable population size' for an island-living primate species indicated by the species–area relationship for the Sunda Shelf may be considerably greater than often suggested for the short term – probably in the thousands rather than in the hundreds. Preliminary information on macaques from that area indicates that genetic diversity is significantly reduced on smaller islands, but it is not yet known whether this is directly linked with the long-term viability of those populations. It is, however, clear that genetic monitoring of both wild and captive populations should be a high priority for the future.

6.3 DNA FINGERPRINTING STUDY OF CAPTIVE GOELDI'S MONKEYS

6.3.1 Introduction

Among other things, it is to be expected that small, isolated populations will experience problems connected with a reduction in genetic diversity. Such problems are particularly acute with captive breeding populations derived from a relatively small number of founders. For this reason, it is important to study the genetic characteristics of small populations both in the wild and in captivity, in order to assess changes and to develop strategies to offset the adverse genetic effects of isolation (see also Wayne *et al.*, Chapter 5). No single factor accounts for the extinction of small, isolated populations, but there are good reasons to believe that a reduction in genetic diversity is at least a contributory factor. In their review of the concept of 'minimum viable populations', Gilpin and Soulé (1986), identified four major causes of extinction:

(a) genetic stochasticity;
(b) demographic stochasticity;
(c) environmental stochasticity;
(d) catastrophes.

Genetic monitoring to increase our understanding of the relationship between genetic variation and population viability is important, and it is clear that genetic monitoring of both captive and wild populations should be an integral part of an effective conservation programme (Lande and Barrowclough, 1987).

As yet, there has been little attempt to investigate genetic differences associated with fragmentation of populations of endangered primate species. One study has, however, investigated the genetic effects of isolation in a lemur species – the black lemur, *Lemur (Eulemur) macaco* – by conducting comparisons between different populations using electrophoretic examination of red cell enzymes (Arnaud *et al.*, 1992; Meier and Rumpler, 1992). In that study, a comparison was made between 43 animals sampled from the mainland population, 46 animals sampled from a small population on an island off the coast of Madagascar (Nosy Ambariovato) and 32 animals from a captive breeding colony. Of the 10 red cell enzymes examined, four proved to be polymorphic (with two or three genotypes per enzyme) in the mainland population. In the sample from the island population, one of these enymes was found to be monomorphic, while for another enzyme the number of phenotypes was found to be only two instead of three. The sample from the captive population showed an even more dramatic difference from the mainland population in that three of the four potentially polymorphic enzymes were found to be monomorphic, while for the remaining enzyme the number of phenotypes was found to be only two instead of three. Although these results are preliminary, they do suggest some reduction in genetic variability in the island population compared with the mainland, and it is fairly clear that there has been a major reduction in genetic variability in the captive breeding programme. Genetic monitoring of wild and captive primate populations is therefore essential, in order to establish baseline values and to provide a proper foundation for long-term interpretation of changes in populations of endangered species (Lande and Barrowclough, 1987).

As a general rule, it is likely that fragmentation of natural populations will be associated with a reduction in genetic variability, and that captive populations can be used as an extreme model of the changes that occur. One threatened primate species which can be taken as an example is Goeldi's monkey (*Callimico goeldii*). This species inhabits a relatively restricted geographical range extending from south-eastern Peru and northern Bolivia to southern Colombia. Within this range, its distribution is sparse and apparently very patchy, such that the species may be naturally subdivided into small semi-isolated subpopulations. A recent field survey in Bolivia has confirmed that *Callimico goeldii* is indeed relatively rare and sparsely distributed there. Further, the species has already disappeared from at least one well-documented area in which it had been studied by previous field investigators (A. Christen, pers. comm.).

Because of its known threatened status, breeding colonies of *Callimico goeldii* have been established in a number of zoos and other institutions and an international studbook has been established (Warneke, 1992). At present 440 animals are held in 75 institutions. The international conservation effort is focused on the establishment of a global management programme for the subdivided captive population. Our work aims to provide the appropriate genetic information necessary for optimal management. The long-term goal is the genetic typing of the entire captive population and present work is focused on the establishment of methods suitable for this purpose. Information about the genetic composition of the fragmented captive population will benefit global management efforts which attempt to preserve genetic variability in the population.

Preliminary studies in our laboratory involved electrophoretic surveys of blood proteins in a small breeding colony held in Zürich. Of the 26 systems tested in 28 individuals, only four are definitely polymorphic, nine are possibly polymorphic (requiring further investigation) and 13 are definitely monomorphic (W. Scheffrahn, pers. comm.). This low level of variability in blood proteins rules out any meaningful investigation of genetic relationships using classical electrophoretic techniques. In what follows, part of the captive population of Goeldi's monkey is therefore taken as an illustration for potential applications of genetic monitoring using DNA-based techniques within a conservation framework.

6.3.2 Methods

(a) Subjects

A breeding colony of Goeldi's monkey *(Callimico goeldii)* is currently maintained at the Anthropological Institute of the University of Zürich, in collaboration with the zoos of Chicago, Jersey and Zürich. DNA samples from 18 individuals in this colony were studied. Of these, 10 animals originated from the Jersey Zoo, five from the Chicago Zoo, and two from the Stuttgart Zoo. One animal was born in Zürich from a mating between a Jersey animal and a Chicago animal.

(b) Extraction of DNA

DNA was purified using a modified version of the salt precipitation method (Signer *et al.*, 1988). The leucocyte fraction from 2 ml fresh or thawed blood was mixed with 3 ml sterile distilled water. After centrifugation at 10 000 rpm and 4°C for 30 minutes, the pellet was resuspended in 1 ml of 0.1% Nonidet P-40 and centrifuged again as above. Proteinase K digestion and ethanol precipitation were carried out according to Signer *et al.* (1988). Precipitated DNA was dissolved, depending on amount, in 100–200 ml of 1xTE (0.01 M Tris–HC1, 0.2 mM EDTA, pH 7.5). Quantitative assessment was carried out through comparisons of samples to λ standards of known concentration on a test gel.

(c) DNA fingerprinting

DNA (4 μg) was digested to completion with Hinf I and then electrophoresed in a 1% agarose gel (1400 VHrs). DNA was blotted onto Pall Biodyne nylon membranes, probed with ^{32}P-labelled human minisatellite probe 33.6 (Jeffreys, Wilson and Thein, 1985). Hybridizations were carried out at 42°C for 16 h with 50% formamide. Membranes were washed at 65°C with 1 × SSC and autoradiographed in an intensifying screen at −70°C for 24 h or longer.

(d) Analysis of DNA fingerprint patterns

To date, approximately 50% of the 153 possible pairwise comparisons among the 18 subjects have been completed. Comparison of banding patterns was carried out visually and was restricted to samples within the same gel.

In order to estimate the level of genetic variability in the study population, the band-sharing value or similarity index (SI) was calculated according to Lynch (1991) as $2N_{AB}/(N_A + N_B)$, where N_{AB} is the number of bands individual A and B have in common, and N_A and N_B are the total numbers of bands possessed by individuals A and B, respectively. SI values calculated within gels were pooled across all gels. Average band-sharing values were computed for animals originating from Chicago and for those originating in Jersey, as well as between the two groups. Furthermore, average band-sharing values were compared between groups of first-degree relatives (parent/offspring and siblings), distantly related individuals (lesser degrees of relationship known from the pedigree in captivity) and unrelated individuals (no known relationship since arrival in captivity).

6.3.3 Results

(a) Confirmation of relationships using DNA fingerprinting

A major practical problem associated with the analysis of relationships within any captive colony is that standard branching family trees typically become extremely complicated when multiple generations are involved. Indeed, it would be practically impossible to produce such a standard tree for the existing *Callimico* studbook (Warneke, 1992). For this reason, a new and simpler type of pedigree was designed in which every individual in a set of breeding records can easily be traced back to the wild-caught founders. Figure 6.4 demonstrates the relationships between the 18 individual Goeldi's monkeys studied, using this revised form of pedigree.

Using DNA fingerprinting, first-degree relationships can be confirmed through comparison of banding patterns. So far, in our studies of *Callimico goeldii*, each individual has been found to have a unique pattern, despite the occurrence of a certain degree of inbreeding in some cases. Thus, DNA fingerprinting may be used for long-term identification of individuals both in captivity and in the wild. In Figure 6.5, DNA fingerprints from two families are shown. In each comparison, every band in the pattern of the offspring (middle lanes) is present in either one or both parents (left and right lanes). Accordingly, DNA fingerprints can also be used for the identification of direct relatives, for example in confirming paternity or in resolving cases where breeding history is unclear.

Figure 6.4 Simplified pedigree for the 18 study subjects in the Zürich breeding population of *Callimico goeldii*, going back to the original founders in captivity. Individuals are listed at the left in descending order of birth date. Each vertical line represents a nuclear family of parents (x) and offpsring (o). Horizontal positioning has no significance relative to breeding history. Boxes indicate study subjects.

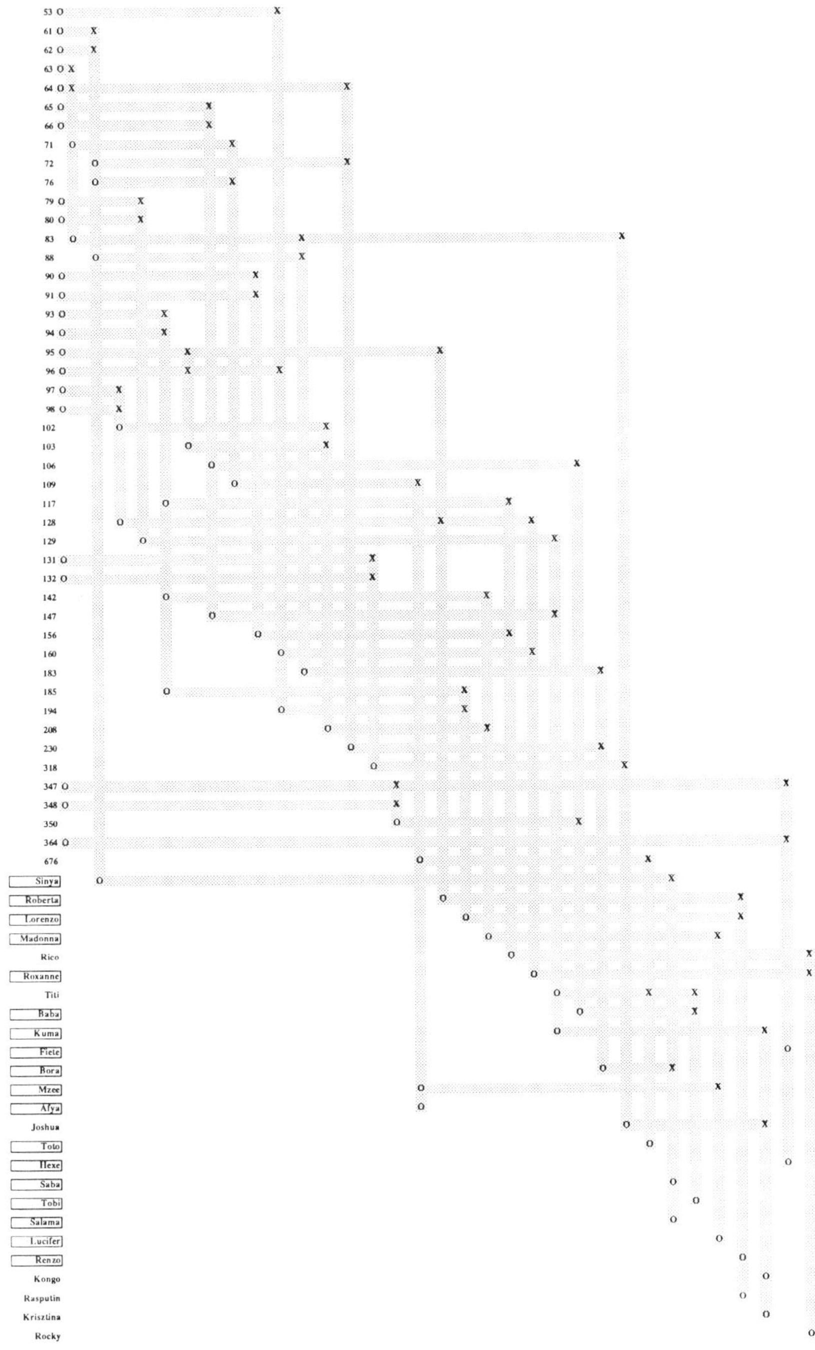
53
61
62
63
64
65
66
71
72
76
79
80
83
88
90
91
93
94
95
96
97
98
102
103
106
109
117
128
129
131
132
142
147
156
160
183
185
194
208
230
318
347
348
350
364
676
Sinya
Roberta
Lorenzo
Madonna
Rico
Roxanne
Titi
Baba
Kuma
Fiete
Bora
Mzee
Afya
Joshua
Toto
Hexe
Saba
Tobi
Salama
Lucifer
Renzo
Kongo
Rasputin
Krisztina
Rocky

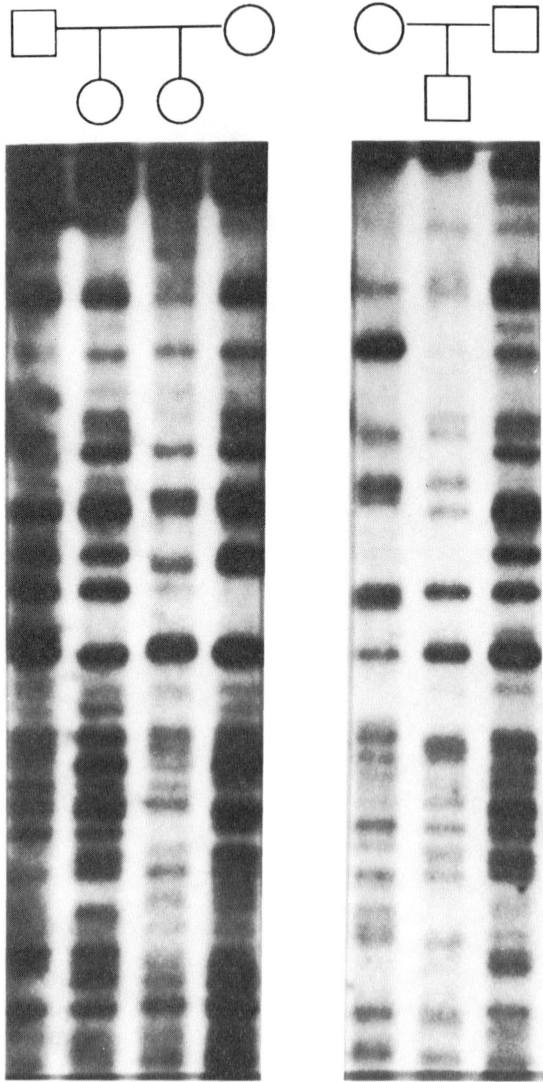

Figure 6.5 DNA fingerprints of two family groups of *Callimico goeldii* from the Zürich colony.

(b) Calculation of the similarity index

In addition to comparisons of band-sharing between individuals, it is also possible to calculate overall levels of band-sharing as an indirect indicator of degree of relatedness. In this way, it is possible to reach an overall assessment of genetic variability within a population and in sub-

Table 6.1 Comparisons of average similarity indices (SI) among subgroups of the study population, based on origin of individuals (X indicates the comparison of SI values given above)

Subgroup	*Jersey–Jersey*	*Jersey–Chicago*	*Chicago–Chicago*
SI	*0.613*	*0.390*	*0.403*
$Z = -2.595$ $p = 0.0095$	X		X
$Z = -5.685$ $p = 0.0001$	X	X	
$Z = -0.658$ $p = 0.5108$		X	X

populations thereof. Similarity index values were accordingly calculated for subgroups representing place of origin and level of relatedness between individuals. The results of these analyses are summarized in Tables 6.1 and 6.2.

On the basis of the pedigree, pairs of Jersey animals were classified as first-degree relatives, distantly related or unrelated, whereas all Chicago animals were distantly related. Band-sharing in the Chicago group as a whole was markedly lower than in the Jersey group (Figure 6.6). Furthermore, band-sharing in the Chicago group (all distantly related) was also lower than in the 'distantly related' subgroup of the Jersey group. In subgroups based on the level of relatedness, band-sharing was highest among first degree relatives (parent and offspring,

Table 6.2 Comparisons of average similarity indices (SI) among subgroups of the study population, based on relationship among individuals (X indicates the comparison of SI values given above)

Subgroup	*First degree relatives*	*Distantly related*	*Unrelated*
SI	*0.708*	*0.485*	*0.449*
$Z = -4.644$ $p = 0.0001$	X		X
$Z = -3.102$ $p = 0.0019$	X	X	
$Z = -1.501$ $p = 0.1344$		X	X

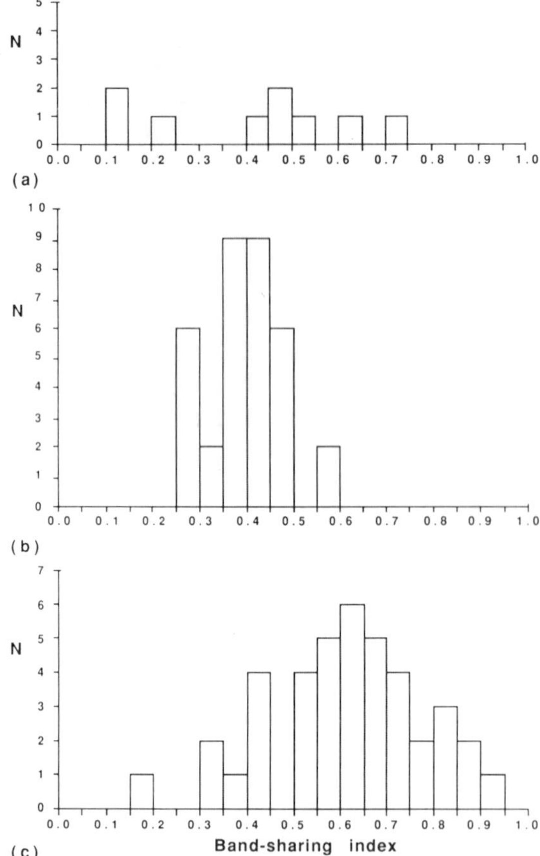

Figure 6.6 Similarity index values for *Callimico goeldii*, showing the distributions for comparisons both within colonies (Chicago/Chicago (a) and Jersey/Jersey (c)) and between colonies (Chicago/Jersey (b)). Note the relatively high values for comparisons within the Jersey colony, contrasting with the values for both Chicago/Chicago and Chicago/Jersey comparisons.

or siblings), as expected, and lowest among distantly related and unrelated individuals (Figure 6.7). All Chicago comparisons were found to be at the low extreme of the distribution for distantly related pairs.

6.3.4 Discussion

It is now generally accepted that persistent breeding between closely related individuals, leading to increased homozygosity in the population, is associated with reduced fitness in most, although not all, species (Lacy, Petric and Warneke, in press). An important goal of captive

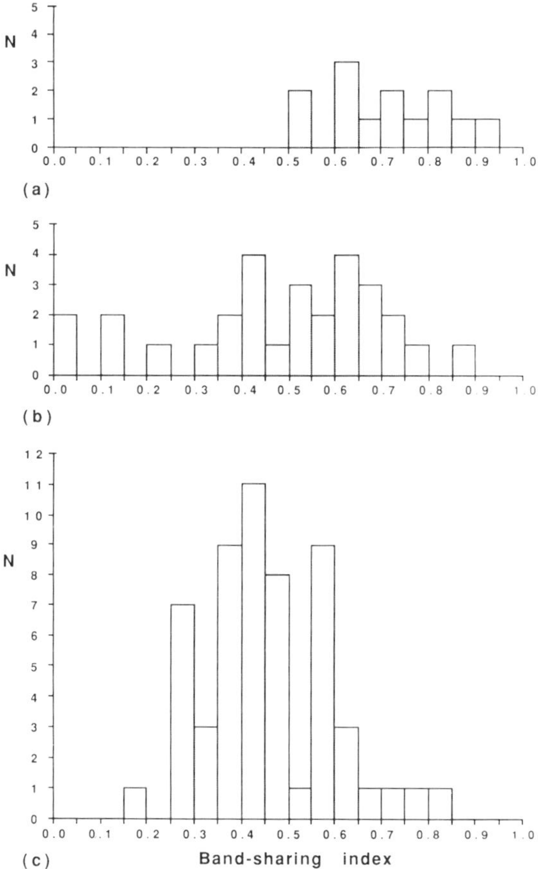

Figure 6.7 Band-sharing index values for *Callimico goeldii*, subdivided into (a) 'direct relatives' (parent–offspring or sibling relationship documented from captive breeding records), (b) 'indirect relatives' (some lesser degree of relationship documented from captive breeding) and (c) 'unrelated' (no known relationship since entering the captive breeding programme).

breeding programmes is therefore the maintenance of genetic variation present in the original founders. In order to avoid the detrimental effects of inbreeding and thus shield threatened species against extinction, captive breeding programmes should incorporate genetic monitoring as an integral feature. An array of genetic typing methods is available for the genetic characterization of captive populations. These include electrophoresis of blood proteins, RFLP analyses, DNA fingerprinting and the amplification of hypervariable genetic loci using the polymerase chain reaction (PCR). The work reported here provides an example of how one of these methods, DNA fingerprinting, may be applied to certain questions relating to the genetic management of a threatened species.

DNA fingerprinting has been applied to problems in genetic management of a variety of endangered species (Hall, Groth and Wetherall, 1992). In the present study, a small breeding colony of *Callimico goeldii* was studied using DNA fingerprinting in order to address two questions relevant to the management of this small group. One goal was to determine if subgroups in this colony originating from different zoos have different levels of genetic variability. A second was to assess the suitability of the degree of band-sharing as an indicator of the level of relatedness of individuals.

With respect to the first question, we found that the level of genetic variability measured by the similarity index is lower among animals derived from the Jersey stock than among animals from the Chicago stock. Although comparisons between these two groups are undoubtedly influenced by a difference in sample size, a restricted comparison of the 'distantly related' subgroup of the Jersey stock with the Chicago stock clearly indicates that animals from Chicago are less similar to one another than the animals from Jersey.

Although band-sharing levels are on average higher for direct relatives than for unrelated individuals, there is no sharp distinction between the two categories, in agreement with previous work (Mace, Pemberton and Stanley, 1992). Unexpectedly, one animal (Baba) from the Jersey group was found to show unusually high values (within the range for first-degree and distant relatives) for band-sharing with reportedly unrelated Jersey animals. The question arises whether an error may have occurred in the pedigree assignment of this individual. Alternatively, the wild-caught ancestors of the animals now showing such unexpectedly high levels of band-sharing may have been closely related. As information about the founders of the captive population of *Callimico* is lacking, the perhaps unjustifiable assumption was made that all founders are unrelated. Further genetic studies will now be carried out to explore the possibility of close relationships between founders.

As the present work illustrates, DNA fingerprinting is suitable for the assessment of the overall level of genetic variability in a small group of individuals. Captive breeding of *Callimico goeldii* necessarily takes place at a large number of individual institutions and attempts are being made to achieve global management of the population. In arranging exchanges of animals between institutions and in choosing partners for breeding pairs it is therefore of interest to characterize the variation in levels of genetic diversity between groups housed at different locations. Further applications of DNA fingerprinting in the context of zoo management may include individual identification for pedigree confirmation and paternity determination (Wayne *et al.*, Chapter 5).

Although DNA fingerprinting can be a useful tool for genetic monitoring, it is important to note that certain disadvantages limit its application in captive breeding programmes. For the study of population genetics, DNA fingerprinting is best confined to investigations of small populations. Because of the complex patterns obtained the evaluation of DNA fingerprints is somewhat subjective, and for this reason it is often necessary to restrict comparisons to pairs of individuals represented within the same gel. Because of this restriction, the study of large populations rapidly becomes impractical. For example, in the case of *Callimico goeldii*, 631 gels (each with 18 samples) would have to be run to study the entire captive population of 440 individuals. A much-cited disadvantage of DNA fingerprinting is associated with its desirable feature of detecting a large amount of variability. DNA fingerprints are so complex that the determination of the allelic relationships between bands is virtually impossible. Methods which detect variation at single loci are preferable because of their ease of interpretation, despite the disadvantage of having to combine information from several loci to achieve adequate resolving power. DNA fingerprinting is also a method that requires a relatively large amount of DNA. This is of concern in studies of endangered species, where it is important to obtain a maximum of information from a minimum amount of tissue, and indeed to avoid invasive methods of sampling whenever possible.

The present study has served as a preliminary step in the genetic characterization of the captive population of *Callimico goeldii*. Further work will focus on expansion of the scope of study to include as much of the captive population as possible. The concerns mentioned above regarding the use of DNA fingerprinting will be addressed, and the primary method of analysis will involve characterization of microsatellite loci using the polymerase chain reaction (PCR).

6.4 CONCLUDING DISCUSSION

The primary aim of this contribution is the demonstration, by way of example, of the necessity for interdisciplinary collaboration in tackling conservation problems. The examples presented happen to be drawn from the fields of evolutionary biology and genetics, but numerous other fields are equally important within the framework of the emerging discipline of conservation biology. The two fields of evolutionary biology and genetics were chosen because both demonstrate the continuing need for university-based research in the development of conservation biology, and because they represent two extremes of the biological spectrum involved in the emergence, transformation and extinction of species.

The concept of the species–area relationship was chosen as an illustration of the special perspective that can be provided by an evolutionary approach. Despite its inevitable coarse-grained nature and the resulting limitations, the evolutionary approach provides the only source of information about long-term changes. It is vital to consider such long-term changes in any overall conservation strategy, as considerable resources may otherwise be devoted to strategies that will postpone but not arrest the abnormal extinction process arising from the conflict between human activities and the natural world. In this specific case, application of the species–area concept to data on primate species suggests that the population size required for long-term survival of primate species may be considerably larger than often proposed. Further, there is some evidence that reduced genetic variability even in quite large isolated natural populations may be a factor contributing to their vulnerability to extinction. In the light of such indications, it is obvious that breeding programmes for endangered primate species such as *Callimico goeldii* (with a total captive population of less than 500 individuals) urgently require attention to genetic aspects.

Loss of genetic variability is a major threat to the survival of any species. Therefore, in addition to the establishment of the optimal size for populations, active management of these populations in order to preserve genetic variability is essential. Based on a knowledge of the genetic composition of a population, optimal management strategies can be devised. These may involve transfers or exchanges of animals between institutions as well as choices of appropriate breeding pairs in captive breeding programmes, in addition to reintroduction programmes and subsequent genetic monitoring in the wild. Such strategies can reduce loss of genetic variability in a population, thereby furthering the chances for species survival.

In summary, the present contribution attempts to demonstrate the ways in which two fields of study as pursued in a university setting – evolutionary biology and genetics – may contribute to conservation of endangered species. In addition to these two fields, numerous others have applications to species conservation. It is up to members of the research community as well as zoos and wildlife reserves to work together and take advantage of the benefits of collaboration for the sake of our common goal, which is the preservation of biodiversity.

ACKNOWLEDGEMENTS

We are particularly grateful to the Jersey Wildlife Preservation Trust (especially Dr Bryan Carroll) and to the Brookfield Zoo, Chicago (especially Dr Ann Baker) for providing the initial stock for the breeding colony of *Callimico goeldii* at the Anthropological Institute in Zürich. The cooperation of the Zürich Zoo (especially Dr Christian Schmidt) is also gratefully ackowledged. Thanks are due to Dr Chris Pryce and to the animal caretakers (Inge Sandmeier, Karin Etter and Ursula Gonsalez) for their skilled care of the animals and for providing blood samples for genetic typing. Thanks are also due to Josiane Tardent for permission to use data on the primates of the Sunda Shelf region from her unpublished semester thesis. The concept for the pedigree representation in Figure 6.4 was contributed by Sandy Rutherford and Daniel Thomas. Research on our breeding colony of *Callimico goeldii* has been generously supported by the Swiss National Science Foundation (grant 3100-27806), by the A.H. Schultz Foundation and by the Julius Klaus Foundation in Zürich.

REFERENCES

Aiello, L.C. (1992) Allometry and the analysis of size and shape in human evolution. *Journal of Human Evolution*, **22**, 127–47.

Arnaud, J., Meier, B., Dugoujon, J.M. *et al.* (1992) Study of the variability of erythrocyte enzymes in captive and wild populations of the black lemur (*Eulemur macaco macaco*). An indispensable preliminary in captive breeding programmes. *Primates*, **33**, 139–46.

Ashton, P. and Ashton, M. (1972) *The Quaternary Era in Malesia*, Department of Geography Miscellaneous Series 13, University of Hull, Hull.

Crockett, C.M. and Wilson, W.L. (1980) The ecological separation of *Macaca nemestrina* and *M. fascicularis* in Sumatra, in *The Macaques: Studies in Ecology, Behavior and Evolution* (ed. D.G. Lindburg), Van Nostrand Reinhold, New York, pp. 148–81.

Darlington, P.J. (1957) *Zoogeography: The Geographical Distribution of Animals*, John Wiley, New York.

Gilpin, M.E. and Soulé, M.E. (1986) Minimum viable populations: processes of species extinction, in *Conservation Biology: The Science of Scarcity and Diversity* (ed. M.E. Soulé), Blackwell Scientific/Sinauer Associates, Oxford/Sunderland, Mass., pp. 19–34.

Hall, G., Groth, C.M., and Wetherall, J. (1992) Application of DNA profiling to the management of endangered species. *International Zoo Yearbook*, **31**, 103–8.

Harvey, P.H. and Mace, G.M. (1982) Comparison between taxa and adaptive trends: problems of methodology, in *Current Problems in Sociobiology* (eds King's College Sociobiology Group), Cambridge University Press, Cambridge, pp. 343–61.

Jeffreys, A.J., Wilson, V. and Thein, S.L. (1985) Individual-specific 'fingerprints' of human DNA. *Nature, London*, **316**, 76–9.

Kawamoto, Y., Ischak, T.M. and Supriatna, J. (1984) Genetic variations within and between troops of the crab-eating macaque (*Macaca fascicularis*) on Sumatra, Java, Bali, Lombok and Sumbawa, Indonesia. *Primates*, **25**, 131–59.

Kawamoto, Y., Nozawa, K. and Ischak, T.M. (1981) Genetic variability and differentiation of local populations in the crab-eating macaque (*Macaca fascicularis*). *Kyoto University Overseas Reports, Studies on the Indonesian Macaque*, **1**, 15–39.

Kawamoto, Y., Nozawa, K., Matsubayasji, K. and Gotoh, S. (1988) A population-genetic study of crab-eating macaques (*Macaca fascicularis*) on the island of Angaur, Palau, Micronesia. *Folia Primatologica*, **51**, 169–81.

Kondo, M., Kawamoto, Y., Nozawa, K. *et al.* (1993) Population genetics of crab-eating macaques (*Macaca fascicularis*) on the island of Mauritius. *American Journal of Primatology*, **29**, 167–82.

Lacy, R.C. (1992) The effects of inbreeding on isolated populations: are minimum viable population sizes predictable?, in *Conservation Biology: The Theory and Practice of Nature Conservation, Preservation and Management* (eds P.L. Fiedler and S.K. Jain), Chapman & Hall, London, pp. 277–96.

Lacy, R.C., Petric, A. and Warneke, M. (in press) Inbreeding and outbreeding in captive populations of wild animal species, in *The Natural History of Inbreeding and Outbreeding*, University of Chicago Press, Chicago.

Lande, R. and Barrowclough, G. (1987) Effective population size, genetic variation and their use in population management, in *Viable Populations for Conservation* (ed. M.E. Soulé), Cambridge University Press, Cambridge, pp. 87–123.

Lynch, M. (1991) Analysis of population genetics structure by DNA fingerprinting, in *DNA Fingerprinting: Approaches and Applications* (eds T. Burke, D. Dolf, A. Jeffreys and R. Wolff), Birkhauser Verlag, Basel, pp. 113–26.

MacArthur, R.H. and Wilson, E.O. (1963) An equilibrium theory of insular zoogeography. *Evolution*, **17**, 373–87.

MacArthur, R.H. and Wilson, E.O. (1967) *The Theory of Island Biogeography*, Princeton University Press, Princeton, New Jersey.

Mace, G.M., Pemberton, J.M., and Stanley, H.F. (1992) Conserving genetic diversity with the help of biotechnology – desert antelopes as an example. *Symposia of the Zoological Society of London*, **64**, 123–34.

Martin, R.D. and Barbour, A.D. (1989) Aspects of line-fitting in bivariate allometric analyses. *Folia Primatologica*, **53**, 65–81.

Meier, B. and Rumpler, Y. (1992) Enzyme variability in island and mainland populations of black lemur (*Eulemur macaco*): a contribution to conservation biology. *Karger Gazette*, **54**, 10–12.

Nilsson, S.G., Bengtsson, J. and Ås, S. (1988) Habitat diversity or area *per se*? Species richness of woody plants, carabid beetles and land snails on islands. *Journal of Animal Ecology*, **57**, 685–704.

Preston, F.W. (1962) The canonical distribution of commonness and rarity. Part I. *Ecology*, **43**, 185–215.

Shafer, C.L. (1990) *Nature Reserves: Island Theory and Conservation Practice*, Smithsonian Institution Press, Washington.

Shaffer, M.L. (1981) Minimum population sizes for species conservation. *Bioscience*, **31**, 131–4.

Signer, E., Kuenzler, C.C., Thomann, P.E. *et al.* (1988) DNA fingerprinting: improved DNA extraction from small blood samples. *Nucleic Acids Research*, **16**, 7738.

Simberloff, D.S. and Abele, L.G. (1976) Island biogeography theory and conservation practice. *Science*, **191**, 285–6.

Soulé, M.E., Bolger, D.T., Alberts, A.C. *et al.* (1988) Reconstructed dynamics of rapid extinctions of chapparal-requiring birds in urban habitat islands. *Conservation Biology*, **2**, 75–92.

Soulé, M.E. and Simberloff, D. (1986) What do genetics and ecology tell us about the design of nature reserves? *Biological Conservation*, **35**, 19–40.

Soulé, M.E., Wilcox, B.A. and Holtby, C. (1979) Benign neglect: a model of faunal collapse in the game reserves of East Africa. *Biological Conservation*, **15**, 259–72.

Tardent, J. (1988) Die Entstehung des heutigen indonesischen Archipels und die daraus resultierende Artenverteilung der Primaten und Spitzhörnchen. University of Zürich Semester Thesis.

Terborgh, J.W. (1976) Island biogeography and conservation: strategy and limitations. *Science* **193**, 1029–30.

Warneke, M. (1992) *International Studbook:* Callimico goeldii, Chicago Zoological Society, Chicago.

Whitmore, T.C. (1975) *Tropical Rain Forests of the Far East*, Clarendon Press, Oxford.

Wilcox, B.A. (1980) Insular ecology and conservation, in *Conservation Biology: An Evolutionary–Ecological Perspective*, (eds M.E. Soulé and B.A. Wilcox), Sinauer Associates, Sunderland, Mass., pp. 95–117.

7

Reproductive technologies

W.V. Holt
Zoological Society of London, UK.

7.1 INTRODUCTION

Basic knowledge about reproductive processes has burgeoned over the past decade, and many new insights have been gained into major areas such as testicular and ovarian function and their control by neuroendocrine mechanisms. Technological developments in many fields have also advanced considerably at the same time, thus making it feasible to intervene at numerous different levels of the reproductive process to assist, manipulate or even prevent animal breeding.

Many of the technological developments have been stimulated by the needs of human clinical medicine, either through research aimed at alleviating infertility or conversely the requirement for more acceptable and effective methods of contraception. It is therefore appropriate that efforts to conserve rare and endangered species should reap some benefits from this work.

The scientist concerned with breeding endangered species cannot, however, simply transfer technologies developed for use in medicine or agriculture directly to the animals of interest. In mammals alone several fundamental reproductive strategies have evolved (polyoestrus, induced ovulation, delayed implantation, sperm storage, etc.) which have led to numerous differences in gamete or hormonal function between species. The investigator is likely to be faced with new problems each time a different species requires attention.

Creative Conservation: Interactive management of wild and captive animals.
Edited by P.J.S. Olney, G.M. Mace and A.T.C. Feistner.
Published in 1994 by Chapman & Hall, London.
ISBN 0 412 49570 8

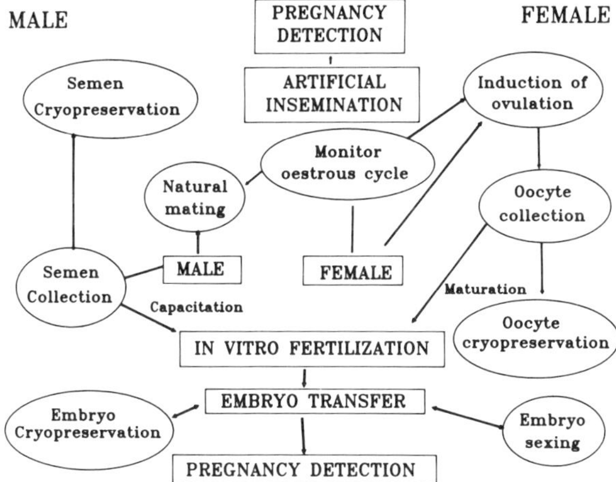

Figure 7.1 The relationships between different techniques available for use in artificial breeding.

Figure 7.1 puts into perspective the various levels at which some form of reproductive intervention or manipulation is currently possible. Unfortunately each aspect of this diagram has its own problems, and the entire set of possibilities can be realized in only a small number of species.

In this review I will aim to discuss the merits of a number of the possibilities offered by reproductive technology, emphasizing the practical limitations imposed when dealing with wild species.

7.2 MONITORING FEMALE REPRODUCTIVE STATUS

Egg development is not a continuous process like spermatogenesis, and ovulation is preceded by the recruitment of one or more Graafian follicles, their development to maturity over a period lasting several days, and the release of oocytes at ovulation. Clearly, for techniques such as artificial insemination or oocyte collection to be performed successfully, accurate knowledge of the timing of egg development and ovulation is required. The various techniques available for monitoring these aspects of reproductive function were recently reviewed in detail by Hodges (1992), and I will provide a general outline with some comments on their practical relevance to wild species.

The different reproductive strategies used by various mammals have a major influence in the approach to studying and monitoring the reproductive cycle. Many mammals, particularly ungulates, primates and many rodents, undergo spontaneous oestrous cycles on a regular basis, with egg development and ovulation recurring regularly. These animals frequently exhibit external signs which can be exploited for the detection of oestrus. Some primates exhibit perineal swelling and coloration change in association with the elevated circulatory oestrogen concentrations which precedes ovulation (Gould and Martin, 1986; Wildt, 1986). However, these signs are not usually accurate enough to be useful except as a guide.

The pheromonal signals produced at oestrus in some species can, however, be recognized by a conspecific male who reacts with characteristic sexual behaviour. Deer and antelope are among the species for which a 'teaser' male (frequently vasectomized) provides a reasonably effective means of detecting oestrus. The effectiveness was tested in a study of artificial insemination techniques in blackbuck (*Antelope cervicapra*; Holt *et al.*, 1988), and oestrus was accurately detected in five out of 10 females tested. The reasons for failing to observe oestrus in the other five animals could have been due to short duration of oestrus behaviour. Some unreliability in behavioural observations has also been found recently in a study of scimitar-horned oryx (*Oryx dammah*) at Marwell Zoological Park (Holt *et al.*, unpublished observations), where a vasectomized male unexpectedly and selectively stopped responding to the oestrous pheromones of certain females, who were known to be undergoing regular oestrous cycles.

More sophisticated approaches to oestrus detection can involve the analysis of blood samples for specific hormones. Circulatory oestrogens provide a direct indication of imminent oestrus, as also does a transient increase in luteinizing hormone concentration. However, these techniques require frequent access to the animals for blood sampling and are often impractical without elaborate holding facilities or tame animals.

Analysis of urine samples offers a more realistic approach to monitoring reproductive status, given two prerequisites. A deceptively simple point is that regular urine samples from individually identified animals must be obtained. The necessary frequency of urine collection is proportional to the length of the oestrous cycle, but daily samples are ideal. The nature of the species has considerable bearing on the success of such sampling procedures; with species which urinate infrequently, especially those such as addax (*Addax nasomaculatus*) which are adapted to dry conditions, there can be a real problem in obtaining urine samples. A practical consideration is that urine collection is difficult if the animals are normally housed on absorbent surfaces.

The second requirement is that the appropriate hormonal metabolite must be identified for analysis. Pregnanediol-3α-glucuronide is suitable for use with a number of ungulates – e.g. blackbuck (Holt *et al.*, 1988), okapi (*Okapia johnstoni*) (Loskutoff, Ott and Lasley, 1982) – but

20α-dihydroprogesterone is the appropriate metabolite for black rhinoceros (*Diceros bicornis*) and white rhinoceros (*Ceratotherium simum*) (Hindle, Möstl and Hodges, 1992). As more progress is made in this field the catalogue of suitable hormonal metabolites for each species will become more comprehensive. I would urge that such information should be maintained within a globally managed and maintained database, such as that proposed by a recent World Conservation Union (IUCN) resolution on Animal Genetic Resource Banking for Species Conservation.

Some species do not ovulate spontaneously, and may require the presence of a male (e.g. wild cavids; Weir, 1971) or actual mating, as in rabbits, for ovulation induction. Members of the cat family exhibit continuous waves of follicular development, but ovulation has to be specifically induced by mating. Clearly, the strategy for timing inseminations or oocyte recoveries has to be adapted to take these factors into account. Moore, Bonney and Jones (1981) performed successful artificial insemination in a puma (*Felis concolor*), having timed the insemination by monitoring oestrogen levels, vaginal cornification and follicular development (by laparoscopy). In this case, ovulation was induced hormonally, using pregnant mare's serum gonadotrophin (PMSG) during dioestrus, followed by human chorionic gonadotrophin (HCG) administered 72 h later. Similar strategies have been used in the domestic cat (Platz, Wildt and Seager, 1978; Goodrowe *et al.*, 1988). A similar type of ovulatory mechanism operates in camelids (e.g. the llama; Bravo *et al.*, 1990); procedures for artificial insemination in this group have not been reliably defined, but the induction of ovulation by HCG is a necessary part of the procedure.

Procedures involving the induction of ovulation allow some degree of control over the timing of reproductive events; this is convenient, especially when a team has to be assembled or when the animal is at some distant zoo. This control is imperfect, however, if the reproductive physiology of the animals in question is poorly understood. Here there is a practical requirement for new advanced sensing techniques, either involving rapid and simplified hormonal assays or the development of entirely novel systems. Urine analyses at present can be undertaken in as little as 3 h, but hour by hour information about the status of an animal would be a considerable advantage.

With some species (see Hodges, 1985; Schiewe *et al.*, 1991) it is possible to synchronize the onset of oestrus using a prostaglandin $F_2\alpha$ analogue, which induces regression of an active corpus luteum and allows a new follicular phase to begin. Alternatively, artificially high levels of circulating progesterone can be maintained with a progesterone-loaded vaginal pessary – then when this is removed a new cycle of follicular growth is initiated (Schiewe *et al.*, 1991).

These methods have their respective advantages; it seems likely that the progesterone pessary technique allows more reproducibility in the timing of return to oestrus, but for wild, nervous animals the pessaries can only be inserted and removed under general anaesthetics. As the technique requires an insertion and a removal step, its use may be inadvisable for practical reasons. The prostaglandin method has the advantage that it requires administration of only small doses of liquid, typically 0.5–1.5 ml depending upon body weight; this can be achieved by the use of dart guns or blowpipes. However, considerable research needs to be undertaken within the species of interest to establish the correct timings for initially administering the prostaglandins, and to determine the interval to the subsequent oestrus period. Experience with scimitar-horned oryx at the Institute of Zoology (London) and Marwell Zoological Park (Winchester, UK) has shown that the interval between prostaglandin administration and return to oestrus can vary from three to six days, which is clearly unacceptable.

Pregnancy detection after artificial inseminations or embryo transfers can be a considerable problem. Measurements of urinary or blood hormones provide accurate and reliable indications of pregnancy, provided the appropriate samples can be obtained. However, actually obtaining the samples can present some difficulties.

Animals in zoos are often intractable, often housed in large paddocks or dens, and often kept in groups; this makes collection of urine samples somewhat difficult. Recent work on the analysis of faecal samples, which are easier to collect, has shown a great deal of promise for pregnancy detection. Measurement of faecal steroids has shown that oestrous cycles can be monitored in addax and scimitar-horned oryx (H. Shaw, unpublished data). It is likely that pregnancy detection will also be possible in these species. As with urinary steroid metabolites it will be important to establish the identity of appropriate candidate steroids for measurement in individual species, and such information should be made available through a database.

The advent of ultrasonic imaging techniques provides a useful solution to the problem of pregnancy detection, direct observation of a foetus being feasible in most species. However, the practical drawback to this technique is that animals must be sedated or restrained before ultrasonic examination is possible. The slight risks of foetal loss or damage caused by the anaesthetic procedures have to be weighed against the benefit of knowing the reproductive status of the animals.

Assuming that the procedures outlined above have allowed accurate identification of the female reproductive status, it is technically possible to carry out either artificial insemination or oocyte collection and *in vitro* fertilization. Considerations for these techniques will be outlined in the following discussion.

7.3 OOCYTE COLLECTION, MATURATION AND *IN VITRO* FERTILIZATION

7.3.1 Oocyte collection and maturation

Only fully mature oocytes have the functional ability to form viable offspring; this competence is gained after germinal vesicle breakdown and resumption of meiosis (Moor and Trounson, 1977). A 6–8 h inductive period within the follicle is required for complete acquisition of developmental potential (Cran and Moor, 1990). Since unfertilized, ovulated eggs retain their developmental capacity for only about 24 h, the collection window for oocytes undergoing normal *in situ* maturation is narrow, and careful monitoring of follicular development is required if useful oocytes are to aspirated from the pre-ovulatory follicle or the Fallopian tube.

In vitro maturation of oocytes offers one method of gaining extra flexibility in the timing of oocyte recovery. Short-term, 18–24 h, incubation of oocytes (and surrounding cumulus cells) prior to performing *in vitro* fertilization maximizes their developmental potential. However, of longer term interest and importance is the culture of primary oocytes together with cumulus and granulosa cells. This enables use of larger numbers of follicles, and also eliminates the need for close hormonal monitoring of individual donor females. Viable offspring have been obtained from primary follicular oocytes, fertilized *in vitro*, of several domestic and laboratory species (Schroeder and Eppig, 1984; Leibfried-Rutledge *et al.*, 1989; Gandolfi and Moor, 1987). Recent studies of the mouse indicate that even pre-antral follicles can undergo successful maturation *in vitro*, and furthermore can be cryopreserved at this early stage (Carroll *et al.*, 1990; Telfer, Torrence and Gosden, 1990).

A further logical refinement of *in vitro* fertilization and oocyte maturation technology would be to establish and choose the sex of embryos for transfer to recipients. Such techniques are currently under development for clinical use, where sometimes the main reason for undertaking *in vitro* fertilization is the avoidance of sex-linked inherited diseases. Karyotyping and identification of the sex chromosomes in individual blastomeres dissected from the embryo by micromanipulation is one possibility (King, 1984), which has recently been modified to make use of fluorescent *in situ* hybridization probes (Griffin *et al.*, 1992). Alternatively, the polymerase chain reaction, which detects and amplifies a particular DNA sequence, can be applied to DNA samples obtained from individual blastomeres (Coutelle *et al.*, 1989; Kunieda *et al.*, 1992). With regard to the breeding of endangered species, it is possible in principle to undertake sex determination on all cryopreserved embryos, then at some future date choose which embryos to recover for further development. In practice, this scenario is at present out of the question for most endangered species, but could be realized for a particular species or group if the appropriate effort and resources were invested.

7.3.2 Capacitation of spermatozoa and the acrosome reaction

The spermatozoa of all eutherian mammals must undergo a capacitation process in the female reproductive tract, or in appropriate culture media, before fertilization can occur. This physiological process typically lasts several hours, and requires that relatively stringent, and species-specific, conditions are met (Yanagimachi, 1988). Capacitation results in molecular and physical changes within the sperm plasma membrane, and permits occurrence of the acrosome reaction, a central process in fertilization.

A major difficulty with *in vitro* fertilization procedures for exotic species is that the optimal conditions for inducing capacitation usually have to be determined on an empirical basis. The amount of work involved in these investigations is unpredictable when a species is being investigated for the first time. Unfortunately, although some laboratory-based indications of capacitation are available, the definitive proof requires demonstration that the spermatozoa can actually fertilize eggs of the homologous species. It is self evident that when dealing with endangered species the availability of oocytes for experimental use will be minimal.

The use and interpretation of functional tests for capacitation and the occurrence of the acrosome reaction is therefore of considerable importance. Various simple but effective methods for assessing acrosomal status have been developed; they mostly require the use of a fluorescence microscope. A monoclonal antibody (Figure 7.2) raised against acrosomal membranes can be used for the immunofluorescent staining of acrosomes in a wide variety of species (Moore *et al.*, 1987). Spermatozoa with intact acrosomes are readily distinguished from acrosome-reacted spermatozoa by the pattern of fluorescence (Moore *et al.*, 1987; Zhang *et al.*, 1990).

The antibody-binding technique is simple to carry out, and the results can be quantified by counting and scoring 100–200 cells. An even simpler, but equivalent procedure, employing fluorescent-labelled lectins isolated from peas (*Pisum sativum*) or peanuts (*Arachis hypogea*), can also be used (Cross *et al.*, 1986; Mortimer, Curtis and Miller, 1987). The lectins bind to carbohydrate groups on the acrosomal membranes or contents, and the structure of the acrosome is thus easily seen.

A more functional assessment of both the acrosome reaction and the membrane fusing capacity of spermatozoa can be achieved by the use

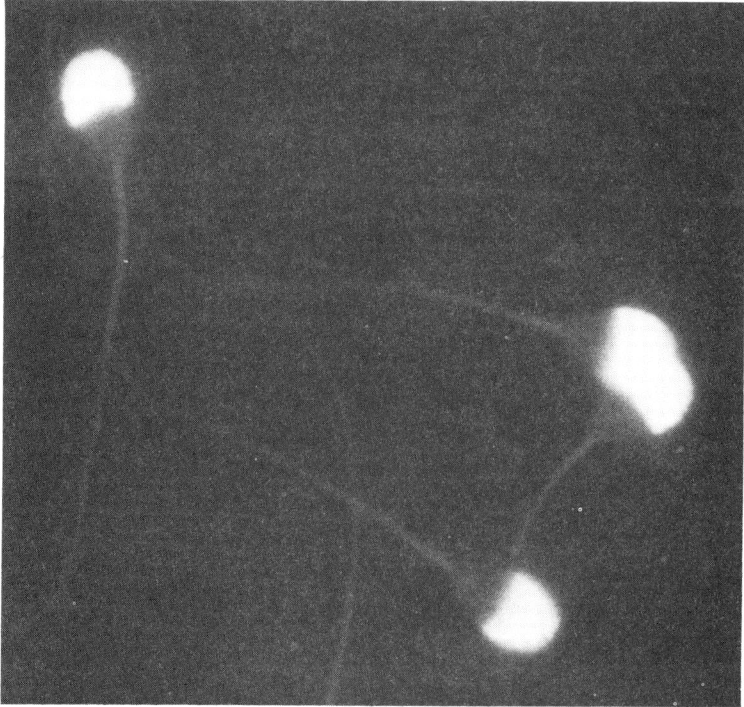

Figure 7.2 Frozen then thawed spermatozoa from scimitar-horned oryx, stained with monoclonal antibody '18.6'. The acrosomes appear as bright semilunar areas at the anterior region of the sperm heads.

of zona-free hamster eggs. These oocytes, denuded of their zona-pellucida, will allow heterologous spermatozoa to bind and fuse with the oolemma (Yanagimachi, Yanagimachi and Rogers, 1976), provided they have undergone the acrosome reaction. To some extent, the difficulties of inducing the physiological acrosome reaction in an unfamiliar species can be circumvented artificially. An early event in the initiation of the acrosome reaction is an intracellular rise in calcium concentration, which acts to trigger the membrane fusions involved in the acrosome reaction. The ionophore A23187, a drug with calcium translocating ability, can bypass the physiological processes of capacitation, and induce the acrosome directly in a number of species (Green, 1976; Hanada, 1985). Other treatments, such as the use of heparin and lysophosphatidylcholine (Parrish *et al.*, 1988) may be of use in this context, but once again are species-specific.

7.3.3 Microinjection of spermatozoa

A relatively novel way to generate fertilized oocytes, and hence embryos, is to bypass penetration of the zona-pellucida by microinjecting spermatozoa into the perivitelline space, or even into the egg cytoplasm. The former technique has been relatively successful in the mouse (Mann, 1988) in producing fertilized eggs capable of developing into offspring, and has been used to assist human conception when sperm numbers are low (Laws-King *et al.*, 1987). This method, however, requires spermatozoa that are capacitated and have undergone, or are capable of undergoing, the acrosome reaction, since spermatozoa in the perivitelline space must still fuse naturally with the oolemma for fertilization to occur. For endangered species, therefore, this technique does not eliminate the need to induce capacitation *in vitro*, and the problems discussed above are still relevant.

The only way to avoid the capacitation problem completely is to use microinjection directly into the egg cytoplasm (Thadani, 1980; Markert, 1983). This generates its own problems, as the act of puncturing the egg plasma membrane can lead to irreversible oocyte damage. The advantages are, however, that there is no requirement for the spermatozoa to be motile or even intact. Fertilized hamster eggs with male pronuclei have been generated by microinjecting isolated sperm heads that had previously been frozen or freeze-dried (Uehara and Yanagimachi, 1976). Although these embryos ceased development at the two-cell stage, there are obvious implications for rescuing important genetic material.

The suggestion that surrogate oocytes could be used to synthesize pronuclei for transfer to activated, homologous oocytes has been raised by observations that human sperm heads formed pronuclei when injected into hamster oocytes (Uehara and Yanagimachi, 1976). If perfected, such techniques could circumvent many of the species-specific aspects of *in vitro* fertilization. However, this approach will have to be explored with care, because although the studies to date have not indicated any genetic abnormalities in offspring produced by sperm injection, bypassing the natural barriers to fertilization may eliminate natural selection mechanisms imposed during normal processes. For example, karyotypic studies of human spermatozoa indicate that about 10% display chromosomal abnormalities (Martin *et al.*, 1983). As there is no way of distinguishing these abnormal cells there is obviously a 10% chance that they will be used in sperm microinjection techniques. Whether this rate is higher than the naturally occurring fertilization rate with genetically abnormal cells is unknown at present.

7.3.4 Embryo cryopreservation

The production of viable embryos can be achieved by several routes including *in vitro* fertilization; oestrus synchronization and natural mating; and superovulation and either natural mating or artificial insemination. One consequence of these techniques is that excess embryos are sometimes produced.

Ever since Whittingham, Leibo and Mazur (1972) demonstrated that mouse embryos could be stored by cryopreservation, with subsequent birth of live offspring after embryo transfer, the storage of embryos from endangered species has been a realistic possibility. The techniques for embryo cryopreservation have been simplified by the demonstration that vitrification can be used successfully (Fahy *et al.*, 1984). This technique involves exposing the embryos to high molar concentrations of cryoprotectants (DMSO, polyethylene glycol, propylene glycol), thereby inhibiting ice crystal formation. On cooling, this mixture forms a glass-like solid. This technique has considerable potential for field use, because there are much less stringent requirements for specialized cooling equipment.

The scientific background and application of embryo cryopreservation with respect to endangered species was recently reviewed by Schiewe (1991a). Embryo cryopreservation has been successful for humans, sheep, cattle and the common marmoset (Summers *et al.*, 1987), but further development of techniques, taking into account species variability, is needed before the methods can be reliably and widely applied. The difficulties experienced in trying to develop protocols for the pig embryo, which is unusually sensitive to temperatures below 10°C, provide a salutary warning.

7.4 SEMEN TECHNOLOGY

7.4.1 Semen collection

There have been relatively few advances in semen collection procedures since the topic was reviewed in detail by Watson (1978). The most practical and generally applicable technique is still electroejaculation of a sedated animal, and there have been numerous instances of successful semen collection in different mammalian groups – e.g. primates (Holt, 1986; Wildt, 1986), wild bovids (Schiewe *et al.*, 1991), giant panda (Moore *et al.*, 1984).

For semi-domesticated species such as deer, various artificial vagina (AV) devices which do not require operator handling have been developed. Krzywinski and Jaczewski (1978) described a device, like a false hindquarter, which could be attached to the back of a female and housed an AV into which ejaculation occurred at mating. An internal AV has recently been developed for use with farmed fallow deer (Jabbour and Asher, 1991), where does are fitted with the device before the onset of oestrus. After mating the AV is removed and the semen aspirated for use. Although this system requires some animal handling for insertion and removal of the AV, it has considerable potential for captive wild deer or antelope if investigated in more detail.

When a genetically important male dies through accidental or natural causes some consideration should be given to conserving its spermatozoa for future use. The techniques for obtaining epididymal spermatozoa are relatively simple. The cauda epididymidis is cleaned of blood and extraneous contaminants, and the spermatozoa are then released by making several incisions across the epididymal tubules. If the epididymis is placed in either a cryopreservation or physiological solution, the spermatozoa will be released into suspension for subsequent storage. Spermatozoa collected in this way are fully viable and capable of fertilization, even though they have never been exposed to seminal plasma. A protocol for the collection and cryopreservation of epididymal spermatozoa from wild cattle was outlined recently by Hopkins (1991).

In some instances epididymal spermatozoa are more robust and suitable for *in vitro* handling than ejaculated cells; this is especially true with certain primates where ejaculates undergo coagulation, thereby trapping the spermatozoa. For example, successful *in vitro* fertilization procedures for the common marmoset (*Callithrix jacchus*) depends upon the use of epididymal spermatozoa; their exposure to seminal plasma during electroejaculation severely impairs their *in vitro* lifespan (L. Wilton, pers. comm.).

The general desirability of separating spermatozoa from their seminal plasma before cryopreservation has to be investigated separately for each new species. For domestic species there are conflicting reports on the effects of freezing ram or bull spermatozoa in the presence of seminal plasma, but there is clear evidence that goat seminal plasma is detrimental. The enzyme phospholipase A, secreted by the Cowper's gland into the seminal plasma, hydrolyses egg yolk lipids to fatty acids and lysolecithin, which are both toxic to spermatozoa (Roy, 1957).

7.4.2 Semen cryopreservation procedures

Semen freezing procedures have changed remarkably little since their initial development in the early 1950s. In general, fresh semen is diluted into a medium containing buffer salts, metabolic substrates, a cryoprotectant such as glycerol and additional cryoprotective agents such as egg yolk or milk proteins. The introduction of the plastic straw and pellet as new methods for the packaging of frozen semen (Cassou, 1964; Nagase and Niwa, 1964), which allowed faster cooling and thawing rates than previously possible, was an important advance at the time and resulted in the improvement of fertility.

Over the last four decades, numerous investigators have experimented with the composition of freezing media, exploring the possible benefits to be gained from different additives and various buffer salts. This empirical approach resulted in the optimization of freezing techniques for application to particular species of agricultural importance. The high conception rates (about 70–80%) obtained with good quality frozen cattle semen underline the success of this research. However, species differences in the responses of spermatozoa to cooling and freezing mean that considerable background research on spermatozoa from an individual species is required prior to initiating a germ plasm cryobanking programme. Schiewe (1991b) recently presented detailed sperm cryopreservation data on wildebeeste (*Connochaetes gnou*), greater kudu (*Tragelaphus strepsiceros*), gaur, (*Bos gaurus*), scimitar-horned oryx and muntjac (*Muntiacus* spp.) which emphasized this point.

Clearly, empirical research of this type has a valuable role, but for significant advances to be made in this field there has to be more investment in basic cryobiological science. A critical appraisal of the results typically obtained with frozen semen demonstrates considerable scope for improvement. Direct comparisons between fertility rates obtained with fresh and frozen spermatozoa are instructive in revealing the degree of damage inflicted by the best current cryopreservation methods for cattle and sheep spermatozoa. I have chosen three examples.

(a) Colas (1984) compared fertility rates after intracervical inseminations in sheep with fresh or frozen semen. Conception rates of 70% were obtained when doses of 100–400 million fresh spermatozoa were inseminated into each ewe; 59% conception rates were obtained using 600–700 million frozen spermatozoa per ewe.
(b) McKelvey (pers. comm.), using laparoscopic inseminations in approximately 1000 sheep, obtained a conception rate of 80% with 12–20 million fresh spermatozoa per uterine horn, and a rate of 62% with 40 million frozen spermatozoa per uterine horn.
(c) Shannon (1978) quoted figures from the New Zealand Dairy Board showing that transcervical cattle inseminations with two million fresh spermatozoa per cow resulted in a 60% conception rate, whilst 25 million frozen spermatozoa per cow produced a 65% conception rate.

If an arbitrary Fertility Index (F) is defined as

$$F = \frac{\text{Conception rate}}{\text{Number of spermatozoa inseminated}}$$

then relative fertilities of fresh and frozen spermatozoa can be examined as ratios

$$\text{Relative fertility} = \frac{\text{F (frozen)}}{\text{F (fresh)}}$$

The data from Example (a), using values of 250 and 650 million fresh and frozen spermatozoa, respectively, show that freezing ram semen depresses its fertility by approximately two thirds (68%). Remarkably close agreement is afforded by the laparoscopic data from Example (b), where a 61% loss in fertility is indicated. The most surprising conclusion comes from Example (c), however, where a relative fertility of only 9% is obtained, representing a loss in fertility of over 90%.

These figures are based on numbers of motile frozen spermatozoa, and do not therefore take into consideration the variable proportion of cells which undergo irreversible damage during the freezing process. This is commonly as high as 50% of the cell population. Why do the frozen spermatozoa exhibit such poor fertility? Mattner, Entwhistle and Martin (1969) demonstrated that frozen spermatozoa were considerably less efficient than fresh spermatozoa in their ability to colonize the ewe reproductive tract. In addition, 24 h after insemination the frozen spermatozoa were no longer present in the uterus and Fallopian tube, whereas numerous (12–30 000) fresh spermatozoa were detected.

Decreased ability to reach the oviduct is probably due to reduced vigour and shortened sperm survival time after thawing. A correlation between survival *in vitro* and actual conception rates was recently demonstrated for frozen human spermatozoa (Holt *et al.*, 1989). The suggestion has also been made that freezing and thawing accelerates the acrosome reaction (Critser *et al.*, 1987; Watson, 1990). These observations point towards the molecular modification of the sperm plasma membrane by chemical or physical influences during cryopreservation.

Whatever the cause of poor fertility after freezing and thawing, it is nevertheless the case that fertilizations do occur, and that amongst the many functionally deficient spermatozoa there is a small, physiologically normal, population. To make significant further advances in semen freezing we should study these normal cells and determine how they differ from the defective majority.

The freeze–thaw cycle to which spermatozoa are subjected during cryopreservation involves several distinct steps. The cells are initially diluted into a prewarmed extending medium, which may or may not contain glycerol as cryoprotectant. They are then cooled slowly to approximately 5°C, at which point glycerol is slowly added if necessary. The spermatozoa are frozen relatively rapidly (20–40°C/min) to approximately −80°C, then plunged into liquid nitrogen for long-term storage. Once frozen at this low temperature (−196°C) biological material is stable for many years. For subsequent use, the straws, pellets or ampoules of semen are thawed rapidly in warm water.

The hazards associated with each step are numerous, and were recently reviewed by Hammerstedt, Graham and Nolan (1990). The initial cooling step causes molecular rearrangements within the sperm membrane lipid bilayers (Holt and North, 1984, 1985, 1986; De Leeuw *et al.*, 1990). Addition of glycerol raises the osmolarity of the extending medium, with a subsequent loss of intracellular water. The glycerol also becomes associated with the cellular membranes, causing further molecular reorganization (Hammerstedt and Graham, 1992). Once the freezing process starts, the solutes present in the unfrozen fraction become highly concentrated as water is converted to ice, a process likely to cause further loss of intracellular water and inward leakage of ions. The sites of high salt concentration occur as channels between ice crystals; steep concentration gradients of extracellular salts run from one side of each channel to the other (Persidsky and Luyet, 1959), and these randomly intersect different regions of the spermatozoon. Rapid thawing causes a sudden decrease in extracellular solute concentration, and water is osmotically driven back into the cell; if the cell membrane has insufficient elasticity it may be disrupted by the increase in volume thereby caused. Removal of cryoprotectant, i.e. lowering of the extracellular glycerol concentration, causes yet another osmotic influx of water, sometimes with lethal consequences. Careful study of membrane permeabilities to water have permitted minimization of osmotic stresses during freezing in other cell types. Such approaches have to be developed for spermatozoa. Duncan and Watson (1992) have initiated this work by evaluating specific parameters relevant to the flux of water across the ram sperm plasma membrane during freezing and thawing. Estimates of cellular surface area, of volume of sperm cell water under isotonic conditions and of hydraulic conductivity were determined experimentally in an effort to model the loss of water across the plasma membrane during freezing. Unfortunately the unusually specialized shape of the spermatozoa makes this type of study very difficult, and furthermore the morphological and biochemical diversity of spermatozoa from different species means that the required parameters may have to be derived on a species by species basis.

Cryomicroscopy offers one method of investigating the cryobiology of spermatozoa by allowing direct observation as they freeze and thaw at controlled rates on the stage of a light microscope. This has the great advantage that individual cells can be observed throughout the process, in contrast to conventional methods in which the origin of deleterious lesions has to be assigned **after** the whole freezing and thawing process has been completed. Although the principles of cryomicroscopy were well understood many years ago (e.g. Chambers and Hale (1932) and references quoted therein), for technical reasons they were difficult to implement satisfactorily. Consequently almost nothing is known about the cryomicroscopy of spermatozoa, apart from observations by Smith, Polge and Smiles (1951) on fowl spermatozoa.

Recent experiments in this laboratory have explored the use of the

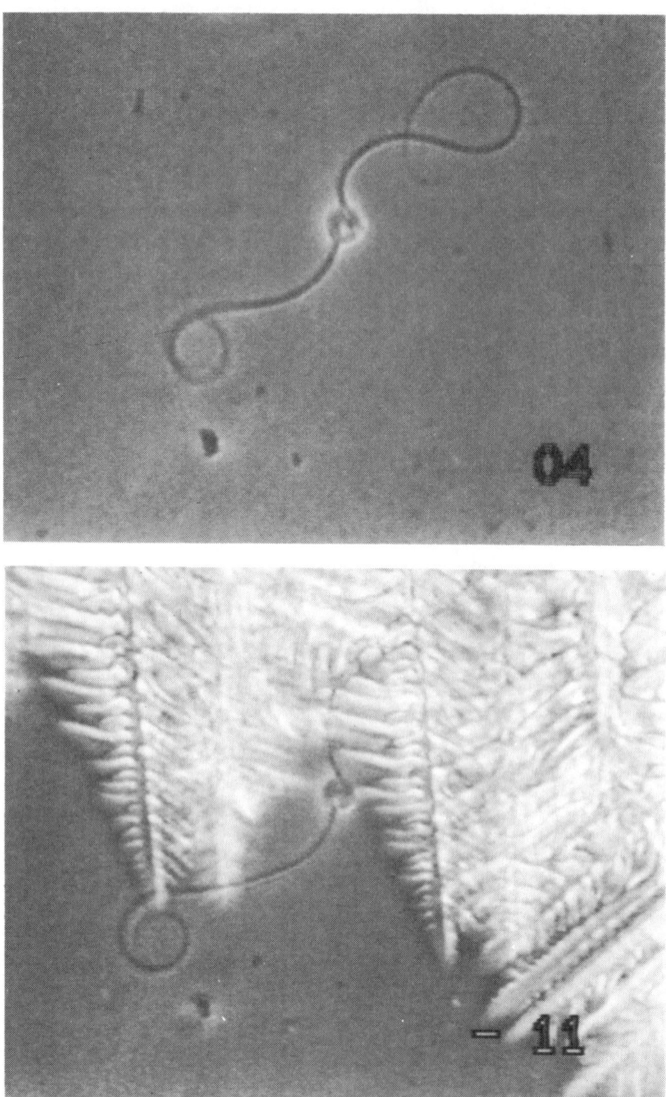

Figure 7.3 Two sequential views of a pair of spermatozoa from a grey, short-tailed opossum (*Monodelphis domestica*), viewed through the cryomicroscope. Upper: spermatozoa seen at +4°C; lower: spermatozoa undergoing freezing at −11°C.

cryomicroscope in conjunction with direct assays of sperm function. Fluorescent probes for monitoring plasma membrane integrity (e.g. fluorescein diacetate; Garner, Johnson and Allen, 1988; Harrison and Vickers, 1990) afford the identification of specific temperature zones and conditions which cause cell damage, whilst objective measurement of sperm motility provides a functional endpoint by which to judge the success of particular cryopreservation protocols. One interesting observation made by this technique is that under certain conditions spermatozoa can survive freezing, but do not tolerate thawing. Membrane integrity, as evidenced by the fluorescent probe, was not lost until post-thaw rewarming; furthermore there was a relationship between minimum cooling temperature and the temperature associated with loss of intracellular fluorescence (Holt, Head and North, 1992).

Besides its use to explore the fundamental cryobiology of spermatozoa, the cryomicroscope is useful in a strictly practical sense for establishing and optimizing protocols for the freezing of semen from particular species, or even individual animals of importance. Several such investigations have been undertaken in this laboratory, and successful cryopreservation of spermatozoa from the laboratory mouse and common marmoset has been achieved. *In vitro* fertilization with frozen–thawed spermatozoa from both species was used to demonstrate the effectiveness of the procedures developed. These two species illustrate the use of this system particularly well, for the investigations were performed using the small quantities of spermatozoa available, only 4 μl of diluted sperm suspension being needed for each trial.

Work is currently in progress to establish a cryopreservation procedure for spermatozoa from the grey short-tailed opossum (*Monodelphis domestica*). Techniques for the cryopreservation of marsupial spermatozoa have never previously been established. As these particular animals store very few epididymal spermatozoa, cryomicroscopy is an ideal way to begin investigating the problem. Although a satisfactory technique has not yet been developed, it has emerged that glycerol in concentrations which would normally be considered excessive (10–15%) is needed for the recovery of motility after freeze–thawing. Figure 7.3 illustrates the appearance of one pair of opossum spermatozoa during the freezing process.

7.4.3 Assessment of spermatozoa

The use of the zona-free hamster oocyte test as a functional assay for testing the ability of spermatozoa to undergo the acrosome reaction and fuse with the oolemma has already been described above. It represents one method by which the survival of spermatozoa during cryopreservation can be assessed, and despite its uncertainties and imperfections provides a good basis for evaluating the quality of the material stored in a cryobank of germplasm.

This question of ensuring that spermatozoa stored in such a cryobank are actually useful for future breeding plans is of central importance. Hence it is worth repeating the point made in the preceding section, that it is not acceptable to assume that spermatozoa from any species can be frozen using techniques designed for cattle artificial insemination.

A working group at the Wild Cattle Symposium, held at Omaha, USA, in June 1991, made some recommendations concerning minimal acceptable standards for wild cattle semen, which included the following points.

(a) For fresh semen there should be at least 70% normal progressive spermatozoa, and 70% normal structural morphology.
(b) At least 25% of spermatozoa should be motile upon thawing, and possess 50% intact acrosomes.
(c) At least 15% of spermatozoa should be motile after 3 h, and possess 30% intact acrosomes.

These ratings are based on the procedures normally used for cattle artificial insemination. Similar guidelines exist for the acceptability of human semen samples for donor insemination.

These guidelines are based on the knowledge that for successful fertilization, a minimum number of spermatozoa are required and that these must be capable of surviving for several hours within the female reproductive tract. The requirement for intact acrosomes recognizes the inability of damaged spermatozoa to undergo a normal acrosome reaction, and fertilize an oocyte. The monoclonal antibody and lectin tests outlined above for checking acrosomal integrity are therefore useful in the assessment of cryopreserved spermatozoa.

Differences between reproductive tract anatomy or fertilization mechanisms between species mean, however, that the exact minimum requirements for sperm quality after freezing and thawing must be tailored for each individual case. In some cases, such as cattle, a relatively small number of spermatozoa (20×10^6) introduced into the uterus through the cervix will produce a high conception rate, whilst difficulties with transcervical access in sheep mean that a much larger number of spermatozoa ($> 200 \times 10^6$) is required to achieve the same degree of success. Pig ejaculates are typically of high volume, 100–300 ml, and artificial insemination requires both this large volume and high sperm numbers (about 1.5×10^9 spermatozoa). Clearly, for endangered species, background information about normal reproductive function is of utmost importance.

7.4.4 Insemination techniques

One way to circumvent the idiosyncratic requirements of different species in terms of sperm numbers required for insemination, or optimal site for deposition, is to attempt to place spermatozoa as near to the ovulated oocyte as possible.

The development of fibreoptic equipment has meant that it is now possible to deliver spermatozoa to the uterus under direct visual control. Sheep artificial insemination provided a stimulus for the development of this technology, as many fewer spermatozoa are required for the achievement of good fertility rates. In practice a small incision in the abdomen allows an operator to insert a fibreoptic endoscope and view the uterine horns directly. Semen in a specially adapted syringe can be injected into the uterine horn, thus bypassing many of the obstacles which impede the normal progress of spermatozoa through the cervix.

With sheep artificial insemination the lambing rates achieved by this method are very high, whilst a relatively small number of spermatozoa is required. Using frozen semen, Maxwell (1986) obtained 62% fertility rates (percentage of inseminated ewes lambing) after insemination with 50 million motile spermatozoa per ewe. This compares well with the 60–70% fertility rates obtained with the older intracervical technique, where typically 400–600 million spermatozoa are required for each ewe. More recently McKelvey (pers. comm.), using fresh spermatozoa, obtained 80% conception rates with 40 million spermatozoa per ewe. The great advantage of the laparoscopic technique is that frozen spermatozoa can be used more sparingly and to greater effect than was possible by the intracervical method. McKelvey (pers. comm.), working with 80 million frozen spermatozoa per ewe, obtained a conception rate of 62% (1000 ewes inseminated). This result is comparable to data obtained by Findlater *et al.* (1988), whose maximal lambing rates achieved with frozen spermatozoa were almost 80%.

The laparoscopic approach will usefully supplement existing insemination techniques for breeding endangered species. Impaired sperm-transport is the principal cause of poor fertility with frozen semen (Salamon and Lightfoot, 1967; Mattner, Entwhistle and Martin, 1969), and therefore direct intrauterine insemination circumvents these problems to some extent; especially when semen quality is doubtful. Laparoscopy was used in 1989 without difficulty to inseminate two female addax at Marwell Zoological Park, and is used commercially for the artificial insemination of red deer.

Traditionally, inseminations are performed by depositing semen within the vagina or cervix, or by manipulating an insemination pipette or catheter through the cervix. In principle the latter technique of intrauterine insemination should be as effective as laparoscopy, but in practice it is often difficult to perform in small animals. Laparoscopic insemination should then be considered.

In some species there is a need for the accumulation of anatomical knowledge of the reproductive tract, in order to design instruments specifically for delivering semen. The elephants are a notable example in which intensive studies of this nature are currently in progress (Balke *et al.*, 1988). The unusual anatomy of the female reproductive tract in this species means that an insemination device must be at least 100 cm in length in order to reach the vagina in some elephants. Other examples include the scimitar-horned oryx, where the uterine horns are isolated from each other by a bifurcated cervix, and the giant panda in which the urethra is easily mistaken for the cervix (Knight *et al.*, 1985).

REFERENCES

Balke, J.M.E., Barker, I.K., Hackenberger, M.K., McManamon and Boever, W.J. (1988) Reproductive anatomy of three nulliparous female asian elephants: the development of artificial breeding techniques. *Zoo Biol.*, **7**, 99–113.

Bravo, P.W., Fowler, M.E., Stabenfeldt, G.H. and Lasley, B.L. (1990) Ovarian dynamics in the llama. *Biol. Reprod.*, **43**, 579–85.

Carroll, J., Whittingham, D.G., Wood, M.J., Telfer, E. and Gosden, R.G. (1990) Extra-ovarian production of mature viable mouse oocytes from frozen primary follicles. *J. Reprod. Fert.*, **90**, 321–7.

Cassou, R. (1964) La méthode des paillettes en plastique adaptée à la généralisation de la congélation. *V Int. Congr. Anim. Reprod. Artif. Insem. (Trento).*, **4**, 540–6.

Chambers, R. and Hale, H.P. (1932) The formation of ice in protoplasm. *Proc. Roy. Soc. Lond. Ser. B.*, **110**, 336–52.

Colas, G. (1984) Semen technology in the ram, in *The Male in Farm Animal Reproduction* (ed. M. Courot), Martinus Nijhoff, Boston, pp. 219–34.

Coutelle, C., Williams, C., Handyside, A. *et al.* (1989) Genetic analysis of DNA from from single human oocytes; a model for preimplantation diagnosis of cystic fibreosis. *Br. Med. J.*, **299**, 22–4.

Cran, D.G. and Moor, R.M. (1990) Programming the oocyte for fertilization, in *Fertilization in Mammals* (eds B. Bavister, J. Cummins and E. Roldan), Serono Symposia, Norwell, USA, pp. 35–47.

Critser, J.K., Arneson, B.W., Aaker, D.V. *et al.* (1987) Cryopreservation of human spermatozoa. II. Post-thaw chronology of motility and of zona-free hamster ova penetration. *Fert. Steril.* **47**, 980–4.

Cross, M.L., Morales, P., Overstreet, J.W. and Hanson, F.W. (1986) Two simple methods for detecting acrosome-reacted human sperm. *Gamete Res.*, **15**, 213–26.

De Leeuw, F.E., Chen, H-C., Colenbrander, B. and Verkleij, A.J. (1990) Cold-induced ultrastructural changes in bull and boar sperm plasma membranes. *Cryobiology*, **27**, 171–83.

Duncan, A.E. and Watson, P.F. (1992) Predictive water loss curves for ram spermatozoa during cryopreservation: comparison with experimental observations. *Cryobiology*, **29**, 95–105.

Fahy, G.M., MacFarlane, D.R., Angell, C.A. and Meryman, H.T. (1984) Vitrification as an approach to cryopreservation. *Cryobiology*, **21**, 407–26.

Findlater, R.C.F., Haresign, W., Curnock, R.M. and Beck, N.F.G. (1988) Effect of timing of intrauterine insemination with frozen-thawed semen on fertility in ewes. *Proc. 11th Int. Congr. Anim. Reprod. Artif. Insem.* (Dublin), **3**, 242.

Gandolfi, F. and Moor, R.M. (1987) Stimulation of early embryonic development in the sheep by co-culture with oviduct epithelial cells. *J. Reprod. Fert.*, **81**, 23–8.

Garner, D.L., Johnson, L.A. and Allen, C.H. (1988) Fluorometric evaluation of cryopreserved bovine spermatozoa extended in egg yolk and milk. *Theriogenology*, **30**, 369–78.

Goodrowe, K.L., Wall, R.J., O'Brien, S.J. *et al.* (1988) Developmental competence of domestic cat follicular oocytes after fertilization *in vitro*. *Biol. Reprod.*, **39**, 355–72.

Gould, K.G. and Martin, D.E. (1986) Artificial insemination of nonhuman primates, in *Primates: The Road to Self-sustaining Populations* (ed. K. Benirschke), Springer–Verlag, New York, pp. 425–43.

Green, D.P.L. (1976) Induction of the acrosome reaction in guinea-pig spermatozoa *in vitro* by the Ca ionophore A23187. *J. Physiol. Lond.*, **260**, 18P–19P.

Griffin, D.K., Wilton, L.J., Handyside, A.H. *et al.* (1992) Dual fluorescent *in situ* hybridisation for simultaneous detection of X and Y chromosome-specific probes for the sexing of human preimplantation embryonic nuclei. *Human Genetics*, **89**, 18–22.

Hammerstedt, R.H. and Graham, J.K. (1992) Cryopreservation of poultry sperm: the enigma of glycerol. *Cryobiology*, **29**, 26–38.

Hammerstedt, R.H., Graham, J.K. and Nolan, J.P. (1990) Cryopreservation of mammalian sperm: what we ask them to survive. *J. Androl.*, **11**, 73–88.

Hanada, A. (1985) *In vitro* fertilization in cattle with particular reference to sperm capacitation by ionophore A23187. *Jap. J. Anim. Reprod.*, **31**, 56–61.

Harrison, R.A.P. and Vickers, S.E. (1990) Use of fluorescent probes to assess membrane integrity in mammalian spermatozoa. *J. Reprod. Fert.*, **88**, 343–52.

Hindle, J.E., Möstl, E. and Hodges, J.K. (1992) Measurement of urinary oestrogens and 20α-dihydroprogesterone during ovarian cycles of black (*Diceros bicornis*) and white (*Ceratotherium simum*) rhinoceroses. *J. Reprod. Fert.*, **94**, 237–49.

Hodges, J.K. (1985) The endocrine control of reproduction. *Symp. Zool. Soc. Lond.*, **54**, 149–68.

Hodges, J.K. (1992) Detection of oestrous cycles and timing of ovulation. *Symp. Zool. Soc. Lond.*, **64**, 73–88.

Holt, W.V. (1986) Collection, assessment and storage of sperm, in *Primates: The Road to Self-sustaining populations* (ed. K. Benirschke), Springer–Verlag, New York, pp. 413–24.

Holt, W.V. and North, R.D. (1984) Partially irreversible lipid phase transitions in mammalian sperm plasma membrane domains: freeze-fracture study. *J. Expl. Zool.*, **230**, 473–83.

Holt, W.V. and North, R.D. (1985) Determination of lipid composition and thermal phase transition temperature in an enriched plasma membrane fraction from ram spermatozoa. *J. Reprod. Fert.*, **73**, 285–94.

Holt, W.V. and North, R.D. (1986) Thermotropic phase transitions in the plasma membrane of ram spermatozoa. *J. Reprod. Fert.*, **78**, 447–57.

Holt, W.V., Moore, H.D.M., North, R.D. *et al.* (1988) Hormonal and behavioural detection of oestrus in blackbuck, *Antilope cervicapra*, and successful artificial insemination with fresh and frozen semen. *J. Reprod. Fert.*, **82**, 717–25.

Holt, W.V., Shenfield, F., Leonard, T. *et al.* (1989) The value of sperm swimming speed measurements in assessing the fertility of human frozen semen. *Human Reprod.*, **4**, 292–7.

Holt, W.V., Head, M.F. and North, R.D. (1992) Freeze-induced membrane damage in ram spermatozoa is manifested after thawing. Observations with experimental cryomicroscopy. *Biol. Reprod.*, **46**, 1086–94.

Hopkins, S.M. (1991) Wild cattle epididymal spermatozoa collection and freezing, in *Proc. Wild Cattle Symposium*, (*June 13–16th, 1991*) (eds D.L. Armstrong and T.S. Gross), Henry Doorly Zoo, Omaha, pp. 79–81.

Jabbour, H.N. and Asher, G.W. (1991) Artificial breeding of farmed fallow deer (*Dama dama*), in *Wildlife Production: Conservation and Sustainable Development* (eds L.A. Renecker and R.J. Hudson), AFES Misc. Publication, University of Alaska, Fairbanks, pp. 485–91.

King, W.A. (1984) Sexing embryos by cytological methods. *Theriogenology*, **21**, 7–17.

Knight, J.A., Bush, M., Celma, M. *et al.* (1985) Veterinary aspects of reproduction in the giant panda (*Ailuropoda melanoleuca*), in *Bongo, Berlin, Proc. Int. Symp. Giant Panda*, pp. 93–126.

Krzywinski, A. and Jaczewski, Z. (1978) Observations on the artificial breeding of red deer. *Symp. Zool. Soc. Lond.*, **43**, 271–8.

Kunieda, T., Xian, M., Kobayashi, E. *et al.* (1992) Sexing of mouse preimplantation embryos by detection of Y chromosome-specific sequences using polymerase chain reaction. *Biol. Reprod.*, **46**, 692–7.

Laws-King, A., Trounson, A., Sathananthan, A.H. and Kola, I. (1987) Fertilization of human oocytes by microinjection of a single spermatozoon under the zona pellucida. *Fert. Steril.*, **48**, 637–42.

Leibfried-Rutledge, M.L., Critser, E.S., Parrish, J.J. and First, N.L. (1989) *In vitro* maturation and fertilization of bovine oocytes. *Theriogenology*, **31**, 1201–7.

Loskutoff, N.M., Ott, J.E. and Lasley, B.L. (1982) Urinary steroid evaluations to monitor ovarian function in exotic ungulates 1. Pregnanediol-3-glucuronide immunoreactivity in the Okapi (*Okapia johnstoni*). *Zoo Biol.*, **1**, 45–54.

Mann, J.R. (1988) Full term development of mouse eggs fertilized by a spermatozoon microinjected under the zona pellucida. *Biol. Reprod.*, **38**, 1077–83.

Markert, C.L. (1983) Fertilization of mammalian eggs by sperm injection. *J. Exp. Zool.*, **228**, 195–201.

Martin, R.H., Balkan, W., Burns, K. *et al.* (1983) The chromosome constitution of 1000 human spermatozoa. *Hum. Genet.*, **63**, 305–9.

Mattner, P., Entwhistle, K.W. and Martin, I.C.A. (1969) Passage, survival and fertility of deep-frozen ram semen in the genital tract of the ewe. *Aust. J. Biol. Sci.*, **22**, 181–7.

Maxwell, W.M.C. (1986) Artificial insemination of ewes with frozen–thawed semen at a synchronized oestrus. 2. Effect of dose of spermatozoa and site of intrauterine insemination on fertility. *Anim. Reprod. Sci.*, **10**, 309–16.

Moor, R.M. and Trounson, A.O. (1977) Hormonal and follicular factors affecting maturation of sheep oocytes *in vitro* and their subsequent developmental capacity. *J. Reprod. Fert.*, **49**, 101–9.

Moore, H.D.M., Bonney, R.C. and Jones, D.M. (1981) Successful induced ovulation and artificial insemination in the puma (*Felis concolor*). *Vet. Rec.*, **108**, 282–3.

Moore, H.D.M., Bush, M., Celma, M. *et al.* (1984) Artificial insemination in the Giant panda (*Ailuropoda melanoleuca*). *J. Zool. Lond.*, **203**, 269–78.

Moore, H.D.M., Smith, C.A., Hartman, T.D. and Bye, J.D. (1987) Visualization and characterization of the acrosome reaction of human spermatozoa by immunolocalization with monoclonal antibody. *Gamete Res.*, **17**, 245–9.

Mortimer, D., Curtis, E.F. and Miller, R.G. (1987) Specific labelling by peanut agglutinin of the outer acrosomal membrane of the human spermatozoon. *J. Reprod. Fert.*, **81**, 127–35.

Nagase, H. and Niwa, T. (1964) Deep freezing of bull semen in concentrated pellet form. I. Factors affecting the survival of spermatozoa. *V Int. Congr. Anim. Reprod. Artif. Insem. (Trento)*, **4**, 410–15.

Parrish, J.J., Susko-Parrish, J., Winer, M.A. and First, N.L. (1988) Capacitation of bovine sperm by heparin. *Biol. Reprod.*, **38**, 1171–80.

Persidsky, M.D. and Luyet, B.J. (1959) Low-temperature recrystallization in gelatin gels and its relationship to concentration. *Biodynamica*, **8**, 107–20.

Platz, C.C., Wildt, D.E. and Seager, S.W.J. (1978) Pregnancy in the domestic cat after artificial insemination with previously frozen spermatozoa. *J. Reprod. Fert.*, **52**, 279–82.

Roy, A. (1957) Egg yolk coagulating enzyme in the semen and Cowper's gland of the goat. *Nature, Lond.*, **179**, 318–19.

Salamon, S. and Lightfoot, R.J. (1967). Fertilisation and embryonic loss in sheep after insemination with deep frozen semen. *Nature, Lond.*, **216**, 194–5.

Schiewe, M.C. (1991a) The science and significance of embryo cryopreservation. *J. Zoo Wildl. Med.*, **22**, 6–22.

Schiewe, M.C. (1991b) Species variations in semen cryopreservation of nondomestic ungulates, in *Proc. Wild Cattle Symposium, (June 13–16th, 1991)* (eds D.L. Armstrong and T.S. Gross), Henry Doorly Zoo, Omaha, pp. 56–64.

Schiewe, M.C., Bush, M., Phillips, L.A. *et al.* (1991) Comparative aspects of estrus synchronization, ovulation induction, and embryo cryopreservation in the scimitar-horned oryx, bongo, eland and greater kudu. *J. Exp. Zool.*, **258**, 75–88.

Schiewe, M.C., Bush, M., de Vos, V. *et al.* (1991) Semen characteristics, sperm freezing, and endocrine profiles in free-living wildebeest (*Connochaetes taurinus*) and greater kudu (*Tragelaphus strepsiceros*). *J. Zoo Wildl. Med.*, **22**, 58–72.

Schroeder, A.C. and Eppig, J.J. (1984) The developmental capacity of mouse oocytes that matured spontaneously *in vitro* is normal. *Devl. Biol.*, **102**, 493–7.

Shannon, P. (1978) Factors affecting semen preservation and conception rates in cattle. *J. Reprod. Fert.*, **54**, 519–27.

Smith, A.U., Polge, C. and Smiles, J. (1951) Microscopic observation of living cells during freezing and thawing. *J. Roy. Micr. Soc.*, **71**, 186–95.

Summers, P.M., Shephard, A.M., Taylor, C.T. and Hearn, J.P. (1987) The effects of cryopreservation and transfer on embryonic development in the common marmoset monkey, *Callithrix jacchus*. *J. Reprod. Fert.*, **79**, 241–50.

Telfer, E., Torrance, C. and Gosden, R.G. (1990) Morphological study of cultured preantral ovarian follicles of mice after transplantation under the kidney capsule. *J. Reprod. Fert.*, **89**, 565–71.

Thadani, V.M. (1980) A study of hetero-specific sperm egg interactions in the rat, mouse and deer mouse using *in vitro* fertilization and sperm injection. *J. Exp. Zool.*, **212**, 435–53.

Uehara, T. and Yanagimachi, R. (1976) Microsurgical injection of spermatozoa into hamster eggs with subsequent transformation of sperm nuclei into male pronuclei. *Biol. Reprod.*, **16**, 315–21.

Watson, P.F. (1978) A review of techniques of semen collection in mammals. *Symp. Zool. Soc. Lond.*, **43**. 97–126.

Watson, P.F. (1990) Cryopreservation of ram semen. V. *Jornadas Int. Repro. Anim. A.I.*, **5**, 95–112.

Weir, B.J. (1971) The evocation of oestrus in the cuis, *Galea musteloides*. *J. Reprod. Fert.*, **26**, 405–8.

Whittingham, D.G., Leibo, S.P. and Mazur, P. (1972) Survival of mouse embryos frozen to −196°C and −269°C. *Science*, **178**, 411–14.

Wildt, D.E. (1986) Spermatozoa: collection, evaluation, metabolism, freezing, and artificial insemination, in *Comparative Primate Biology: Vol. 3, Reproduction and Development* (eds W.R. Dukelow and J. Erwin), A.R. Liss, New York, pp. 171–93.

Yanagimachi, R. (1988) Mammalian fertilization, in *The Physiology of Reproduction* (eds E. Knobil and J.D. Neil), Raven Press, New York, pp. 135–85.

Yanagimachi, R., Yanagimachi, H. and Rogers, B.J. (1976) The use of zona-free animal ova as a test system for the assessment of the fertilizing capacity of human spermatozoa. *Biol. Reprod.*, **15**, 471–76.

Zhang, J., Boyle, M.S., Smith, C.A. and Moore, H.D.M. (1990) Acrosome reaction of stallion spermatozoa evaluated with monoclonal antibody and zona-free hamster eggs. *Molec. Reprod. Dev.*, **127**, 152–8.

8

The role of environmental enrichment in the captive breeding and reintroduction of endangered species

D. Shepherdson
Metro Washington Park Zoo, Portland, Oregon, USA.

8.1 INTRODUCTION

Environmental enrichment is an increasingly popular method for improving the well-being of animals in zoos. Research has shown that simple and eminently practical changes to the way zoo animals are kept can have highly beneficial effects on their behaviour and physiology. Carlstead, Seidensticker and Baldwin (1991), for example, found that providing bears with hidden food and manipulable objects greatly increased activity and exploration at the expense of repetitive stereotyped behaviours. Similarly Kastelein and Wiepkema (1989) observed that a walrus provided with realistic foraging opportunities, namely searching for food items amongst concrete blocks, greatly reduced the amount of stereotyped swimming. Providing a kinkajou with food that required exploratory and manipulatory behaviour, by hanging whole fruit from branches as an alternative to chopped fruit in a bowl, resulted in greatly increased rates of locomotion, exploration and foraging behaviour which corresponded with reductions in the levels of stereo-

Creative Conservation: Interactive management of wild and captive animals.
Edited by P.J.S. Olney, G.M. Mace and A.T.C. Feistner.
Published in 1994 by Chapman & Hall, London.
ISBN 0 412 49570 8

typed behaviour (Shepherdson, Brownback and Tinkler, 1990). These and other successful techniques share an important characteristic: they restore to the captive animal the contingency between the performance of behaviour (foraging for example) and the appropriate consequences (such as finding food). This contingency or control is an integral characteristic of the interactions between a wild animal and its environment but one that is frequently lost in captivity with serious consequences for psychological well-being.

As wild habitats continue to shrink and human pressures on wildlife continue to rise, zoos are becoming increasingly concerned with the conservation and well-being of animal groups (populations and species) as well as individuals. Environmental enrichment has an important role to play in this endeavour also, but one that may differ in some respects from the predominantly individual-oriented approach to welfare of the past. These new and interrelated roles for enrichment include:

(a) conservation of culturally determined repertoires of behaviour for posterity, research and education;
(b) facilitating the reproduction and maintenance of genetically viable captive populations;
(c) reintroduction of captive bred animals to the wild.

8.2 CONSERVING BEHAVIOUR

One of the putative aims of captive breeding is to preserve in captivity populations of animals as representatives of their wild ancestors, for posterity, research and education. There is more to preserving an animal species, however, than looking after its genes. An elephant's rich array of diverse individual and social behaviour is as much a part of what makes it an elephant as is its trunk and its huge feet. Since behaviour is a consequence of interactions between both genetic and environmental factors, failure to reproduce an environment that is at least functionally equivalent to that of the wild will inevitably result in the loss of many forms and patterns of natural behaviour. In many cases behavioural repertoires may be recovered even after generations of absence simply by recreating the correct environmental stimuli. However this will not be the case for behaviours that are learnt by individuals and then passed on from generation to generation. The diversity of these culturally transmitted behaviours (which concern activities as diverse as infant rearing and food preference) has the potential to be lost much faster than does genetic diversity (May, 1991), and, unlike genes, behaviour cannot be preserved frozen in a test tube.

8.3 CAPTIVE BREEDING

Strictly speaking the environmental requirements for maintaining genetically viable captive populations and for reintroduction should be the same, since we can never be certain that a captive population may not, at some future date, be a candidate for reintroduction (Konstant and Mittermeier, 1982). In view of this and the arguments above, Snowdon (1989) has suggested that, as with reintroduction, the criterion for successful captive breeding should be the potential to survive in the wild. In practice, however, limited resources are likely to impose more fundamental criteria. The less space required to maintain each captive population the greater the total number of populations that can be maintained. Enrichment can usefully be thought of as a way of increasing the psychological space available to captive animals (Chamove, 1989). Specifically, enrichment techniques can be used to optimize the levels of social and physical stimulation provided by captive environments to maximize reproduction and ensure normal behavioural development.

Chronic stress is known to result in high levels of pituitary–adrenal activity which can have inhibitory effects on reproduction as well as reduced immune response, growth and digestion (Sapolski, 1989; Moburg, 1991). Chronic stress may be responsible for the failure of some zoo populations to achieve their reproductive potential. That simple forms of environmental enrichment can reduce the levels of stress-related hormones in zoo animals has been demonstrated by Carlstead, Brown and Seidensticker (1991). Providing leopard cats with enriched enclosures containing places to hide and climb resulted in reduced urinary cortisol levels. Abnormal behaviours such as stereotypic behaviour, coprophagy, inactivity, and excessive aggression have been attributed to the stress of captivity and may, in some cases, directly interfere with reproduction by reducing the diversity of social behaviour. The zoo literature contains many examples of enriched environments that have successfully reduced the incidence of these behaviours, presumably by reducing or eliminating the source of this stress (e.g. Bowen, 1980; Ackers and Schildekraut, 1985; Kastelein and Wiepkema, 1989; Shepherdson, Brownback and Tinkler, 1990; Carlstead, Seidensticker and Baldwin, 1991).

Stressful stimuli over which animals have no control are likely to be far more damaging than those to which they can respond appropriately (Wiess, 1968). The kinds of stress that animals face in the wild are nearly always those for which an appropriate behavioural response exists. For example when faced by a predator, a wild animal will usually be able to perform an appropriate predator avoidance behaviour. Environmental enrichment can provide the substrates necessary for appropriate behavioural responses. A potential source of uncontrollable stress for all captive animals lies in their interactions with human caretakers (Hemsworth and Barnett, 1987) and visitors (Glatston *et al.* 1984; Chamove, Hosey and Schatzel, 1988). In support of this Mellen (1991) found a positive correlation between the amount of friendly keeper interaction with small exotic cats and reproduction. This finding implies that by habituating or 'taming' animals the stress of human interaction is reduced. This is an area deserving of more research and is also one area in which the needs of captive breeding may conflict with those of reintroduction.

Paradoxically, under-stimulation may also be a cause of reproductive inhibition. Short periods of acute stress may not only be benign, they may be a necessity for normal behavioural and physiological development (Weiss, Sundar and Becker, 1989). Complete lack of stress in early life can reduce subsequent ability to cope with stress and may even lead to death through fatal shock syndrome (Snyder, 1977; Wood-Gush, 1973). Moodie and Chamove (1990) provide evidence that short threatening events are indeed beneficial to the well-being of captive tamarins. In fact a certain minimum level of hormonal arousal is necessary for the activation of many of the physiological and behavioural mechanisms of reproduction. This may explain why previously non-reproductive gorillas moved to enriched enclosures sometimes become reproductively active (Maple and Stine, 1982), and Eaton and Craig (1973) have proposed that the stimulation of catching and killing live prey facilitates breeding in captive cheetah.

It is well known that providing the appropriate social environment is of critical importance, not only for reproduction to occur but also for the development of reproductively normal adults. For many species, notably primates, it is necessary for juveniles to learn parenting and social skills from other group members if they are not to become behaviourally and reproductively compromised as adults (e.g. Harlow and Harlow, 1962; Suomi, 1986; Beck and Power, 1988; Mellen, 1992; Berman, 1990; Hannah and Brotman, 1990). Removing juveniles from their family group prematurely can result in higher rates of reproduction in the short term but may not fulfil the long-term goals of captive breeding. The physical environment plays an important role in determining the quantity and quality of social interactions. Factors such as visual barriers (Erwin *et al.*, 1976) and feeding methods (Chamove *et al.*, 1982) can have far reaching effects on the social dynamics of captive animal groups.

8.4 REINTRODUCTION

The skills required for survival in the wild have been categorized and described by Box (1991):

(a) orientation and locomotion skills;
(b) feeding and foraging;
(c) obtaining suitable places to rest and sleep;
(d) interspecies interaction including predator avoidance;
(e) intraspecific interaction.

Clearly animals born and reared in the wild have superior survival skills to those reared in captivity. Wild-born animals reintroduced to the same sites as captive-born animals adapt quicker and survive longer. This has been shown for both golden lion tamarins (Beck *et al.*, 1991) and Siberian polecats translocated in lieu of black footed ferrets (Miller *et al.*, Chapter 27). This is further supported by the fact that survivorship of the wild-born offspring of reintroductees is frequently greater than that of their captive-born parents (Beck *et al.*, 1991), and by the relatively high success rates of ungulate translocations (Gordon, 1991). The failure of captive-bred game animals to survive in comparison with wild-born animals also underlines the importance of the pre-release environment to the acquisition of behaviours necessary for survival in the wild (e.g. Schladweiler and Tester, 1972).

Two interrelated factors are likely to influence post release survival of reintroduced animals: the possession of specific survival skills gained from previous experience, and the capacity to learn new and flexible behavioural strategies and skills in response to a new, dynamic environment. Both these factors are influenced by the quality of the pre-release environment.

8.4.1 Skill training

Most reintroductions are conducted using 'soft' release techniques whereby animals undergo a period of acclimatization directly pre-release, post-release or a combination of the two (see review in Stanley Price, 1989). Sometimes referred to as a training period this might more accurately be thought of as a period during which animals can learn the skills necessary for survival in the wild, although apes have been actively trained by the example of human companions (e.g. Brewer, 1978). While there seems to be little doubt that some form of acclimatization is necessary at the release stage of a reintroduction, the role of the captive environment prior to that point is less clear. Few studies have compared the post-release survival of animals reared in different captive environments (although when acclimatization periods are long this distinction becomes somewhat blurred).

Golden lion tamarins given free range of a large wooded area at the National Zoo, Washington, DC (Bronikowski, Beck and Power, 1989) showed enhanced behavioural adaptation upon release; however, previous attempts to train the animals using smaller enriched environments failed to confer any enduring advantages, suggesting the enriched cage environments failed to accurately reproduce the challenges of the wild (Beck *et al.*, 1987). Miller *et al.* (Chapter 27) compared the post-release survivorship of Siberian ferrets raised in cages, ferrets raised in an enclosed semi-natural prairie dog colony and a third group of wild-caught translocated ferrets. The results suggested that the translocated ferrets survived longest but were closely followed by the semi-natural group. The animals reared in conventional breeding cages survived for the shortest period of time. These examples suggest that sufficiently enriched environments do enable captive animals to learn valuable survival skills. They also suggest, however, that providing a sufficiently enriched environment for significant gains in post-release survival may be very difficult.

Clearly some of the stimuli that wild animals have to cope with are stressful and, in some cases, harmful. In order for captive animals to learn useful survival skills in captivity they must experience the unpleasant consequences of making mistakes. If a golden lion tamarin gets separated from its group it may have to spend a cold, wet and frightening night out on its own. A red wolf needs to know that contact with human beings is an unpleasant experience and black-footed ferrets need to know that hanging around on the surface of a prairie dog colony is asking for trouble.

Miller *et al.* (1990) have used mild aversive stimuli (being shot by a rubber band) presented in tandem with predator models to enhance predator avoidance in Siberian polecats; and every effort is made to ensure that any contact red wolves have with humans is unpleasant (Moore and Smith, 1991). It has been suggested that this is the point at which the well-being of the individual is being subordinated to that of the group or species (Beck, 1991). In some cases this will undoubtedly be the case. Allowing infant offspring to die in order that the parent(s) shall learn how to rear young successfully is obviously compromising the well-being of the offspring; and of course providing predators with live prey is not always in the best interests of the prey (and is illegal in some countries, such as the UK). But, as the experiment with Siberian polecats described above demonstrates, it is often possible to simulate aversive events without causing any long-lasting injury. In other cases it may be our perception of animal well-being that needs to change. As we have already seen, maximizing well-being does not necessarily mean removing all stress from the lives of captive animals; rather it requires providing the optimal level of appropriate stimulation.

For many animals it may not be possible, for both ethical and practical reasons, to provide captive environments realistic enough for the animals to learn from experience the skills necessary for survival in the wild. In these cases, and they may well be in the majority, we must rely on providing a captive environment that is at least sufficient to allow animals to retain the ability to learn and adapt to new environments, and hope that these abilities will allow them to acquire necessary survival skills during the acclimatization phase of release.

8.4.2 Behavioural flexibility

Laboratory research, mainly on rodents, suggests that environmental enrichment can have far reaching physiological and behavioural consequences. These consequences have been reviewed by Widman, Abrahamsen and Rosellini, (1992) and include changes in brain morphology, such as increased cortical thickness and weight; increases in the size, number and complexity of nerve synapses; increases in the amounts of acetylcholinesterase and cholinesterase; and increased ratio of RNA to DNA. A number of important behavioural changes are also associated with enriched environments: fewer errors in maze-type orientation problems; quicker learning of operant tasks; reduced emotionality; and quantitative and qualitative increases in exploratory behaviour. Renner (1987) suggests that these changes in exploratory behaviour may 'broaden the range of possible information that may be available during exploration. This resulting difference in environmental knowledge could in turn, have functional significance under conditions of challenge (e.g. predation).' Research has also shown that animals kept in socially and physically impoverished environments tend to develop rigid, unvarying behaviour patterns (stereotypic behaviour) rather than remaining exploratory and alert to stimuli in the environment (reviewed by Mason, 1991).

These kinds of generalized effects clearly have great significance for reintroductions since they would undoubtedly influence the efficiency and speed with which survival skills are learned by reintroduced animals, regardless of the method of release. Most of these experiments compared groups of animals in normal laboratory caging with larger enriched environments containing fixed and manipulable objects which were regularly changed. It could be argued that by this definition most zoo enclosures are already enriched. This may be true in many cases but, as we have already seen, zoo animals do develop rigid, repetitive behaviours (e.g. Kastelein and Wiepkema, 1989; Carlstead, Seidensticker and Baldwin, 1991; Carlstead, 1991; Wechsler, 1991), and while these may not necessarily indicate reduced well-being (Mason, 1991) they are not suggestive of environments that are rich in behavioural opportunity or stimulus diversity. Furthermore, as the demands for space in which to house endangered populations continue to rise (Maguire and Lacey, 1990) enclosures are likely to become smaller and, potentially, less enriched. Different species may require different lengths of exposure to enriched environments for these benefits to be acquired and this would be a useful topic for research. Evidence that these generalized effects are relevant to real captive breeding and re-introduction endeavours is provided by Miller *et al.* (Chapter 27) who found that black-footed ferrets reared in an enriched environment (but with no previous prey-killing experience) killed hamsters more efficiently than animals reared in standard pens.

Finally there is another way in which the developmental environment of captive animals can influence the attainment of future survival skills. Many animals need to learn specific skills at certain critical periods in their development. Vargas (pers. comm.) for example, found that black-footed ferrets established adult dietary preferences during their third month of life, and Pfaffenberger and Scott (1976) suggest that guide dogs for the blind exposed to an enriched environment (living in a country as opposed to a city environment) between the ages of three and 12 weeks coped better with city hazards as adults. In fact dogs kept in a sterile kennel environment for the first 12 weeks of life suffered shock on introduction to a more complex environment, from which they never recovered. Failure to identify these periods and provide the appropriate environmental stimuli may severely effect subsequent reintroduction survivorship.

8.5 CONCLUDING COMMENTS

Environmental enrichment techniques have several important roles to play in the breeding of endangered species in captivity. Enrichment techniques can help to preserve the behavioural diversity of captive populations and ensure the growth of behaviourally viable populations. Enrichment of captive rearing environments can help ensure the success of reintroduction programs by providing animals with the opportunity to learn skills necessary for life in the wild, and by optimizing the capacity of animals to learn from, and adapt to, new environments. Experimental establishment of free-ranging captive-born groups of animals such as those described by Price *et al.* (1989) and Chamove and Rohruber (1989) for cotton-top tamarins, Bronikowski, Beck and Power (1989) for golden lion tamarins and Miller *et al.* (Chapter 27) for Siberian polecats, offer an excellent opportunity to test these ideas.

Resources for captive breeding and reintroduction are always scarce, and enriched environments requiring more resources, notably space, will reduce the potential size and productivity of captive populations and hence the number of animals available for reintroduction. The benefits of enriched captive environments (in terms of increased post-release survival) must be balanced against the cost of having fewer animals available for release.

Conflicts of interest may arise between the needs of captive breeding, reintroduction and animal well-being. In some cases this is inevitable and tough decisions will have to be made, but others may be resolved by adopting a broader view of well-being in which challenges that may be stressors in the short term lead, in the long term, to increased behavioural adaptability, responsiveness and, consequently, well-being.

ACKNOWLEDGEMENTS

My thanks to the Metro Washington Park Zoo, the Friends of Metro Washington Park Zoo, the Universities Federation for Animal Welfare and the Zoological Society of London for funding and support. My thanks also to the keeping staff of M.W.P.Z. and Z.S.L., and to Dr J. Mellen, Dr K. Carlstead, J. Barker, Dr B. Beck, Dr A. Feistner, Dr B. Miller, R. Smith, Dr T. Thorne, A. Vargas and Wyoming Game and Fish Dept. for information and discussion.

REFERENCES

Akers, J.S. and Schildekraut, D.S. (1985) Regurgitation/reingestion and coprophagy in captive gorillas. *Zoo Biology*, **4**, 99–109.

Beck, B. (1991) Managing zoo environments for re-introduction, in *AAZPA Annual Conference Proceedings*, Wheeling, WV, pp. 436–40.

Beck, B.B., Castro, I., Kleiman, D. *et al.* (1987) Preparing captive-born primates for re-introduction. *International Journal of Primatology*, **8**, 426.

Beck, B.B., Kleiman, D.G., Dietz, J.M. *et al.* (1991) Losses and reproduction in the reintroduced golden lion tamarins *Leontopithecus rosalia. Dodo, Journal of the Jersey Wildlife Preservation Trust*, **27**, 50–61.

Beck, B. and Power, M.L. (1988) Correlates of sexual and maternal competence in captive gorillas. *Zoo Biology*, **7**, 339–50.

Berman, C.M. (1990) Intergenerational transmission of maternal rejection rates among free-ranging rhesus monkeys. *Animal Behaviour*, **39**, 329–37.

Bowen, R.A. (1980) The behaviour of three hand-reared lowland gorillas *Gorilla g. gorilla* with emphasis on the response to a change in accommodation. *Dodo, Journal of the Jersey Wildlife Preservation Trust*, **17**, 63–79.

Box, H.O. (1991) Training for life after release: simian primates as examples. *Symposium of the Zoological Society of London*, **62**, 111–23.

Brewer, S. (1978) *The Forest Dwellers*, Collins, London.

Bronikowski, J., Beck, B. and Power, M. (1989) Innovation, exhibition and conservation: free-ranging tamarins at the National Zoological Park, in *AAZPA Annual Conference Proceedings*, Wheeling, WV, pp. 540–6.

Carlstead, K., Brown, J.L. and Seidensticker, J. (1991) Assessing stress in captive environments for leopard cats (*Felis bengalensis*), in *AAZPA Annual Conference Proceedings*, Wheeling, WV, pp. 319.

Carlstead, K., Seidensticker, J and Baldwin, R. (1991) Environmental enrichment for zoo bears. *Zoo Biology*, **10**, 3–16.

Carlstead, K. (1991) Husbandry of the fennec fox *Fennecus zerda*: environmental conditions influencing stereotypic behaviour. *International Zoo Yearbook*, **30**, 202–7.

Chamove, A.S. (1989) Environmental enrichment: a review. *Anim Technology*, **40** (3), 155–78.

Chamove, A.S., Anderson, J.R., Morgan-Jones, S.C. and Jones, S.P. (1982) Deep woodchip litter: hygiene, feeding, and behavioural enhancement of eight primate species. *Int. J. Stud. Anim. Prob.* **3**, 308–18.

Chamove, A.S., Hosey, G.R. and Schatzel, P. (1988) Visitors excite primates in zoos. *Zoo Biology*, **7**, 359–69.

Chamove, A.S. and Rohruber, B. (1989) Moving callitrichid monkeys from cages to outside areas. *Zoo Biol.* **8**, 151–63.

Eaton, R.L. and Craig, S.J. (1973) Captive management and mating behaviour of the cheetah, in *The World's Cats. Vol. 1, Ecology and Conservation* (ed. R.L. Eaton), World Wildlife Safari and Institute for the Study and Conservation of Endangered Species, Winston, Oregon, pp. 216–54.

Erwin, J., Anderson, B., Erwin, N. *et al.* (1976) Aggression in captive groups of pigtail monkeys: effects of provision of cover. *Perceptual Motor Skills*, **42**, 219–24.

Glatston, A.R., Geilvoet-Soetman, E., Hora-Pecek, E. and Van Hooff, J.A.R.A.M. (1984) The influence of the zoo environment on the social behaviour of groups of cotton topped tamarins, *Saguinus oedipus oedipus*. *Zoo Biology*, **3**, 241–53.

Gordon, I.J. (1991) Ungulate re-introductions: the case of the scimitar-horned oryx. *Symposium of the Zoological Society of London*, **62**, 217–40.

Hannah, A.C. and Brotman, B. (1990) Procedures for improving maternal behaviour in captive chimpanzees. *Zoo Biology*, **9**, 233–40.

Harlow, H.F. and Harlow, M.K. (1962) Social deprivation in monkeys. *Scientific American*, **207**, 137–46.

Hemsworth, P.H. and Barnett, J.L. (1987) Human–animal interactions, in *Veterinary Clinics of North America: Food Animal Practice, Vol. 3*, pp. 339–56.

Kastelein, R.A. and Wiepkema P.R. (1989) A digging trough as occupational therapy for Pacific Walruses (*Odobenus rosmarus divergens*) in human care. *Aquatic Mammals*, **15**, 9–17.

Konstant, W.R. and Mittermeier, R.A. (1982) Introduction, re-introduction and translocation of neotropical primates: past experiences future possibilities. *International Zoo Yearbook*, **22**, 69–77.

Maguire, L.A. and Lacey, R.C. (1990) Allocating scarce resources for conservation of endangered species: partitioning zoo space for tigers. *Conservation Biology*, **4**(2), 157–70.

Maple, T.L. and Stine, W.W. (1982) Environmental variables and great ape husbandry. *American Journal of Primatology*, Supp. 1, 67–76.

Mason, G.J. (1991) Stereotypies and suffering. *Behavioural Processes*, **25**, 103–15.

May, R. (1991) The role of ecological theory in planning the re-introduction of endangered species. *Symposium of the Zoological Society of London*, **62**, 145–63.

Mellen, J.D. (1991) Factors influencing reproductive success in small exotic felids (*Felis* spp.): a multiple regression analysis. *Zoo Biology*, **10**, 95–110.

Mellen, J.D. (1992) Effects of early rearing experience on subsequent adult sexual behavior using domestic cats (*Felis catus*) as a model for exotic small felids. *Zoo Biology* **11**, 17–32.

Miller, B., Biggins, D., Wemmer, C. *et al.* (1990) Development of survival skills in captive-raised Siberian polecats (*Mustela eversmanni*) II: predator avoidance. *Journal of Ethology*, **8**, 95–104.

Moburg, G.P. (1991) How behavioral stress disrupts the endocrine control of reproduction in domestic animals. *Journal of Dairy Science*, **74**, 304–11.

Moodie, E.M. and Chamove, A.S. (1990) Brief threatening events are beneficial for captive tamarins. *Zoo Biology*, **9**, 275–86.

Moore, D.E. and Smith, R. (1991) The red wolf as a model for carnivore re-introductions. *Symposium of the Zoological Society of London*, **62**, 263–78.

Pfaffenberger, C.J. and Scott, J.P. (1976) Early rearing and testing, in *Guide dogs for the blind: their selection, development and training. (Developments in animal and veterinary sciences, 1)*, Elsevier, Amsterdam.

Price, E.C., Feistner, A.T.C., Carroll, J. and Young, J.A. (1989) Establishment of a free ranging group of cotton-top tamarins *Saguinus oedipus* at the Jersey Wildlife Preservation Trust. *Dodo, Journal of the Jersey Wildlife Preservation Trust*, **26**, 60–9.

Renner, M.J. (1987) Experience dependant changes in exploratory behavior in the adult rat (*Rattus norvegicus*): overall activity level and interactions with objects. *Journal of Comparative Psychology*, **101**(1), 94–100.

Sapolski, R. (1989) Physiological perspectives on non-human primate well-being. *Scientists Centre For Animal Welfare Newsletter* (Bethesda, Maryland), **11** (3), 4–8.

Schladweiler, J.L. and Tester, J.R. (1972) Survival and behaviour of hand-reared mallards released in the wild. *Journal of Wildlife Management*, **36**(4), 1118–27.

Shepherdson, D.J., Brownback, T. and Tinkler, D. (1990) Putting the wild back into zoos: enriching the zoo environment. *Applied Animal Behaviour Science*, **28** (3), 300.

Snowdon, C.T. (1989) The criteria for successful captive propagation of endangered primates. *Zoo Biology*, Supp. 1, 149–61.

Snyder, R. (1977) Putting the wild back into the zoo. *Int. Zoo News*, **24**(4), 11–18.

Stanley Price, M.R. (1989) *Animal Re-introductions: The Arabian Oryx in Oman*, Cambridge University Press, Cambridge, UK.

Suomi, S.J. (1986) Behavioural aspects of successful reproduction in primates, in *Primates: the road to self-sustaining populations* (ed. K. Benirschke), Springer-Verlag, New York, pp. 331–40.

Wechsler, B. (1991) Stereotypes in polar bears. *Zoo Biology*, **10**, 177–88.

Weiss, J.M. (1968). Effects of coping responses on stress. *Journal of Comparative Physiology and Psychology*, **65**, 251–60.

Weiss, J.M., Sundar, S.K. and Becker, K.J. (1989) Stress-induced immunosuppression and immunoenhancement: cellular immune changes and mechanisms, in *Neuroimmune Networks: Physiology and Diseases* (eds E.J. Goetzl and N.H. Spector), Alan R. Liss, New York, pp. 193–206.

Widman, D.R., Abrahamsen, G.C. and Rosellini, R.A. (1992) Environmental enrichment: the influences of restricted daily exposure and subsequent exposure to uncontrollable stress. *Physiology and Behaviour*, **51**, 309–18.

Wood-Gush, D.G.M (1973) Animal welfare in modern agriculture. *British Veterinary Journal*, **129**, 164–74.

9

Disease risks associated with wildlife translocation projects

M.H. Woodford
IUCN/SSC Veterinary Specialist Group, Washington, DC, USA
and
P.B. Rossiter
Kenya Agricultural Research Institute, Kikuy, Kenya.

9.1 INTRODUCTION

The translocation of wild animals by man for the establishment of new populations is a very ancient practice. There is evidence in the writings of Pliny that the Romans artificially extended the natural range of the rabbit (*Oryctolagus cuniculus*) from southern Spain to most of western Europe (although the Normans are credited with the introduction of the species into Britain). The Romans, too, are said to have moved the fallow deer (*Dama dama*) from its ancestral range in the Mediterranean forests to the newly conquered territories of Gaul and Britannia. The motive for most of these artificial animal movements was then, as now, to establish an easily accessible supply of food 'on the hoof' and to provide a suitable quarry for hunting. Rarely was the motive anything other than practical, although Pliny does mention the translocation of a species of croaking frogs from the Continent to Cyrene to liven up the local frogs which were said to be dumb. Translocation of wild animals as a wildlife management tool thus has many precedents, and the vast

Creative Conservation: Interactive management of wild and captive animals.
Edited by P.J.S. Olney, G.M. Mace and A.T.C. Feistner.
Published in 1994 by Chapman & Hall, London.
ISBN 0 412 49570 8

majority of movements of wild animals whether they are introductions, reintroductions or reinforcements are still made for sporting purposes.

A recent survey (1973–1986) of **intentional** releases of indigenous birds and mammals into the wild in Australia, Canada, Hawaii, New Zealand and the United States shows that on average nearly 700 translocations were conducted each year. Native game species, many of them birds, comprised 90% of these movements (Griffith *et al.*, 1989). Many similar releases, usually for sporting purposes, are made each year in Europe and on a smaller scale elsewhere in the world.

The translocation of rare and endangered species has become an important conservation technique, especially for those species which possess limited dispersal ability – since these often find themselves confined in shrinking and disconnected habitat fragments where early extinction can be predicted (Wilson and MacArthur, 1967).

Translocations of rare species can be expensive (Cade, 1988; Kleiman, 1989) and often attract great public attention (Booth, 1988). Several factors have been associated with the outcome of such enterprises. These have been discussed and evaluated by Griffith *et al.* (1989), but surprisingly the possibility that disease might present a negative influence on the efforts of founder animals to establish themselves was not considered. However, other authors (Aveling and Mitchell, 1982; Brambell, 1977; Caldecott and Kavanagh, 1983) have drawn attention to the problem.

9.2 TYPES OF DISEASE RISK

Most of the diseases likely to be of importance in translocation projects are infectious, and the risks, whatever the objective of these operations may be, will depend on a variety of factors. Amongst these are the epidemiological situations at the source from which the animals derive and those at their destination or release site.

The source is often a zoo, a ranch or an extensive captive breeding establishment, and it is sometimes located on a distant continent. Occasionally, however, a supply of suitable wild animals for translocation can be found in another area within the country where the release is planned.

The release site may be a national park or other protected area where contact with other wild species may occur, or it may be an undesignated area of suitable habitat where contact with domestic livestock and humans is an added hazard.

Animals born and bred in a zoo or in a captive breeding facility in a distant country will acquire local infections, and in some cases will become symptomless carriers of disease agents. Zoo-bred stock is often exposed to exotic pathogens brought in from foreign countries and to infections transmitted to them by their attendants and visitors. Furthermore, captivity subjects some species to continuous stress resulting in immunodepression and increased susceptibility to infection. Tuberculosis, usually of the bovine type in ungulates and of human origin in primates, is common in zoological collections, and unfortunately, at least for wild ungulates, the current diagnostic tests for this disease are unreliable and difficult to interpret (Philip, 1989; Flamand, 1990).

Ranch-raised animals, usually herding ungulates, are often exposed to the common pathogens of local domestic animals. Brucellosis, bluetongue and various tick-borne haemoparasites are likely examples, while equids will be exposed to the endemic diseases of horses and donkeys in the country where they are raised.

9.3 DISEASES INTRODUCED BY TRANSLOCATED ANIMALS

There is a risk, therefore, that both zoo-bred and ranch-raised animals, as well as wild-caught stock, may bring new pathogens into a release area and that these may cause disease amongst coexisting immunologically naïve wild and domestic animals. As has been aptly remarked, each translocated animal is not a representative of a single species but is rather a biological package containing a selection of viruses, bacteria, protozoa, helminths and arthropods (Nettles, 1988).

There have been many instances in which captive-bred (and sometimes wild-caught) animals have brought with them pathogens which have had a severe impact on wild and domestic stock in and around the release site (Table 9.1). Some of these cases, like the introduction of African horse sickness into Spain by two zebra (*Equus burchelli*) from Namibia in 1987 are well documented (OIE, 1987). Unfortunately others are only anecdotal or described in rudimentary and unconfirmed detail.

It should be borne in mind that capture and quarantine themselves may both place a high level of stress on wild animals, especially those caught in the wild, and that there is a great deal of variation in the response of species and age classes. Stress can lead to the clinical recrudescence of latent infectious diseases. For example, Cape buffaloes (*Syncerus caffer*) transported a long distance by lorry have excreted foot and mouth disease virus in sufficient quantities to infect nearby cattle (Hedger and Condy, 1985) and fatal cases of trypanosomiasis have arisen in black rhinoceros (*Diceros bicornis*) captured and translocated while harbouring inapparent trypanosome infections (McCulloch and Achard, 1969).

A classic case has been the outbreak of tuberculosis in Arabian oryx (*Oryx leucoryx*) in Saudi Arabia (Heuschele, 1990; Asmode and Khoja, 1989; Kock and Woodford, 1988). The late King Khaled of Saudi Arabia was in the habit of turning loose in a large, fenced enclosure outside

Table 9.1 Diseases introduced into release areas by translocated wildlife[a]

Species translocated	*Source*	*Disease introduced*	*Release area*	*Species affected*	*Reference*
Zebra	Namibia, w/c	African horse sickness	Spain	Domestic equids	OIE, 1987
Arabian oryx	Riyadh, Saudi Arabia, c/b	Tuberculosis	Taif, Saudi Arabia	Translocates	Heuschele, 1990; Asmode and Khoji, 1989; Kock and Woodford, 1988
Raccoon	Texas, w/c	Parvoviral enteritis	West Virginia	Local raccoons	Allen, 1986
Raccoon	Florida, w/c	Rabies	West Virginia	Skunks, local raccoons	Anon, 1990
Reindeer	Norway, c/b	Warble flies, nostril flies	Greenland	Caribou	Rosen, 1955; Rosen, 1958; Thing and Thing, 1983
Mojave desert tortoise	Petshops in California	'Respiratory disease'	Mojave Desert	Wild tortoises	Jacobsen *et al.*, 1991
Bighorn sheep	Arizona, w/c	'Viral pneumonia'	New Mexico	Local bighorns	–
Plains bison	Montana, c/b	Tuberculosis, Brucellosis	Canada	Wood bison	Environmental Assessment Panel, 1990
Hares	Hungary, Czechoslovakia, w/c	Brucellosis	Switzerland; Italy	Domestic animals, people	Pastoret *et al.*, 1988
Rainbow trout	USA, c/b	'Whirling disease'	UK	Trout	Trust, 1986

[a]w/c = wild caught; c/b = captive bred

Riyadh, all the herbivorous diplomatic gifts that he received. Amongst these assorted animals, most of which had been zoo-bred in the country of their donors, must have been some carrying sub-clinical tuberculosis. When later it was decided to prepare a herd of Arabian oryx for release into the wild in Saudi Arabia, 57 of these desert antelopes were separated out from the mixed herd of herbivores and relocated by air to the National Wildlife Research Centre at Taif, some 400 miles away. Within a few weeks of this translocation, some of the oryx became sick and died of acute tuberculosis, a disease of which there had been no signs in the mixed antelope herd at Riyadh. It was considered likely that the stresses of capture, crating and translocation probably played a part in precipitating the clinical disease. Plains bison (*Bison bison*), translocated in 1907 from Montana to Canada brought with them tuberculosis and brucellosis, two infections which have resulted in a recent decision to slaughter 3200 infected animals in an attempt to eliminate the diseases from the Wood Buffalo National Park in Canada, where they threaten the relict herd of wood bison (*Bison bison athabascae*). However, public pressure has now caused the slaughter decision to be rescinded.

In 1985, raccoons (*Procyon lotor*), translocated from Texas to West Virginia to augment the local raccoon stock for hunting purposes are believed to have brought with them parvoviral enteritis, a serious disease previously not present in West Virginia and now enzootic in the local raccoon population (Allen, 1986). A similar case occurred when 'a few thousand' raccoons were translocated from Florida to West Virginia in 1977, again for hunting purposes. This particular relocation is still blamed for the current rabies epizootic in raccoons and skunks (*Mephitis mephitis*) in Pennsylvania, Virginia and Maryland (Anon, 1990). Rabies is said to have been far less common in these states before the raccoon translocation from Florida took place. It is, of course, possible that the ongoing rabies outbreak may be a reflection of the increase in the local raccoon population in the area.

In September 1952, 225 domestic reindeer (*Rangifer tarandus*) were translocated by ship from northern Norway to western Greenland. The object of this introduction was to provide a new livelihood for the local Inuit (Rosen, 1955). Unfortunately the reindeer brought with them a warble fly (*Oedemagna tarandi*) and a nostril fly (*Cephenemyia trompe*), both of which at that time did not occur in Greenland (Rosen, 1958). In due course these parasites transferred their attention to the indigenous wild Greenland caribou (*Rangifer tarandus groenlandicus*). Both these flies severely harass the caribou throughout the relatively warm long summer days to the extent that they are unable to feed sufficiently to build up the fat stores upon which they depend to survive the Arctic winter. This translocation was a disaster for the Greenland caribou, which are now greatly reduced in numbers due to severe winter mortality. All the caribou in western Greenland are now infested with both parasites (Thing and Thing, 1983).

Sick specimens of the endangered Mojave Desert tortoise (*Xerobates agassizii*), unwanted by their owners, have been released back into the desert and are believed to have infected the wild tortoises with a fatal upper respiratory tract infection, probably acquired on pet shop premises (Jacobsen *et al.*, 1991).

Large numbers of hares (*Lepus europaeus*) are imported annually from Hungary and Czechoslovakia into Switzerland and Italy for sporting purposes. Some of these animals are infected with *Brucella suis*, an important pathogen which can seriously affect domestic livestock (Pastoret *et al.*, 1988).

9.4 DISEASES ENCOUNTERED BY TRANSLOCATED ANIMALS AT THE RELEASE SITE

Animals born and bred on a distant continent in a very different epidemiological environment, may have encountered diseases endemic in that area (and may sometimes have become symptomless carriers of the pathogens), but will also inevitably lack acquired immunity or resistance to infections that are equally exotic (to them) which will challenge them at the release site. Many diseases and parasites are very local in their distribution as a result of the highly specific ecological requirements of the pathogens and their vectors, and even a short translocation of wild-caught animals from one ecozone to another can result in their exposure to unsuspected disease problems. Some examples are given in Table 9.2.

About half of 50 koalas (*Phascolartos cinereus*) translocated from a tick-free area in Victoria, Australia to a nearby range infested with the Ixodid vector of tick paralysis, rapidly succumbed, (S. McOrist, pers. comm.). Adult Kafue Flats lechwe (*Kobus leche*) – a large Zambian antelope – were translocated into an area infested with the tick vector of heartwater and where this disease was enzootic in nearby domestic stock. Within two months 56 lechwe had died; autopsies revealed heavy tick infestations (*Amblyomma variegatum*) and characteristic lesions of heartwater (Pandey, 1991). Had either of these two diverse species been born in the areas infested by the tick vectors they would probably have acquired resistance to the respective diseases and the translocations would have been a success. Caribou and moose (*Alces americana*) are seriously affected by the 'meningeal worm' (*Pneumostrongylus tenuis*) of white tailed deer (*Odocoileus virginianus*), and attempts to translocate these ungulates into the white-tailed deer range in north-eastern America have been frustrated by cerebrospinal nematodiasis acquired by the

Table 9.2 Diseases encountered at release areas by translocated wildlife[a]

Species translocated	*Source*	*Disease encountered at release site*	*Release area*	*Source of pathogen*	*References*
Koala	Victoria, Australia, w/c Tick paralysis, vector: *Ixodes* spp.		Victoria, Australia	Toxic agent in tick saliva	–
Lechwe	Zambia, w/c	Heartwater, vector: *Amblyomma* spp.	Zambia	Enzootic in local wild ruminants	Pandey, 1991
Caribou	Eastern USA, w/c	Cerebrospinal nematodiasis	Ontario	White-tailed deer	Anderson, 1971
Caribou	Quebec, w/c	Cerebrospinal nematodiasis	Nova Scotia	White-tailed deer	Anderson, 1971

Arabian oryx	USA, c/b	Botulism	Oman	Enzootic in Oman	Stanley Price, 1989
Muskrat	Canada, w/c	Tularaemia	USSR	Water voles	Pavlovsky, 1966
Golden lion tamarin	USA, c/b	'Disease'	S.E. Brazil	?	May and Lyles, 1987
Hawaiian goose	UK, c/b	Avian pox vector: mosquitoes	Hawaii	Local birds	Kear, 1977; Kear and Berger, 1980; Berger, 1978
Brush-tailed possum	Tasmania, w/c	Bovine tuberculosis	New Zealand	Domestic cattle	Hickling, 1991
Black rhinoceros	Highland Kenya, w/c	Trypanosomiasis	Lowland Kenya	Tsetse flies, local wildlife and domestic stock	Mihok *et al.*, 1992

[a] w/c = wild caught; c/b = captive bred.

accidental ingestion of infected terrestrial molluscs while grazing (Anderson, 1971).

Botulism is enzootic in sheep and goats in Oman, and Arabian oryx bred in captivity under zoo conditions in the USA have died of this disease when they were released into the desert in Oman (Stanley Price, 1989).

When muskrats (*Ondatra zibethicus*) were translocated into the Soviet Union in the 1930s to augment the population of local fur-bearing rodents, they multiplied rapidly. Tularaemia was, at the time, enzootic in the local water voles (*Arvicola amphibius*) which shared the wetlands with the translocated muskrats. Muskrats are highly susceptible to tularaemia and there ensued a massive epizootic of the disease in their expanding population. Tularaemia is an unpleasant zoonosis and it was not long before the disease affected the human muskrat trappers. In this case, a disease which previous to the muskrat translocation and population explosion had been enzootic amongst the water voles and sporadic in the human population, assumed epizootic proportions and became a serious health hazard to humans and wild animals alike (Pavlovsky, 1966).

Failure of the Hawaiian goose (*Branta sandvicensis*) to increase in numbers after reintroduction may have been due to its occupancy of only the higher altitudes of its historical range. Breeding had formerly occurred at lower, more suitable altitudes but these were subsequently infested with introduced mosquitoes, vectors of avian pox virus to which the goose is very susceptible (Kear, 1977).

Black rhinoceros translocated from tsetse fly-free highland Kenya to tsetse fly-infested areas in lowland Kenya have become infected with trypanosomiasis due to *Trypanosoma brucei* and *T. congolense* (Mihok *et al.*, 1992).

Many years ago brush-tailed possums (*Trichosurus vulpecula*) from Tasmania were introduced into New Zealand to establish a new species of fur-bearer. The animals found an empty niche and in spite of heavy trapping for the fur trade, multiplied greatly. Then the demand for possum fur collapsed and it became no longer economic to trap the animals. There are now over 70 million possums in New Zealand and between 3% and 30% of them are said to be infected with bovine tuberculosis, a disease which does not occur in possums in Tasmania (Hickling, 1991). The economic consequences for New Zealand's livestock industry, including deer farming, need hardly be stressed.

On occasion, an avirulent, inapparent infection, when translocated along with a wild animal host, becomes virulent on passage through a new host in a new environment. An example of this occurred when some wild rabbits (*Lepus cuniculus*) were sent from England to South Africa. A small number (5.7%) of rabbits in England carry an avirulent strain of toxoplasma. On reaching South Africa these animals infected the local rats (*Rattus natalensis*) and the passaged toxoplasma assumed a virulent form.

9.5 MINIMIZING THE RISKS

The success of potentially expensive translocation projects depends to a large extent on the ability of wildlife biologists and their veterinary advisers to assess and evaluate the suitability of the release site chosen, and the ability of the translocated animals to colonize the area and establish a viable breeding population.

In this context the veterinary implications of translocation projects are proving to be of great importance, and failure to carry out adequate preliminary investigations has already resulted in several expensive fiascos. Worse still, the introduction of destructive parasites and disease agents into often naïve resident populations has had profound negative consequences. It is therefore important that systematic veterinary investigations be carried out prior to selecting the release site so that all ecological risks can be assessed in advance and if necessary appropriate modifications in the reintroduction plans can be made.

9.5.1 Veterinary intervention at the source of the founder stock

It is extremely important to establish a dialogue at an early stage in the planning of the wildlife translocation project with the director of veterinary services (DVS) of the recipient country. Many DVSs, especially those of countries where domestic stock raising is an important economic activity and where export markets for beef are dependent on stringent disease control regulations, are becoming very reluctant to permit the import of wild animals from outside their region. African countries like Zimbabwe, Namibia, Botswana and South Africa are now free of many of the diseases which until recently severely constrained their livestock industries. In some cases the whole country may be free of certain diseases, and in others disease-free zones are assiduously maintained. The DVSs of these countries will make a risk assessment of each application to translocate wild animals into their areas of jurisdiction. They will base this assessment on their knowledge of the diseases likely to be carried by the species concerned, its country or area of origin and the standards and ability of the veterinary services in that country to report and control disease outbreaks. For this reason it is wise to consult the veterinary authorities at a very early stage so that their attitude to the proposed translocation can be ascertained. Some DVSs may require the imported animals to be confined in fenced enclosures after release and may not allow any possibility of stray animals reaching commercial farming areas. Some DVSs may permit the import of wild species only after stringent quarantine has been undergone in the country of origin. In other cases the DVSs may not only require that the quarantine premises be approved by their departments but that any biological samples shall be collected under the supervision of their officers, and that subsequent diagnostic tests shall be conducted at their departments' laboratories. Furthermore it is becoming a common practice for the DVS to require a further period of quarantine, during which some of the diagnostic tests may be repeated, after the imported animals arrive in the destination country.

These regulations and requirements are likely to become more stringent as international animal disease control policies develop. Their impact on the future prospects for intercontinental animal movements will be considerable. Even within countries, the distribution of disease agents is irregular and the local movement of animals, both wild and domestic, is often subject to strict veterinary regulation.

The criteria for the selection of founders for a reintroduction attempt have been discussed by Stanley Price (1989). In this review, disease risk is considered to be a major factor in assessing the wisdom of population reinforcements and the case is cited where at least one release of rehabilitated orang utans (*Pongo pygmaeus*) in Indonesia was cancelled because some were found to be infected with human tuberculosis (Jones, 1982). Tuberculosis and human herpesvirus have recently been found to infect orang utans translocated from Taiwan, where they had been held captive, to Malaysia, where it was planned to release them back into the wild. These animals are now (1992) being held in isolation until an appropriate solution is found.

After selection of the founders has taken place, the group must be placed in quarantine for a minimum of 30 days (some countries may require 60 days). The site of the quarantine station may be at the zoo, wild animal park or ranch where the animals were raised, or it can be at a distance if animals from several separate sources are to be combined to make up the herd or group. The main requirement is that the group should be completely isolated from its own and related species (if possible from all species), with the necessary exception of a limited number of human attendants.

The veterinary import requirements of the receiving country should be known so as to ensure that the period of quarantine and any special screening requirements can be met.

The quarantine premises must be appropriate for the species concerned and for the epidemiological situation, e.g. mosquito-netted enclosures will need to be provided where vector-borne diseases are to be excluded. The facilities must allow adequate visual access and permit clinical examination, sampling and, if necessary, chemical immobilization. Isolation from all possible sources of infection must be absolute and human attendants must be screened for transmissible diseases, e.g. tuberculosis. Early in the quarantine period disease screening procedures such as the following must be instituted.

(a) The breeding history and clinical records of the founder stock (if available) will be examined. Since founders gathered at a quarantine station may have their origins in several zoos or captive breeding establishments this may be a tedious process, but it is important because some zoos have a poor health record, especially for some infectious diseases such as tuberculosis, and this will influence the interpretation of subsequent tests.
(b) The stock should be given a thorough visual clinical examination during which the veterinarian will look for aberrant behavioural traits as well as clinical signs. When the animals are released and exposed to the selective forces of their wild environment any physical or behavioural abnormality will reduce their chances of survival. This is particularly important when the founders are derived from captive-bred stock. Only very rarely are animals that are habituated to humans suitable for a reintroduction project.
(c) Local and regional disease patterns in the source area must be assessed, if necessary referring to the Office International des Epizooties (OIE), and the Food and Agriculture Organization (FAO)'s *Animal Health Yearbook*, and for Europe, *Faune Sauvage d'Europe, Informations Techniques des Services Veterinaires*, for details of disease occurrence in the country concerned. The advice of the national DVS and of local veterinarians, wildlife biologists and livestock farmers should also be actively sought. On the basis of these enquiries and on a knowledge of the common pathogens of the founder species, the veterinarian will prepare a screening protocol that will also take into account any disease problems which may have been associated with previous translocation of the taxon and any special pathogens stipulated by the international or national veterinary authorities in the release country.
(d) Before undertaking the screening procedures all founder animals should be permanently marked by ear tag, tattoo and/or electronic micro-chip.
(e) The collection of biological samples from the founder animals is best carried out early in the day so as to allow for same-day processing and swift dispatch to the laboratory.
(f) The samples should be sent to internationally recognized reference laboratories. These institutions have the requisite expertise and will be using the most up-to-date tests. It may be an advantage to send replicate samples to different laboratories and to ask the laboratories for their interpretation of the results.

(g) Laboratory procedures for the direct detection of evidence of infection will include:
- clinical haematology;
- blood smears for haemoparasites (for trypanosomes the buffy coat of a microhaematocrit tube may be examined);
- bacterial cultures;
- viral cultures;
- faecal egg counts;
- faecal larval culture (especially for lungworm);
- urinalysis (and examination of urine for kidney worm eggs);
- molecular biological techniques, especially nucleic acid hybridization with specific probes, are robust, highly sensitive and increasingly available.

(h) Laboratory procedures for the indirect detection of evidence of infection will include demonstration of antibodies. The presence of antibodies to a specific pathogen in a single serum sample at a determined level of dilution (titre) reveals that the animal has been exposed to infection by that pathogen. This indicates either natural infection or, in the case of zoo or captive-bred stock, the possibility of prior vaccination. Serial serum samples showing an active rising antibody titre are required to determine whether an animal is incubating or recovering from an infection. Depending on the pathogen concerned, further tests may be needed to differentiate whether an animal whose serum contains specific antibodies is likely to be an infectious carrier of the disease.

In some cases it may be desirable to recommend a total ban on the importation (into a 'clean' country) of certain species. For example, Greenwood and Cooper (1982) drew attention to the likely danger presented by prairie falcons (*Falco mexicanus*) and great horned owls (*Bubo virginianus*), both of which are commonly infected with falcon herpesvirus and suggested that the importation of these species from the USA into the UK should be restricted or banned.

9.5.2 Veterinary intervention at the proposed release site

Where possible, the veterinarian should visit the proposed release site during the planning phase (and well in advance of the translocation) to carry out the following.

(a) Consult with the directors of veterinary and wildlife services of the recipient country. This is important to avoid misunderstandings and to find out whether there are any special tests or certifications required by these directorates.

(b) Determine which diseases of wild and domestic animals and what vectors and reservoirs are enzootic in contiguous territories. A study of local seasonal wildlife migrations and domestic livestock trade routes is also relevant.

(c) Consult with regional veterinary authorities about the prevalence and incidence of local diseases.
(d) Check with the local diagnostic laboratory on the results of recent domestic animal disease surveys.
(e) Ascertain the degree of contact likely between the translocated animals and domestic stock, wildlife and humans.
(f) Decide if vaccination programmes of the national veterinary department are adequate to provide a protective immune barrier for susceptible wildlife against such diseases as rinderpest and foot and mouth disease for ungulates, and rabies for carnivores.
(g) Survey the proposed release site and assess the likely presence of disease vectors and disease foci. The overall suitability of the release site may be greatly influenced by these factors.
(h) If possible, survey and sample selected wild species and domestic stock at the release site and screen for appropriate suspected pathogens.
(i) Investigate availability of suitable vaccines to protect founder stock against local diseases identified during these investigations.
(j) Check if any mineral or trace element deficiencies or excesses occur in the release area. If supplementary feeding is planned, foodstuffs, especially hay, may have to be transported over long distances from a different ecozone and should be checked for toxic substances (e.g. selenium), for ticks and other parasites and for agricultural pesticides.
(k) Check the release area for the presence of poisonous plants which may be unfamiliar to translocated animals, especially in the dry season.
(l) Find out why the species to be translocated is absent from the area. Have previous translocations failed? If so, was disease a reason?
(m) Prepare a report and send copies to the directors of veterinary and wildlife services of the recipient country.

9.5.3 Pre-release planning

The well-tried strategy of confining an integrated social group in a large open enclosure near or actually in the release area for a considerable period before release has a number of advantages (Stanley Price, 1989).

(a) Social groupings can be adjusted and become established.
(b) The period of confinement acts as an extended quarantine during

which the clinical signs of infectious or contagious disease may be detected.

(c) Contact with domestic stock and other wildlife can be excluded, thus reducing the chance of disease transmission.

(d) The foundation stock can become acclimatized to local conditions.

(e) Handling can be minimized but if necessary can be achieved with little stress if the enclosure is equipped with well-designed facilities for restraint.

Once the founder stock is released there will be little chance of protecting them against the challenge of disease agents. Nevertheless, successful attempts have been made in the Kruger National Park to vaccinate valuable roan antelope (*Hippotragus equinus*) against anthrax by darting them from a helicopter (de Vos, Van Rooyen and Kloppers, 1973), and the success of oral rabies vaccine baits for foxes (*Vulpes vulpes*) in Europe is now well known (Blancou *et al.*, 1988; Pastoret *et al.*, 1988).

9.6 INTERPRETATION OF SURVEY AND SCREENING RESULTS

The results of the investigation of the disease history of the source of the animals will help in deciding if any biological tests should be given priority. The finding that zoo-bred animals have been vaccinated against a certain disease, for example, leptospirosis, may be evidence that this disease is a problem in the zoo supplying the animals. Tuberculin tests for animals originating in a zoo where tuberculosis is enzootic need to be interpreted more carefully than where the disease is unknown. Bluetongue and epizootic haemorrhagic disease sera may cross-react, and virus isolation tests may be necessary to determine whether positive serology indicates carrier status for either virus. Zebra and wild African suids are symptomless carriers of African horse sickness and African swine fever viruses, respectively, and are therefore subject to strict international movement regulations. Elephants (*Loxodonta africana*) and aardvarks (*Orycteropus afer*) may have antibodies to these respective disease agents but their significance is unclear.

So great is the variety of mammals, birds, reptiles, amphibians and even invertebrates which may in future become the subject of translocation projects, that specialist advice will almost certainly be required for the interpretation of results of the investigations carried out at both the source of the founder species and the release site. International disease monitoring organizations such as the International Office of Epizootics (OIE); the Food and Agriculture Organization of the United Nations (FAO) and the World Health Organization (WHO) may all be consulted, as may the Veterinary Group (VG) of the Species Survival Commission (SSC) of the World Conservation Union (IUCN) and the veterinary services of the countries concerned.

9.7 VACCINATION OF FOUNDERS

The decision to vaccinate and the choice of vaccines will depend on the interpretation of the results of the surveys and investigations described in sections 9.5.1 and 9.5.2.

If there is a significant risk of founder stock being infected by serious pathogens at the release site then it may be advisable to vaccinate them. Two main types of vaccines are available: modified live vaccines which have reduced virulence for one or more species, and a range of 'inactive' vaccines which contain no viable organisms. The advantages of the live vaccines are that they grow in the host after inoculation, and usually induce a solid immunity similar to that found in animals which have recovered from the particular pathogen – if such an immunity occurs. Consequently vaccination with a live vaccine usually requires only one inoculation, which is a great advantage when dealing with wildlife, especially those that need restraint.

The disadvantage of live vaccines is that whilst they may have reduced virulence for some species they may remain pathogenic for others. For instance, live canine distemper virus vaccines have induced disease in susceptible mustelids, including the endangered black-footed ferret (*Mustela nigripes*). Live feline parvovirus vaccine (feline enteritis or panleucopaenia), which is innocuous for domestic cats, is thought to have caused disease in other felines. Although safe in most breeds of cattle, a live rinderpest vaccine prepared in rabbits caused significant mortality when tested in a variety of African artiodactyls (Brown and Scott, 1960). Fortunately the current cell-culture-derived rinderpest vaccine has proved both safe and efficacious in the wild species tested to date, including buffalo, eland and warthog, which are highly susceptible, and impala and oryx, which are less so (Rossiter, unpublished findings; Plowright, 1982; Bengis and Erasmus, 1988). The control of rabies in western Europe and elsewhere has been radically improved through the use of live vaccines administered orally to foxes and other wildlife in 'baits'. There is considerable variation in the susceptibility of different species to live rabies virus and an important concern was the risk that live vaccines, although innocuous for the fox or other target species, could prove virulent for other species which might incidentally scavenge the baits.

Thus for vaccination of founder stock live vaccines offer several advantages so long as they have been proven to be safe in the chosen species. Care should be taken to use the same strain of a particular vaccine that has been successfully tested in the species, since different manufacturers may use different strains. Unless they have been tested with the vaccine, it is best to avoid vaccinating pregnant females as some live vaccines, especially viral, can cause foetal death and/or abortion. Similarly some live vaccines may cause disease in very young animals and the manufacturer's instructions should be sought on this matter. If females have been vaccinated previously, then maternal immunity may reduce the immunogenicity of live vaccines administered to young animals (carnivores less than three to four months old, herbivores less than six months old).

Inactive vaccines carry little risk of infecting the host, although some toxoids can induce local reactions at the site of inoculation. However, since they do not replicate in the host, these vaccines usually require more than one initial application and may require frequent boosting. Whether this will prove to be the case with the inactivated canine distemper virus vaccines currently being investigated for the protection of harbour seals (*Phoca vitulina*) against phocid distemper remains to be seen (Visser *et al.*, 1989). These vaccines are often incorporated into alum or other 'adjuvants', which help stimulate the immune response but can also cause significant local reaction at the inoculation site.

A good example of the use of an inactivated viral vaccine to protect a captive-bred species prior to exposure to infection in the wild is afforded by the case of the whooping crane (*Grus americana*). The whooping crane, although never abundant in North America, became endangered largely because of habitat modification and destruction. To help recovery, a captive propagation and reintroduction program was initiated in 1966 at the Patuxent Wildlife Research Centre (PWRC) at Laurel, Maryland. However, in 1984, seven of 39 whooping cranes at PWRC died from infection by eastern equine encephalitis virus (EEE), an arbovirus that infects a wide variety of indigenous bird species, although mortality is generally restricted to introduced birds. Following identification of the causal agent, surveillance and control measures were implemented, including serological monitoring of both wild and captive birds for EEE viral antibody and assay of locally trapped mosquitoes for the virus. In addition, an inactivated EEE virus vaccine developed for use in humans was evaluated in the captive whooping cranes. Results so far suggest that the vaccine will afford protection to susceptible birds.

As part of the effort to restore whooping crane populations in the wild, three geographically isolated areas in the eastern USA are being considered as possible reintroduction sites. Since EEE viral transmission occurs annually in parts of eastern USA it will be important to survey for EEE virus and its mosquito vectors before selecting the release sites.

Although the die-off of whooping crane due to EEE virus was a set back for the captive breeding programme, it is now known that this disease must also be considered in establishing wild crane populations (Carpenter, Clark and Watts, 1989).

If there is an urgent need to vaccinate stock of unknown susceptibility to a live vaccine then an inactive vaccine should be used, if available. Failing this, the live vaccine could be tested in one or two founder stock, so long as these are isolated from the rest of the group.

Protective immunity may take from a few days to several weeks to develop, so immunization schedules should be carried out well in advance of translocation. This has the added benefit of acting as a further quarantine period lest the animals be incubating a field infection or be excreting the live vaccinal agent administered to them. It should be remembered that some live vaccines induce persistent infections, the vaccinates in effect becoming carrier animals which may pass on infection to vectors or susceptibles, although fortunately this is rare.

9.8 POST-RELEASE HEALTH MONITORING

Regular, systematic monitoring of the health and reproductive performance of translocated animals may provide an early warning of incipient disease problems. If the causes of disease or reduced productivity can be determined, a change in management may be indicated for the current or future translocations.

Provision for long-term monitoring is often included in translocation project plans but is seldom carried out once the released animals appear to be surviving in the wild. Post-release monitoring of the Hawaiian goose was recommended during the project design (Kear and Berger, 1980) but little took place. After the release of 1244 birds over 16 years on Hawaii and a further 391 on Maui Island, the status of the reintroduced population was unknown and the reasons for its limited success (believed to be associated with avian pox virus infection) could only be guessed (Berger, 1978).

9.9 DISEASE TRANSMISSION HAZARDS WITH CRYOPRESERVED GERMPLASM

In the not-too-distant future, translocation of wild animal gametes and embryos may become an accepted technique for minimizing the effects of inbreeding in small isolated populations. When genetic resource banking programmes become established, the disease implications of these activities will have to be addressed.

The diseases of domestic stock transmissible by semen and embryo transfer have been reviewed by Hare (1985) and Singh (1988). Disease transmission between different populations can be greatly reduced by embryo transfer, because an intact embryo collected from a diseased mother is often free from the bacterial and viral disease agent and so does not transmit the disease to the foster mother. The surrogate mother, too, may confer passive immunity to the offspring which develops from a transferred embryo. This passive immunity may be transplacental or via the colostrum, and can prove of value in producing offspring passively immune to local enzootic diseases.

If the results of research on domestic embryos can be extrapolated to wild animals, it would seem that the potential of embryos to transmit infectious disease is considerably less than that of semen or live animals. In fact, it remains to be established whether under field conditions disease transmission will ever occur via embryo transfer. Sufficient research has been carried out with embryos from bovine leukaemia virus-infected donors and foot and mouth disease virus donors to determine that these viruses will not be transmitted via embryos, provided they are handled properly (Singh, 1988). Research is at present in progress to determine whether similar conclusions can be reached for other pathogens.

Little, if any, research has been done on the disease transmission potential of cryopreserved germplasm derived specifically from wild animals. Semen clearly has great potential for the spread of infectious disease, but semen also provides an excellent means of disease control when it is collected and transferred artificially using strict aseptic techniques, because direct animal contact is avoided, the health status of the donor can be predetermined and, if the semen is frozen, aliquots can be tested for the presence of microorganisms.

Embryos have less potential for the transmission of disease than does semen. At the developmental stage at which they are usually transferred embryos are protected by a relatively thick capsule, the zona pellucida. For the embryo to transmit infectious disease it has either to be infected with a minimum infective dose or to act as a carrier through association of the disease agent with the zona pellucida. Preliminary evidence suggests that while many disease agents are removed or inactivated by washing or other treatments, a few are not. Current regulations to control disease transmission by embryos of domestic animals are based largely on the philosophy that if the donors are free from specified diseases, their embryos will be similarly free. Evidence to date suggests that this approach may be unnecessarily restrictive with regard to a number of diseases (Hare, 1985).

It is expected that the principles summarized here will be applicable to cryopreserved wildlife germplasm, but that while the results of current research on diseases of domestic animals transmissible by semen and embryo transfer techniques can probably be extrapolated to include cryopreserved wildlife genetic material, care will be needed to ensure that specific wildlife diseases are not transferred.

9.10 DISCUSSION

Many endangered species have been reintroduced into the wild after captive propagation. In at least six cases – Père David's Deer, Przewalski's horse (*Equus przewalski*), red wolf (*Canis rufus*), Arabian oryx, Guam kingfisher (*Halcyon cinnamomina*) and Guam rail (*Rallus owstoni*) – these species were extinct in the wild at the time of the reintroduction: in the case of Père David's deer, extinct for 800 years! A number of other species, notably the California condor (*Gymnogyps californianus*) and the black-footed ferret, are likely to be reintroduced into their native habitat as soon as captive numbers are considered adequate.

In the majority of translocation projects few systematic attempts are made to quantify and minimize the veterinary risks attending the translocation of the founder animals. Notable exceptions to this generalization have been the careful screening and preparation of the black-footed ferret for return to the wild, and the development of stringent veterinary protocols for the Arabian oryx destined for release in the deserts of Oman. The latter have now become so strict that the captive source (a zoological collection in the USA), from which founder and reinforcement stock have been obtained, is no longer able to comply.

The 'rescue', rehabilitation and return to the wild of sick and injured wild animals and birds is an activity which attracts the attention and money of the general public. These activities are often carried out by enthusiastic amateurs and can, in some cases, be hazardous. Animals which are sick may wander long distances from their home range, and birds on migration may alight many miles from their destination. If such specimens, when apparently restored to health, are released back into the wild, they may become serious foci of infection for local animals. Even more dangerous is the return of sick strays to their distant natural habitat after a period in captivity during which they may have been in contact with diseased animals of the same or related species from other sources.

Members of the veterinary profession with experience and expertise in wildlife disease need to be involved in the timely planning of translocation projects to ensure that appropriate quarantine and screening are carried out. In special cases the quarantine period may need to be extended, e.g. if tuberculosis is perceived as a special hazard. The disease conditions affecting the great range of taxa which may be the subject of future translocation projects are so varied that no simple guidelines can be drawn to cover all exigencies. Each case must be separately evaluated, taking into account all biological, ecological, geographical and epidemiological circumstances. Only then will the inherent risks in moving potential 'disease packages' across the world be minimized, and the chances of failing to establish a healthy new wild population significantly reduced.

In the longer term, the establishment of a database on infectious agents and diseases of wildlife would be a very important supporting activity. An expanding source of information, publishing bulletins of wildlife disease occurrence on a global scale, will assist veterinarians to evaluate the disease risks which may attend a translocation proposal.

ACKNOWLEDGEMENTS

Permission to reproduce this paper is given by the Director General of OTE.

REFERENCES

Allen, T.J. (1986) Evaluation of movements, harvest rate, vulnerability and survival of translocated raccoons in southern West Virginia. *Trans. Northeast. Sect. Wildl. Soc.*, **43**, 64.

Anderson, R.C. (1971) Lungworms, in *Parasitic Diseases of Wild Animals* (Eds J.W. Davis and R.C. Anderson), Iowa State University Press, Ames, Iowa, pp. 81–126.

Anon (1990) A bridge too far. *The Economist*, May 12, 1990, 48–9.

Aveling, R. and Mitchell, A. (1982) Is rehabilitating orang utans worthwhile? *Oryx*, **16**, 263–71.

Asmode, J.F. and Khoja, A.R. (1989) Arabian oryx captive breeding and re-introduction in Saudi Arabia, in *Proc. Captive Breeding Specialist Group Aridland-Antelope Workshop, San Antonio, Texas* (eds U.S. Seal, K. Sausman and J. Mikolai), pp. 109–25.

Bengis, R.G. and Erasmus, J.M. (1988) Wildlife diseases in South Africa: a review. *Rev. Sci. Tech. Off. Int. Epiz*, **7** (4), 807–21.

Berger, A.J. (1978) Reintroduction of Hawaiian geese, in *Endangered Birds: Management Techniques for Preserving Threatened Species* (ed. S.S. Temple), Univ. of Wisconsin Press, London, pp. 339–44.

Blancou, J., Pastoret, P.-P., Brochier, B., Thomas, I and Bogel, K. (1988) Vaccinating wild animals against rabies. *Rev. Sci. Tech. Off. Int. Epiz.*, **7**, (4), 989–1003.

Booth, W. (1988) Reintroducing a political animal. *Science*, **24**, 156–9.

Brambell, M.R. (1977) Reintroductions. *Int. Zoo. Ybk*, **17**, 112–16.

Brown, R.H. and Scott, G.R. (1960) Vaccination of game with lapinised rinderpest virus. *Vet. Rec.*, **72**, 1232.

Cade, T.J. (1988) Using science and technology to reestablish species lost in nature, in *Biodiversity* (eds E.O. Wilson and F.M. Peter), National Academy Press, Washington, DC, pp. 279–88.

Caldecott, J.O. and Kavanagh, M. (1983) Can translocation help wild primates? *Oryx*, **17**, 135–9.

Carpenter, J.W., Clark, G.G. and Watts, D.M. (1989) The impact of Eastern equine encephalitis on efforts to recover the endangered whooping crane, in *Disease and Threatened Birds* (ed. J.F. Cooper), ICBP Technical Publications No. 10, pp. 115–20.

de Vos, V., Van Rooyen, G.L. and Kloppers, J.J. (1973) Anthrax immunisation of free-living roan antelope (*Hippotragus equinus*) in the Kruger National Park. *Koedoe*, **16**, 11–25.

Flamand, J. (1990) An outbreak of tuberculosis in a herd of Arabian oryx. Diagnosis and management, in *Abstracts: VI Int. Conf. on Wildl. Dis.*, Berlin, p. 21.

Greenwood, A.G. and Cooper, J.E. (1982) Herpes virus infections in falcons. *Vet. Rec.*, **11**, 514.

Griffith, B., Scott, J.M., Carpenter, J.W. and Read, C. (1989) Translocation as a species conservation tool: status and strategy. *Science*, **245**, 477–80.

Hare, W.C.D. (1985) Diseases transmissible by semen and embryo transfer. OIE Tech. Series No. 4, Office International des Epizooties, Paris.

Hedger, R.S. and Condy, J.B. (1985) Transmission of foot and mouth disease from African buffalo virus carriers to bovines. *Vet. Rec.*, **117**, 205.

Heuschele, W.P. (1990) Tuberculosis in captive Arabian oryx, in *Abstracts: VI Int. Conf. on Wildl. Dis.*, Berlin, p 31.

Hickling, G.J. (1991) The ecology of brush-tailed possum populations infected with tuberculosis, in *Proc. Symp. Tuberculosis. Publication 132* (ed. R. Jackson), Massey Univ., Palmerston North, New Zealand.

Jacobsen, E.R., Gaskin, J.M., Brown, M.B. *et al.* (1991) Chronic upper respiratory tract disease of free-ranging desert tortoises (*Xerobates agassizii*). *J. Wild. Dis.*, **27** (2), 296–316.

Jones, D.M. (1982) Conservation in relation to animal disease in Africa and Asia. *Symp. Zoo. Soc. Lond.*, **50**, 271–85

Kear, J. (1977) The problems of breeding endangered species in captivity. *Int. Zoo Ybk.*, **17**, 5–14.

Kear, J. and Berger, A.J. (1980) *The Hawaiian Goose: An Experiment in Conservation.* T. & A.D. Poyser, Calton.

Kleiman, D.G. (1989) Reintroduction of captive mammals for conservation. *Bioscience*, **39**, 152–69.

Kock, R.A. and Woodford, M.H. (1988) Reintroduction of Père David's deer (*Elaphurus davidianus*), scimitar-horned oryx (*Oryx dammah*) and the Arabian oryx (*Oryx leucoryx*) to their native habitats – a veterinary perspective, in *Proc. Joint Conf. Amer. Assoc. Zoo Vets and Amer. Assoc. Wildl. Vets*, Toronto, Ontario, pp. 143–4.

May, R.M. and Lyles, A.M. (1987) Living Latin binomials. *Nature*, **334**, 642–3.

McCulloch, B and Achard, P.L. (1969) Mortalities associated with capture, translocation, trade and exhibition of black rhinos. *Int. Zoo. Ybk.*, **9**, 184–91.

Mihok, S., Munyoki E., Brett R.A. *et al.* (1992) Trypanosomiasis and the conservation of black rhinoceros (*Diceros bicornis*) at Ngulia Rhino Sanctuary, Tsavo West National Park, Kenya. *Afr. J. Ecol.*.

Nettles, V.F. (1988) Wildlife relocation: disease implications and regulations. *Proc. 37th Ann. Conf. Wildl. Dis. Assoc. Athens*, Georgia, p. 52.

Environmental Assessment Panel (1990) Northern Diseased Bison, Report of the Environmental Assessment Panel, 510–750 Cambie St, Vancouver, B.C. V6B 2P2, Canada.

OIE (1987) Epizootiological Information, Nos. ESP/87/1/117, ESP/87/2/119, ESP/87/3/121, ESP/87/4/127, ESP/87/5/142 and ESP/87/6/145, Office International des Epizooties, Paris.

Pastoret P.-P., Thiry E., Brochier B. *et al.* (1988) Diseases of wild animals transmissible to domestic animals. *Rev. Sci. Tech. Off. Int.* Epiz., **7**, (4), 705–36.

Pandey, G.S. (1991) Heartwater (*Cowdria ruminantium*) with special reference to its occurrence in Zambian wildlife. *Newsletter of the Centre for Tropical Veterinary Medicine (Edinburgh)*, **52**, 6.

Pavlovsky, E.N. (1966) The natural nidus of a disease as a pathobiocenose, in *Natural Nidality of Transmissible Diseases*, Univ. of Illinois Press, Urbana, pp. 13–15.

Philip, P. (1989) Tuberculosis in deer in Great Britain. *State Vet. J.*, **43**, 125.

Plowright, W. (1982) The effects of rinderpest and rinderpest control on wildlife in Africa. *Symp. Zool. Soc. Lond.*, **50**, 1–28.

Rosen, J. (1955) Norwegian domesticated reindeer in Greenland, In *Polarboken*, Norwegian Polar Institute, Oslo, pp. 147–59.

Rosen, J. (1958) Reindeer flies and reindeer lice. *Gronlandsposten*, **26**, 24.

Singh, E. (1988) Potential of embryos to control transmission of disease: a review of current research. Animal Disease Research Inst., Nepean, Ontario, Canada.

Stanley Price, M.R. (1989) *Animal Re-introductions: The Arabian Oryx in Oman.* Cambridge Univ. Press, Cambridge, UK, pp. 188, 119–201, 222–38.

Thing, E.M. and Thing, H. (1983) Cow–calf behaviour in West Greenland caribou on Sdr. Stromfjord summer range. *Acta Zoologica Fennica*, **175**, 113–15.

Trust, T.J. (1986) Pathogenesis of infectious diseases of fish. *Ann. Rev. Microbiol*, **40**, 479–502.

Visser, I.K.G., van der Bildt, M.W.G., Brugge H.N. *et al.* (1989) Vaccination of harbour seals (*Phoca vitulina*) against phocid distemper with two different inactivated canine distemper virus (CDV) vaccines. *Vaccine*, **7**, 521–6.

Wilson, E.O. and MacArthur, R.H. (1967) *The Theory of Island Biogeography* (ed. E.O. Wilson), Princeton Univ. Press, Princeton, NJ, pp. 203.

10

Legalities and logistics of meta-population management

A.M. Dixon
Zoological Society of London, UK.

10.1 INTRODUCTION

Much of the philosophy propounded for meta-population management, and the long-term plans for species survival depend fundamentally on the ready transfer of individuals, gametes or embryos between populations or breeding groups, both wild and captive. But the question must be asked whether it will actually be possible to execute such transfers on the scale envisaged, or whether the legal requirements and practical logistics inherent in the process will become so cumbersome and expensive that the process can only be realistically undertaken for a limited number of species.

There tends to be a presumption that, when a transfer is clearly based on scientific and conservation principles, it should be a straightforward matter to obtain the necessary official sanctions. In fact the situation is not so simple. The movement of animals and their products is regulated to control potentially deleterious trade in species and to protect domestic livestock and people from disease. Both objectives are valid but sometimes the practice is found to hamstring legitimate work, causing delays which are frustrating and expensive for the institutions involved, not to mention detrimental to the goals of the meta-population programme.

Creative Conservation: Interactive management of wild and captive animals.
Edited by P.J.S. Olney, G.M. Mace and A.T.C. Feistner.
Published in 1994 by Chapman & Hall, London.
ISBN 0 412 49570 8

10.2 CITES

The Convention on International Trade in Endangered Species (CITES) concluded in Washington, DC on 3 March 1993, entered into force in July 1975. (For text of CITES, see EC, 1982.) Over 100 states are now parties, and it is undoubtedly one of the most effective international wildlife conservation treaties in existence. However, the Convention is only as strong as its implementation by the individual member states and in turn this implementation depends on the enactment and enforcement of appropriate national legislation. This imposition of this legislation, as well as the fulfilment of the requirements of CITES, is the responsibility of the government-appointed Management Authority acting with advice from a similarly appointed Scientific Authority. With few exceptions, the Management Authority will be asked to approve every single international transfer of a threatened species or its product (i.e. sperm, blood, ova, etc.); understanding the constraints under which they operate is therefore an essential part of any long-term species plan.

CITES itself is not intended to eliminate trade; it is essentially based on the premise of sustainable utilization made possible by the continued survival of species. Trade is defined as 'export, re-export, import and introduction from the sea' (Article 1(c), see EC, 1982). Appendix I (see EC, 1982) includes all those species 'threatened with extinction which are or may be affected by trade' (Article II.1). Trade in these species 'must only be authorized in exceptional circumstances'. Fortunately, captive-bred specimens of Appendix I species are treated as Appendix II species (Article VII.5) which affords a less stringent degree of protection (Article II.2), and there is a recognition that trade in Appendix I species can be demonstrated to be of benefit to the species involved if it is part of a coordinated breeding programme. The difficulties arise when dealing with wild-caught specimens for which documentary proof is required that the transfer is not only beneficial to the species but, more problematic, not deleterious to the species in the wild. Counterpart management authorities across the world are often under-resourced, may not share a common language and simply may not be able to provide the detailed documentation required. The permit process can then become a protracted nightmare of misunderstandings, delays and expense.

As an example, all five species of rhino are listed on Appendix I, and the rigorous permit procedure which is aimed at eliminating the illegal trade in their horns applies with equal force to the movement of one live animal for breeding purposes. The shipment of a single male northern white rhino (*Ceratotherium simum cottoni*) from Sudan to the USA in 1990 took two years to accomplish despite the very clear objective of removing this animal from a solitary existence and placing him where he could be expected to reproduce. In part, the delays were due to the sheer logistical difficulties of getting an animal of that size out of Khartoum, but substantial problems were encountered in the permitting procedure. As one of only about 40 survivors of this subspecies in the world, the animal in question was of critical importance to the breeding programme, and its transfer was endorsed by the Captive Breeding Specialist Group of the World Conservation Union Species Survival Commission (IUCN/SSC). But it was wild-caught, Sudan was in the throes of civil war and drought, English is not their first language and the legalities of shipping one rhino to the USA were not a high priority for a small struggling management authority. In the end the rhino was moved only because the receiving institution was prepared and able to make the necessary commitment of money and time to send people to Khartoum to sort out the paperwork.

In fairness it needs to be appreciated that the remit of CITES is extensive, within which the movement of animals or their derivatives as part of an organized meta-population programme forms only a small part. If CITES is to be effective in controlling international deleterious trade, there can be only a limited degree of flexibility in its interpretation, and both sides must learn to work within those interpretive boundaries.

10.2.1 National legislation

In addition to the text of the Convention, it is important to understand the legislation of the individual parties enacted to implement the Convention within national borders. Probably the most extreme example is the European Community where the Commission in Brussels draws up the regulations to apply throughout the Community but then each member must enact its own laws to implement them.

Although Regulation EEC 3626/82 (EC, 1982) as last amended by Regulation EEC No. 197/90 (EC, 1990) is still in operation, a proposal for a new Regulation laying down provisions with regard to possession and trade in specimens of species of wild fauna and flora is currently under review by the member states (EC, 1992). This Regulation would be the legislative instrument for the enforcement of CITES within the Community, but its strictures may apply to other species which for one reason or another are considered by the member states to be of concern. Annex A accordingly proposes including all CITES Appendix I species and 'any species which is or may be in demand for utilization in the Community or for international trade and which is either threatened with extinction or so rare that any level of trade would imperil the survival of the species' and 'species which under the provision of other

Community legislation on the conservation of wild fauna and flora benefit from a prohibition on trade or taking'. Annex A therefore includes protected European species, including most bats. But it also includes a number of primates and felid species which are listed in CITES Appendix II, wild dogs (not on CITES at all), pygmy hippos (CITES II), 12 species of falconiformes which are in CITES Appendix II and all Cetacea – just to name a few differences from CITES. In theory, these species are listed in response to the definitions cited above, in short to get a grip on any trade that may threaten them. In practice, however, other less-defined considerations are often at work.

In particular, concern for animal welfare exerts a pervasive political pressure, the results of which are increasingly to be seen in CITES and CITES-related legislation despite the fact that CITES is not a welfare treaty. For example, Article 16 of the new Regulation would require that 'any movement within the Community of a live specimen of a species listed in Annex A . . . shall require prior authorization from a management authority of the Member State in which the specimen is registered'. It states that 'Such authorization will only be given when the competent scientific authority . . . has advised in writing that the intended accommodation is adequately equipped and suited to the biological and, in the case of an animal, behavioural needs of the species' (EC, 1992). This kind of terminology causes concern because although there is agreement in principle, there is also an inherently dangerous vulnerability to subjective interpretations of behavioural needs. Endless and fruitless debate over whether monkeys need to have bark on their swings can in the meantime hold up a permit necessary for a whole breeding programme.

This is especially true since, unfortunately, increased regulations often do not result in increased resources to the responsible government authorities. Often permit offices lack necessary skills and experience to fulfil their responsibilities efficiently and with confidence. The result can be a serious loss of time and money, either of which can have unfortunate implications for a breeding programme. The provisions proposed currently do not apply to specimens of species 'that are commonly bred in captivity', nor is it proposed that they apply to species listed on the other annexes (EC, 1992), but the trend for increasing the legal obligation of the regulatory authorities to oversee the practical care and management of indivdual species is clearly established, and one would be well advised to recognize the inherent long-term implications now.

It would be logical to assume that good zoos would be in a strong position to advise on the welfare of exotic species. However, in Europe (and increasingly in the USA) there is a forceful anti-zoo movement which prefers to ignore the practical realities of an animal's existence – in the wild, in a zoo or in a factory farm. On an international scale CITES is being used to accomplish welfare objectives not only with regard to animal accommodation but also shipment. The stated objective is to reduce mortality, and while the responsible captive breeding community concurs wholeheartedly with this philosophy, the details of the regulations do no necessarily reflect the best interests of the animals themselves, nor is there any guarantee that they will be appropriately enforced; there are inumerable laws regarding the transport of domestic stock but the pages of the local animal protection society magazines in Europe, the USA, Australia and increasingly in Asia illustrate how these are ignored and how they are open to interpretation at the discretion of a customs or ministry inspector at the local dock or airport.

10.3 HEALTH REGULATIONS

Conservation-based legislation is only a part of the legislative process involved in animal movements – the other pertains to health. In the long run, it may well be that it is questions of disease control which cripple meta-population management plans. With the substantial backing of massive domestic livestock industries and human health concerns, agricultural ministries are responsible for the enforcement of a formidable number of health regulations which, while largely aimed at domestic stock, nevertheless encompass exotic species. Efforts to control swine fever in domestic pigs, for example, mean that no members of the Suidae family can at present be imported into the UK. By maintaining the animals under strictly controlled conditions, the sanitary requirements can be met but once again this takes time and money – as demonstrated by New Zealand's requirement that all animals imported for zoological gardens must 'have been domiciled in the zoological garden during the 12 months immediately preceding shipment or since birth' (New Zealand Ministry of Agriculture and Fisheries, pers. comm.).

Nor should the mistake be made of thinking that the health regulations apply only to wild-caught animals or animals coming from 'less developed' countries. Recent plans to send Arabian oryx back to Oman as part of a reintroduction programme were severely undermined because individuals did not clear for bluetongue, despite many different sets of animals being tried, each one compromising further the ideal genetic set for reintroduction. Although enough animals were eventually identified, this kind of set-back can delay if not totally destroy a reintroduction programme and thereby call into question the whole concept of individuals, gametes, or embryos free-flowing between populations. This limitation urgently needs to be taken into consideration right at the beginning of any species plan and serious questions need to be asked about whether meta-population programmes based on genetic, demographic even behavioural grounds are not going to der on the rocks of health regulations. If the Arabian oryx pro-ıme with all its resources and financial support had such difficulties, what will happen in instances involving less glamorous but equally valid species in less affluent circumstances?

10.4 CONCLUSIONS

In fact, there are several things that need to be done. The first is to ensure that the captive breeding community is part of the consultative process right from the beginning of any legislation. The practical consequences of many of the regulations are simply not appreciated by the health and conservation regulatory authorities.

The second is to recognize that the zoo and captive breeding community are the experts – in animal welfare, disease and transport, in short on the practical management of species for conservation. This means active and constant participation in such fora as CITES and IATA, and with regional and national regulatory authorities.

The third is for active conservationists to follow legislative developments closely, and to comment and object where appropriate. This requires comprehensive and up-to-date information but it is particularly important that the views of the responsible modern captive breeding world be registered.

The fourth, and most important, is to incorporate a recognition of logistic and legislative limitations in every species plan from the outset. This means realistically assessing exactly what is going to be possible in terms of disease controls, movement of animals, time and costs. Without this sort of assessment, the scientifically elegant theories of meta-population management promulgated in the conference halls and laboratories may well prove impractical on the ground.

REFERENCES

EC (1982) Official Journal of the European Communities. Council regulation (EEC) No. 3626/82 of 3 December 1982 on the implementation in the Community of the Convention on international trade in endangered species of wild fauna and flora. No. L 384, vol. 25, 31.12.1982, Office for Official Publications of the European Communities, Luxembourg.

EC (1990) Official Journal of the European Communities. Council Regulation to amend (EEC) No. 3626/82. No. L29, 31.1., Office for Offical Publications of the European Communities, Luxembourg.

EC (1992) Official Journal of the European Communities. Information and Notices Proposal for a council regulation laying down provisions with regard to possession of and trade in specimens of species of wild fauna and flora. COM(91)448final-SYN 370. 3 February 1992, Office for Official Publications of the European Communities, Luxembourg.

11

Training in zoo biology: two approaches to enhance the conservation role of zoos in the tropics

D.R. Waugh
Jersey Wildlife Preservation Trust, Channel Islands, UK
and
C. Wemmer
Conservation and Research Center, Front Royal, Virginia, USA.

11.1 INTRODUCTION

Twenty years ago, at the First World Conference on Breeding Endangered Species in Captivity, Gerald Durrell made a particular observation about the need to institutionalize training of zoo personnel so as to cultivate higher standards and create a world-wide professionalism in the management of captive animals (Martin, 1975). His observation was made truly in the spirit of wanting to transfer resources from the 'haves' to the 'have-nots' in the common interest of preventing extinctions and maintaining global biodiversity. Its genesis lay in Durrell's many years' experience of the virtual absence of any goals in the keeping of wild animals in captivity in less-developed countries. It should be emphasized that inadequate standards of captive animal management in any part of the world would not have been spared his

Creative Conservation: Interactive management of wild and captive animals.
Edited by P.J.S. Olney, G.M. Mace and A.T.C. Feistner.
Published in 1994 by Chapman & Hall, London.
ISBN 0 412 49570 8

criticism. However, Durrell did keenly sense that zoos in developing nations were handicapped, not just by lack of funds but, as importantly, because they were isolated from the accumulated knowledge forming the foundation of an emerging professionalism across the zoo community of the developed countries. In other words, the discipline of zoo biology was virtually absent from the tropics, and Durrell considered it essential that zoos in the richer nations should make the efforts and bear the costs to provide the training necessary to effect a transfer of knowledge and skills for improved animal management.

Returning to the kernel of his argument though, the need was to place the captive breeding of endangered species at the forefront of the training initiative, thereby to simultaneously increase both the sense of responsibility and the effectiveness of government departments mandated to protect indigenous wildlife. Thus, the contenders for training might be personnel from forestry and wildlife departments with specialized captive breeding stations or wildlife confiscation centres, in addition to the traditional zoo employee. People in these institutions would largely have been starved of the widely dispersed and diverse literature on zoos and captive animal management, not just the more obscure and specialized but also the classic treatments of progressive zoo operations (Hediger, 1970; Durrell, 1976). Even texts giving general statements on the current state of zoos, e.g. Bendiner (1981), UFAW (1988), and the benefits that can be derived from them, have not been widely available to people working with captive animals in the tropics. Despite a bias towards the zoo scene in the developed nations, these works could be expected to provide cues for self-improvement, but in their absence the training imperative merely becomes stronger.

This is clearly reinforced by more recent initiatives amongst zoos, prevalently in the high-income nations to date, to define priorities for conservation of species and to rapidly expand cooperation to make practical gains. As an example, the Captive Breeding Specialist Group (CBSG) of the World Conservation Union Species Survival Commission (IUCN/SSC) has a pivotal role in linking the work of zoos with field conservation efforts in the IUCN and elsewhere, in addition to helping bring zoo closer to zoo (Foose, 1991). Foose (1988) has spoken of the world as a megazoo, and of the importance of the CBSG in helping to focus the efforts of individual zoos onto agreed conservation priorities. To encompass the world's zoos in a coherent strategy implies a preparedness of the individual institutions, in which training of personnel has a critical part to play.

11.2 TRAINING INITIATIVES: CAPTIVE AND FIELD

In the developed nations, the perceived need to provide systematic training for at least the zoo keeper level has long been in existence – e.g.

at the Calgary Zoo (Karsten, 1974) – but expressed with varying degrees of enthusiasm. Certainly, in-house training of keepers and staff in other positions has been perceived as an investment for the future in North America (Aum, 1982; Carr, 1990; Curtis, 1982; Kohn, 1990; Koontz and Hutchins, 1990) and Europe (Jones, 1985; T. Kauffels, pers. comm.; C. Schmidt, pers. comm.). Waugh (1988) has reported elsewhere on several training programmes with similar goals.

However, what of training efforts outside North America and the other developed nations? Mittermeier (1986) stated one of the recommendations of the IUCN/SSC Primate Specialist Group as 'To assist in developing training programmes in which zoo staff teach essential zoo technologies to our colleagues and counterparts from developing countries.' This was certainly a reinforcement of what had been proposed by Gerald Durrell (Martin, 1975; Durrell, 1976), and was made at a time when training help from zoos in developed countries varied from one person/single visit efforts, to coordinated and repeatable programmes. Two examples of the latter are the training programmes offered by Wildlife Preservation Trust (WPT) (Waugh, 1983) and the National Zoological Park Front Royal Conservation and Research Centre (NZP) (Wemmer, 1990; Wemmer, Pickett and Teare, 1990), and these are further reported below. The WPT and NZP programmes in particular have helped to spawn within-country and regional training courses and workshops in developing nations, which indeed should be the desired outcome of any North–South aid.

Other instances of training oriented to developing countries have been reported by Calvo (1989), Desai (1992), Harris (1992), Krausman and Johnsingh (1990), Luthin (1987), Vecchio (1990), Walker (1987) and B. Hastings (pers. comm.). These are examples which fit into a long-standing goal of the World Conservation Strategy (IUCN/UNEP/WWF, 1980) and a more recently stated CBSG goal to establish an international conservation and captive breeding training programme network (Hage, 1992). They can be aided by single-institution initiatives, such as accords with government wildlife departments (Mallinson, 1988), bilateral or sister-zoo arrangements (e.g. Christman, 1989) and specific support groups (e.g. Anderson, 1989; Anderson, 1990; Skrei and Lieberman, 1990). The major thrust of all these initiatives is to broaden access to sources of training, and to emphasize that training programmes should be designed to improve local capacities to solve problems (Freese and Saavedra, 1991; Rudran, Wemmer and Singh, 1990; IUCN/UNEP/WWF, 1991).

11.3 TRAINING AND BIODIVERSITY

Why have this emphasis on the provision of training and transfer of resources to zoos in tropical countries? This seems to ask the obvious but would bear further examination. Demonstrably there is no shortage of current literature dealing with the concentration of biological diversity within the equatorial latitudes, e.g. Wilson (1988), and further still a dissection of which countries hold most species, the so-called megadiversity countries (Mittermeier, 1988). Mittermeier (1988) states that 50% of the world's total biodiversity is found in only six countries and 80% in only 12, and he lists Brazil, Colombia, Indonesia, Madagascar, Mexico and Zaire as a starting point. At the same time, however, there is compelling evidence of the attrition of this biodiversity, and informed comment on the negative social and economic effects to human communities which are often already impoverished (Myers, 1979; Ehrlich and Ehrlich, 1981; Norton, 1986; McNeely, 1988). These and many other commentators have also suggested measures to maintain biological diversity and to mitigate the worst effects of environmental misuse.

If we hope to have well-run zoos and captive breeding centres as a result of training, and expect those institutions to participate in conservation and recovery programmes for native fauna, those in the megadiversity countries take on a particular importance. This is amplified if the role of the zoo in educating the visitors about native fauna and conservation is taken into account, coupled with a knowledge of the annual visitor intake and size of the potential visitor pool. Table 11.1 gives some crude indicators of the relative potential of well-run zoos to educate people, especially in relation to the number of threatened species that might require captive breeding programmes. Although somewhat dependent on the efficiency of the listing authorities, the IUCN Red Data Book mammals (IUCN, 1990) gives a reasonable indication of the zoos' potential to do some good. This is clearer when taken together with primate diversity (Macdonald, 1985; Mittermeier and Oates, 1985), primates being chosen because they are invariably popular as exhibits in zoos. It is recognized that visitor attendance is related to several factors, not least cultural differences and relative mobility of the population. However the gross national product figures are presented not in this context but are intended to reflect how much money might be available to the zoos, given especially that the majority in the low income nations are government funded. This difference is shown clearly in Table 11.2 and, relative to government-owned zoos, there are significantly fewer in non-government ownership in developing countries ($X^2 = 114.2$, df = 1, $P < 0.001$).

It is the city zoos that assume importance in conservation education and awareness to large numbers of people. In just the Chinese zoos, which participated in a 1990 NZP Zoo Biology Training Course, alone the potential audience for conservation education programmes exceeds 40 million people a year (Beehler, 1990). Sinclair (1988) has presented some useful indicators on the potential of city zoos to respond positively to training of personnel (Table 11.3). More people visit the government city (municipal) zoos which hold more species, but with the encouraging sign that trained staff might individually have responsibility for fewer specimens. The breeding record of IUCN threatened mammals and birds is the lowest in these zoos, further indicating the potential value of training to help improve management and breeding. There is also evidence that mortality rate is sometimes significantly higher for certain taxa than in zoos in developed countries, which suggests deficiencies in nutrition and husbandry.

11.4 TRAINING IN ADVERSITY

The foregoing hints at the adverse conditions that beset zoos in developing countries. But for the problems to be most effectively addressed by training they must first be clearly defined. They should also not be taken in isolation from the obstacles that most wildlife departments face, because many of the problems are the same and some zoos are run by these departments anyway. The challenges facing these zoos and wildlife departments have been discussed in some detail elsewhere (Waugh, 1988; Rudran, Wemmer and Singh, 1990; Wemmer, 1990; Wemmer, Picket and Teare, 1990) and are reviewed here.

11.4.1 Lack of financial support

This is an obvious and pervasive problem which is predominantly related to a lack of appreciation by decision-makers of the immense potential of good zoos, even simply in a recreational sense, let alone with regard to nature and natural resources. Such a paucity of funds places a limitation on all operations, and in particular the prospect of a very low salary means that potentially valuable graduate staff will seek employment elsewhere. Allied to this, financial impoverishment has been seen to dissuade upper management from initiating in-house training programmes, and to shy away from offers to train staff overseas without an accompanying guarantee of external funding (Table 11.4).

11.4.2 Poor communication and access to information

There are recent encouraging signs that more regional zoological parks associations are being formed, and existing ones becoming more active. Developments such as these are essential to improve communication and information exchange within countries and regions, to promote learning from one another, to expand animal exchange, to develop cooperative breeding programmes and so on. Just as vital is to improve communication with zoos and wildlife departments in developed coun-

Table 11.1 Comparative statistics on human population, wealth and zoo attendance, biodiversity and rarity[a]

Country/region	*City (and zoo)*	*Annual attendance (thousands)*[d]	*City population (thousands)*[d]	*Per capita GNP (US$)*	*Number of IUCN/RDB mammal taxa*	*Number of primate species in country/region*
Argentina	Buenos Aires	3 000	10 728	2 640	23	4
Brazil	Sao Paulo	2 551	10 998	2 280	44	51
China	Beijing	11 000	6 800	330	34	16
Colombia	Medellin	650	2 121	1 240	31	27
Cuba	Havana (National)	–	2 096	–	10	0
Egypt	Alexandria	1 840	2 917	650	9	0
India	Bombay	2 999	8 243	350	38	11
Indonesia	Jakarta (Ragunan)	2 013	6 504	430	51	27–30
Madagascar	Antananarivo	–	406	180	63	28
Malaysia	Kuala Lumpur	1 050	920	1 870	28	17
Mexico	Mexico City (Chapultepec)	12 540	9 191	1 820	33	3
Morocco	Rabat (National)	321	1 287	750	12	1
Nepal	Kathmandu	–	235	170	17	2
Nigeria	Ibadan	247	847	290	20	23 } 33–4
Cameroon	Douala[e]	0	1 030	1 010		28–29 } 33–4

Peru	Lima	1 332	5 008	1 440	31	27
Philippines	Manila	–	1 631	630	11	2
West Africa	Principal cities[b]	–	532[b]	404[b]	26	16
Ghana	Kumasi	114	376	380	–	–
Zaire	Kinshasa[e]	0	2 444	170	32	29–32
Australia	Sydney (Taronga)	912	3 431	12 390	47	0
Europe	Principal cities[c]	928[c]	1 442[c]	13 263	18	0
France	Paris (Parc Zoologique)	878	2 189	17 830	3	0
Japan	Tokyo (Ueno)	6 120	8 156	23 730	8	1
South Africa	Pretoria	709	443	2 290	24	5
Spain	Barcelona	1 000	1 704	7 740	5	0
Former USSR	Moscow	3 509	8 967	–	18	0
USA	Washington, DC	3 300	617	19 780	65	0
	New York (Bronx)	2 078	7 000			

[a]Sources – mammals: IUCN (1990); primates: Macdonald (1985), Mittermeier and Oates (1985); city populations and GNP: Europa (1991); annual zoo attendances: Olney and Ellis (1989). Bars indicate no data available or compiled.
[b]Mean values calculated from 10 countries in West African Zoogeographical region.
[c]Mean values calculated from 25 European countries.
[d]Numbers rounded to the nearest thousand.
[e]Not listed as having zoos.

Table 11.2 Distribution of government and non-government zoos (source; Olney and Ellis, 1989)

Country	*Ownership*		
	Government	*Non-government*	*Both*
High income	327	283	610
Low income	229	20	249
Both	556	303	859

tries, and to gain better access to the kinds of information that western zoos take for granted. Diaz Matalobos and Barraza (1992), for example, have highlighted this continuing need for zoos in Latin America, and it is incumbent on the zoos and zoo associations in high-income nations to initiate and sustain the information flow. Much more of the technical literature presently only available in English should be translated to other languages, in particular Spanish, French and Portuguese.

11.4.3 Insufficient knowledge of applied ecology and zoo biology fundamentals

Applied ecology, zoo biology and zoo animal management, as scientific disciplines, are still insufficiently recognized and understood, even among zoo personnel of high rank. In the zoos, this is reflected in many ways, but especially in the paucity of properly planned captive breeding efforts, regionally managed cooperative breeding programmes and active participation in international species recovery plans. Unawareness of the importance of applied ecology and zoo biology among university graduates can bias them against seeking employment in zoos and wildlife departments, and thus helps to perpetuate the insufficiency. This is compounded by a more general misconception that the existence of these zoos is in no way connected to wildlife management and conservation in the wild. It is therefore important to demonstrate that a valid scientific connection exists, namely the integration of zoo biology and applied ecology. In particular, decision-makers and politicians need to be made aware of the application and significance of these disciplines to the improvement of resource management and economic advantages for local communities.

11.4.4 Politics and bureaucracy

Rarely are these two factors entirely absent anywhere, but too often they can be a serious encumbrance to progress in zoos and wildlife depart-

Table 11.3 Comparative statistics on city zoos of different ownership and type[a]

		N	*Annual attendance*	*City population*	*Number of species held*	*Number of specimens per staff member*	*Number of specialist staff*	*IUCN Red Data Book mammals/birds as percentage of their total stock*	
								*Held**	*Bred*
Government	Municipal	60	737 904	835 603	247	20.8	2.5	5.1	3.1
Non-government	Private	33	241 546	219 280	129	38.2	1.1	3.9	3.4
	Non-profit	29	331 131	619 441	201	21.3	2.7	5.1	4.7

[a]Modified from Sinclair (1988). Mean values are given; all are significantly different (ANOVA, $P < 0.05$) except *.

Table 11.4 Factors that have or would prevent decision-makers from absenting staff for training (N = 27) (unpublished data from JWPT survey 1990/91)

	Level of concern (%)		
Factor to consider	*Never*	*Possibly*	*Frequently*
Lack of funds	11.1	18.5	70.4
Central treasury will not fund	22.2	44.4	25.9
No time to find scholarship funds	22.2	40.8	37.0
Staff without qualifications	29.6	44.4	26.0
Staff without job experience	37.0	44.4	18.6
No prospects for staff to use training	55.6	44.4	0.0
Other staff discontented	85.2	7.4	7.4
Trained staff promoted to inappropriate job	63.0	33.3	3.7
Staff request training only for promotion	74.1	22.2	3.7
Trained staff take other jobs	77.8	18.5	3.7
Trained staff too demanding	74.1	22.2	3.7
No replacement for staff in training	40.7	40.7	18.6
No training meets your requirements	48.1	48.1	3.7
Training gives no 'recognized' qualification	59.3	37.0	3.7
Training too short or long	44.4	40.7	11.1
Curriculum not same as usual for staff	25.9	51.8	18.5
Lack of information to assess suitability of training	25.9	59.2	14.8
Lack of advance notice	25.9	63.0	11.1

ments of developing countries. There can be frequent lack of continuity at the director level, occasioned by decisions made remote from the institution which is directly and negatively affected. These decisions are based on changes of government or, for example, the officer career structure of the forestry service, without regard to the value of existing and accumulated knowledge, interest and commitment. Political appointees may be interested, but could well be entirely unqualified to run these zoological institutions. Communication between zoos, and necessary management decisions within zoos can also be hampered, postponed or discouraged by political interference or a bureaucratic command structure where responsibility is a weaker partner to authority.

11.4.5 Credibility and public image problems

Because of financial difficulties and lack of information in particular, many zoos in less-developed countries (and to some extent also in the western countries) do not have a good record on breeding, health, mortality, education programmes and so on. The consequent inability to demonstrate real achievements goes no way to convincing already sceptical field biologists and decision-makers that these zoos have valuable biological and cultural roles to play. The official response may be at best to vote no increase in the annual operating budget, thereby reinforcing the downward trend. Of the millions of visitors annually to Third World zoos, probably not many have a distinctly negative image of these institutions, but more likely have a neutral attitude. However, to foster a positive public image will of course require the funding, political backing and training to ensure the visitors see animals in good conditions and are provided with good educational opportunities.

11.4.6 Legal restrictions

Laws in many developing countries have been formulated, or are interpreted, in such a way as to make barely possible or even illegal some functions which should be considered essential to the running of zoos. Laws may prohibit sale or purchase of animals, or exclude any transactions of endangered species, thereby reducing the scope for these zoos to become involved in the conservation of indigenous wildlife. Regulations might prevent the direct receipt of funds from any source other than the designated authorities, so discouraging initiatives to raise extra money badly needed for improvements to exhibits. Legislation on the importation of drugs can rightly be severe, but again interpreted in such a way as to make procurement laborious and disheartening. In most cases where illegally held wildlife is confiscated by the appropriate authority, the zoos are compelled to receive and accommodate animals in sub-standard facilities, and in the extreme may not subsequently be permitted to move them elsewhere.

11.4.7 Curriculum deficiences and lack of in-house training

In universities and colleges, it is essential to establish applied ecology and wildlife management much more widely and firmly in existing curricula, perhaps along the lines of what we see at the Wildlife Institute of India (Krausman and Johnsingh, 1990). Eventually, this needs to extend into the zoos via graduate input into new and expanded in-house training programmes targeted at various levels. Certainly from the current standpoint of zoos in the tropics, this route to improvement will need to engage the recurring areas of weakness – educational signs and graphics, policy and practice on animal acquisition and disposition, immobilization as a feature of improved veterinary care, the marking of animals, and record keeping. The combination of these negative elements might suggest an insuperable barrier to change, but the substantial advances made in zoos with trained personnel clearly contradict any pessimistic view, and no real progress can be made unless all the problems are confronted anyway. Thus, some of the challenges outlined can be met more immediately with training, and positive results gained relatively quickly. Others are evidently more intractable, with more inertia to overcome, and require longer term remedial action through training at an upper management level, plus increased exposure of high-level officials to the value of zoos and wildlife departments, and their need for greater input of funds and improved access to information.

11.5 SETTING THE SCENE FOR TRAINING

With the intention that training should contribute to continued improvements in zoos and captive breeding centres, both immediate and long term, it is important to examine the necessary preliminaries to any programme, and worthwhile to set these in the context of broader collaboration.

11.5.1 Learn the cast of characters

In proposing training of any form or level, it is essential not only to identify the appropriate candidates for a course, but also to be informed of where they fit into the overall command structure of the organizations involved. In dealing with selection of applicants, funding and other practicalities, it is important to interact with those who hold the responsibility for these decisions. Figure 11.1 presents a simplified view of zoo administrative organization as an indication of the levels of interaction and for the targeting of training.

11.5.2 Learn the conservation arena

To be most effective, the training needs to:

(a) relate to the wider conservation issues of the region concerned and in particular to which species are identified as threatened or endangered, and which are already in captivity or are definite contenders for captive breeding programmes;
(b) identify the major conservation problems, the socio-economic forces which cause them, and what existing government or non-governmental conservation action is taking place.

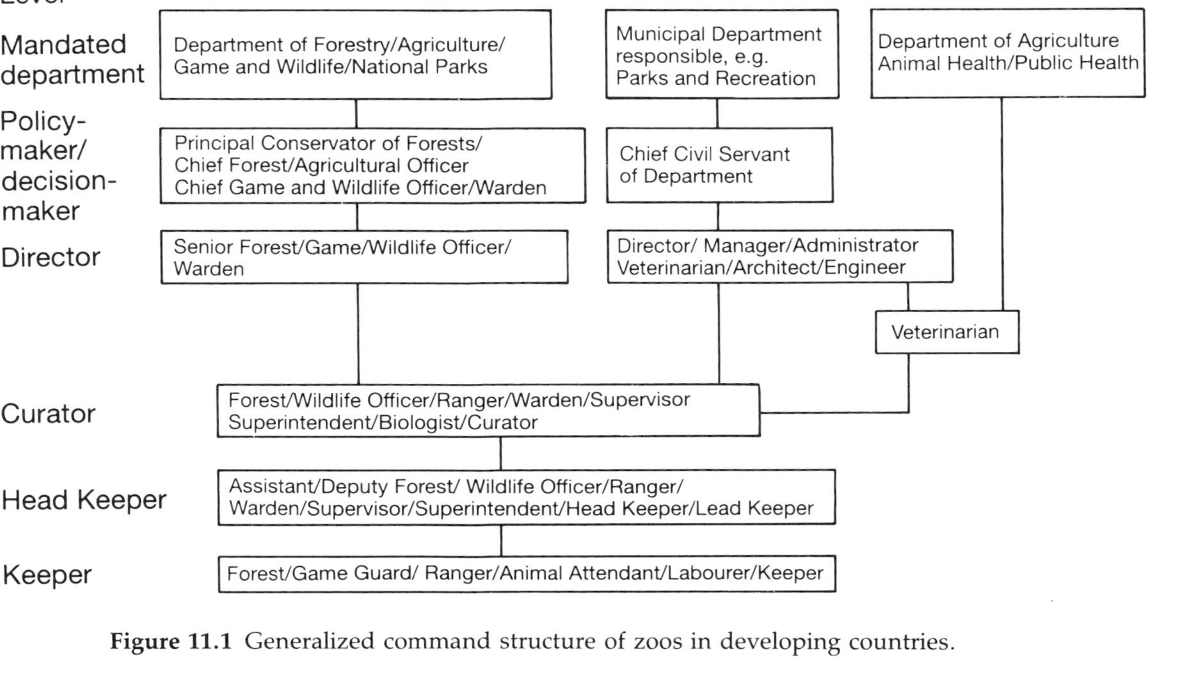
Level
Mandated department
Policy-maker/decision-maker
Director
Curator
Head Keeper
Keeper
Department of Forestry/Agriculture/
Game and Wildlife/National Parks
Municipal Department
responsible, e.g.
Parks and Recreation
Department of Agriculture
Animal Health/Public Health
Principal Conservator of Forests/
Chief Forest/Agricultural Officer
Chief Game and Wildlife Officer/Warden
Chief Civil Servant
of Department
Senior Forest/Game/Wildlife Officer/
Warden
Director/ Manager/Administrator
Veterinarian/Architect/Engineer
Veterinarian
Forest/Wildlife Officer/Ranger/Warden/Supervisor
Superintendent/Biologist/Curator
Assistant/Deputy Forest/ Wildlife Officer/Ranger/
Warden/Supervisor/Superintendent/Head Keeper/Lead Keeper
Forest/Game Guard/ Ranger/Animal Attendant/Labourer/Keeper

Figure 11.1 Generalized command structure of zoos in developing countries.

11.5.3 Explore in-country sources of funding

To help fund training and conservation programmes arising from it, one should identify which funding agencies are operating within the country. In many cases multilateral aid-giving agencies are potentially available for funding zoo-related initiatives, particularly in the area of training.

11.5.4 Identify the problems

Within the organization which is targeted to receive training, the existing problems should be identified and agreement reached on which ones, in which order of priority, can be tackled by a particular level of training. An important aspect of this process can be the concerns that decision-makers/directors might have about possible negative implications of training, especially in the temporary loss of staff to attend regional or international programmes. Table 11.4 shows the results of a recent (1990/91) survey of decision-makers and gives an indication of the relative importance they attach to various aspects of staff secondment for training.

11.5.5 Develop the collaborative agenda

A sense of true collaboration oils the wheels of progress, and this will be a strong feature during a programme of training *in situ*, where the roles of the host institution and the trainers can be defined to maximum advantage. The agenda can extend beyond a first-time training programme to include collaboration in refresher courses, and joint efforts in captive and field conservation campaigns over the long term. Three important elements to such collaboration are:

(a) to have counterpart investigators;
(b) to cultivate employees of lower status who might show great enthusiasm which deserves to be nurtured;
(c) to establish a system of rewards and incentives for genuine efforts and results in making improvements within the institution.

11.5.6 Follow-up activities and reinforcement

A continuity of collaboration after training will rely on a pattern of follow-up events designed to augment initial experiences, stimulate further enquiry, provide support for host institution programmes, and assess the effectiveness of the training based on evidence of implementation and feedback from participants and those in charge. Follow-up activities are further discussed below in relation to how western zoos can increase help to zoos in developing countries.

11.6 QUALITIES OF THE CANDIDATE FOR TRAINING

The criteria to be used in selecting participants for training will obviously relate to the employee level at which it is aimed. In all cases, however, the best end-result is likely to occur where the trainer's requirements coincide with the qualities that the decision-maker takes into account when allocating these same people to particular staff positions and duties. Remembering that 'craftsmanship involves both attitude and skill' (Braham, 1988), the range of qualities to assess can be extensive. MacKinnon *et al.* (1986), in dealing with issues of staff selection for national parks, state that the warden may be in the lucky position of selecting his own staff, but more often he 'inherits' them. They go on to list the factors that the warden will have to take into account in allocating staff to positions and duties, and these are: the existing level of training and education, skills and abilities; work aptitude, ability to follow orders and ability to assume responsibility; capacity to improvise in new situations; trustworthiness, honesty and personal courage; ability to work together with other staff members; past work performance and seniority in service; promotion opportunities and guarantees; status in the local community and relationships with key personages (village heads, officers of other government divisions, headquarters staff); home ties, reasons to stay in the area; and professional appearance and image. The preceding considerations of the challenges to be met and suggestions on how to structure a response, have relevance to any type of training programme. However, since 1986 the essentially complementary approaches of two training programmes have encompassed all these factors and a description of each follows.

11.7 WILDLIFE PRESERVATION TRUST TRAINING PROGRAMME

11.7.1 Wildlife Preservation Trust

The trio of WPT organizations work in concert to support training: the Jersey Wildlife Preservation Trust (JWPT), Jersey, British Isles; Wildlife Preservation Trust International (WPTI), Philadelphia, USA; and Wildlife Preservation Trust Canada (WPTC), Toronto. This support is provided in the context of the other conservation work undertaken by WPT, which is focussed on endangered species recovery efforts involving captive breeding programmes, field and captive research, reintroductions, habitat purchase, *ex situ* and *in situ* education work, and collaboration with governments and NGOs.

11.7.2 The Training Programme

The International Training Programme in Conservation and the Captive Breeding of Endangered Species has been in existence since 1978. It has been designed to cater for the requirements of people working directly with captive animals, the extent of practical involvement in their work places largely being dictated by employment position. From the outset, the programme has not been confined to the training of zoo personnel only, and the title alludes to the need to train staff from government departments such as forestry, wildlife and national parks, as well as from conservation NGOs. Whatever the organization, if part of its remit is to have endangered species captive breeding and recovery efforts, and attempts to achieve this are evident, then the training of key staff is a priority for WPT. The very first participant of the programme, the government of Mauritius's conservation officer, is a good example of a forestry department employee being directly involved with the captive breeding of endangered endemic fauna.

A further priority has been to train people from certain geographical and geopolitical areas, based on the overall extent of existing cooperation and collaboration that WPT has with governments and NGOs. These areas have been further prioritized based on endemicity, biodiversity and extent of threats to the fauna. These priorities have been set so as to get the best conservation return for WPT investment but, nevertheless, the training programme has been able to cater for many people from other geographical areas, to the extent that 330 people from 70 countries had been trained by March 1992. It should also be noted that more of the participants from the tropics have been zoo employees: for a subset of trainees from developing countries, Table 11.5 classifies them by occupation. Several cases elude an occupational category simply because the person involved seems to be doing virtually everything, and the captive breeding effort rests on the enthusiasm, commitment and knowledge of that individual.

11.7.3 Aims of the programme

The principal aim could always be stated thus: 'To give training in the captive breeding of endangered species to individuals with differing needs, in order for them to advance the cause of endangered species work and conservation in their own countries. To give basic practical training in modern husbandry methods and breeding techniques for captive animals'. However, given the spread of occupations evident in Table 11.5, a widening of the aims has occurred to provide additional training in broader animal management and zoo issues founded on the correct practical basics. Riding close behind is the intent to improve the

Table 11.5 Occupations of WPT trainees from developing countries

Occupation	*Zoo*		*Non-zoo*		*Total*	
	N	%	*N*	%	*N*	%
Chief forest officer	–	–	1	1.6	1	0.7
Director/asst. dir./senior officer forests-wildlife	10	11.0	8	12.7	18	11.7
Veterinarian	20	22.0	4	6.4	24	15.5
Curator/superint./biologist/forest-wildl. officer	26	28.6	7	11.1	33	21.4
Head keeper/keeper/asst. forest-wildl. officer	29	31.8	15	23.8	46	28.6
Researcher	2	2.2	16	25.4	18	11.7
Educator	4	4.4	6	9.5	10	6.5
Graduate	–	–	6	9.5	6	3.9
Total	91	100.00	63	100.00	154	100.00

success rate of specialized captive breeding centres, to increase the positive visitor impact of developing countries' zoos and to improve collaboration by fostering an information network of motivated zoo people.

11.7.4 Training site and resources

The programme is run at the JWPT, with the zoological collection forming the core resource for training. All participants have direct access to the staff working areas on every animal section, and access under supervision to off-view breeding and public exhibition enclosures. Other key areas at their disposal in the zoo are the veterinary centre, laboratory, egg incubation and bird rearing units, animal records office and education centre. At the International Training Centre (ITC), there are purpose-built facilities in the form of a lecture hall, reference library, computer work-stations, graphics studio and photographic dark room, audio-visual equipment, and museum with adjoining specimen preparation area. This technical wing is adjacent to the ITC residential area, which accommodates all trainees plus occasional researchers and other visitors. The ITC was purchased and equipped with foundation support, and an additional important funding requirement has been the provision of scholarships to trainees from less-developed countries who otherwise would not have the opportunity to attend. The years 1986 and 1987 were high in relation to more scholarship trainees on shorter training periods, but otherwise the trend is one of increase (Table 11.6) and this is set to continue. The support of JWPT, WPTI and foundations sustains three full-time employees to administrate the programme, provide instruction and coordinate the other practical training given by different staff members of JWPT.

11.7.5 Programme structure and curriculum

Since the inception of the programme, the duration of training has varied from person to person according to their work-site circumstances and requirements. Therefore some tailoring to needs has taken place, although most trainees have attended the programme for 10 or 16 weeks' duration. The intention is to keep the programme duration flexible, and also to maintain maximum flexibility for trainee subject specialization, again related to home priorities. In the core practical part of the training programme, where each trainee works on various animal sections alongside the experienced keepering staff, this flexibility can be achieved by allocating trainees to appropriate sections, and for more or less time according to need. The number of trainees simultaneously on one section has to be limited so as not to compromise standards of

Table 11.6 Yearly pattern of WPT training scholarship recipients

Year	*1978*	*1979*	*1980*	*1981*	*1982*	*1983*	*1984*	*1985*	*1986*	*1987*	*1988*	*1989*	*1990*	*1991*	*All*
Number of scholarships	(1)	1	2	1	5	6	9	9	16	17	12	12	13	15	199
Proportion of total trainees (%)	(50.0)	4.2	9.1	6.3	25.0	33.3	31.0	31.0	51.6	48.6	46.2	44.4	59.1	71.4	37.0

animal care, and the tried and tested way to deal with this is to have a rolling programme whereby as one person completes, the next arrives to begin. In this way, there are usually six trainees in the programme at any one time.

Lectures, workshops and practical demonstrations form the other part of the programme, these theoretical aspects being presented by the training officer and his assistant, other staff members and occasional visiting scientists and speakers. The curriculum of practical work and theory covers a range of topics essential for an understanding of modern zoo management (listed as an appendix to this chapter). All topics have accompanying printed materials and these, and the instruction, are in English. There is also time allotted for trainees to carry out research projects or other further study, which may involve nutrition, animal health, behavioural observation, education, or use of the Trust's animal records. During a complete period of training, there is constant access to material in the Trust's library for all trainees to relate their practical work at the Trust to what has been done previously and what is happening elsewhere. Trainees are assessed on their performance as they work through the various sections, and they receive a final report after completion of the course.

An important consideration is the appropriateness of the training, in terms of the few resources with which many trainees will have to work when they return to their own countries. The emphasis often needs to be on the simple but effective, in the knowledge especially of what the home zoo budget is likely to be for the foreseeable future. An up-coming development to better deal with this is a two-track programme, with participants higher in the decision-making tree having more emphasis on aspects of fund raising, for example.

A proportion of the training programme participants are funded by WPT to extend their training within other zoological institutions which might better fulfil a particular requirement. Additionally, a Canada's New Noahs scheme funds one Canadian per year to extend the training programme with six months' field training involving endangered species recovery programmes in Mauritius.

11.7.6 The Diploma in Endangered Species Management

Since 1986, entrants to the training programme have had the option to combine this with a dissertation by research to attain a Diploma in Endangered Species Management (DESMAN). This University of Kent (England) diploma, awarded via the Durrell Institute of Ecology and Conservation (DICE), involves the participant in full – or part-time – investigation, usually in the home country or at the JWPT. DESMAN provides an opportunity for trainees to research in some detail the species and subjects of importance to their home situation, and for the results to have a practical conservation application. It also provides welcome academic recognition of their captive breeding and conservation work, and this is expected to have a positive influence on decision-maker attitudes. Investigations can be captive or field, or a combination of both. Recent examples include captive breeding and population modelling of the northern helmeted curassow (*Pauxi pauxi*) in Venezuela; captive breeding and behaviour of maned wolves (*Chrysocyon brachyurus*) in Brazilian zoos; captive management, breeding and conservation of white-winged wood ducks (*Cairina scutulata*) with special reference to Thailand; and evaluation of field study and capture methods for carnivores in Mexico.

11.7.7 Other courses

There are three other WPT initiatives to provide training and improve awareness, communication and information flow.

(a) JWPT Summer School

Since 1980, this annual three-week intensive course on breeding and conservation of endangered species has provided an introduction to many practical aspects of management and investigation of captive animals and field conservation. Each course has had 24 participants from a range of backgrounds, including zoo people from director to keeper, wildlife wardens, field biologists, college students, environmental journalists and so on. The value of this course has been to expose more people to the scientific basis of modern zoo management and its relevance to conservation, and it has been accessible to people with little time available for training. Invited instructors and speakers augment the teaching team, and increasingly there is WPT scholarship assistance for applicants from low-income countries.

(b) Policy-makers' Course

1991 saw the first initiative to expose people at the policy-maker and decision-maker level (see Figure 11.1) to the conservation potential of zoos and zoo personnel in the tropics. Entitled 'Wildlife Economics and Management: Policy and Practice', this three-week course drew together 21 senior administrative staff and seasoned policy-makers of departments and institutions with responsibility for wildlife conservation and utilization. The course examined wildlife conservation from the perspective of using effective methods to manage wildlife as a renewable resource successfully, in order to yield an economic return. This economic perspective encompassed zoos, including how to run them better, encourage and utilize staff most effectively and increase their conservation contribution. Organized by The British Council, the course was run at DICE and JWPT with an international instruction team of high repute. It is intended that this course will run annually.

(c) Zoo Educator Training Course

Also held for the first time in 1991 was this course, aimed particularly at zoo educators, with the goal of showing them how to create or improve conservation education programmes in their zoos. With the experimental curriculum developed by an international team of experienced educators, the six-week course was run by the JWPT education officer with the assistance of educators from developing countries. The 12 participants came from South America, Africa and Asia, each employed in a zoo or otherwise working with captive animals. Curriculum refinement is taking place and the improved version is intended for use throughout the world, irrespective of resource level but again with a focus on zoos in the tropics.

11.7.8 Training follow-up

To support the work of trainees, post-training has always been an important feature of any form of WPT training, and has been enhanced since 1987 with improved financial backing. Follow-up measures have aspired to:

(a) assess the effectiveness of training through feedback and direct observation of implementation;
(b) assess the continuing needs of trainees in the work place to determine the options for further support.

WPT has thus developed follow-up activities on a broad front, with training effectiveness and continuing needs being assessed through questionnaire analysis, and WPT staff visits to see trainee work-practices on-site and, where possible, to reinforce dialogue with decision-makers.

Ex-trainees have been supported financially to expand their involvement in integrated field conservation projects, for example with golden-headed and black lion tamarins (*Leontopithecus chrysomelas* and *L. chrysopygus*, respectively) in Brazil. Their captive breeding efforts have been bolstered by supply of equipment and building of breeding enclosures, as for the St Vincent parrot (*Amazona guildingii*) on its native island. Support is regularly given to help trainees organize their own in-house or regional workshops and training, the Kumasi and Accra Zoos in Ghana being a good example from 1991. Opportunities are fostered to gain further experience elsewhere, by attendance at workshops and conferences, or even to witness reintroductions. An example of the latter was WPT support for two Mexicans and a Brazilian to join in with the recent black-footed ferret (*Mustela nigripes*) releases in the USA. The final part of follow-up is to provide a continuing means of communication between trainees, and between them and WPT, and this is achieved through a post-training newsletter.

11.8 NATIONAL ZOOLOGICAL PARK ZOO BIOLOGY TRAINING PROGRAMME

The National Zoological Park – bureau of the Smithsonian Institution (SI) – commenced its Zoo Biology Training Programme in 1987 with seed grant support from the institution's Office of International Programmes. The course was developed in response to increasing requests for zoo-related information received during the Wildlife Conservation and Management Training Course, which since 1981 has been presented annually at the National Zoo's Conservation and Research Center in Front Royal, Virginia. While the Wildlife Conservation Course is designed almost exclusively for *in situ* wildlife and conservation managers from developing countries, it became clear from numerous queries about capture and restraint, transport, captive propagation, and husbandry that there is a need for specific curricula on broad issues of zoo biology. The Zoo Biology Course is now one of 10 training curricula that target developing country nationals, and that are organized and presented by the Smithsonian Conservation Training Council (SICTC). The council was organized in 1990 to coordinate and promote all Smithsonian training in the realm of biodiversity conservation.

11.8.1 The training programme

The National Zoo's Zoo Biology Course shares with the JWPT a commitment to improving zoos in the developing world through instruction and the implementation of new practices. Inherent in the method is an emphasis on the biological understanding of the problems of maintaining wild animals in captivity. The programme differs operationally from the JWPT programme in several details. To begin with, all courses have been presented abroad to student bodies representing mid-level management of several zoos in the country or, in some cases, the region. This group includes primarily curatorial staff, but also senior supervisory keepers, veterinarians, and, in some cases, even assistant-directors and directors. The breadth of the audience represents more the attitude of the participants than the comprehensiveness of the curriculum.

Each three-week course has been taught by an instruction team, initially employees of the National Zoo, but more recently, curators and veterinarians of other zoos in the USA, Canada and Europe. In every course, one senior instructor with past teaching experience in the programme serves as the leader and official representative. The guidelines and sequence for lectures is contained in a manual of zoo biology written by C. Wemmer, Charles Pickett and J. Andrew Teare, which has now been translated into Mandarin, Spanish, Bahasa Indonesia and French. A revised and expanded English language version of the manual will be published by the Smithsonian Press.

The course is generally conducted in a large zoo, often in the capital city. The hosting zoo is responsible for providing logistical support and making local arrangements, including the invitation of eligible participants from its own and other zoos in the country or region. Conducting the course in a developing country's zoo has several advantages. Ironically, most of these advantages are the very limitations that inevitably confront the instructors. These are the same limitations that beset most zoos and other organizations in those regions where biological diversity is under impending threat. Solutions must be sought using local materials and talents, and the message is that 'so-called' high-tech solutions are in most cases simply unrealistic.

Recently, the course has to some extent been franchised in that the format and materials are supplied by the National Zoo, while most instructors are provided and partially funded by other zoos. This has added benefits. It offers zoo personnel a unique educational experience, increases appreciation of cultural diversity and promotes a more sophisticated understanding of the challenges confronting developing country zoos and international conservation in general. It also allows the contributions of many people to improve upon the materials and methods in various zoological disciplines.

11.8.2 Aims of the programme

Achieving the long-term goals of the programme – modern zoos which engender respect for wildlife, and understanding the dependency of the human condition on environmental welfare – hinges to a great extent on various kinds of reinforcement, and a steady stream of opportunities to those in the developing world who are committed to the zoological profession. This responsibility is obviously beyond the means of any single institution, and the programme is dependent on related activities of other zoos and zoo groups – such as the Madagascar Fauna Interest Group (Anderson, 1990), and the Zoo Conservation Outreach Group (Anderson, 1989), the CBSG, and NGOs which can complement the work. Effective collective effort requires communication and the degree of coordination necessary to focus attention and resources.

Several activities have been important in sustaining the efforts of certain alumni of the programme. During each course attempts are made to identify outstanding students, usually those individuals most interested in extra-curricular projects, such as research and studbook development. Several short study tours have been funded by grants from the Smithsonian's Office of International Programs, and the Visiting Research Programs. Equally important, however, have been the matching contributions from North American zoos which have provided instructors to the programme. These zoos have hosted the visitors for several days at a time, have funded their housing and local transport or have made local arrangements for room and board. Each visitor's itinerary has been designed with respect to specific goals, such as the husbandry of a particular taxon, environmental education programmes, or studbook development. In two instances now, developing country zoo employees have drafted proposals to fund in-country initiatives, and received advice on the topic from North American NGOs.

11.8.3 Training site and resources

The hosting zoo for the training programme serves as the training site, and each site has differed in its strengths and weaknesses. Deficiencies of course serve as lessons to the instructors, and beg solutions, which in many cases are beyond financial reach. The most pronounced deficiencies have been in exhibits, which may be poorly constructed of inappropriate materials, or may be structurally sound but inappropriately designed. Examining the exhibits of host zoos has been an illuminating experience for course instructors, who have discovered large cats behind chain link barely sufficient to hold a macaque, or iguanas contained in cement-walled and moated fortresses. Workable, economic, 'keeper-friendly', and aesthetically pleasing exhibit designs are not widespread in the developing world, but outstanding existing designs should be communicated. This recurrent problem of zoo exhibit design is not unique to the developing world, but the lack of permanent and trained personnel among the mid-level ranks of management is partially responsible for these problems.

Husbandry, nutrition, and veterinary care are often areas of deficiency because of inadequate staff expertise and/or insufficient equipment and supplies. The latter deficiency pertains in particular to veterinary care, where there is still a large unaddressed need.

The programmes and facilities of any zoo embody its priorities. Special facilities for visitors, even guest houses, auditoriums and animal shows, are the dominating recreational themes in many zoos of means. These are complex problems requiring sustained exposure to a wide range of disciplines.

11.8.4 Programme structure and curriculum

Prior to each course, materials such as a species inventory and a list of special and recurrent problems are sought from the hosting zoo to prepare the instructors for the experience. Presenting the course during the first two years also gave us the opportunity to modify the scope of topics and activities, and this led to the development of the manual which has served as a guideline for lectures. While lectures are the backbone of the course, they are supplemented with several other activities. Competitive problem-solving exercises have tackled common dilemmas in captive animal management, such as crating and transport procedures. A workshop on capture and restraint has been a popular recent innovation with festive overtones, and quizzes have proven a useful diversion from classroom routine and an important means of gauging individual knowledge of various areas of zoo biology. Class projects involving small teams were conducted during the early courses, but have been discontinued because of insufficient time. Experience has also taught us that adaptability is a key requisite for successful instruction in different cultural contexts. The existence of a range of educational activities, from lectures to demonstrations, discussions, quizzes and projects has armed us with a menu of options. Some of these are better suited to certain circumstances than others, but the decision of which is best suited to a particular situation can only be made by experienced instructors.

As the interests and appetite in subject matter differ between regions and zoos, so do the assessments of knowledge needed by students differ between students and instructors.

11.8.5 Other needs for training in zoo biology and management

This training course has exposed us to many significant needs of zoos in the developing world, some of which could be assisted by training. The following thoughts are offered in light of the growing interest of European and North American zoos in playing a role of international assistance.

In 1990 the American Association for Zoological Parks and Aquariums (AAZPA) organized a workshop for zoo associations emphasizing those of the developing world. Clearly, many of the challenges to, and perceptions of, those zoos and aquariums can be changed only through a high level of concerted and organized action. The zoo and aquarium associations of economically developed countries have much to offer their sister organizations of the developing world, and a concensus recommendation was that information gateways be established to regularly communicate information and printed materials from rich to poor countries. Unfortunately, the recommendation has not been heeded to the extent needed.

In zoos and wildlife departments of the developing world the demand for wildlife veterinary skills is great, and the extent of local expertise limited. To our knowledge, wildlife medicine is not taught in Africa, Asia or Latin America, and on many occasions we have been called upon to deal with special health problems and the capture and transport of wild animals. The Wildlife Veterinary Information Service Asia South (VISA) being developed by the Zoological Society of London holds promise, and if expanded could effectively address the need. In addition, regionally tailored workshops on specific veterinary problems would prove beneficial. Directors and keepers have training needs as well. As already mentioned, the directors are often short-term employees without relevant experience in management and zoological matters, while the keepers are often labourers, many of limited literacy. Until the time that high-ranking zoo positions move out of the realm of political entitlement, there is a need to provide the ever changing pool of directors with orientation into the philosophy of captive animal management and other responsibilities of the profession. While such a course would be best given to a mixed group of directors in Europe or North America, keeper instruction is a need that is best done in the home zoo during the training period. Several model programmes serve as examples of what has been done, and they could be modified to serve the purposes of developing country zoos (Koontz and Hutchins, 1990).

Lastly, most zoos of the developing world have financial difficulties. Zoological societies and volunteer or friends groups can generate significant funds and labour for their parent organization, but these mechanisms seem to be under-utilized in many regions we have visited. While there may be legal obstacles to establishing such groups in some zoos, others are well positioned to take advantage of such organizational aids. A workshop involving a small number of society leaders and developing country zoo directors interested in the topic could begin a movement to improve zoo conditions by these means.

11.9 DISCUSSION

The above accounts of the WPT and NZP training programmes demonstrate their essentially counterpart nature in trying to achieve the common goal of helping zoos in the tropics to improve and to have a conservation input. Both programmes can demonstrate heartening examples of improvements in captive animal management through the efforts of people given the knowledge and stimulus from a period of training. These efforts have most often been made in the face of the palisade of obstacles already mentioned, and thus we salute these people, plus untold others, whose enthusiasm for doing a good job remains undimmed. Much of what they have to deal with would provide a sobering experience for those associated with zoos in developed nations, and perhaps an increased stimulus for change through assistance. It is difficult to envisage that all of the problems besetting zoos in the tropics can be dealt with on the same time-scale. Some zoo functions taken for granted in the west are likely to remain as continuing needs for training long-term, and policy formulation and implementation, will be more difficult still. The one certainty is that zoos in less-developed nations will remain in existence, the number could possibly even increase, and therefore there is a continuing imperative for the western zoos to help. A substantial proportion of the following suggestions cannot be tackled by the existing training opportunities, and demand increased scope and capacity.

The priority continuing need is, of course, increased funding, and a straightforward response can be for zoos of developed nations to donate directly to their boot-strapped cousins. For example, a recent analysis of budgets and priorities for zoos in high-income countries (Anon, 1991a) showed the annual collective budget to approximate US$1 billion. The suggestion was made that an allocation to off-site conservation activities of only 1% would net $10 million, of which $1 million (0.1%) could be used as a North/South venture to support developing countries' zoos (Adopt-A-Zoo). This could represent a large expansion of the sister-zoo concept as described for the La Aurora (Guatemala) and Dallas zoos (Christman, 1989). Additional to this, 0.25% was suggested as an allocation to Adopt-A-Park programmes, and Tilson (1992) has already described how the Minnesota Zoo is contributing in this way to the preservation of critical habitat. Providing a mechanism to involve in-country zoos, preferably nearest the adopted park, would show sensitivity to their aspirations, confer credibility and give a stimulus for management improvements.

However, previous sections of this paper intimate that administrative changes in developing-country zoos will frequently be necessary for foreign funds to be successfully received and disbursed. This is not to deny the ingenuity and inventiveness that are frequently used to make the most of meagre existing resources. Instituting the right framework, and then operating within it, are skills likely to need development through training of personnel at higher levels of management. This implies an expansion of specialist help for policy-makers, and the creation of a suitable programme for zoo directors. The AAZPA Management School would be one option, but only feasible if large subsidies could be procured. Another option could be the collaboration of several zoos within a region (e.g. North American, European, Australasian) to run directors' courses, or otherwise it might be a feasible undertaking for a single large zoo and would certainly earn that institution approbation. Often the donation of materials is a possibility when direct funding is not. Trained personnel at various levels in developing-country zoos may not be able to implement all their acquired and developed skills due to the lack of equipment. Direct donations of items such as immobilization and capture equipment, endoscopes and other veterinary apparatus, specified drugs, marking equipment and tags, computer hardware and software, video and audio-visual equipment for in-house training and schools education programmes, books, periodicals, and scientific reprints are all going to help.

Once again, for zoos in the tropics properly to solicit such materials, training of personnel in the appropriate skills will make a difference. Much of this discussion comes back to the issue of changing existing policies and how they are implemented. Amendment of legislation unsympathetic to modern zoo operations is a related issue, as is the need to alter the system of political appointments so as to extend stability and experience at the director level. These are points of some sensitivity, but highlight the need to expose policy-makers to those things that handicap zoo development, and for them to better understand why the directors make particular decisions or requests. For example, where the wildlife department is making regular confiscations of illegally held wildlife, and no quick release is possible, an advisable policy would be to allocate these animals fairly among the zoos according to available space, rather than threaten overcrowding in one or two. This is likely to promote a better relationship between the zoos and the department.

In turn, the zoos need the goodwill of the public. In zoos of the developed countries, this is viewed as a core aspect of public relations and marketing, and is assiduously courted (Montagu, 1983). The tradition of zoos in the tropics has, with some notable exceptions, left these activities poorly developed, one result being no clear focus as to future direction. Since tropical zoos are largely not in a position to employ public relations personnel, a directors' training programme would appear to be the vehicle whereby market research, visitor surveys, redefining of objectives and future planning can become more commonplace. If a zoo and its managing authority are seen to be more responsive to visitors and the local community, this must surely fall within a broad definition of marketing 'profit'. A directors' training programme, or possible specialized course for zoo fund-raisers, should also deal with the subject of how to circumvent government institutional restrictions on the receiving of external funds.

For example, the tactful development of parallel supporting groups, such as 'friends of the zoo', to facilitate the raising of funds for specific projects, and especially to attract business sponsors and patronage of the well-connected. The animal management section of a directors' training curriculum, would need emphasis on the development of a master plan or collection plan, on how to arrive at species selection priorities and how to institute effective acquisition and disposition policies. This prepares the ground for successful captive breeding programmes, and the subsequent international recognition that is going to be increasingly necessary for developing-country zoos to attain. In advance of any changes to legislate in favour of minimum zoo standards, zoos in the tropics would benefit from stronger national and regional zoo associations, with accreditation in the style seen via AAZPA, and with enhanced information exchange via meetings, newsletters and other circulars. Strong zoo associations can be a more effective lobby to government and also to international aid agencies whose remit does not permit cooperation with individual zoos. Benefits might accrue within-country or even extra-limital, as the recent US Agency for International Development Noah Project initiative suggests in its recommendation of training programmes for developing-country scientists (Anon, 1991b). Fledgling zoo associations will benefit from the networking and assistance of more mature, well-functioning ones, and training plus funding should be contemplated for the staff of an eventual zoo association office. One possibility is for western zoos to provide two to three years' funding for full-time regional conservation officers, which then offers a chance for a strong zoo association to become firmly rooted. For example, Lorena Calvo, the Guatemala-based coordinator for AMAZOO has been funded in this way by WPTI. Although their role and influences vary considerably between institutions, the zoo educators can be very much identified with zoo association activities, visitor surveys and marketing. The WPT Zoo Educator Training Course dealt with these items, as should all future educator courses. In common with keeper and mid-management programmes, the zoo educator training will also profit from instruction in the diplomatic skills of handling decision-makers! Given that signs and graphics in many zoos are woefully inadequate, the elements of presentation and interpretation in the educator training curriculum need to be central. However, as often as not the lack of signs is because there is no zoo educator position, a problem that the policy-makers urgently need to consider.

In more general terms of animal management in tropical zoos, training of the WPT/NZP type should continue, but also the development of outreach programmes and follow-up should be improved. Many of the zoos could benefit from increased manpower, under experienced leadership, for short-term specific projects such as exhibit refurbishment or for longer-term attachment. The utilization of an organization such as Earthwatch has been previously mooted (Wemmer, Pickett and Teare, 1990), as has a zoo version of Peace Corps, perhaps called Zoo Corps.

Certainly, for experienced keepers from zoos in developed nations, being seconded to help directly in areas of high biodiversity would represent an excellent career incentive and achieve real advances where most needed. At the administration level, business-sector secondment of executives is also a possibility never to be undervalued.

Finally, we want to emphasize the point about networking and providing linkages, because the training process invariably involves a joyful exchange of experiences where the remarkable commitment of people working in less fortunate circumstances shines through. As Diaz Matalobos and Barraza (1992) have stated, bridging the gap is all important.

11.10 CONCLUSIONS

(a) There is need for substantial improvements in the functioning of zoos of less-developed countries and a demand for training to effectively address the problems.
(b) The Wildlife Preservation Trust Conservation and Captive Breeding of Endangered Species training programme, and the National Zoological Park Zoo Biology training programme, are two tried-and-tested programmes, which feedback demonstrates are helping to meet the demand in complementary manner.
(c) The values of the kind of training discussed above are to help develop the philosophy of zoo biology, to stimulate scientific curiosity, to provide incentives for professional growth and, finally, to increase linkages.

11.11 APPENDIX: TOPICS IN THE WPT TRAINING PROGRAMME

General principles of conservation: conservation in the wild, who conserves and how, ecopolitics, captive breeding and conservation, zoo cooperation; research basics: methods of behavioural study in captivity; record-keeping: studbooks, use of computers for ARKS and SPARKS animal record keeping and analysis, use of records in research; wildlife reserves: some aspects of design and management; methods of field study; selection of species for captive breeding; the zoo as an education resource; zoo design and cage design: methods of exhibition; environmental enrichment; animal collecting, transport and quarantine; animal restraint: immobilization for examination, medication, surgery, drug types and applications; a preventive approach to good animal health: hygiene, quarantine screening and prophylaxis; nutrition and diets: propagation of live food, feeding, food handling and storage; aspects of bird rearing: artificial incubation; hand-rearing; recent techniques for endangered species: artificial insemination and embryo transplants;

distribution: dispersal of stock and why distribution is necessary; reintroduction of captive-bred animals to the wild; population genetics with relevance to captive breeding programmes and the design and management of nature reserves; legislation: environmental laws, legislation governing captive animals, international animal trade regulations, CITES; safety in the zoo; funding for zoo requirements and field projects: zoo administration and budgeting; calculating costs for captive breeding programmes.

ACKNOWLEDGEMENTS

David Waugh wishes to thank WPTI and WPTC and all donors to the WPT International Training Programme via those organizations. All direct donors to JWPT are likewise thanked, especially the Whitley Animal Protection Trust, Lady McNeice and British Airways Assisting Nature Conservation. The JWPT staff role is gratefully acknowledged, and the friendship, good humour and irrepressible spirit of the trainees greatly admired and appreciated.

REFERENCES

Anderson, D.E. (1989) The next step: Zoo Conservation Outreach Group. *AAZPA Ann. Conf. Proc.*, 20–3.

Anderson, D.E. (1990) Formation of the Madagascar Fauna Captive Propagation Group. *AAZPA Ann. Conf. Proc.*, 37–40.

Anon. (1991a) The conservation role of zoos: budgets and priorities. *CBSG News*, **2** (4), 23–4.

Anon. (1991b) Noah Project news. *CBSG News*, **2** (1), 11–12.

Aum, E. (1982) Training through staff exchange programs. *Animal Keepers' Forum*, December, 1982, 375–80.

Beehler, B.A. (1990) Zoo Biology Training Course: China. *CRC Newsletter*, **2** (2), 4–5.

Bendiner, R. (1981) *The Fall of the Wild: The Rise of the Zoo*, Dutton, New York.

Braham, M. (1988) The ecology of education, in *New Ideas in Environmental Education* (eds S. Briceno and D.C. Pitt), IUCN/Croom Helm, London, pp. 3–32.

Calvo, L. (1989) The current state of zoos in Central America. *AAZPA Ann. Conf. Proc.*, 15–6.

Carr, B.L. (1990) AAZPA Conservation Academy. Broadening perspectives in a shrinking world. *AAZPA Ann. Conf. Proc.*, 398–401.

Christman, J. (1989) The sister zoo project: La Aurora Zoo and the Dallas Zoo. *AAZPA Ann. Conf. Proc.*, 17–19.

Curtis, L. (1982) Husbandry of mammals, in *Zoological Park and Aquarium Fundamentals* (ed. K. Sausman), AAZPA Wheeling, West Virginia, pp. 245–55.

Desai, J.H. (1992) Captive breeding, rehabilitation, and reintroduction of endangered species in India: a status report. *CBSG News*, **3** (1), 31–3.

Diaz Matalobos, M. and Barraza, L. (1992) Bridging the gap – Latin American zoos and the world zoo community. *Int. Zoo News*, **39** (1), 11–13.

Durrell, G. (1976) *The Stationary Ark*, Collins, Glasgow.

Ehrlich, P.R. and Ehrlich, A.H. (1981) *Extinction: the Causes and Consequences of the Disappearance of Species*, Random House, New York.

Europa (1991) *Europa World Yearbook* vols I and II, Europa Publications, London.

Foose, T.J. (1988) Globalization of Species Survival Plan programs – the world as megazoo. *Proc. 5th World Conf. on Breeding Endangered Species in Captivity*, 383–91.

Foose, T. (1991) CBSG Captive Action Plans. *CBSG News*, **2** (2), 5–7.

Freese, C.H. and Saavedra, C.J. (1991) Prospects for wildlife management in Latin America and the Caribbean, in *Neotropical Wildlife Use and Conservation* (eds J.G. Robinson and K.H. Redford), University of Chicago Press, Chicago, pp. 430–44.

Hage, S. (1992) CBSG education and training report. *CBSG News*, **3** (1), 30.

Harris, J. (1992) New International Guest House and Training Center. *The ICF Bugle*, **18** (1), 3–4.

Hediger, H. (1970) *Man and Animal in the Zoo: Zoo Biology*, Routledge & Kegan Paul, London.

IUCN (1990) *1990 IUCN Red List of Threatened Animals*, IUCN Gland, Switzerland and Cambridge UK.

IUCN/UNEP/WWF (1980) *World Conservation Strategy*, IUCN, Gland, Switzerland.

IUCN/UNEP/WWF (1991) *Caring for the Earth. A Strategy for Substainable Living*, IUCN, Gland, Switzerland.

Jones, D.M. (1985) The care of exotic animals, in *Advances in Animal Conservation* (eds J.P. Hearn and J.K. Hodges), Symp. Zool. Soc. London, **54**, 89–101.

Karsten, P. (1974) The apprentice zookeeper training program. *Dinny's Digest*, **2** (10), 3–20.

Kohn, F.B. (1990) American Association of Zoo Keepers: Growth through education and training materials. *AAZPA Reg. Proc.*, 633–6.

Koontz, F. and Hutchins, M. (1990) An investment in the future: Keeper training at the New York Zoological Park. *AAZPA Ann. Conf. Proc.*, 390–7.

Krausman, P.R. and Johnsingh, A.J.T. (1990) Conservation and wildlife education in India. *Wildlife Soc. Bull.*, **18** (3), 342–7.

Luthin, C.S. (1987) Brehm Fund contracts. *Garrulax*, **3**, 6–7.

Macdonald, D. (ed.) (1985) *The Encyclopaedia of Mammals*, vol.1. George Allen & Unwin, London.

MacKinnon, J., MacKinnon, K., Child, G. and Thorsell, J. (comps.) (1986) *Managing Protected Areas in the Tropics*, IUCN, Gland, Switzerland.

Mallinson, J.J.C. (1988) Collaboration for conservation between the Jersey Wildlife Preservation Trust and countries where species are endangered. *Int. Zoo Yb.*, **271**, 176–91.

Martin, R.D. (ed.) (1975) *Breeding Endangered Species in Captivity*, Academic Press, London.

McNeely, J.A. (1988) Ecology is not enough. *IUCN Bull.*, **19**, 2–3.

Mittermeier, R.A. (1986) Who will pilot the ark?, in *Primates: The Road to Self-sustaining Populations* (ed. K. Benirschke), Springer–Verlag, New York, pp. 985–7.

Mittermeier, R.A. (1988) Primate diversity and the tropical forest: case studies from Brazil and Madagascar and the importance of the megadiversity countries, in *Biodiversity* (ed. E.O. Wilson), Nat. Acad. Sciences/Smithsonian Inst., Washington, DC, pp. 145–54.

Mittermeier, R.A. and Oates, J.F. (1985) Primate diversity: the world's top countries. *Newsletter/J. IUCN/SSC Primate Specialist Group*, **5**, 41–8.

Montagu, Lord (1983) *Britain's Zoos: Marketing and Presentation – The Way Forward to Viability*, English Tourist Board, London.

Myers, N. (1979) *The Sinking Ark*, Pergamon Press, Oxford.

Norton, B.G. (1986) *The Preservation of Species: The Value of Biological Diversity*, Princeton University Press, Princeton, New Jersey.

Olney, P.J.S. and Ellis, P. (eds) (1989) *Int. Zoo Yb.* **28**, Zool. Soc. London.

Rudran, A., Wemmer, C.M. and Singh, M. (1990) Teaching applied ecology to nationals of developing countries, in *Race to Save the Tropics: Ecology and Economics For a Sustainable Future* (ed. R. Goodland), Island Press, Washington, DC, pp. 125–40.

Sinclair, C.E. (1988) Financial and conservation differences among various kinds of zoos and recommendations for a proposed new zoo for Milan, Italy. *Unpublished Training Report*, Jersey Wildlife Preservation Trust.

Skrei, S. and Lieberman, G.A. (1990) Zoo Conservation Outreach Group, training opportunities for zoo professionals in Tropical America: experiences to date. *AAZPA Ann. Conf. Proc.*, 402–8.

Tilson, R.L. (1992) Preserving critical habitat. The Minnesota Zoo's Adopt-A-Park Program. *CBSG News*, **3** (1), 20–1.

Towne, D. (1990) Northwest Center for Wildlife Conservation. *Species*, **15**, 7.

UFAW (1988) Why Zoos?, *UFAW Courier* No. 24., Universities Federation for Animal Welfare, Potters Bar, England.

Vecchio, T. (1990) Zoo training programs: a valuable export. *AAZPA Reg. Proc.*, 631–2.

Walker, S. (1987) Zookeeper and other staff training in India: some pioneer efforts. *Animal Keepers' Forum*, December 1987, 432–5.

Waugh, D.R. (1983) The Wildlife Preservation Trust training programme. *Dodo, J. Jersey Wildl. Preserv. Trust*, **20**, 12–16.

Waugh, D.R. (1988) Training in zoo biology, captive breeding and conservation. *Zoo Biol.*, **7**, 269–80.

Wemmer, C. (1990) International Zoo Biology Training: a method, progress and revelations. *AAZPA Ann. Conf. Proc.*, 409–15.

Wemmer, C., Pickett, C. and Teare, J.A. (1990) Training zoo biology in tropical countries: a report on a method and progress. *Zoo Biol.* **9**, 461–70.

Wilson, E.O. (ed.) (1988) *Biodiversity*, National Acad. Sciences/Smithsonian Inst., Washington, DC.

Part Two

Reintroduction and Captive Breeding

12

Reintroduction as a reason for captive breeding

A.C. Wilson and M.R. Stanley Price
IUCN/SSC Reintroduction Specialist Group, Nairobi, Kenya.

12.1 INTRODUCTION

If, as Colin Tudge (1991) asserts, 'the proper end point of captive breeding is reintroduction', then it is timely that we examine the status of the science and art of reintroduction. Restoring extirpated species to their natural environment is not new, but the recent upsurge in interest in reintroductions is in part due to a perceived need for zoos to play a more active role in the conservation of endangered species. The fact that the World Conservation Union/Species Survival Commission (IUCN/SSC) has set up a Reintroduction Specialist Group (RSG) is in part a reflection of this interest, but also of the fact that reintroductions are becoming an accepted item in the conservation tool-box. Despite evidence that wild-to-wild translocations have historically been more successful than releases of captive-bred animals (Griffith *et al.*, 1990), the use of captive-bred animals is often preferable where translocations might further endanger a remnant wild population.

12.2 DEFINITIONS

The following definitions are used in the context of this paper: a 'reintroduction' is an attempt to establish a species (or subspecies) in an

[a]And African Wildlife Foundation, Nairobi, Kenya.

Creative Conservation: Interactive management of wild and captive animals.
Edited by P.J.S. Olney, G.M. Mace and A.T.C. Feistner.
Published in 1994 by Chapman & Hall, London.
ISBN 0 412 49570 8

area which was once part of its historical range, but from which it has disappeared; a 'translocation' involves movement of wild-born individuals or populations from one part of their range to another. When the translocation is to an area where the species has become extinct, it constitutes a reintroduction. A 'conservation introduction' or 'benign introduction' is an attempt to establish a species, for the purpose of conservation, outside its recorded range but within an appropriate habitat and eco-geographical area. Where the conservation introduction is to an island or otherwise isolated habitat, it is sometimes referred to as 'marooning'. An 'extirpation' refers to the disappearance of a species from a limited part of its range.

In this paper we give an overview of the information on historical, current and proposed reintroduction projects which are on the database of the RSG. The information is gathered from the literature (both scientific and 'popular') and from personal communications with RSG members. We also look at recommendations for reintroductions made by members of the IUCN/SSC's Specialist Groups and published in the species Action Plans. We attempt to answer the following questions. What species have been, or are going to be, reintroduced? Where? Are they locally or globally threatened species? Have they come from captive-bred stock? Can we predict what future role reintroduction might play in the conservation of threatened species? Are there groups of organisms for which we might anticipate reintroduction needs? Are there groups of animals for which reintroductions will not be a likely option? What role is captive breeding likely to play? In which geographical areas will reintroductions be needed, or indeed viable? Who decides which species to reintroduce?

Because the RSG database as yet includes little information on reintroductions of taxa other than birds and mammals, we concentrate on those two groups for our analysis of past and proposed reintroductions.

An objective scrutiny of reintroduction as an **effective** tool in conserving endangered species is not yet possible. The reasons for this are elaborated by Beck *et al.* (Chapter 13) but can be summarized under three broad categories. First, the establishment of viable, self-sustaining populations in the wild is a long-term proposition and most projects simply have not been running long enough to assess their success or failure. Second, for each species reintroduction poses unique problems, and for most groups of animals (with the probable exception of raptors) we are still in the process of refining release strategies and post-release monitoring techniques. Third, many past reintroductions were *laissez-faire* exercises which suffered from poor planning and record-keeping and lack of post-release monitoring. Most reintroductions these days are performed more conscientiously. Several publications (e.g. IUCN, 1987; Black, 1991) have already outlined reintroduction procedure, and it is hoped that the Guidelines for Reintroductions (RSG, 1992), which the RSG is developing in conjunction with the Captive Breeding Specialist Group (CBSG) and Veterinary Specialist Group (VSG), will form a practical framework for reintroductions.

12.3 THE CURRENT STATUS OF REINTRODUCTION PROJECTS

Tables 12.1 and 12.2 show past, current and proposed bird and mammal reintroductions which are in the RSG project database. The database is not exhaustive and for some projects we have very little information other than the fact that they exist. However, it serves to illustrate the range of species being reintroduced. The database includes reintroductions of both captive-bred and wild-caught individuals. In Chapter 13 Beck *et al.* analyse reintroductions of exclusively captive-born animals in more detail.

In Tables 12.1 and 12.2 we have included a column indicating the principal original cause(s) of extirpation or decline (where known) of each species. (1) involves direct killing or removal (overhunting for food, sport, trophies, hides, medicines; elimination because of competition with or predation on livestock; overcollection for the pet or falconry trade). Species which have been most affected by these factors include most large mammals, turtles, parrots and raptors. (2) encompasses the effects of environmental toxins. Vulnerable species include those at the top of the food chain, and aquatic species particularly amphibians and fish. (3) involves introduction of exotic competitors or predators. This is a major cause of decline of island species and those in Australia, New Zealand and Hawaii. (4) includes species with small fragmented populations vulnerable to natural disasters which affects a broad range of species, in particular those inhabiting tropical forests, islands and small wetlands. (5) is habitat loss. Three other causes of species declines – accidental killing (e.g. road deaths, agricultural practices), hybridization, and diseases introduced by closely related species – have been grouped under (3).

Tables 12.1 and 12.2 show that of 149 bird and mammal reintroductions involving 121 species, 53 are not species listed as globally threatened in the Red Data Book (IUCN, 1990) and these projects represent the replacement of locally extirpated populations. Almost half the bird reintroduction projects involve birds of prey (Strigiformes and Falconiformes), while 26% of mammal projects are of Carnivora. Of all the 54 projects involving predatory species, an overwhelming 76% are in Europe and North America. For birds the reasons for this are not hard to find: first, the main reasons (overhunting and pollution) that have caused declines of the Falconiformes and Strigiformes in the industrialized countries are beginning to be controlled; second, release techniques

Table 12.1 Bird reintroduction projects with regard to group, conservation status, country of release, origin of release stock and cause of decline

Species	*Country*	*RDB status*	*Origin*	*Cause of decline*	*Comments*
Struthioniformes					
Ostrich *Struthio camelus*	Israel		CB	1/3/5	Proposed
Ciconiiformes					
Puffin *Fratercula arctica*	USA		WW		
Yellow-crowned night heron *Nyctanassa violacea*	Bermuda		WW		
Milky stork *Mycteria cinerea*	Malaysia	V	CB	1/5	
Eastern white stork *Ciconia c. boyciana*	Japan		CB		Proposed
N. bald ibis *Geronticus eremita*	Turkey	E	CB		Failure
N. bald ibis *G. eremita*	Spain	E	CB		Proposed
N. bald ibis *G. eremita*	Morocco	E	CB		Proposed
S. bald ibis *G. calvus*	South Africa	R	?		Proposed
Anseriformes					
Hawaiian goose *Branta sandvicensis*	USA	E	CB	1/3/4/5	
Aleutian Canada goose *B. canadensis leucopareia*	USA	V	CB + WW	1/3	
Light-bellied brent goose *B. berniclahrota*	?Norway		CB		
Lesser white-fronted goose *Anser erythropus*	Sweden	R	CB	1/2	
New Zealand blue duck *Hymenolaimus malacorhyncus*	New Zealand	E			
Mandarin duck *Aix galericulata*	Far East		CB	1/2/5	
New Zealand brown teal *Anas aucklandica*	New Zealand	E		1/3/5	
Hawaiian duck *A. wyvilliana*	USA	E	CB	1/3/4/5	
Mexican duck *A. platyrhyncos diazi*	USA		CB		Completed
New Zealand scaup *Aythya novae-seelandiae*	New Zealand	V			Completed
N. American wood duck *Aix sponsa*	USA		CB		Completed

Marbled teal *Marmaronetta angustirostris*	Pakistan	V	CB	1/5	
White-headed duck *Oxyura leucocephala*	Spain	V	CB	1/5	
White-headed duck *O. leucocephala*	Hungary	V	CB	1/5	Failure
Falconiformes					
California condor *Gymnogyps californianus*	USA	Ex	CB	1/5	
Andean condor *Vultur gryphus*	Columbia		CB		
Andean condor *V. gryphus*	Venezuela		CB		
Lappet-faced vulture *Torgos tracheliotus*	Israel				Proposed
Bearded vulture *Gypaetus barbatus*	Israel		CB		
Bearded vulture *G. barbatus*	European Alps				
Griffon vulture *G. fulvus*	France				
Griffon vulture *G. fulvus*	Israel		CB		Proposed
Black vulture *Aegypius monachus*	France	V			
Bald eagle *Haliaeetus leucocephalus*	USA (New York)			2/5	
Bald eagle *H. leococephalus*	USA (south)			1/2/5	
Imperial eagle *Aquila heliaca*	Spain	R		1/5	
White-tailed eagle *Haliaeetus albicilla*	Germany	R	CB	1/2/5	
White-tailed eagle *H. albicilla*	UK (Scotland)	R	WW	1/2/5	
White-tailed eagle *H. albicilla*	Ireland	R	WW	1/2/5	
White-tailed eagle *H. albicilla*	Israel	R	CB	2/5	Proposed
Bonelli's eagle *Hieraaetus fasciatus*	France		CB		
Aplomado falcon *Falco femoralis*	USA		CB		
Bat falcon *F. rufigularis*	Guatemala				
Peregrine falcon *F. peregrinus*	Sweden		CB		
Peregrine falcon *F. peregrinus*	Canada		CB		
Peregrine falcon *F. peregrinus*	USA (west)		CB + WW		
Peregrine falcon *F. peregrinus*	USA (east)		CB + WW		
Mauritius kestrel *F. punctatus*	Mauritius	E	CB	3/5	
Lesser kestrel *F. naumanni*	Spain	R	CB		
Lanner *F. biarmicus*	Israel		CB		

Table 12.1 (*continued*)

Species	*Country*	*RDB status*	*Origin*	*Cause of decline*	*Comments*
Harris's hawk *Parabuteo unicinctus*	USA		CB		
Osprey *Pandion haliaetus*	USA		WW		
Red kite *Milvus milvus*	UK (Wales)		WW	1/2/5	
Red kite *M. milvus*	UK (Scotland)		WW	1/2/5	
Red kite *M. milvus*	UK (England)		WW	1/2/5	
Galliformes					
Cheer pheasant *Catreus wallichi*	Pakistan	V	CB	1/5	?Failure
Masked bobwhite *Colinus virginianus*	USA		CB		
Malleefowl *Leipoa ocellata*	Australia	K		3/5	
Gruiformes					
Whooping crane *Grus americana*	USA	E	CB	1/5	Proposed
Siberian crane *G. leucogeranus*	Russia				
Eastern sarus crane *G. antigone*	Thailand		CB		
Guam rail *Rallus owstoni*	USA	E	CB	3/5	Marooning
Takahe *Notornis mantelli*	New Zealand	E	CB	3/5	
Lord Howe Is. rail *G. sylvestris*	Australia	E	CB	3/5	
Great bustard *Otis tarda*	Europe	R	CB + WW	1/5	
Houbara bustard *Chlamydotis undulata*	UAE	V	CB	1/5	
Purple swamphen *Porphyrio porphyrio*	Spain		WW		
Charadriiformes					
Black stilt *Himantopus novaezelandiae*	New Zealand	E	CB	3/5	
Columbiformes					
Pink pigeon *Nesoenas mayeri*	Mauritius	E	CB	1/3/5	
Socorro dove *Zenaida graysoni*	Mexico	E	CB	3/4	

Psittaciformes					
Bahama parrot (*Amazona leucocephala bahamensis*)	Bahamas				
Orange-bellied parrot *Neophema splendida*	Australia	R	5		
Thick-billed parrot *Rhyncopsitta pachyrhyncha*	USA	V	1/5		
Strigiformes					
Barn owl *Tyto alba*	UK		WW	2/5	
Elf owl *Micrathene whitneyi*	USA		CB		
Burrowing owl *Speotyto cunicularia*	Canada		WW	3/5	
Eagle owl *Bubo bubo*	Sweden		CB		
Eagle owl *B. bubo*	Europe				
Pygmy owl *Glaucidium passerinum*	Germany				
Passeriformes					
Bali starling *Leucopsar rothschildi*	Indonesia	E	CB	1/5	
Helmeted honeyeater *Lichenostomus melanops cassidix*	Australia		CB		
White-eyed vireo *Vireo griseus*	Bermuda		WW	5	
Black robin *Petroica traversi*	New Zealand	E	WW	3/5	Marooning
Hawaiian crow *Corvus hawaiiensis*	USA	E	CB	3/5	Proposed
San Clemente shrike *Lanius ludovicianus*	USA		CB		
Mauritius fody *Foudia rubra*	Reunion		?CB		Marooning/ Proposed

(Abbreviations: RDB = IUCN Red Data Book; Ex = extinct in wild; E = Endangered; V = Vulnerable; R = Rare; I = Indeterminate; K = Insufficiently known; CB = captive bred; WW = wild to wild; 1 = hunting/collecting; 2 = pollution; 3 = competition/predation; 4 = small, fragmented populations; 5 = habitat loss.)

Table 12.2 Mammal reintroduction projects with regard to group, conservation status, country of release, origin of release stock and cause of decline (Abbreviations as in Table 1.2; status in brackets = Specialist Group opinion)

Species	*Country*	*RDB status*	*Origin*	*Cause of decline*	*Comments*
Marsupialia					
Burrowing bettong *Bettongia lesueur*	Australia	R(E)	WW	3/5	
Brush-tailed bettong *B. penicillata*	Australia	E	CB	3/5	
Rufous hare-wallaby *Lagorchestes hirsutus*	Australia	R(E)	CB	3/5	
Brush-tailed phascogale *Phascogale tapoatafa*	Australia		CB	3/5	Experimental
Eastern barred bandicoot *Parameles gunnii*	Australia	(V)	WW	3/5	
Sugar glider *Petaurus breviceps*	Australia			3/5	
Greater bilby *Macrotis lagotis*	Australia	E(V)	?	3/5	
Numbat *Myrmecobius fasciatus*	Australia	E	CB + WW	3/5	
Primates					
Black and white ruffed lemur *Varecia v. variegata*	Madagascar	V	CB		Proposed
Black-handed spider monkey *Ateles geoffroyi panamensis*	Panama	V	WW		
Golden lion tamarin *Leontopithecus rosalia*	Brazil	E	CB	1/5	
Barbary macaque *Macaca sylvanus*	Morocco	V			
Lion-tailed macaque *M. silenus*	India	E			
Orang-utan *Pongo pygmaeus*	Indonesia	E		1	
Rodentia					
Red squirrel *Sciurus vulgaris*	UK			3/5	Failure
Greater sticknest rat *Leporillus conditor*	Australia	R	CB		
Jamaican hutia *Geocapromys brownii*	Jamaica	I	CB		
Common dormouse *Muscardinus avellanarius*	UK				

Carnivora					
Grey wolf *Canis lupus*	USA (Yellowstone)	V		1	Proposed
Red wolf *C. rufus*	USA (NC)	E	CB	1/3/5	
Red wolf *C. rufus*	USA (TN)	E	CB		
African wild dog *Lycaon pictus*	South Africa	E	WW	1	Several projs.
African wild dog *L. pictus*	Namibia	E		1	Failure
Swift fox *Vulpes velox*	Canada		CB + WW		
N. American river otter *Lutra canadensis*	USA		WW		
European otter *Lutra l. lutra*	UK	V	CB	1/2/5	Two projs.
Sea otter *Enhydra lutris*	USA/Canada		WW	1	
Black-footed ferret *Mustela nigripes*	USA	Ex	CB	1/3/5	
Cheetah *Acinonyx jubatus*	South Africa	V	WW	1/5	
Lynx *Felis lynx*	Switzerland		WW	1/5	
Lynx *F. lynx*	Yugoslavia		WW	1/5	
Lynx *F. lynx*	France		WW	1/5	
Lynx *F. lynx*	USA		WW	1	
Pardel lynx *F. pardina*	Spain/Portugal	E	WW	1/5	Proposed
Panda *Ailuropoda melanoleuca*	China	E	CB	1/5	Proposed
Perissodactyla					
Onager *Equus hemionus onager*	Israel		CB	1/3/5	Diff. ssp.
Onager *E. h. onager*	Saudi Arabia		CB	1/3/5	Diff. ssp.
Przewalski's horse *E. przewalskii*	Siberia	Ex?	CB	1/3/5	Proposed
Przewalski's horse *E. przewalskii*	Mongolia	Ex?	CB	1/3/5	Proposed
Przewalski's horse *E. przewalskii*	China	Ex?	CB	1/3/5	
Great one-horned rhino *Rhinoceros unicornis*	India	E	CB	1/5	

Table 12.2 (*continued*)

Species	*Country*	*RDB status*	*Origin*	*Cause of decline*	*Comments*
Artiodactyla					
Barbary deer *Cervus elaphus barbarus*	Algeria	V	CB	1/?	
Formosan sika deer *C. nippon taiouanus*	Taiwan	E	CB	1/5	
Père David's deer *Elaphurus davidianus*	China	E		1/5	
Heumul *Hippocamelus bisculeus*	Chile	E	CB + WW		
Roe deer *Capreolus capreolus*	Israel		CB	1/3/5	Failure
Blackbuck *Antilope cervicapra*	Pakistan		CB		
Woodland caribou *Rangifer tarandus*	USA/Canada		WW		
Moose *Alces americana*	USA		WW		Proposed
European bison *Bison bonasus*	Poland/Russia	V	CB	1/5	
Addax *Addax nasomaculatus*	Tunisia	E	CB	1/5	
Addax *A. nasomaculatus*	Niger	E	CB	1/5	Proposed
Cuvier's gazelle *Gazella cuvieri*	Algeria	E	CB	1/5	
Mountain gazelle *G. gazella*	Saudi Arabia	E	CB	1/5	
Dama gazelle *G. dama*	Senegal	E	CB	1	
Sand gazelle *G. subgutterosa marica*	Saudi Arabia		CB	1	
Arabian oryx *Oryx leucoryx*	Oman	E	CB	1	
Arabian oryx *O. leucoryx*	Jordan	E	CB	1/5	
Arabian oryx *O. leucoryx*	Israel	E	CB	1/5	Proposed
Scimitar-horned oryx *O. dammah*	Tunisia	E	CB	1	
Abbruzzo chamois *Rupicapra pyrenaica ornata*	Italy	V	CB	1/5	
Nubian ibex *Capra ibex*	Jordan		CB	1/5	
Alpine ibex *C. ibex*	European Alps		CB + WW		
Desert bighorn *Ovis canadensis*	USA		WW	1	Several projs.

for birds of prey are probably better developed than for any other group; and third, as a group they are 'charismatic' animals with tremendous public appeal.

Whether most of these reintroductions will succeed (for a definition of 'success' see Beck, *et al.*, Chapter 13) remains to be seen; no amount of public goodwill or government support will compensate for the loss of habitat most of these species have undergone in the past few centuries. However, as the peregrine (*Falco peregrinus*) reintroductions in North America show, as long as sufficient habitat remains there is a good chance of success (Cade *et al.*, 1988). This might not be the case for carnivorous mammals which pose a real or imagined threat to livestock; some lynx (*Felis lynx*) reintroductions in Europe have been less than successful because of opposition, and sometimes armed action, by farmers (Breitenmoser and Breitenmoser-Wursten, 1990). In the USA, reintroductions of red wolves (*Canis rufus*) in North Carolina and lynx in New York State have been preceded and accompanied by intense public relations exercises, coupled with laws exonerating hunters and trappers from punishment arising from accidental killing of the carnivores (Phillips, 1990; Brocke, Gustafson and Major, 1990); however in neither of these cases was a threat to livestock a major concern. A proposed reintroduction of grey wolves (*C. lupus*) into the Yellowstone ecosystem, however, is meeting fierce opposition from sheep and cattle ranchers which may preclude the project even though there is support from the general public.

The group having the largest number (23 or 36%) of mammal reintroductions is the Artiodactyla, and most of these projects do involve globally endangered species. Again, antelope, deer and sheep are charismatic species and the success of the Arabian oryx (*Oryx leucoryx*) project in Oman (Stanley Price, 1989; Spalton, 1992) has provided additional encouragement to attempt further reintroductions in the Middle East. For most artiodactyls the original cause of decline was overhunting, but even if habitat loss was not a major factor in their extirpation from the wild, it may still become a major obstacle to their successful reintroduction. Rising populations of humans and their livestock, and changes in lifestyle such as possession of vehicles may mean that most arid-land antelope will have to be released into closed reserves rather than being truly free-ranging wild populations.

12.4 GEOPOLITICAL CONSIDERATIONS

Hunting and collecting for the pet trade are theoretically relatively easy to control once legislation is in place. However, this presupposes adequate law enforcement and a public with a supportive attitude made possible through public relations exercises and education. A reintroduction may stand a higher chance of success if made into a country which has a relatively high standard of living (and its frequent corollary, political stability). Currently, there are several reintroduction projects (e.g. Przewalski's horse [*Equus przewalskii*] into Russia and Mongolia, and addax [*Addax nasomaculatus*] into Niger) which are 'on hold' due to economic or political problems in the intended country of release. As Beck *et al.* (See Chapter 13) point out, reintroductions are long-term, expensive undertakings and economic factors are undoubtedly the reason that the great majority of them have been carried out in North America, Europe and Australia/New Zealand.

12.5 THE ROLE OF HABITAT RESTORATION

For sound reasons, few reintroductions have been attempted where habitat destruction has been the major cause of a species' decline, although habitat loss has played a role in almost all species declines. Restoration ecology is a fairly new discipline and the economic, social and technical prerequisites for habitat reconstruction have limited the process to areas such as small prairies, abandoned mines, botanical gardens, small wetlands and islands (Cairns, 1988), with most restorations being attempted in industrialized countries. In North America, tallgrass prairie communities at most a few hundred hectares in extent are being restored, a process that requires thousands of hours of volunteer labour (Jordan and Packard, 1989). In the past, only vegetative restoration was attempted, on the assumption that animals would 'find their way' into restored areas. There are various reasons why this does not happen (Jordan, 1988). Moreover, most restored areas are too small to support viable populations of large vertebrates. As bison *(Bison bison)* must have been major architects of the vegetative composition of the original prairies, the absence of these herbivores poses a problem for purists who wish to return to original conditions (S.Packard, pers. comm.). However, restoration of small areas does provide opportunities for reintroductions of some invertebrates and some vertebrates, such as fish, herpetofauna, birds and small mammals; reintroductions of these types of animals into restored habitats is now becoming popular in the United States (e.g. Stolzenburg, 1992). Captive breeding of small extirpated animals should be encouraged, perhaps ideally in smaller zoos close to the reintroduction area.

Habitat restoration is expensive but it has been attempted in less-developed countries, and even on a much larger scale; the restoration of Guanacaste National Park (Janzen, 1986) is a case in point. However, given the social and economic restraints, it seems unlikely that restoration of large areas of tropical ecosystems and reintroduction of extirpated species to them is a realistic option in the foreseeable future.

Removal of introduced herbivores and predators from islands such as those in New Zealand (Veitch and Bell, 1989) is also a form of habitat restoration, but this elimination is costly and labour-intensive. In Australia and New Zealand it has been successful mainly on small islands or fenced areas such as peninsulas (Brown, 1991). Reintroductions of species into these areas has been successful in some cases (Brown, 1991), although a theoretical problem of this exercise is the limited scope for expansion of the reintroduced species, so that even at carrying capacity, population sizes may not be strictly viable.

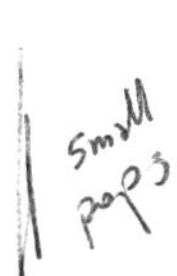

12.6 CONSERVATION INTRODUCTIONS

Marooning of endangered species onto 'clean' islands (or otherwise protected areas) outside their historical range is becoming increasingly popular. The Guam rail *(Rallus owstoni)* has been introduced onto the nearby island of Rota which is free of the predators which were a major factor in its extinction on Guam. There are proposals to establish the Mauritius fody (*Foudia rubra*) onto Reunion Island, and discussions about the potential of introducing the Rodrigues fruit bat (*Pteropus rodricensis*) to an island in the western Indian Ocean outside the cyclone belt (see Mickleburgh and Carrol, Chapter 19). Not reintroductions in the strict sense, these are sometimes called 'benign' or 'conservation' introductions. Care should, however, be exercised; the introduction of the endangered rodent Bahamas hutia (*Geocapromys ingrahami*) onto a small island where it had not been recorded previously, resulted in a massive perturbation of plant community structure. This included the possible extinction of eight plant species which presumably had not coevolved adequate defense mechanisms against grazing by the rodents (Campbell, Lowell and Lightbourn, 1991). Coevolution of plants and animals that eat or pollinate them is poorly documented and should be taken into account when attempting any introduction, no matter how 'benign' the intentions are.

12.7 FUTURE REINTRODUCTION NEEDS

Table 12.3 summarizes data from the 24 species Action Plans which had been completed by IUCN Specialist Groups by the end of 1992.

Of the 2409 species (a tiny proportion of the world's estimated 3–5 million species!) covered by the Red List (IUCN, 1990), over 660 are listed as Extinct, Endangered, Rare, Vulnerable, Indeterminate or Insufficiently Known or are considered by the authors of the Action Plans to be under threat. Reintroduction is recommended for only 68 (10.3%) of these, of which marsupials, crocodiles and otters (a total of 38 species) make up the major proportion. It is interesting to note that

Table 12.3 Comparison between IUCN Action Plans for number of species included and number of species recommended for reintroductions (after Stuart, 1991, with additions)

Species Action Plan	*Total No. species included*	*No. threatened species*	*No. species recommended for reintroduction*	*Percentage of threatened species recommended for reintroduction*
African insectivores and elephant shrews	192	>26	0	0.0
African primates	63	30	0	0.0
Asian primates	63	30	1	3.3
Mustelids and viverrids	123	38	1	2.6
Foxes, wolves, jackals and dogs	34	18	4	22.2
Dolphins, porpoises and whales	77	21	0	0.0
African elephants and rhinos	3	3	2	66.7
Asian rhinoceros	3	3	2	66.7
East African antelopes	59	15	0	0.0
Southern African antelopes	36	2	0	0.0
West and Central African antelopes	44	8	4	50.0

Kouprey	1	1	1	100.0
African forest birds	409	102	0	0.0
Tortoises and freshwater turtles	263	121	4	3.3
Rabbits, hares and pikas	78	17	5	29.4
Australasian marsupials and monotremes	149	27	19	70.3
Crocodiles	23	15	12	80.0
South American camelids	4	2	0	0.0
Otters	13	8	7	87.5
Asian elephant	1	1	0	0.0
Swallowtail butterflies	573	75	2	2.6
Equids	7 (21ssp.)	5 (7ssp.)	2 (3ssp.)	40.0 (42.8)
Old World fruit bats	161	>69	1*	1.5
Lemurs	30 (28ssp.)	27	1 (2ssp.)	3.7
Totals	2409	>664	68	10.2

*A conservation introduction rather than a reintroduction.

although captive breeding is recommended for a great many species by the authors of the Action Plans, reintroduction as the ultimate goal of this exercise is rarely mentioned.

Table 12.4 lists the 68 species recommended for reintroduction in the Action Plans. For a majority (45) of these, captive breeding is also recommended. Most are already being bred with varying degrees of success in zoos; only 14 are reported to have international studbooks (Olney and Ellis, 1991), although others may have regional studbooks. Of the remaining 23 species for which translocation (to areas where they have been locally extirpated) is recommended, sufficient numbers remain elsewhere in the wild (e.g. some marsupials and otters). In some cases, translocation is an option where there are fears that numbers needed to establish viable zoo populations would jeopardize remnant wild populations (e.g. Javan rhino). Otters and fruit bats appear to be particularly difficult to breed in captivity (Foster-Turley, Macdonald and Mason, 1990; Mickleburgh, Hutson and Racey, 1992; but see Mickleburgh and Carroll, Chapter 19), but the Action Plans for these groups clearly make reference to the need for renewed efforts. Not all captive-bred stock need be from zoos; reintroductions of saltwater and mugger crocodiles (*Crocodylus porosus* and *C. palustris*) in India and Pakistan have taken place using surplus stock from commercial farming operations, but there is still a need for zoos to breed the rarer species. Commercial farms are concerned with production rather than genetic purity, and in the case of Siamese crocodiles (*C. siamensis*), interbreeding with *C. porosus* has rendered one farmed population unsuitable for reintroduction (Messel, King and Ross, 1992).

However, locally as well as globally, endangered species will be candidates for reintroductions. Almost half the birds and mammals currently being reintroduced do not appear in the Red List (and there are no doubt more projects of which RSG is not aware). Many of these reintroductions of locally extirpated species are from captive-bred stock; even when translocations would be feasible, captive breeding is desirable where translocations would involve animals from genetically distinct sub-populations, or when wild populations would not sustain a large offtake.

Why is reintroduction considered a viable option for some species, or groups of species, but not for others? Do the Action Plan recommendations reflect the preference or experience of the authors, and are they based on sound criteria? As Stuart (1991) points out, habitat protection (including establishment and management of protected areas) is the most desirable method of conservation for the majority of endangered species. Of the 68 species so far being recommended for reintroduction it must be assumed that sufficient habitat exists. However, 40 of these species are native to areas outside Europe, North America, Japan and

Table 12.4 Species proposed for reintroduction in the IUCN Action Plans, origin of release stock and status of projects (Abbreviations as in Table 12.1; S = studbook; status in brackets = status proposed by authors of Action Plans)

Species	*RDB status*	*Location*	*Origin*	*Reintroduction under way?*
Orang-utan *Pongo pygmaeus* (S)		Sabah/Brunei	CB	
Black-footed ferret *Mustela nigripes* (S)	Ex	USA	CB	Yes
Maned wolf *Chrysocyon brachyurus* (S)	V	South America	?	
Simien jackal *Canis simensis*	E	Ethiopia	CB	
Red wolf *C. rufus* (S)	E	USA	CB	Yes (North Carolina, Tennessee)
Grey wolf *C. lupus*	V	Mexico/N.Am./Eur.	WW	
White rhinoceros *Ceratotherium simum* (S)	E	South Africa	CB	Yes
Black rhinoceros *Diceros bicornis* (S)	E	African sanctuaries	CB + WW	Yes
Indian rhinoceros *Rhinoceros unicornis* (S)	E	India/Nepal	CB	Yes
Javan rhinoceros *R. sondaicus*	E	Sumatra	WW	
Przewalski's horse *Equus przewalskii* (S)	Ex	China/Russia/Mongolia	CB	Yes (China)
Onager *E. hemionus onager*/*E.h.khur*	E	Iran/India		
Addax *Addax nasomaculatus* (S)	E	Sahel	CB	Yes (Tunisia)
Scimitar-horned oryx *Oryx dammah*	E	Sahel	CB	Yes (Tunisia)
Dama gazelle *Gazella dama* (S)	E	Sahel	CB	Yes (Senegal)
Slender-horned gazelle *G. leptoceros* (S)	E	Sahel/Sahara		
Kouprey *Bos sauveli*	E	Indochina	?	
African spurred tortoise *Geochelone sulcata*		Sudan	CB	
Hermann's tortoise *Testudo hermanni*	V	France	CB	Yes
Black softshell tortoise *Trionyx nigricans*	R	Bangladesh	?	
Western swamp turtle *Pseudemydura umbrina*	E	Australia	CB	

Table 12.4 (*continued*)

Species	*RDB status*	*Location*	*Origin*	*Reintroduction under way?*
Hispid hare *Caprolagus hispidus*	E	India/Nepal	CB	
Sumatran rabbit *Nesolagus netscheri*	I	Sumatra	CB	
Tehuantepec jackrabbit *Lepus flavigularis*	E	Mexico	CB	
Volcano rabbit *Romerolagus diazi*	E	Mexico	CB	
Riverine rabbit *Bunolagus sardus*	E	South Africa	CB	
Numbat *Myrmecobius fasciatus*	E	Australia	CB + WW	Yes
Chuditch *Dasyurus geoffroii*		Australia	CB	
Red-tailed phascogale *Phascogale calura*	I (E)	Australia	WW	
Golden bandicoot *Isodon auratus*	(E)	Australia	CB + WW	
Western barred bandicoot *Parameles bougainville*	R (E)	Australia	WW	
Eastern barred bandicoot *P. gunnii*	(V)	Australia	CB + WW	Yes
Greater bilby *Macrotis lagotis*	E (V)	Australia	CB + WW	Yes
N. hairy-nosed wombat *Lasiorhinus krefftii*	E	Australia	WW	
Burrowing bettong *Bettongia lesueur*	R (E)	Australia	WW	Yes
Brush-tailed bettong *B. penicillata*	E	Australia	CB + WW	Yes
Rufous hare-wallaby *Lagorchestes hirsutus*	R (E)	Australia	CB	Yes
Banded hare-wallaby *Lagostrophus fasciatus*	R (E)	Australia	CB + WW	
Tammar wallaby *Macropus eugenii*	(V)	Australia	WW	
Bridled nailtail wallaby *Onychogalea traenata*	E	Australia	WW	
Black-footed rock wallaby *Petrogale lateralis*	(V)	Australia	WW	
Brush-tailed rock wallaby *P. penicillata*	(V)	Australia	WW	
Mountain pygmy possum *Burramys parvus*	(E)	Australia	CB + WW	
Western ringtail *Pseudocheirus occidentalis*	(V)	Australia	WW	
Leadbetter's possum *Gymnobelideus leadbeateri* (S)	(V)	Australia	CB	

Chinese alligator *Alligator sinensis* (S)	E	China	CB	
Broad-snouted caiman *Caiman latirostris*	E	Argentina/Brazil	CB	
Yacare caiman *C. yacare*		Argentina	CB	
Black caiman *Melanosuchus niger*	E	Bolivia	CB	Yes
American crocodile *Crocodylus acutus*	E	Venezuela/Columbia	CB	Yes (Venezuela)
Orinoco crocodile *C. intermedius*	E	Venezuela	CB	Yes
Philippines crocodile *C. mindorensis*	E	Phillippines	CB	
Mugger crocodile *C. palustris*		India/Pakistan	CB	Yes
Saltwater crocodile *C. porosus*		India	CB	Yes
Cuban crocodile *C. rhombifer*	E	Cuba	CB	
Siamese crocodile *C. siamensis*	E	Thailand	CB	
Gharial *Gavialis gangeticus*	E	India/Nepal/Pakistan	CB	Yes (India/Nepal)
Sea otter *Enhydra lutris*		North America	WW	Yes (USA/Canada)
N. American river otter *Lutra canadensis*		North America	WW	Yes (USA)
Eurasian otter *Lutra lutra*	V	Eur./Hong Kong/Japan	CB + WW	Yes (Europe)
Giant otter *Pteronura brasiliensis*	V	Brazil	?	
Hairy-nosed otter *Lutra sumatrana*	V	South-east Asia	?	
Marine otter *L. felina*	V	South America	?	
Asian small-clawed otter *Aonyx cinerea* (S)	K	Hong Kong	?	
Japanese luehdorfia *Luehdorfia japonica*	I	Japan	CB	
Schaus' swallowtail *Papilio aristodenius ponceanus*	(E)	USA	CB	
Black and white ruffed lemur *Varecia v. variegata*	(E)	Madagascar	CB	
Red ruffed lemur *Varecia v. rubra*	(E)	Madagascar	CB	

Australia; some of the social, legal and political difficulties facing reintroductions in less developed countries have been touched on earlier. Of the remaining 28 species, 20 are Australian, a country where reintroductions are not only subject to strict methodology but must be endorsed by state and territory governments. Of the eight endangered species (black-footed ferret [*Mustela nigripes*], red wolf, grey wolf, Hermann's tortoise [*Testudo hermanni*], sea otter [*Enhydra lutris*], North American river otter [*Lutra canadensis*], Eurasian otter [*L. lutra*], Schaus' swallowtail [*Papilio aristodenius ponceanus*]) native to North America or Europe, seven are already being reintroduced.

This leads to another question: who decides which species to reintroduce? Is the process being driven by zoos, by conservationists who write Action Plans, or by governments? Beck *et al.* (Chapter 13) indicate that, in the USA at least, state and federal agencies perform the majority of reintroductions, and this is certainly true in Australia (Kennedy, 1992). The necessary financial resources behind such governmental infrastructures are only to be found in the richer nations. Should zoos be citing 'reintroduction as a reason for captive breeding' of tropical species if it is unlikely that these reintroductions can take place in the foreseeable future? For tropical species which have declined due to habitat loss, the near-term outlook is bleak indeed. Human populations continue to grow rapidly, and the chances are low of substantial new protected areas being established in most of the developing world. In India and Africa even some existing protected areas are under threat from land hunger and civil unrest. Flesness and Foose (1990) point out that 'captive breeding buys time – time to stop human over-exploitation . . . to practise restoration ecology', but sadly this may condemn many endangered species to a very long term in captivity.

However, captivity is arguably better than extinction, and perhaps cryogenic preservation of gametes or embryos for eventual reintroduction is a further option (Dresser, 1988). The lesson for all of us is to preserve biodiversity *in situ*, before it is too late.

REFERENCES

Black, J.M. (1991) Reintroduction and restocking – guidelines for bird recovery programmes. *Bird Conservation International*, **1**, 329–34.

Breitenmoser, U. and Breitenmoser-Wursten, C. (eds) (1990) *Status, Conservation Needs and Reintroduction of the Lynx* (Lynx lynx) *in Europe*, Council of Europe, Strasbourg.

Brocke, R.H., Gustafson, K.A. and Major, R.A. (1990) Restoration of the lynx in New York: biopolitical lessons. *Trans. 55th N.Am. Wildl. Nat. Res. Conf.*, Wildlife Management Institute, Washington, DC, pp. 590–8.

Brown, P. (1991) Re-introductions in Australia: a brief overview. *Re-introduction News* (Newsletter of the IUCN/SSC Re-introduction Specialist Group), **2**, 2–3

Cade, T., Enderson, J. Thelander, C. and White, C. (eds) (1988) *Peregrine Falcon Populations: Their Management and Recovery*, The Peregrine Fund, Boise, Idaho.

Cairns, J. (1988) Increasing diversity by restoring damaged ecosystems, in *Biodiversity* (ed. E.O. Wilson), National Academy Press, Washington, DC , pp. 333–43.
Campbell, D.G., Lowell, K.S. and Lightbourn, M.E. (1991) The effect of introduced hutias (*Geocapromys ingrahami*) on the woody vegetation of Little Wax Cay, Bahamas. *Conservation Biology*, **4** (5), 536–41.
Dresser, B.L. (1988) Cryobiology, embryo transfer, and artificial insemination in *ex situ* animal conservation programs, in *Biodiversity* (ed. E.O. Wilson), National Academy Press, Washington, DC, pp. 296–308.
Flesness, N.R. and Foose, T.J. (1990) The role of captive breeding in the conservation of species. Foreword to 1990 *IUCN Red List of Threatened Animals*, IUCN, Gland, Switzerland and Cambridge, UK, pp xi–xv.
Foster-Turley, P., Macdonald, S. and Mason, C. (eds) (1990) *Otters: An Action Plan for their Conservation*, IUCN/SSC, Gland, Switzerland.
Griffith, B., Scott, J.M., Carpenter, J.W. and Reed, C. (1990) Translocations of captive-reared terrestrial vertebrates 1973–1986. *Endangered Species Update*, **8** (1), 10–14.
IUCN (1987) *The IUCN position statement on translocation of living organisms*, IUCN, Gland, Switzerland.
IUCN (1990) *IUCN Red List of Threatened Animals*, IUCN, Gland Switzerland, and Cambridge, UK.
Janzen, D.H. (1986) *Guanacastle National Park: Tropical, Ecological and Cultural Restoration*, Editorial Universidad Estata a Distancia, San Jose, Costa Rica.
Jordan, W.R. (1988) Ecological restoration: reflections of a half-century of experience at the University of Wisconsin-Madison Arboretum, in *Biodiversity* (ed. E.O. Wilson), National Academy Press, Washington, DC, pp. 311–16.
Jordan, W.R. and Packard, S. (1989) Just a few oddball species; restoration practice and ecological theory, in *Biological Habitat Reconstruction* (ed. G.P. Buckley), Bellhaven Press, London, pp. 18–26.
Kennedy, M. (ed.) (1992) *Australasian Marsupials and Monotremes: An Action Plan for their Conservation*, IUCN/SSC, Gland, Switzerland.
Messel, H., King, F.W., and Ross, J.P. (eds) (1992) *Crocodiles: An Action Plan for their Conservation*, IUCN/SSC, Gland, Switzerland.
Mickleburgh, S.P., Hutson, A.M. and Racey, P.A. (eds) (1992) *Old World Fruit Bats: An Action Plan for their Conservation*, IUCN/SSC, Gland, Switzerland.
Olney, P.J.S. and Ellis, P. (eds) (1991) *International Zoo Yearbook*, **30**, Zoological Society of London.
Packard, S. (in prep.) Restoration as discovery: tallgrass savannah and woodland, in *From Specimen to Habitat Management*, Proceedings of a Conference held at Royal Botanic Gardens, Kew, 1991.
Phillips, M.K. (1990) Measures of the value and success of the red wolf reintroduction project. *Endangered Species Update*, **8** (1), 24–6.
RSG (1992) Draft Guidelines for Reintroductions. *Reintroduction News* (Newsletter of the IUCN/SSC Reintroduction Specialist Group), **4**, 2–3.
Spalton, A. (1992) Arabian oryx in Oman. *Reintroduction News* (Newsletter of the IUCN/SSC Reintroduction Specialist Group), **5**, 6.
Stanley Price, M.R. (1989) *Animal Re-introductions: The Arabian Oryx in Oman*, Cambridge University Press, Cambridge, UK.
Stolzenburg, W. (1992) Silent Sirens. *Nature Conservancy News*, May/June 1992, 8–13.
Stuart, S.N. (1991) Re-introductions: to what extent are they needed?, in *Beyond Captive Breeding: Re-introducing Endangered Mammals to the Wild, Symp. Zool. Soc. Lond.*, No. 62 (ed. J.H.N. Gipps), Clarendon Press, Oxford, pp. 27–37.
Tudge, C. (1991) *Last Animals At The Zoo*, Hutchison Radius Press, London.

Veitch, C.R. and Bell, B.D. (1990) Eradication of introduced animals from the islands of New Zealand, in *Ecological Restoration of New Zealand Islands* (eds D.R. Towns, C.H. Daugherty and I.A.E. Atkinson), Department of Conservation, Wellington, New Zealand.

13

Reintroduction of captive-born animals

B.B. Beck
National Zoological Park, Smithsonian Institution Washington, DC, USA
L.G. Rapaport
University of New Mexico, New Mexico, USA
M.R. Stanley Price
IUCN/SSC Reintroduction Specialist Group, Nairobi, Kenya
and
A.C. Wilson
IUCN/SSC Reintroduction Specialist Group, Nairobi, Kenya

13.1 INTRODUCTION

This paper explores the extent to which reintroduction of captive-born animals is being used as a conservation strategy, the extent to which zoos are participating, the success of reintroduction, and some characteristics of these reintroduction programmes as they relate to success. This paper does not provide guidelines for reintroduction; see Kleiman, Stanley Price and Beck (Chapter 14) for guidelines.

Creative Conservation: Interactive management of wild and captive animals.
Edited by P.J.S. Olney, G.M. Mace and A.T.C. Feistner.
Published in 1994 by Chapman & Hall, London.
ISBN 0 412 49570 8

13.2 METHODS AND DEFINITIONS

Reintroduction as used here refers to the intentional movement of captive-born animals into or near the species' historical range to re-establish or augment a wild population. Translocations of wild animals, and reintroductions for primarily recreational or commercial purposes are not included. Thus our definition is narrower than that of 'translocation' as used in the World Conservation Union (IUCN) position statement (IUCN, 1987) and by Griffith *et al.* (1989) because it does not include movement of wild animals. Our definition is also narrower than that of 'reintroduction' and 'restocking' in the IUCN statement because it additionally excludes movement of animals for recreational (e.g. sport fishing) or commercial (e.g. game ranching) purposes. We also exclude introductions of animals outside their species' historical range.

Our data were collected in 1991 and 1992 from surveys of the published literature; bibliographies compiled by Comly *et al.* (1989) and by Kenyon (1992); returns of standardized questionnaires that we sent to reintroduction managers (included as an appendix to this chapter); and/or personal communications deriving mainly from the IUCN Species Survival Commission's Reintroduction Specialist Group (SSC/RSG) and the Reintroduction Advisory Group (RAG) of the American Association of Zoological Parks and Aquariums (AAZPA).

13.2.1 Project

We counted the number of reintroduction projects and the number of species reintroduced. A project is an administratively distinct reintroduction, where captive-born animals have actually been released anywhere in the world between 1900 and the present. Our database thus differs in two additional ways from that of Griffith *et al.* (1989), which included only projects conducted in Canada, the USA (including Hawaii), Australia and New Zealand, between 1973 and 1986. Our database includes as separate projects reintroductions of the same species into distinct populations, e.g. the reintroduction of Arabian oryx (*Oryx leucoryx*) in Oman, Jordan and Saudi Arabia. Our database thus contains more reintroduction projects than reintroduced species. Additions to the same population by successive reintroductions, even if carried out by different administrative authorities, are not counted as separate projects. In a few cases we were not able to distinguish between different projects reintroducing the same species to different populations. For example the bald eagle (*Haliaetus leucocephalus*) was reintroduced in at least 25 states of the USA and Canadian provinces, but we were unable to ascertain which effort released captive-born eagles and which and how many were administratively distinct: we counted all of these as one project. We similarly compressed the number of projects for white-tailed sea eagles (*Haliaetus albicilla*), peregrine falcons (*Falco peregrinus*), lammergeiers (*Gypaetus barbatus*), wild turkeys (*Meleagris gallepavo*), and Indian gharials (*Gavialus gangeticus*). These compressions result in our underestimating the number of reintroduction projects but do not affect the number of species.

13.2.2 Species

We used project managers' determinations of species and did not attempt reclassification.

13.2.3 Captive-born animals

A project was counted if there was evidence that at least one captive-born animal was reintroduced; this underestimates the number of projects and species since in 16 projects we were unable to find evidence confirming our impression that captive-born animals were reintroduced. Note that a project introducing captive-born animals may also have released rehabilitated or translocated wild-born individuals, but the number of reintroduced animals in this paper refers only to the number of captive-born animals. In 10 projects we confirmed that some captive-born individuals were reintroduced, but no estimate of numbers was found; these projects and species are included in the database but are not used in calculations involving the number of animals reintroduced. In five other projects only the minimum number of captive-born animals was stated (e.g. 'at least 60' or 'more than 100'); in these cases we used the lowest number provided. All of these sources of error tend to underestimate the number of captive-born individuals reintroduced.

Animals, e.g. sand lizards (*Lacerta agilis*) (Spellerberg and House, 1982) and Kemp's ridley sea turtle (*Lepidochelys kempi*) (Burchfield, 1985), born or hatched in enclosures in the natural habitat are considered captive born. An individual of an oviparous species is said to be captive born from an egg laid and hatched in captivity or an egg laid in the wild and hatched in captivity. Individuals from eggs laid in captivity and hatched in the wild are considered wild born. Thus a crane (*Grus* spp.) hatched in captivity and placed as a hatchling under a foster parent in a wild nest is considered captive born, but it is considered wild born if hatched under a foster parent in a wild nest from an egg laid in captivity. Infants hatched in the wild and raised or 'head-started' in captivity are considered wild born.

13.2.4 Zoo involvement

A zoo is said to be involved in a reintroduction project if at least one of the reintroduced captive-born animals, or at least one of its documented ancestors, lived in a zoo. We defined 'zoo' to include zoos, aquaria, wildlife parks, arboretums, living museums, nature centres and other institutions for which public exhibition is the principal function. Animals were counted as having come from a zoo even if they were kept only in the off-exhibit facilities of such an institution. Animals that were kept in research stations, game farms, and hatcheries were not counted as having come from a zoo, even though the public may visit such institutions.

13.2.5 Tropics

A reintroduction is said to have occurred in the tropics if the animals were released between the Tropics of Cancer and Capricorn (23°, 27′ N and S, respectively).

13.2.6 Threatened taxa

A reintroduced species or subspecies is considered threatened if it appears in the *1988 IUCN Red List of Endangered Animals* (IUCN, 1988).

13.2.7 Pre-release training

We tried to determine if the captive-born animals were trained, acclimatized, or medically or genetically screened before release. Pre-release training includes such measures as inducing golden lion tamarins (*Leontopithecus rosalia*) to search for hidden and spatially distributed food and to move around on natural vegetation in their cage (Beck *et al.*, 1991); inducing black-footed ferrets (*Mustela nigripes*) to find and kill prairie dogs (*Cynomys ludovicianus*) in large outdoor enclosures (Oakleaf *et al.*, 1992; A. Vargas, pers. comm.); and inducing thick-billed parrots (*Rhynchopsitta pachyrhyncha*) to handle pine cones (a primary food source) and encouraging them to fly in pre-release cages (Wiley, Snyder and Gnam, 1992).

Nile crocodiles (*Crocodylus niloticus*) were fed live fish which they had to catch in pre-release holding pools (Morgan-Davies, 1980). Black lights were placed in the outdoor cages of elf owls (*Micrathene whitneyi*) to attract insects so that the owls could learn to catch and eat insects (J. Lithicum and B. Walton, pers. comm.). Prior to reintroduction, masked bobwhite quail (*Colinus virginianus ridgwayi*) were harassed by humans, trained dogs and hawks; the quail were allowed to escape, presumably having acquired fear of potential predators (Carpenter, Gabel and Goodwin, 1991). The same programme later grouped naïve masked bobwhite quail with wild Texas bobwhite quail (*C. v. texanus*) which occupy similar habitat and are not threatened. The Texas bobwhites demonstrated food-finding and anti-predator behaviour to the masked bobwhites in acclimatization cages and after reintroduction (the Texas bobwhite were sterilized to prevent hybridization).

This is not a complete list of pre-release training efforts but serves to illustrate the range of techniques. While housing captive-born reintroduction candidates with skilled conspecifics to demonstrate behaviours crucial to survival was considered to be training, simply providing foster parents or conspecifics for companionship was not. Hand-rearing captives, e.g. California condors (*Gymnogyps californianus*) (Wallace, 1990) and Mississippi sandhill cranes (*Grus canadensis pulla*) (Horwich, 1989) with puppets or costumes solely to prevent imprinting on humans was not considered training, but the surrogate's directing the bird to natural food was. A reintroduction project is said to have used training if at least one, but not necessarily all, captive-born reintroductees was trained.

13.2.8 Acclimatization

A project is said to have used acclimatization if at least one of the reintroduction candidates was held at or near the release site in a cage, pen, corral, artificial nest, scaffold, tower or other man-made structure for at least 24 hours before release, in order to allow the animal to become familiar with climatic conditions, landmarks, natural foods or other features of the natural environment. Training may also have taken place during acclimatization if project personnel actively presented environmental features or actively induced the animals to perform specific behaviours. For example, Hispaniolan parrots (*Amazona ventralis*) were held in a field aviary for 9–12 days before release; this by our definition is acclimatization. But they were provisioned with fruits and seeds of naturally occurring plants; thus by our definition they were also trained (Snyder, Wiley and Kepler, 1987). We would not have considered the parrots trained if they simply ate food that occurred naturally in the acclimatization aviary.

13.2.9 Medical screening

A project is said to have used medical screening if the choice of reintroduction candidates was based at least in part on medical considerations (Woodford and Rossiter, Chapter 9). This includes certification of freedom from certain communicable diseases as indicated by quarantine, vaccination requirements, deparasitization requirements, freedom from debilitating injury or deformity, weight or age minima/maxima, reproductive viability and other health-related criteria. Before release, red wolves (*Canis rufus*) were given injectable parasiticides, and were vaccinated against rabies, distemper, canine parvovirus, hepatitis, leptospirosis, corona-virus and parainfluenza (Phillips, 1990). Prior to shipment to Oman, Arabian oryx (*Oryx leucoryx*) chosen for reintroduction had to be certified to be free from tuberculosis, brucellosis, leptospirosis, rinderpest, parasitic mange, foot and mouth disease, anthrax, and bluetongue, and additionally had to be vaccinated against foot and mouth disease, anthrax, clostridial diseases and pasteurellosis (Stanley Price, 1989). Before shipment to Brazil for reintroduction, golden lion tamarins were given antihelminthics, vaccinated against rabies, radiographed to screen for extreme diaphragmatic thinning, and determined to be free of antibodies to callitrichid hepatitis (Bush, Beck and Montali, 1993; Montali and Bush, 1992). Medical monitoring or treatment only after release does not qualify a project as having used medical screening.

13.2.10 Genetic screening

A project is said to have used genetic screening if the choice of reintroduction candidates was based at least in part on genetic or pedigree considerations. These include descent from founders over-represented in the captive population, e.g. scimitar-horned oryx (*Oryx dammah*; Gordon, 1991), sufficient unrelatedness to produce minimally inbred offspring after release, e.g. Arabian oryx in Oman (Spalton, 1991), descent from lineages with or without specific genetically influenced phenotypes, e.g. diaphragmatic thinning in golden lion tamarins (Bush, Beck and Montali, 1993), or descent from non-hybridized lineages. A project is said to have used genetic screening if reintroductees were chosen from lineages originating in the specific area of the planned reintroduction, i.e. from the same putative subspecies or local population, e.g. Chocowhatchee beach mice (*Peromyscus polionotus allophrys*; Wood and Holler, 1990) and Puerto Rican crested toads (*Peltophryne lemur*; Johnson and Paine, 1989; Miller, 1985). In a programme to re-establish the swift fox (*Vulpes velox*) in Canada, Herrero, Schroeder and Scott-Brown, (1986) chose founders of a captive breeding population from the geographically closest population in the USA; the programme was thus said to have used genetic screening. These authors stated further that post-release selection would recreate locally adapted gene complexes that might be absent in stock originating from more southerly locations.

A project is said not to have used genetic screening if pedigrees of the source captive population were known, e.g. Lord Howe Island woodhen (*Tricholimas sylvestris*) (Miller and Mullette, 1985), or the source population was genetically managed, but genetic considerations were not used to select reintroduction candidates.

13.2.11 Post-release training

We also tried to determine if the reintroduced animals were trained, provisioned or monitored after release. Post-release training consists of, for example, active presentation of natural foods, e.g. grasshoppers to golden lion tamarins, or inducing the monkeys to move over natural vegetation (Beck *et al.*, 1991). Post-release training is also exemplified by demonstration of food-gathering and nest-building skills to reintroduced chimpanzees (*Pan troglodytes*) (Carter, 1981).

13.2.12 Provisioning

A project is said to have utilized provisioning after release if the animals were given food, water, and/or shelter-boxes or nest-boxes at least once after release. The animals need not have consumed the food or water or used the boxes. Reintroduced Mauritius kestrels (*Falco punctatus*) were trained before release to come to a whistle to get mice or chicks. This allowed them to be provisioned easily after release, and additionally allowed fully independent kestrels to be called in for identification after several years (Jones *et al.*, 1991). These kestrels were also given artificial nesting boxes after release. Reintroduced American burying beetles (*Nicrophorus americanus*) were given carrion on which to lay eggs (Amaral and Morse, 1990), and reintroduced Puerto Rican crested toads were given water, not for drinking but as a medium to prevent dehydration (R. Johnson, pers. comm.); we consider these both to be provisioning.

13.2.13 Monitoring

Post-release monitoring is an active attempt to determine the size of the reintroduced population, the occurrence of births, the occurrence of deaths, causes of death, and/or the behaviour of individuals. At its simplest, post-release monitoring is an attempt to determine the existence of a population after reintroduction, i.e. to determine if any of the released animals survived. Monitoring could involve direct visual contact with the animals; determining location and activity by radiotelemetry; or inferring survival, population size or activity from faeces, prey remains, or nests. Extraordinary ingenuity has produced transmitters for such diverse species as Guam rails (*Rallus owstoni*) (Meadows, 1992) and pine snakes (*Pituophis melanoleucus*) (Burger and Zappalorti, 1988). Individual identification through tattoos, tags or bands, natural or applied body marks, transponders and/or transmitters, was used in many projects but was not a requisite for monitoring. The ease of monitoring affects the precision of estimates of post-release survivorship, reproduction and behaviour. Transmitter-equipped animals returning for provisioned food, e.g. Jamaican hutias (*Geocapromys brownii*) (Oliver *et al.*, 1986) are easily monitored while black-footed ferrets which disperse widely and quickly, shed radiocollars easily, and live secretively underground are more difficult (Oakleaf *et al.*, 1992). Frequency of post-release monitoring also influences precision. If a project used monitoring we tried to determine if monitoring attempts took place on 1–12 days per year, 13–100 days, or more than 100 days. Carter (1981, 1988) literally lived with reintroduced chimpanzees, observing them daily at very close range; her monitoring results are thus very precise and detailed. In contrast, herds descending from reintroduced wood bison (*Bison b. athabascae*) live in remote and inaccessible areas and move widely and unpredictably; periodic monitoring from aircraft yields coarser grained information (Hoefs and Reynolds, 1989; H.P.L. Kiliaan, pers. comm.).

13.2.14 Local employment

We tried to determine if reintroduction projects provided local employment or professional training opportunities, or if they had a community education programme. Local employment means providing salaries to people living in the area of the reintroduction project in exchange for working on the project. The workers need not have worked with the reintroduced animals directly; many projects have employed local people to build acclimatization enclosures (e.g. plains bison, *Bison bison*) (Loring, 1906; Sanborn, 1908), build fire breaks, restore habitat, participate in community education programmes and tourism, and serve as rangers and guards. Indirect economic benefits to local people, e.g. to merchants and hotel owners, are not counted as employment.

13.2.15 Professional training

A project offers professional training opportunities if graduate students pursued research for master's or doctoral theses at the reintroduction site. The students need not study reintroduced animals, but the subject of the research must be directly related to the conservation programme. Undergraduate participation, internships, non-thesis research collaborations and professional career development programmes are not counted as professional training.

13.2.16 Community education

A reintroduction project is said to have had a community education programme if project personnel presented lectures or slide programmes; distributed posters, T-shirts, hats or pins; participated in fairs, parades, community meetings or other cultural or civic events; visited schools, clubs or individual households; or helped to prepare stories and releases for radio, television or newspapers. The golden lion tamarin programme made the community education effort a subject of research, providing evidence of improved local support for and understanding of conservation (Dietz and Nagagata, 1986; Kleiman *et al.*, 1986; Dietz, Dietz and Nagagata, Chapter 2). The conservation program for the Bali starling (*Leucopsar rothschildi*) complements reintroduction with a community education programme that both raises awareness and retrieves genetically valuable starlings that have been stolen from the wild for the pet trade; captive-bred starlings from over-represented lineages are traded to pet owners for the genetically valuable birds and the owners are given amnesty (Seibels, 1991).

13.2.17 Release years

We tried to determine the number of calendar years in which animals were actually released in each project. We were unable to make a confident estimate for 22 projects, and for six others we could only determine a minimum number of release years. Note that there may have been many years of preparation and captive breeding before the releases, years between releases, and years of post-release monitoring, but we count only actual release years as a measure of project longevity and effort.

13.2.18 Success

A reintroduction project is counted as successful if the wild population subsequently reached at least 500 individuals which are free of provisioning or other human support, or where a formal genetic/demographic analysis (e.g. Population Viability Analysis or Population and Habitat Viability Analysis) of the sort advocated by Foose (1991) predicts that the population will be self-sustaining. The reintroduction itself need not be the sole factor contributing to population growth, and indeed other measures may have been more instrumental in population recovery. For example Rees (1989) suggests that habitat protection, predator control, regulation of hunting, and public education seem to have contributed more than reintroduction to the recovery of the Aleutian goose (*Branta canadensis leucopareia*).

13.3 RESULTS

We can document 145 administratively distinct projects in which 13 275 295 documented individual captive-born animals of 126 species were reintroduced. Table 13.1 shows these data arrayed by class, with

Table 13.1 The frequency of reintroduction projects, reintroduced species, and individual reintroduced captive-born animals, by class

	Projects	*Species*	*Individuals*
Mammals	46 (32%)	39 (31%)	2317
Birds	65 (45%)	54 (43%)	39 054
Reptiles and amphibians	23 (16%)	22 (17%)	31 483
Fish	9 (6%)	9 (7%)	13 201 050
Invertebrates	2 (1%)	2 (2%)	1391
	145 (100%)	126 (100%)	13 275 295

reptiles and amphibians combined. Foose *et al.* (1992) calculate that mammals constitute about 8% of all vertebrate species, birds about 19%, reptiles and amphibians about 21%, and fish about 52%. But mammals constitute about 32% of all reintroduced captive-born vertebrate species, birds about 44%, reptiles and amphibians about 18%, and fish about 7%. Thus reintroduction of captive-born mammals and birds is used more frequently than would be predicted from species abundance, and reintroduction of captive-born fish is less frequent. Further, since there are at least 300 000 species of invertebrates compared to 47 500 species of vertebrates, it would seem that reintroduction of captive-born invertebrates is very uncommon compared with species abundance. Of course it is endangerment rather than abundance that should drive the frequency of use of any recovery method, but we could not identify a representative measure of endangerment of various classes.

Table 13.2 shows the average number of reintroduced individuals per project and species, arrayed by class. Note that the number of reintrodu-

Table 13.2 The average number of individual captive-born animals reintroduced per project and species, by class

	Project	*Species*
Mammals	50	59
Birds	620	737
Reptiles and Amphibians	1431	1499
Fish	3 200 263	3 200 263
Invertebrates	696	696

ced individuals was not available for two bird projects (one species), one herpetological project (one species), and five fish projects (five species). The differences between classes may reflect differences in litter sizes and growth rates, as well as in relative ease of handling and transporting vertebrate eggs, fry, tadpoles and adults. Ounsted (1991) has already noted that the preponderance of bird over mammal reintroductions is due in part to the ease of manipulating and fostering eggs. The two invertebrate reintroduction projects were largely pilot feasibility studies; because of the ease of handling eggs, larvae and even the small-bodied adults, we would anticipate large average numbers of individuals per project if reintroduction becomes a widely used recovery tool for invertebrates.

13.3.1 Zoo involvement

Zoo-bred animals or the captive-bred descendants of zoo animals were reintroduced in 76 (59%) of the 129 reintroduction projects for which we could confidently determine whether zoo-born animals or their descendants were used. As there are about 350 zoos and aquaria in the developed world, on average only about one in five has been involved in reintroduction. This is a rough estimate since some projects used animals from several zoos while some zoos have been solely responsible for several projects. We did not count as zoo-born one chimpanzee born in a road-side zoo, removed from its mother on the day of birth, and sold to a private party on the next day. This animal was ultimately reintroduced to the wild (Carter, 1981), but this did not seem to reflect meaningful zoo involvement in this project. Zoo-bred mammals were reintroduced in 37% of the 76 projects, birds in 41%, reptiles and amphibians in 20%, fish in 3%, and invertebrates in none (the Cincinnati Zoo is now involved in captive breeding of North American burying beetles but no zoo beetles have yet been reintroduced, but see Pearce-Kelly, Chapter 17). These proportions approximate the distribution of animal classes in all reintroductions (Table 13.1), suggesting that a zoo-bred species of any class is equally likely to be reintroduced.

Since some projects released zoo-bred and non-zoo-bred (but captive-born) animals, we could not determine precisely how many zoo-born animals have been reintroduced. But a minimum of 20 849 animals (1958 mammals, 8271 birds, 10 620 reptiles and amphibians) were released in these projects. This is the equivalent of the collections of only three or four major North American or European zoos combined.

Of course, zoos participate in reintroduction in ways other than providing zoo-born animals for release. For example, the Gladys Porter Zoo provides husbandry expertise in the Kemp's ridley sea turtle reintroduction but does not provide zoo-born turtles. The Frankfurt Zoological Society is a major financial supporter of the golden lion tamarin reintroduction, but has provided only two zoo-born tamarins. Many zoos feature reintroduction in their public education programmes.

Nevertheless, it does not appear that zoos are the primary proponents, animal providers, funders, or managers of reintroduction programmes. State and federal wildlife agencies are involved in a vast majority of reintroductions, including many of those where zoo-born animals were released, and appear to be the major driving force.

13.3.2 Tropics

Only 30 (21%) of 144 projects released animals in the tropics (we could not determine the precise release location of one project). The vast majority of reintroductions have been sited in temperate North America, Europe, and Australia/New Zealand. This too is consistent with the conclusion that state and federal wildlife agencies manage reintroduction projects, since these organizations are particularly well funded and well staffed in the developed temperate world. Mammals were released in 11 (37%) of the 30 tropical projects, birds in nine (30%), reptiles and amphibians in 10 (33%), and fish and invertebrates in none. Thus zoo-bred reptiles and amphibians are more likely than expected to be reintroduced in the tropics, and birds and fish less likely, based on their proportions of all reintroduction projects (Table 13.1).

13.3.3 Threatened taxa

Seventy (48%) of the 145 reintroduction projects released species or subspecies listed in the IUCN Red Data Book as threatened. Additionally, the Aleutian goose was listed as threatened when its reintroduction began, and in all likelihood wood bison, plains bison and wood ducks would also have been listed as threatened. Père David's deer (*Elaphurus davidianus*) is extinct in the wild and therefore not listed as threatened. Including these, 27 of 46 (59%) mammal projects released a threatened taxon, as did 21 of 65 (32%) bird projects, 16 of 23 (70%) reptile and amphibian projects, nine of nine (100%) fish projects, and two of two (100%) invertebrate projects.

13.3.4 Some characteristics of reintroduction projects

Table 13.3 summarizes some characteristics of reintroduction projects. Pre-release acclimatization was more frequently used (76% of all projects) than pre-release training (35%). This was true for mammals, birds, and reptiles and amphibians, probably reflecting the greater costs of pre-release training, and the limited evidence for its effectiveness (Kleiman

Table 13.3 Characteristics of reintroduction projects

Factor	*Proportion of projects (%)*[a]			
	All[b]	*Mammals*	*Birds*	*Reptiles and amphibians*
Pre-release training	35	36	48	7
Acclimatization	76	82	83	56
Medical screening	46	60	47	31
Genetic screening	37	35	34	46
Post-release training	12	12	19	0
Provisioning	63	69	84	13
Monitoring	96	97	98	87
Local employment	53	50	64	54
Professional training	56	52	64	54
Community education	70	59	76	77
Release years[c]	6.51	3.03	6.09	7.50

[a]Due to lack of information, overall percentages are based on between 59 and 104 projects.
[b]Fish and invertebrate projects are included in the overall percentages but are not shown separately because of small sample sizes.
[c]Average number of years in which animals were released.

et al., 1986; Snyder, Wiley and Kepler, 1987). Post-release training (12% of all projects) was less commonly used than pre-release training (35%), probably reflecting the logistical difficulty of training free-ranging, often widely dispersed, reintroductees.

Pre-release training was used more frequently for mammals (36% of projects) and birds (48%) than for reptiles and amphibians (7%). Likewise, acclimatization was used more frequently for mammals (82%) and birds (83%) than for reptiles and amphibians (56%). Despite comparable levels of post-release monitoring, there was more post-release training for mammals (12%) and birds (19%) than for reptiles and amphibians (0%), and there was more post-release provisioning for mammals (69%) and birds (84%) than for reptiles and amphibians (13%). The same general trends are apparent in the fish and invertebrate subsamples. All of these trends probably result from the conclusion that foraging, locomotor, and anti-predator behaviours, and other behaviours essential for survival, are more heavily dependent on learning and specific environmental experience in mammals and birds. Reptiles and amphibians might thus be expected to require less pre-release preparation and post-release support. But there are no significant differences in reintroduction success rate among the three class samples (see below).

There are fewer release years for mammals (3.03 per project) than for birds (6.09) and reptiles and amphibians (7.50). To the degree that the number of release years is positively correlated with reintroduction success (Griffith *et al.*, 1989), this may retard the success of mammal reintroduction projects. But again there are no significant differences in success rates between mammals, birds and reptiles and amphibians.

Kleiman, Stanley Price and Beck, (Chapter 14) provide a comprehensive compilation of guidelines for reintroduction. These include stringent pre-release veterinary screening and genetic screening, and strong community relations programmes. Careful consideration must be given to pre-release training and acclimatization, and post-release support. Our data suggest that to date reintroduction managers have largely met guidelines regarding acclimatization (76% of projects), post-release provisioning (63%), post-release monitoring (96%) and community relations (53% offered local employment and 70% offered community education programmes). But there appear to be serious shortfalls with regard to pre-release training (35% of projects), medical screening (46%) and genetic screening (37%). In view of the potentially catastrophic effects of communicable disease on a remnant population of conspecifics or on a natural ecosystem (Bush, Beck and Montali, 1993), the low frequency of medical screening is of great concern. Indeed, in addition to our frequencies for medical and genetic screening being low, they are based on fewer than half of the known projects since we could not even determine from available sources whether there had or had not been screening.

13.3.5 Is reintroduction successful?

We can find evidence that only 16 (11%) of the 145 reintroduction projects were successful as defined above. These 16 projects reintroduced captive-born wood bison, plains bison (two projects), Arabian oryx (Oman), Alpine ibex (*Capra ibex*), bald eagle, Harris' hawk (*Parabuteo unicinctus*), peregrine falcon, Aleutian goose, bean goose (*Anser fabalis*), lesser white-fronted goose (*Anser erythropus*), wood duck (*Aix sponsa*), gharial (in India), Galapagos iguana (*Conolophus subcristatus*), pine snake (*Pituophis melanoleucus*), and Galapagos tortoise (*Geochelone elaphantopus*).

This estimate of success is provisional, since our measure of success is very general and conservative. Further, the other 89% are not all failures. In some projects there is encouraging progress toward a self-sustaining population. Some projects are in their infancy and techniques are being improved. There are also indirect benefits of reintroductions where a self-sustaining wild population may never be established, namely increased public awareness and support for conservation, professional training, enhanced habitat protection, and increased scientific knowledge (Kleiman, Stanley Price and Beck, Chapter 14; Lindberg, 1992).

Griffith *et al.* (1989), in their analysis of projects involving the reintroduction of captive-bred animals and translocation of wild-born animals, and using project managers' judgments of success of their own projects, estimated that 38% of projects reintroducing captive-bred animals were successful, and 75% of projects translocating wild animals were successful.

13.3.6 Some characteristics of successful reintroduction projects

Of the 16 successful projects, five reintroduced mammals (11% of all mammal projects), seven reintroduced birds (11% of all bird projects) and four reintroduced reptiles or amphibians (17% of all herpetological projects). None of the successful projects reintroduced fish or invertebrates. Thus herpetological reintroductions are slightly more likely to be successful, and fish reintroductions slightly less likely to be successful than expected, but sample sizes are small. At present there is no evident relationship between class and success.

We have reasonably complete documentation on 14 of the 16 successful reintroduction projects, and on 62 other projects. By 'reasonably complete' we mean confirmed responses to at least six of the 11 characteristics shown in Table 13.3. Of the 62 'other' projects (not necessarily unsuccessful but not yet proven successful by our definition), 20 (32%) reintroduced mammals, 26 (42%) reintroduced birds, 11 (18%) reintroduced reptiles or amphibians, three (5%) reintroduced fish, and two (3%) reintroduced invertebrates. These proportions closely represent the class distribution of the database as a whole (see above); thus the sample of 'other' projects is representative in a taxonomic sense. However, the data from this sample are not representative of reintroduction projects as a whole since one might expect more confirmable responses from better organized and better documented projects.

Table 13.4 compares characteristics of the 'successful' and 'other' projects. Sample sizes for any one characteristic ranged from 11 to 13 for successful projects and 47 to 62 for other projects. There were five notable differences. Successful projects used medical screening and post-release provisioning less often than the other projects. This is a counter-intuitive outcome which can be a chance occurrence in a small sample, or it may mean that these measures are not essential for success. Successful projects more frequently provided local employment and community education programmes. These outcomes confirm a conclusion of the oldest project in our database. In his report on the suitability of the Wichita Buffalo Range for the New York Zoological Society's

Table 13.4 Characteristics of successful and other reintroduction projects

Factor	*Proportion of projects (%)*	
	Successful	*Other*
Pre-release training	50	32
Acclimatization	75	68
Medical screening	17	49
Genetic screening	25	35
Post-release training	8	11
Provisioning	42	63
Monitoring	100	94
Local employment	75	47
Professional training	58	51
Community education	100	62
Release years[a]	11.8	4.7

[a]Average number of years in which animals were released.

reintroduction of plains bison, J. Alden Loring wrote:

> 'In establishing this range everything possible should be done to foster good feeling between the Government and the public. To a large extent this may be done by giving employment to persons living on or near the range. These people should be made to feel that it is to their interest to watch over the animals in the range, and report everything that should be brought to the attention of the forester.'
>
> (Loring, 1906, p. 198.)

Finally, successful projects released animals in an average of 11.8 calendar years (n = 15) while other projects released animals in an average of 4.7 years (n = 61); this difference is significant (Mann–Whitney U Test, U = 222.5, z = 3.06, $p < 0.002$, two-tailed). This conforms to differences in the number of captive-born animals released. An average of 726 animals (n = 16) were released in successful projects compared to an average of 336 (n = 56) in other projects (for this calculation fish and invertebrate projects were not included in the sample of other projects since both tend to reintroduce large numbers of individuals and there were no fish or invertebrate projects in the successful sample); this difference is significant (Mann–Whitney U Test, U = 180.5, z = 3.62, $p < 0.001$, two-tailed). Griffith *et al.* (1989) found that successful translocation programmes were of longer duration and released more animals than unsuccessful programmes.

13.4 CONCLUSIONS

Reintroduction is a common conservation strategy in the temperate, developed world but is used rarely in the tropics, where the loss of biodiversity is accelerating. Only about 50% of reintroduction projects have released threatened species or subspecies, suggesting that the potential for reintroduction as a recovery strategy has yet to be realized. This shortfall is especially marked for bird reintroductions, where only 32% of projects released threatened taxa.

Some zoos and some zoo critics exaggerate the importance of reintroduction as a zoo conservation function. Reintroduction is one of a balanced suite of zoo contributions to conservation (Mallinson, 1991), but our results suggest revision of any claims about the primacy of zoos in reintroduction or the primacy of reintroduction in zoos. State and federal wildlife agencies are the major proponents and managers of reintroduction.

In two years of intense searching we were able to acquire reasonably complete information on less than 50% of projects known to have reintroduced captive-born animals. Written information documenting reintroduction procedures and post-release outcomes for over 13 million individual animals fills less than one file drawer. Both Griffith *et al.* (1989) and Kleiman, Stanley Price and Beck (Chapter 14) argue strongly for documentation and published descriptions of techniques and outcomes.

With this paucity of documentation we were forced to use very general criteria for reintroduction success. Had we used more specific criteria the sample of projects allowing confident determinations of success would have been too small to allow meaningful comparisons. In addition to providing better documentation, we need to develop specific criteria of reintroduction success and apply them vigorously and impartially.

Whether the success rate for reintroduction is 11% (this study) or 38% (Griffith *et al.*, 1989), we need to improve reintroduction techniques. Retrospective studies such as the present study will continue to suggest correlates of success. These need to be confirmed by controlled research. Research is also needed in areas where the results are ambiguous, e.g. the importance of pre-release training, medical screening, and post-release provisioning. Only with research-based improvement in techniques will reintroduction attain the success and applicability to threatened taxa needed to gain credibility as a recovery strategy.

We do know that successful reintroduction projects seem to extend over many years and release large numbers of animals. Successful projects invest heavily in the involvement of local people through employment opportunities and community education. All of these characteristics are resource-intensive, suggesting that prospects for adequate, long-term funding must be good if reintroduction is to be undertaken responsibly.

13.5 APPENDIX: REINTRODUCTION DATABASE QUESTIONNAIRE

Common name: ____________
Family: ____________
Genus: ____________
Species: ____________
(Subspecies: ____________)

Country and/or continent on which reintroduction(s) occurred: ________
Park or locale in which reintroduction(s) occurred: ________

In what year(s) were animals released? ________
Total number released: ________
Out of the total number released, how many were captive-born? (For egg-laying animals, 'captive-born' includes individuals that were removed from the wild as eggs and then hatched in captivity.)

____

Out of the total number released, how many were wild-born? ('Wild-born' includes individuals that were removed from the wild, spent time in captivity, and were then released back to the wild.)

____

PRE-RELEASE
Were releasees trained in any way prior to release? Y/N
Please describe if yes: ________
Were the animals acclimatized to the release area prior to release? Y/N
Were potential releasees medically assessed? Y/N
Please describe if yes (for example, fecal samples, blood samples taken): ________
Were potential releasees genetically assessed? Y/N
Please describe if yes: ________

POST-RELEASE
Were releasees trained after release? Y/N
Please explain if yes: ________
Were releasees provisioned after release? Y/N
Were releasees monitored after release? Y/N
If yes, were they monitored more than 100 times per year? Y/N
    12–100 times per year? Y/N
    Less than 12 times per year? Y/N

SURVIVAL

What is the total number of releasees that are known to have died after release? ________

How many releasees died during the first year following release?

____

Out of those, how many died during the first month? ________

How many releasees are known to have died **after** the first year of release? ________

What is the total number of releasees that are known to have disappeared after release? ________

How many releasees disappeared during the first year following release? ________

Out of those, how many disappeared during the first month?

____

How many releasees disappeared **after** the first year of release?

____

What is the total number of releasees that were removed after release?

____

How many releasees were removed during the first year following release? ________

Out of those, how many were removed during the first month? ________

How many releasees were removed **after** the first year of release?

____

What is the total number of births to released animals? ________

How many births occurred during the first year following release?

____

Out of those, did any occur during the first month? ________

How many births were there **after** the first year of release? ________

Mortality and birth information is current as of: ________

Was there a public awareness program associated with the reintroduction project? Y/N

Please describe if yes: ________________________________

Were there student research projects (e.g. Master's or Ph.D. dissertations or internships) initiated as a result of the reintroduction program? Y/N

Were there local employment opportunities created as a result of the reintroduction program? Y/N

Are you willing to be listed as the person to contact for further information regarding this reintroduction project? Y/N

If so,
your name:
address:

phone:
FAX:

Comments:

REFERENCES

Amaral, M. and Morse, L. (1990) Reintroducing the American burying beetle. *Endangered Species Technical Bulletin*, **XV**(10), 3.

Beck, B.B., Kleiman, D.G., Dietz, J.M. *et al.* (1991) Losses and reproduction in reintroduced golden lion tamarins *Leontopithecus rosalia. Dodo, Journal of the Jersey Wildlife Preservation Trust*, **27**, 50–61.

Burchfield, P.M. (1985) Gladys Porter Zoo's role in Kemp's ridley sea turtle conservation. *Proceedings of the Annual Meeting of the American Association of Zoological Parks and Aquariums*, pp. 157–61.

Burger, J. and Zappalorti, R.T. (1988) Habitat use in free-ranging pine snakes, *Pituophis melanoleucus*, in New Jersey Pine Barrens. *Herpetologica*, **44**(1), 48–55.

Bush, M., Beck, B.B. and Montali, R.J. (1993) Medical considerations in reintroduction, in *Zoo and Wildlife Medicine III* (ed. M.E. Fowler), W.B. Saunders, Philadelphia, pp. 24–6.

Carpenter, J.W., Gabel, R.R. and Goodwin, J.G., Jr. (1991) Captive breeding and reintroduction of the endangered masked bobwhite. *Zoo Biology*, **10**, 439–49.

Carter, J. (1981) A journey to freedom. *Smithsonian*, **12** (1), 90–101.

Carter, J. (1988) Freed from keepers and cages, chimps come of age on Baboon Island. *Smithsonian*, **19** (3), 36–49.

Comly, L.M., Griffith, B., Scott, J.M. and Carpenter, J.W. (1989) *An Annotated Bibliography of Wildlife Translocations*. Department of Wildlife, College of Forest Resources, University of Maine (unpublished).

Dietz, L.A. and Nagagata, E.Y. (1986) Community conservation education program for the golden lion tamarin, in *Building Support for Conservation in Rural Communities – Workshop Proceedings*, vol. 1, (ed. J. Atkinson), QLF-Atlantic Center for the Environment, Ipswich, Massachusetts, pp. 8–16.

Foose, T.J. (1991) Viable population strategies for reintroduction programmes, in *Beyond Captive Breeding: Re-introducing Endangered Mammals to the Wild*, (ed. J.H.W. Gipps), Clarendon Press, Oxford, pp. 165–72.

Foose, T., Flesness, N., Seal, U. *et al.* (1992) *Ark into the 21st Century. A Global Conservation Strategy for Zoos and Aquaria* (unpublished).

Gordon, I.J. (1991) Ungulate re-introductions: the case of the scimitar horned oryx, in *Beyond Captive Breeding: Re-introducing Endangered Mammals to the Wild*, Symposia Zoological Society of London No. 62 (ed. J.H.W. Gipps), Clarendon Press, Oxford, pp. 217–40.

Griffith, B., Scott, J.M., Carpenter, J.W. and Reed, C. (1989) Translocation as a species conservation tool: status and strategy. *Science*, **245**, 477–80.

Herrero, S., Schroeder, C. and Scott-Brown, M. (1986) Are Canadian foxes swift enough? *Biological Conservation*, **36**, 159–67.

Hoeffs, M. and Reynolds, H. (1989) Management Plan for Wood Bison in the

Yukon. Report to the Canadian Wildlife Service, Canadian Parks Service and Yukon Department of Renewable Resources (unpublished).

Horwich, R.H. (1989) Use of surrogate parental models and age periods in a successful release of hand-reared sandhill cranes. *Zoo Biology*, **8**, 379–90.

IUCN (1987) *Introductions, Re-introductions and Re-stocking. The IUCN position statement on translocation of living organisms*, IUCN, Gland, Switzerland.

IUCN (1988) *1988 IUCN Red List of Endangered Animals*. IUCN, Gland, Switzerland.

Johnson, B. and Paine, F. (1989) The release of captive bred Puerto Rican crested toads: captive management implications and the cactus connection. *Proceedings of Regional Meetings of the American Association of Zoological Parks and Aquariums*, pp. 962–7.

Jones, C.G., Heck, W., Lewis, R.E. *et al.* (1991) A summary of the conservation management of the Mauritius kestrel *Falco punctatus* 1973–1991. *Dodo, Journal of the Jersey Wildlife Preservation Trust*, **27**, 81–99.

Kenyon, K. (1992) *Reintroduction of Captive-Bred Animals into Their Native Habitat: A Bibliography*, National Zoological Park, Smithsonian Institution (unpublished).

Kleiman, D.G., Beck, B.B., Dietz, J.M. *et al.* (1986) Conservation program for the golden lion tamarin: Captive research and management, ecological studies, educational strategies, and reintroduction, in *Primates: The Road to Self-Sustaining Populations* (ed. K. Benirschke), Springer–Verlag, New York, pp. 960–79.

Lindberg, D.G. (1992) Are wildlife reintroductions worth the cost? *Zoo Biology*, **11**, 1–2.

Loring, J.A. (1906) The Wichita buffalo range. *Tenth Annual Report of the New York Zoological Society*, New York Zoological Society, New York, pp. 181–200.

Mallinson, J.J.C. (1991) Partnerships for conservation between zoos, local governments and non-governmental organizations, in *Beyond Captive Breeding: Re-introducing Endangered Mammals to the Wild* (ed. J.H.W. Gipps), Clarendon Press, Oxford, pp. 57–74.

Meadows, R. (1992) The Guam rail – a second chance for survival. *Zoogoer*, **21** (1), 11–15.

Miller, T.J. (1985) Husbandry and breeding of the Puerto Rican toad (*Peltophryne lemur*) with comments on its natural history. *Zoo Biology*, **4**, 281–6.

Miller, B. and Mullette, K.J. (1985) Rehabilitation of an endangered Australian bird; the Lord Howe Island woodhen *Tricholiminas sylvestris* (Salsler). *Biological Conservation*, **34**, 55–95.

Montali, R.J. and Bush, M. (1992) Some diseases of golden lion tamarins acquired in captivity and their impact on reintroduction. *Proceedings Joint Conference American Association of Zoo Veterinarians and American Association of Wildlife Veterinarians*, pp. 14–16.

Morgan-Davies, A.M. (1980) Translocating crocodiles. *Oryx*, **15**, 371–3.

Oakleaf, B., Luce, B., Thorne, E.T. *et al.* (1992) Black-footed ferret reintroduction in Wyoming: project description and 1992 protocol, Wyoming Game and Fish Department, Laramie (unpublished).

Oliver, W.R., Wilkins, L., Kerr, R.H. and Kelly, D.L. (1986) The Jamaican hutia (*Geocapromys brownii*) captive breeding and reintroduction programme – history and progress. *Dodo, Journal of the Jersey Wildlife Preservation Trust*, **23**, 32–58.

Ounsted, M.L. (1991) Re-introducing birds: lessons to be learned for mammals, in *Beyond Captive Breeding: Re-introducing Endangered Mammals to the Wild, Symposia Zoological Society of London No. 62* (ed. J.H.W. Gipps), Clarendon Press, Oxford, pp. 75–85.

Phillips, M.K. (1990) Reestablishment of red wolves in the Alligator River National Wildlife Refuge, North Carolina. Progress Report No. 4, United States Fish and Wildlife Service, Washington, DC.

Rees, M.D. (1989) Aleutian Canada goose proposed for reclassification. *Endangered Species Technical Bulletin*, **XIV** (11–12), 8–9.

Sanborn, E.R. (1908) The national bison herd. *Zoological Society Bulletin* No. 28, New York Zoological Society, New York, pp. 400–12.

Seibels, R.E. (1991) Bali mynah status. *Captive Breeding Specialist Group News*, **2** (3), 8.

Snyder, N.F.R., Wiley, J.W. and Kepler, C.B. (1987) *The Parrots of Luquillo: Natural History and Conservation of the Puerto Rican Parrot*, Western Foundation of Vertebrate Zoology, Los Angeles.

Spalton, A. (1991) Recent developments in the re-introduction of the Arabian oryx (*Oryx leucoryx*) to Oman. *Captive Breeding Specialist Group News*, **2** (1), 8–10.

Spellerberg, I.F. and House, S.M. (1982) Relocation of the lizard *Lacerta agilis*: an exercise in conservation. *British Journal of Herpetology*, **6**, 245–8.

Stanley Price, M.R. (1989) *Animal Re-introductions: the Arabian Oryx in Oman*, Cambridge University Press, Cambridge.

Wallace, M. (1990) The California condor: current efforts for its recovery. *Endangered Species Update*, **8** (1), 32–5.

Wiley, J.W., Snyder, N.F.R. and Gnam, R.S. (1992) Reintroduction as a conservation strategy for parrots, in *New World Parrots in Crisis* (eds S.R. Beissinger and N.F.R. Snyder), Smithsonian Institution Press, Washington, DC, pp. 165–200.

Wood, D.A. and Holler, N.R. (1990) Of men and mice. *Florida Wildlife*, **44** (4), 37–9.

14

Criteria for reintroductions

*D.G. Kleiman**
National Zoological Park, Washington, DC, USA
M.R. Stanley Price
IUCN/SSC Reintroduction Specialist Group, Nairobi, Kenya
and
B.B. Beck
National Zoological Park, Washington, DC, USA.

14.1 INTRODUCTION

The recent surge of interest in the use of reintroduction, especially of captive-bred animals, as a conservation tool, has resulted in two recent reviews of the subject (Jones, 1990; Gipps, 1991) and the formation of a Re-introduction Specialist Group (RSG) of the Species Survival Commission (SSC) of the World Conservation Union (IUCN) and a Reintroduction Advisory Group (RAG) for the American Association of Zoological Parks and Aquariums (AAZPA).

While it is advisable to consult the RSG and RAG prior to initiating a reintroduction, there is currently no formal global mechanism for application to a national/international body before a reintroduction is planned and execution implemented. The purpose of this paper is to review the necessary criteria that should be met prior to a reintroduction, and to give additional guidelines and recommendations. Also we will provide

* Corresponding author.

Creative Conservation: Interactive management of wild and captive animals.
Edited by P.J.S. Olney, G.M. Mace and A.T.C. Feistner.
Published in 1994 by Chapman & Hall, London.
ISBN 0 412 49570 8

both a framework for the initial decision-making as well as a set of issues to be addressed as the process proceeds. Not all guidelines and recommendations will be relevant for all species or all conditions, but planners need to address the criteria none the less.

Reintroduction has the following goals or objectives (from Draft Guidelines for Re-introductions, IUCN/SSC Re-introduction Specialist Group):

(a) to enhance the long-term survival of the species;
(b) to re-establish a keystone species (in the ecological or cultural sense);
(c) to increase or maintain biodiversity;
(d) to provide long-term economic benefits to local people;
(e) or to achieve a combination of all of the above.

Reintroduction is an attempt to re-establish a species in an area which was once part of its historical geographical range, but in which it has become extinct. Sources of stock may be from another region within the species' range or from a captive population. Reintroduction using captive animals is an expensive approach to conserving a species, and is not the most cost-effective conservation strategy for the majority of rare and endangered species (Kleiman, 1989; Chivers, 1991; Stanley Price, 1991).

The success of a reintroduction can only be measured in terms of its goals and objectives. However, in most cases, success requires that there is eventually a self-sustaining viable population of the species at the reintroduction site (see Beck *et al.*, Chapter 13); this means that there must also be an intact functioning ecosystem.

Terms used by authors to define the release of animals or plants into their natural areas vary (Konstant and Mittermeier, 1982; Cade, 1986; Kleiman, 1989; Stanley Price, 1989). In the current definition used by IUCN (1987) 'translocation' is the movement of living organisms from one area with free release in another. The three main classes of translocation are defined as follows.

(a) 'Introduction' of an organism is the intentional or accidental dispersal by human agency of a living organism outside its historically known native range.
(b) 'Re-introduction' of an organism is the intentional movement of an organism into a part of its native range from which it has become extirpated in historic times as a result of human activities or natural catastrophe.
(c) 'Re-stocking' is the movement of numbers of plants or animals of a species with the intention of building up the number of individuals of that species in an original habitat.

IUCN (1987) does not distinguish between different origins of organisms, e.g. in the wild, in captivity, wild-bred or captive-bred, although these differences may have a major impact on the potential for success of a translocation. Also, many individuals use the term 'reintroduction' when they are releasing animals into already occupied habitat; the terms 'reinforcement' or 'restocking' may be more appropriate in such circumstances. However, with populations of some species very fragmented and the potential dispersal distance for the individuals of a species not being known, it may be impossible to distinguish between restocking and reintroduction.

The reintroductions considered most effective and successful have all been comprehensive efforts involving a large team and many resources. An evaluation and consideration of the criteria presented in this paper should also contribute to the success of individual reintroduction programmes.

There appear to be 13 criteria which should be considered prior to implementing a reintroduction. Subsumed under these criteria are additional guidelines and recommendations which will enhance the likely success of the effort. While these criteria can be considered for both animals and plants, they are most relevant for animals. Specific guidelines for the reintroduction of plants are currently being drafted by the IUCN/RSG; the issues of concern are presented in Falk (1990) and Maunder (1992).

In this paper, we will first review the criteria for reintroduction and then present several examples of how the criteria can be used in decision-making, especially regarding proposals to use captive-bred animals for a reintroduction. The zoo community is particularly entranced with the idea of using reintroduction as a conservation tool to save endangered species that are bred successfully in captivity. The criteria can be lumped into four discrete categories: condition of the species; environmental conditions; biopolitical conditions; and biological and other resources. Kleiman (1990) presented 10 criteria in evaluating the appropriateness of reintroducing three species of lion tamarins (*Leontopithecus*); the additional three criteria presented here involve an expansion of the number of biopolitical criteria.

14.2 CONDITION OF THE SPECIES

14.2.1 Need to augment the wild population

There should be a need to augment the size or genetic diversity of the wild population. It may be self-evident, but a reintroduction is not warranted unless there is a need to increase the number of individuals of a species, the number of separate populations of the species, or unless some genetic manipulations are needed to increase the genetic diversity.

Additionally, an evaluation of the environmental conditions, the local biopolitical considerations, the available resources and information, and the species biology will indicate whether reintroduction is the most cost-effective species recovery effort (Kleiman, 1989; Stanley Price, 1989; MacKinnon and MacKinnon, 1991). The attractiveness of reintroduction as a public relations tool should not result in cheaper and more efficient methods of species conservation and recovery being ignored.

14.2.2 Available stock

There should be available appropriate stock for use in a reintroduction. A reintroduction can use existing wild animals that are transplanted from another area to the reintroduction site. (For some biologists, the use of wild animals by definition means that they are involved with a translocation although translocation is the umbrella term used by the IUCN (1987) for any movement of organisms with free release in a second area, and thus encompasses introductions, reintroductions, and restocking.) Or, captive animals (either wild-born or captive-born) may be used. In either case, the individuals selected for release should be as close genetically to the original population as is feasible (Campbell, 1980; Caldecott and Kavanagh, 1983; Cade, 1986; IUCN, 1987; Wemmer and Derrickson, 1987; Kleiman, 1989; Stanley Price, 1989) to minimize chances of interbreeding between the original population and the newly released animals resulting in outbreeding depression (Shields, 1982; Templeton, 1986). Additionally, since the original stock was adapted to the particular environmental conditions present at the release site, genetically closely-related individuals will have a better chance of surviving (Brambell, 1977; Stromberg and Boyce, 1986).

If captive animals are to be used, the individuals must be surplus to the genetic and demographic needs of the captive population. However, reintroduction is not a mechanism for removing surplus animals from a captive population or a method of population management. There must be a viable self-sustaining well-managed captive population with a surplus from which to draw animals for release (Konstant and Mittermeier, 1982; Kleiman, 1989). A reintroduction should never jeopardize the captive population unless there is strong evidence that the release will save the species' habitat from further destruction or if prospects in the wild are better than in captivity, e.g. if there are severe husbandry problems. The explosive growth in the scientific management of breeding programmes in zoos since 1980 has meant that using captive-born animals has become more feasible.

14.2.3 No jeopardy to the wild population

The reintroduction should not prejudice the existing wild population. The release of captive or translocated animals into or near an already existing wild population, no matter how small the latter, can be extremely dangerous and even have the reverse effect of that desired. While we cannot achieve absolute certainty that our activities will not be prejudicial, no reintroduction should jeopardize the original wild population through disease, genetic swamping, or social disruption (Brambell, 1977; Campbell, 1980; Konstant and Mittermeier, 1982; Harcourt, 1987; Chivers, 1991). Ultimately, the protection of the original wild stock in its original habitat should have the highest priority (Caldecott and Kavanagh, 1983; IUCN, 1987; Kleiman, 1989). Especially when using captive animals, stringent pre-release veterinary screening is necessary to ensure, as much as possible, that the candidates for release are free of contagious disease, genetic defects, and other medical problems which could be transmitted to native animals (Brambell, 1977; Caldecott and Kavanagh, 1983; Wemmer and Derrickson, 1987; Woodford and Kock, 1991; Bush, Beck and Montali, 1993; Woodford and Rossiter, Chapter 9). In some cases, vaccinations against common diseases may be recommended.

14.3 ENVIRONMENTAL CONDITIONS

14.3.1 Causes of decline removed

Biotic and abiotic causes for the species' decline should have been eliminated or controlled. A reintroduction is not warranted unless the reasons for a species' reduction are known and have been removed (Brambell, 1977; Konstant and Mittermeier, 1982; Caldecott and Kavanagh, 1983; Cade, 1986; Griffith *et al.*, 1989; Stanley Price, 1989; Chivers, 1991). These may include, but are not restricted to, habitat destruction, hunting (or taking), predation by non-native forms (see Carey, 1988), introduced disease and environmental pollutants.

14.3.2 Sufficient protected habitat

Sufficient appropriate habitat in species' original range should be protected and secure. Without adequate habitat being available, a reintroduction is not warranted. Additionally, this habitat must be protected from further degradation so that the released individuals and their descendants will be secure (Aveling and Mitchell, 1982; Borner, 1985; MacKinnon and MacKinnon, 1991). The site must be large enough eventually to house a viable population, i.e. there must be sufficient carrying capacity (Brambell, 1977).

There has been considerable discussion about the release of individuals into habitat not previously occupied by the species. While most authors would restrict a reintroduction to ecologically suitable habitat within the original range (Brambell, 1977; Griffith *et al.*, 1989; Kleiman, 1989; Stanley Price, 1989), a few authors (e.g. Cade, 1986) have contrasting views.

Since the ultimate goal of any reintroduction is actually the maintenance of biodiversity and the potential for further evolution, a reintroduction can be a major incentive for a habitat restoration effort, including both basic research and actual implementation. Restoration of natural habitats is a relatively new discipline which is becoming more important as human population explosion and greed lead to global habitat destruction. Since many species' extinctions derive from severe habitat degradation, having a habitat restoration component in a reintroduction programme is likely to become increasingly common. Habitat management and restoration may involve very aggressive activities (Kleiman, 1989; Stanley Price, 1989).

14.3.3 Unsaturated habitat

There should be low densities (or none) of the species existing in the available habitat. It is usually preferable to release specimens into habitat which has no members of the species (reintroduction) or a very sparse population (restocking) unless the purpose of the release is to increase genetic diversity within a limited region. The release of individuals into saturated habitat can result in severe intraspecific conflict between the newly-released individuals and the stable residents. Competition for resources and social disruption (Brewer, 1978; Carter, 1981; Borner, 1985; Hannah and McGrew, 1991) are two possible outcomes while the transmission of an unknown pathogen to which the wild population is susceptible is a third (Brambell, 1977; Aveling and Mitchell, 1982; Caldecott and Kavanagh, 1983; Woodford and Kock, 1991; Bush, Beck and Montali, 1993). As already indicated, protection of the native wild population should always be given a higher priority than of individuals targeted for release.

14.4 BIOPOLITICAL CONSIDERATIONS

14.4.1 No negative impact for locals

The costs and benefits to the local human population shall have been assessed, and there should be no major negative impact. No reintroduction can succeed if any segment of the local community will be seriously hurt by the effort. It is self-evident that a release should cause minimal risk to human life and property; releasing a 'man-eating' tiger is clearly a non-productive strategy (Stanley Price, 1989).

It is also non-productive if a large segment of the local population will be severely hurt economically by the reintroduction, without the possibility of reasonable recompense or alternative employment. In many cases, there will be a minimal negative impact on the local community which can be absorbed, but which appears more serious due to local publicity by special-interest groups. Appropriate education may result in a reduction of the negative perceptions about the reintroduction.

In other cases, the reintroduction programme itself might add to the economic base of a community through eco-tourism or simply additional employment and the people will change their negative attitudes after assessing the costs and benefits.

14.4.2 Community support exists

Local community support for the reintroduction (and conservation more generally) should be achieved through economic benefits, conservation eduction, or other means. This is the positive side of the cost-benefit analysis. Local support is essential for a successful reintroduction and can be achieved in a variety of ways. First, if a reintroduction can provide economic benefits to the local community, there is likely to be a more positive rapport (Stanley Price, 1989; Kleiman, 1989). Secondly, a community conservation education programme that explains the aims and objectives of the reintroduction, in a manner that is relevant to the local people, is important for gaining support for the effort (Carley, 1981; Cade, 1986; Dietz and Nagagata, 1986; Kleiman, 1989; Phillips, 1990; Chivers, 1991). This is especially important in international programmes where there may be significant cultural differences between the local community and those involved with the reintroduction.

Community conservation education programmes can involve giving lectures while attending meetings of adult groups and visiting schools, but also developing public relations efforts that target the entire community or particular members of the local population. Using the media (television, radio, newspapers, magazines) to keep the broader public informed of the reintroduction's activities is extremely helpful; this form of public relations can reach the local community but also be distributed nationally and internationally (Phillips, 1990). Although media attention is generally viewed in a positive light, there are some situations when secrecy is warranted, especially about specific details of the reintroduction. For example, it may be preferable not to publicize the site of the actual release. Additionally, a reintroduction can produce materials that are distributed locally such as notebooks, T-shirts, posters and pins, as well as staging specific events such as parades and plays, which have an educational as well as recreational function (Dietz and Nagagata, 1986).

A reintroduction programme can gain considerably more local support if it includes a component of professional education, that is, if the budget has resources for supporting higher education for some individuals. This commitment also provides convincing evidence that 'outsiders' intend to assist seriously in local development and even turn over management of the reintroduction effort to the community.

14.4.3 GOs and NGOs are supportive/involved

All relevant governmental/non-governmental organizations (GOs and NGOs) should support the effort; where appropriate, they should be part of the decision-making process. Most reintroductions are likely to involve multiple organizations and agencies, both in the approval and the implementation process. All such agencies need to support the effort strongly since officials with negative attitudes can prove very obstructive. It is therefore best that the 'recovery' team include representatives from most relevant government and non-governmental agencies (Stanley Price, 1989; Kleiman *et al.*, 1991). While extremely complex for a multinational programme, the institutional collaboration and cooperation is essential to a smooth operation. Additionally, it provides the appropriate framework and basis for eventually turning over the programme to the local community.

All reintroduction programmes require that there be a clearly defined structure for making decisions, and that the authority and responsibilities of different members of the 'recovery' team are delineated (Clark and Westrum, 1989). Policies for interventions, talking with the press, hiring and firing staff and other sensitive issues need clarification long before a release is attempted. For example, it is inappropriate to develop a policy on interventions as individuals disappear from view after release. Members of the team can have vastly different opinions about how and when to provide support to released specimens; in times of crisis there is too much emotion for objective decision-making (Kleiman, 1989). Conflicts among the 'recovery' team members can only be avoided when the structure of the team and procedures for decision-making are laid out well in advance. As an additional guideline, it is a good idea to have a contingency plan to discontinue or suspend the reintroduction, if there are unanticipated problems or serious mortality.

14.4.4 Conformity with all laws/regulations

The relevant provincial, national and international legislation, regulations, treaties and guidelines should be incorporated into the action plan and not form a barrier to the reintroduction effort. It seems self-evident that a reintroduction programme should comply with all the appropriate legislation and regulations, but these factors are not always addressed. The lack of a single completed and signed form can stop a release from proceeding regardless of how much time, energy and resources have been put into the preparation and organization.

14.5 BIOLOGICAL AND OTHER RESOURCES

14.5.1 Reintroduction technology known/in development

There should be knowledge of techniques for reintroduction for this (or a closely related) species – or the protocol should incorporate methods for evaluating and improving the techniques. The 'science' of reintroduction is new and the techniques used will vary across species, habitats and other local conditions. There is currently no standardized approach, even within closely related taxa. Thus, a reintroduction programme must always contain a 'research and development' component. There needs to be incorporated into the protocol a mechanism both for evaluating the 'success' of the effort and for changing (improving) the methodology. This might include using a formal experimental design and statistical analysis of the resulting relative rates of survivorship (or whatever measure is used to indicate 'success') (see Beck *et al.*, 1991 for an example of this process). In any case, a reintroduction programme must specify a clearly defined method of evaluating the results.

Reintroduction protocols depend on the application of basic biological information. The degree of pre-release acclimatization or adaptation to local (biological and non-biological) conditions, and pre-release preparation or training will depend on the circumstances (Temple [1978] reviews this area extensively for birds). Obviously, it is better to prepare candidates for release as much as possible by exposing them to the appropriate environmental conditions and providing training (for examples of extensive training see Brewer, 1978; Carter, 1981, 1988; Cade, 1986; Beck *et al.*, 1991). This might include holding them in cages at the reintroduction site before release for recuperation from transport and local acclimatization. The release of totally naïve animals into a situation that is known to be difficult would be inhumane to say the least, especially in a species that requires considerable learning and experience (however, see Griffith *et al.* [1989] and AAZPA-RAG [1992] for some of the negative aspects of pre-release training). For captive-born animals, imprinting or dependence on humans should be avoided. Whether a species should be reared to avoid all humans will depend on the degree to which humans threaten it.

The degree of post-release life support and supplementation provided for each individual depends on a variety of factors peculiar to the species and local conditions. In general 'hard' releases do not involve much supplementation and support – and often minimal monitoring. Once released, the animals are expected to survive without much further human involvement. Alternatively, in 'soft' releases the individuals are first prepared/trained and then supported by humans for considerable time – through the provision of food, medical care and shelter and a period of pre-release acclimatization to the local environment (Scott-Brown, Herrero and Mamo, 1986; Kleiman, 1989; Stanley Price, 1989). Each reintroduction will have a different policy concerning the degree of life support proposed for individuals after release.

Numbers of individuals to be released simultaneously and numbers of successive cohorts to be released depend on the project's goals. In general, the more individuals that are released, the more likely that a viable population will be established, assuming that the conditions are appropriate.

Reintroduction techniques may be developed using non-native animals (Wallace, 1990; Miller *et al.*, 1991) but must include mechanisms for the protection of the native ecosystem and the recapture of the animals to prevent establishment of an unwanted population (IUCN, 1987; Stanley Price, 1989).

14.5.2 Knowledge of species' biology

Sufficient information should exist about the species' (or closely related taxa's) biology (e.g. demography, genetics, behaviour, reproduction, and ecology) in the wild (and/or captivity) to evaluate the success of the reintroduction.

This criterion is critical. Developing the preparation, release, and post-release management of a species requires the ability to answer questions about basic biology and critical resource needs in advance. Thus, well before the target date for the reintroduction the 'recovery' team needs to have access to information such as preferred types of habitat; typical size and structure of social groupings; required territory size and home range per individual/group; and amount and distribution of critical requirements such as water, den or shelter sites, or particular plants or animals. The existence and potential impact of competitors and/or predators needs evaluation (Konstant and Mittermeier, 1982; Cade, 1986; Wemmer and Derrickson, 1987; Griffith *et al.*, 1989, Kleiman, 1989; Stanley Price, 1989; Chivers, 1991; AAZPA-RAG, 1992).

Biological knowledge will help make multiple decisions about the reintroduction, e.g. choosing the best date, season, or time of day for a release or determining how many specimens to release over what period of time. This information is also important during the preparation phase, e.g. developing appropriate social groups before release (Jeffries *et al.*, 1985; Oliver *et al.*, 1986; Stanley Price, 1989; Hannah and McGrew, 1991; Beck *et al.*, 1991).

The problems and challenges posed by a natural environment for a newly released captive animal strongly suggest that for humane reasons no individual should be released with injuries, genetic defects, or other characteristics that might reduce their chances of survival (Harcourt, 1987; Stanley Price, 1989; Bush, Beck and Montali, in press).

14.5.3 Sufficient resources exist for the programme

Sufficient resources and mechanisms should exist to conduct the reintroduction, monitor the results of the effort, and do a cost–benefit analysis of the techniques. There have been several reintroductions where little was learned because there was insufficient post-release monitoring, and evaluation protocols were not developed, e.g. the early releases of orang-utans, *Pongo pygmaeus* (Aveling and Mitchell, 1982; Borner, 1985). Release of specimens into the wild, regardless of origin, is a wasted effort without a simultaneous commitment to understand which animals survived and why. This requires an intensive post-release monitoring programme during which the behaviour of permanently identified individuals is regularly observed, and causes of disability and death determined (Beck *et al.*, 1991). Monitoring should include behavioural observations of individuals' feeding, ranging, social interactions, activity cycles, and any other parameters which are relevant to the measurement of successful adaptation. Adequate monitoring cannot be accomplished without a secure source of funding (Brambell, 1977; Campbell, 1980; Cade, 1986; Scott and Carpenter, 1987; Griffith *et al.*, 1989; Kleiman, 1989; Stanley Price, 1989; Chivers, 1991). Beck *et al.* (Chapter 13) have found that long-term funding is significantly correlated with success of a reintroduction.

14.6 EXAMPLES OF THE USE OF CRITERIA

In 1990, a Population Viability Analysis (PVA) Workshop for the genus *Leontopithecus* was held in Brazil. For this, Kleiman applied the criteria for reintroduction to the three major forms of lion tamarins. Table 14.1 (adapted from Kleiman, 1990) presents the criteria and indicates where each species is with respect to the recommendation to initiate or carry out a reintroduction programme. In some cases, we have insufficient

Table 14.1 Translocation/reintroduction of lion tamarins: do appropriate conditions exist? (scale 5 = best) (adapted from Kleiman, 1990)

	Golden	*Golden-headed*	*Black*
Condition of species			
1. Need to augment wild popn.	Yes	No	Yes (?)
2. Available stock	Yes	Yes	No
3. No jeopardy to wild popn.	?	?	?
Environmental conditions			
4. Causes of decline removed	?	No	No
5. Sufficient protected habitat	Yes?	No	Yes
6. Unsaturated habitat	Yes	Yes(?)	?
Biopolitical conditions			
7. No negative impact for locals	No	?	?
8. Community support exists	5	2	4
9. GOs/NGOs supportive/involved	Yes	Yes	Yes
10. Conformity with all laws/regulations	Yes	?	?
Biological and other resources			
11. Reintroduction technology known/in development	4	3	3
12. Knowledge of species' biology	5	1.5	3
13. Sufficient resources exist for programme	Yes	No	No
Recommended reintroduction/translocation?	Yes	No	No

information to determine whether a criterion is met; e.g. for all three species, we do not know whether a reintroduction directly into already occupied habitat will jeopardize the wild population.

Overall, a consideration of the 13 criteria suggests that a reintroduction or translocation would be appropriate for the golden lion tamarin (*L. rosalia*), but not for the golden-headed (*L. chrysomelas*) and black (*L. chrysopygus*) lion tamarins. In the case of the black lion tamarin, the chief problem is the lack of a large, genetically diverse, captive, self-sustaining population. For the golden-headed lion tamarin, probably the major factor against a reintroduction/translocation is the lack of a need to augment the numbers of the wild population within the available habitat. It appears that there are still sufficient numbers of golden-headed lion tamarins in the wild although severe habitat destruction continues, there is not yet sufficient protected habitat, and the causes of the decline have certainly not been controlled.

Table 14.2 presents a comparable analysis for the giant panda (*Ailuropoda melanoleuca*), based on discussions with experts in the manage-

Table 14.2 Translocation/reintroduction of giant pandas (*Ailuropoda melanoleuca*): do appropriate conditions exist? (scale 5 = best)[a]

Condition of species	
1. Need to augment wild popn.	Yes
2. Available stock	No
3. No jeopardy to wild popn.	?
Environmental conditions	
4. Causes of decline removed	No
5. Sufficient protected habitat	No
6. Unsaturated habitat	No
Biopolitical conditions	
7. No negative impact for locals	No?
8. Community support exists	2
9. GOs/NGOs supportive/involved	Yes?
10. Conformity with all laws/regulations	?
Biological and other resources	
11. Reintroduction technology known/in development	1
12. Knowledge of species' biology	2.5
13. Sufficient resources exist for programme	No
Recommended reintroduction/translocation?	No

[a]Developed during Workshop 'Pandas: A Conservation Initiative', 2–8 June, 1991, Front Royal, VA, with input from numerous experts in giant panda biology.

ment and behaviour of giant pandas during a Workshop on Pandas held in June, 1991 at Front Royal, Virginia (Kleiman amd Roberts, 1991). A major block to a consideration of a reintroduction/translocation programme is the lack of a viable self-sustaining captive population of giant pandas from which candidates for reintroduction can be chosen. Additionally, it is not clear whether there is a need to augment the size or genetic diversity of the wild population, and the causes of the giant panda's decline have not been controlled. Finally, we have too little information on the species' biology and on reintroduction/translocation techniques. Although pandas have been taken into captivity and released, there has been insufficient monitoring of released animals (or even free-ranging wild animals) to develop a reintroduction technology. Thus, at this stage, it would seem unproductive to recommend a reintroduction programme for giant pandas. Indeed, given the status of the species, individuals which are brought into captivity for medical or other treatment should be released as soon as is feasible into the location from which they were taken. Giant pandas should not be brought into and then kept in captivity without a very strong justification.

14.7 CONCLUSIONS

Beck *et al.* (Chapter 13) present the results of a survey of reintroduction projects using captive animals. Additionally, they attempt to evaluate those characteristics of reintroductions that lead to success (their definition of success requires that the wild population reaches 500 individuals free from provisioning or other human support or a PVA suggests that the population is self-sustaining). Successful projects tend to extend over many years, and involve the local community through conservation education programmes and local employment. It is still unclear whether pre- and post-release acclimatization, training, or extensive post-release provisioning have positive effects. Based on Beck *et al.* (Chapter 13), it would appear that Criteria 7, 8, 9 and 13 (Tables 14.1 and 14.2) need special attention before a reintroduction is planned. Their preliminary findings suggest therefore that biopolitical conditions and long-term funding are as or more important to the success of a reintroduction as knowledge of species' biology or reintroduction technology, a tentative result which needs to be examined in much more detail. Hopefully, with greater documentation of future reintroductions and a more expanded database we can test the effectiveness of different approaches more fully, as recommended by Beck *et al.* (Chapter 13).

ACKNOWLEDGEMENTS

Kleiman and Beck would like to thank the following organizations for their support of the reintroduction programme for the golden lion tamarin: Smithsonian Institution (International Environmental Sciences Programme), Friends of the National Zoo, World Wildlife Fund, Wildlife Preservation Trust International, Frankfurt Zoological Society, and the many zoos who have contributed time, money and animals to the programme. Table 14.2 is based on information supplied by Claudio and Susanna Padua, Anthony Rylands, Cristina Alves, James and Lou Ann Dietz, Faiçal Simon, Jeremy Mallinson and Jon Ballou.

REFERENCES

AAZPA-Reintroduction Advisory Group (1992) Guidelines for reintroduction of animals born or held in captivity. Bethesda, Maryland.

Aveling, R. and A. Mitchell (1982) Is rehabilitating Orang utans worthwhile? *Oryx*, **16**, 263–71.

Beck, B.B. (1991) Managing zoo environments for reintroduction. AAZPA 1991 Proceedings Annual Conference, 436–40.

Beck, B.B., D.G. Kleiman, I. Castro, B. *et al.* (in press) Preparation of captive-born golden lion tamarins for release into the wild, in *A Case Study in Conservation Biology: the Golden Lion Tamarin* (ed. D.G. Kleiman), Smithsonian Institution Press, Washington, DC.

Beck, B.B., D.G. Kleiman, J.M. Dietz, I. *et al.* (1991) Losses and reproduction in reintroduced golden lion tamarins, *Leontopithecus rosalia. Dodo, Journal of the Jersey Wildlife Preservation Trust,* **27**, 50–61.

Borner, M. (1985) The rehabilitated chimpanzees of Rubondo Island. *Oryx,* **19**, 151–4.

Brambell, M.R. (1977) Reintroduction. *International Zoo Yearbook,* **17**, 112–16.

Brewer, S. (1978) *The Chimps of Mt. Asserik.* Alfred Knopf, New York.

Bush, M., B.B. Beck, and R.J. Montali (1993) Medical considerations of reintroduction, in *Zoo and Wild Animal Medicine,* 3rd edn (ed. M.E. Fowler), W.B. Saunders, Philadelphia, pp. 24–6.

Cade, T.J. (1986) Reintroduction as a method of conservation. *Raptor Research Reports* No. 5, 72–84.

Caldecott, J.O. and M. Kavanagh (1983) Can translocation help wild primates? *Oryx,* **17**, 135–9.

Campbell, S. (1980) Is reintroduction a realistic goal? in *Conservation Biology* (eds. M.E. Soulé and B.A. Wilcox), Sinauer Associates, Sunderland, MA, pp. 263–9.

Carey, J. (1988) Massacre on Guam. *National Wildlife,* **26**, 13–15.

Carley, C.J. (1981) Red wolf experimental translocation summarized. *Wild Canid Survival and Research Center Bulletin.,* Part 1, 4–7, Part 2, 8–9.

Carter, J.A. (1981) A journey to freedom. *Smithsonian Magazine,* **12**, 90–101.

Carter, J.A. (1988) Survival training for chimps. *Smithsonian Magazine,* **19**, 34–48.

Chivers, D.J. (1991) Guidelines for re-introductions: procedures and problems, in *Beyond Captive Breeding: Re-introducing Endangered Mammals to the Wild. Symposia Zoological Society of London No. 62,* (ed. J.H.W. Gipps), Clarendon Press, Oxford, pp. 89–99.

Clark, T.W. and R. Westrum (1989) High performance teams in wildlife conservation: a species reintroduction and recovery example. *Environmental Management,* **13** (6), 663–70.

Dietz, L.A. and E. Nagagata (1986) Community conservation education program for the golden lion tamarin, in *Building Support for Conservation in Rural Areas – Workshop Proceedings,* vol. 1 (ed. J. Atkinson), QLF-Atlantic Center for the Environment, Ipswich, MA, pp. 8–16.

Falk, D.A. (1990) Integrated strategies for conserving plant genetic diversity. *Ann. Missouri Bot. Gdn.,* **77**, 38–47.

Gipps, J.H.W. (ed.) (1991) *Beyond Captive Breeding: Re-introducing Endangered Mammals to the Wild. Symposia Zoological Society of London No. 62* (ed. J.H.W. Gipps), Clarendon Press, Oxford.

Griffith, B., J.M. Scott, J.W. Carpenter, and C. Reed (1989) Translocation as a species conservation tool: status and strategy. *Science,* **245**, 477–80.

Hannah A.C. and McGrew, W.C. (1991) Rehabilitation of captive chimpanzees, in *Primate Responses to Environmental Change* (ed. Hilary Box), Chapman & Hall, London, pp. 167–86.

Harcourt, A.H. (1987) Options for unwanted or confiscated primates. *Primate Conservation,* **8**, 111–13.

IUCN (1987) Translocation of living organisms: introductions, re-introductions, and re-stocking. *IUCN Position Statement,* IUCN, Gland, Switzerland.

Jeffries, D.J., P. Wayre, R.M. Jessop, A.J. *et al.* (1985) The composition, age, size and pre-release treatment of the groups of otters *Lutra lutra* used in the first releases of captive-bred stock in England. *Otters – Journal of the Otter Trust* 1984, 11–16.

Jones, S. (ed.) (1990) Captive propagation and reintroduction: A strategy for preserving endangered species? *Endangered Species Update,* **8** (1) 1–88.

Kleiman, D.G. (1989) Reintroduction of captive mammals for conservation. *BioScience*, **39** 152–61.

Kleiman, D.G. (1990) Decision-making about a reintroduction: do appropriate conditions exist? *Endangered Species Update*, **8** (1), 18–19.

Kleiman, D.G., B.B. Beck, J.M. Dietz, and LA. Dietz. (1991) Costs of a re-introduction and criteria for success: accounting and accountability in the Golden Lion Tamarin Conservation Program, in *Beyond Captive Breeding: Re-introducing Endangered Mammals to the Wild. Symposia Zoological Society of London No. 62* (ed. J.H.W. Gipps), Clarendon Press, Oxford, pp. 125–42.

Kleiman, D.G. and M.S. Roberts (eds) (1991) *The Pandas: A Conservation Initiative*, Giant Panda and Red Panda Conservation Workshop, Working Group Reports, National Zoological Park, Washington, DC.

Konstant, W.R. and R.A. Mittermeier (1982) Introduction, reintroduction, and translocation of Neotropical primates: past experiences and future possibilities. *International Zoo Yearbook*, **22**, 69–77.

MacKinnon, K. and J. MacKinnon (1991) Habitat protection and re-introduction programmes, in *Beyond Captive Breeding: Re-introducing Endangered Mammals to the Wild. Symposia Zoological Society of London No. 62* (ed. J.H.W. Gipps), Clarendon Press, Oxford, pp. 173–98.

Maunder, M. (1992) Plant reintroduction: An overview. *Biodiversity and Conservation*, **1**, 51–61.

Miller, B., D. Biggins, L. Hanebury, C. *et al.* (1991) Rehabilitation of a species: the black-footed ferret (*Mustela nigripes*). *Wildlife Rehabilitation*, **9**, 183–92.

Oliver, W.L.R., L. Wilkins, R.H. Kerr, and D.L. Kelly (1986) The Jamaican hutia *Geocapromys brownii* captive breeding and reintroduction programme – history and progress. *Dodo. Journal of the Jersey Wildlife Preservation Trust*, **23**, 32–58.

Phillips, M.K. (1990) Red wolf: recovery of an endangered speces. *Endangered Species Update*, **8** (1), 79–81.

Scott, J.M. and J.W. Carpenter (1987) Release of captive-reared or translocated endangered birds: what do we need to know? *Auk*, **104** (3), 544–5.

Scott-Brown, J.M., S. Herrero, and C. Mamo (1986) Monitoring of released swift foxes in Alberta and Saskatchewan. Final report, 1986. Unpublished Report to the Canadian Fish and Wildlife Service.

Shields, W.M. (1982) *Philopatry, Inbreeding, and the Evolution of Sex*, SUNY Press, Albany.

Stanley Price, M.R. (1989) *Animal Re-introductions: the Arabian Oryx in Oman*, Cambridge University Press, Cambridge.

Stanley Price, M.R. (1991) A review of mammal re-introductions, and the role of the Re-introduction Specialist Group of IUCN/SSC, in *Beyond Captive Breeding: Re-introducing Endangered Mammals to the Wild. Symposia Zoological Society of London No. 62* (ed. J.H.W. Gipps), Clarendon Press, Oxford, pp. 9–25.

Stromberg, M.R. and M.S. Boyce (1986) Systematics and conservation of the Swift fox, *Vulpes velox*, in North America. *Biological Conservation*, **35**, 97–110.

Temple, S.A. (1978) *Endangered Birds: Management Techniques for Preserving Threatened Species*, University of Wisconsin Press, Madison.

Templeton, A. (1986) Coadaptation and outbreeding depression, in *Conservation Biology* (ed. M. Soulé), Sinauer Associates, Sunderland MA, pp. 105–16.

Wallace, M. (1990) The California condor: current efforts for its recovery. *Endangered Species Update*, **8**, 32–5.

Wemmer, C. and S. Derrickson (1987) Reintroduction: The zoobiologists' dream. *Annual Proceedings of the American Association of Zoological Parks and Aquariums*, pp. 48–65.

Woodford, M.H. and R.A. Kock (1991) Veterinary considerations in re-introduction and translocation projects, in *Beyond Captive Breeding: Reintroducing Endangered Mammals to the Wild. Symposia Zoological Society of London No. 62* (ed. J.H.W. Gipps), Clarendon Press, Oxford, pp. 101–10.

15

Development of coordinated genetic and demographic breeding programmes

L.E.M. de Boer
National Foundation for Research in Zoological Gardens and EEP Executive Office, Amsterdam, The Netherlands.

15.1 INTRODUCTION

There are some early examples of the potential of captive breeding in zoos for preventing extinction of highly endangered species. Species such as Przewalski's horse (*Equus przewalskii*), Père David's deer (*Elaphurus davidianus*), Hawaiian goose (*Branta sandvicensis*) and a number of others would no longer exist but for the fact that their populations were maintained and propagated for many decades in captivity. However, the concept that zoos could serve the goal of preserving biological species through propagation of captive stock at a more than incidental scale is much more recent. The zoo world, aroused by the rapidly accelerating destruction of the world's habitats and increasing loss of biological diversity, started to explore this idea and its consequences only some 20 years ago. This resulted in the organization of the first cooperative and coordinated breeding programmes in the late 1970s and early 1980s. Since then, considerable progress has been made with respect to theoretical considerations, practical implications

Creative Conservation: Interactive management of wild and captive animals.
Edited by P.J.S. Olney, G.M. Mace and A.T.C. Feistner.
Published in 1994 by Chapman & Hall, London.
ISBN 0 412 49570 8

and organizational structures required. Simultaneously the number of captive breeding programmes has increased exponentially. It can now be stated beyond doubt that captive propagation potentially constitutes one of the important tools in conservation of biological diversity (e.g. Foose, Seal and Flesness, 1987).

15.2 GOALS AND PREREQUISITES

The ultimate goal of captive, or *ex situ*, breeding programmes as we see them now, is to provide support for the survival of species, subspecies or populations in natural environments (Seal, 1986; Foose, 1986). To achieve this goal, the primary objectives of *ex situ* programmes are:

(a) to propagate and manage *ex situ* populations of highly endangered taxa to prevent their immediate extinction;
(b) to employ these programmes as parts of conservation strategies that interactively manage *ex situ* and *in situ* populations to ensure ultimate survival of these taxa in the wild;
(c) to develop stable captive populations of rare and endangered taxa for education programmes which are beneficial to the survival of conspecifics in the wild.

The foundations for the achievement of these objectives for every breeding programme can be defined on three levels, as follows.

(a) At the level of individuals, sufficient longevity and physical and psychological well-being should be ensured. Animals should live long lives of acceptable quality.
(b) At the level of breeding pairs and colonies, sufficient reproduction should occur, to guarantee continuity over the generations.
(c) At the level of populations, the conservation of genetic population structures that resemble the natural ones as closely as possible should be assured, while demographically stable population structures should be established as well.

It is axiomatic that preservation of a population's genetic structure is crucial to its survival (de Boer, 1989a, b; Foose and Ballou, 1988). The preservation of such genetic structure determines the probability of success when the *ex situ* population is used to re-establish or reinforce wild populations that are exposed to natural selection at some stage in the future. Similarly, it is self-evident that the achievement of a demographically stable population structure is crucial to a population's survival. Demographic instability affects genetic structure and hampers optimal harvesting from the population for reintroduction, while demographic imbalance may even cause a population's extinction. It is these modern population biological insights that demand coordinated management of *ex situ* populations. Obvious reasons for this are:

(a) that *ex situ* populations generally are much smaller than original, unthreatened wild populations, rendering them subject to severe risks of genetic and demographic stochasticity;
(b) that they are nearly always extremely fragmented, scattered among many dozens of zoos all over the world, which considerably increases the risk of inbreeding and extinction in the subunits;
(c) that in the *ex situ* situation they are subject to risks of unnatural, artificial selection.

The effects of these risks can be summarized as degeneration, domestication and vulnerability to extinction. Coordinated management is necessary to minimize these risks. Such population management procedures greatly rely on three conditions:

(a) development of a sound theoretical basis with respect to genetics and demography of small and fragmented populations;
(b) development of 'tools' to analyse breeding programme populations in the light of the available theoretical knowledge;
(c) development of organizational structures to implement the management proposals.

Major progress has been made in these three areas over the past 20 years.

15.3 RECENT ADVANCES

Many theoretical population geneticists and demographers have been working hard in recent years on the development of the theoretical background for small population management. Currently it is fairly clear how the negative effects of inbreeding and loss of genetic variability can be prevented, and how healthy populations can be maintained over many generations without degeneration and signs of domestication (e.g. Foose *et al.*, 1986). On the basis of these general insights refinements of theory are currently being developed for species-specific situations.

Tools for population management have been developed in the form of the ISIS (International Species Information System) database, the ARKS (Animal Records Keeping System) software and the studbook keeping and population management software packages (Jones, 1990; Flesness and Mace, 1988). In addition to these, various extremely helpful software tools have been developed for specific population analyses and complex calculations (e.g. Lacy, 1989; Ballou, 1990).

Much research has resulted in the availability of technical tools for genetic analyses of populations, for instance with respect to the identification of subspecific status of breeding programme animals, assessment of genetic variability, paternity and maternity, etc. (Moore, Holt and Mace, 1992; see Wayne *et al.*, Chapter 5 and Holt, Chapter 7).

Organizational structures to bring the theory of small population management into practice have been set up as well. The regional breeding programme organizations are the basis for this structure (de Boer, 1992). The early examples set by the North American Species Survival Plans (SSP), Australian Species Management Plans (ASMP) and British Joint Management of Species Programme (JMSG) have been followed by the European Endangered Species Programme (EEP), the Japanese, African and Indian programmes, while South-East Asia, South America and China are close to initiating such programmes as well. Soon all major regions of the world will have organized regional breeding programmes.

Global coordination of population management for species which have programmes in more than a single region has also been developed. Species coordinators of the various regions, together with representatives of the Captive Breeding Specialist Group (CBSG) of the Species Survival Commission (SSC) of the World Conservation Union (IUCN) develop global coordination plans for such species (Seal, 1992; Seal, Foose and Ellis-Joseph, Chapter 16). Moreover, regular contacts between the conservation coordinators of the breeding programme regions and between these and CBSG ensure increasing global coordination and cooperation (CBSG, 1992). Obviously, the International Union of Zoological Gardens (IUDZG), which is increasingly developing into a truly global zoo organization, also plays an important role in this respect (IUDZG, in press).

To date, all these activities have resulted in the establishment of over 300 breeding programmes in the various regions, involving over 200 different endangered animal species, and involving several hundred species coordinators and studbook keepers, not to mention the countless others who serve as members of species committees and other accessory groups (CBSG, 1992).

In the beginning, in all of the breeding programme regions, species for management programmes were chosen more or less on an *ad hoc* basis. Personal interests, the availability of enthusiastic potential coordinators and of an initial population of sufficient size played important roles in choosing species for new programmes. In the meantime structures have been set up that enable a more adequate system of designating species for management programmes, taking into account the specific potentials of the zoo world as well as the conservation needs. On the one hand systems of Taxon Advisory Groups (TAGs) have been developed in various regional organizations (e.g. Hutchins *et al.*, 1991). On the other hand CBSG-developed techniques such as Population and

Habitat Viability Assessments (PHVAs) and Conservation Assessment and Management Plans (CAMPs) to systematically assess the need for support by captive management of a wide variety of vertebrate taxa (Clark, Backhouse and Lacy 1990; Mace and Lande, 1991; Seal, Ballou and Padua, 1990; Seal, Foose and Ellis-Joseph; Chapter 16). On the basis of a combination of these two factors – assessment of zoo potentials and assessment of conservation priorities – newly initiated regional breeding programmes in the near future will be parts of global strategies.

Much progress has also been made in recent years with respect to the ultimate goal of *ex situ* breeding programmes – the lending of support to the survival of species in their natural environments. Many breeding programmes have already shown their educational value, on the basis of which public and governmental interest in habitat protection and protection of remnant wild populations of the breeding programme species dramatically increased. Methods have been explored for interactive management of wild and captive populations. In fact the vision grows that captive and wild should be regarded as two components of the same population, and that the captive component at an early stage should be fully supportive to the survival of the wild component (Foose, 1986; Foose, Seal and Flesness, 1987).

Biotechnological methods of artificial reproduction (Moore, Holt and Mace, 1992; Holt, Chapter 7) and cryopreservation have been explored, and the usefulness of germplasm banks in combination with these technologies has been investigated as a third major component to species conservation (CBSG, 1991). This third component may be developed as a powerful link between captive and wild management.

Much experience has also been gained with respect to reintroduction in the past years (see Beck *et al.*, Chapter 13 and Kleiman, Stanley Price and Beck, Chapter 14). An increasing number of reintroduction projects have shown that bringing captive-born animals back to natural habitats is possible, provided that enough research is done, that projects are well prepared and that cooperation is sought with local inhabitants, nature reserves, conservation organizations and governments. With respect to the last item, a number of zoos have set excellent examples of how cooperation with governments, organizations and people of the countries of origin of breeding programme species can be established, and how beneficial such cooperation is for the survival of such species in the wild.

15.4 THE FUTURE

Summarizing, it can be stated that in the past five to 10 years – thanks to a tremendous amount of work of many persons, institutions and organizations – considerable progress has been made with respect to theoretical backgrounds and practical implications of breeding programmes. It can also be stated that in the past few years many, if not nearly all, aspects which are necessary to make *ex situ* breeding programmes truly effective components of species survival and nature conservation have been developed to a great extent. Currently a major project is underway: the development of the 'World Zoo Conservation Strategy' by IUDZG and CBSG (IUDZG/CBSG, 1993). In this document – which will outline the role of the zoos and aquaria of the world in global conservation – all aspects and components of captive breeding programmes that have been explored and developed in the past years, will be put together in a large framework, a global strategy in support of the rescue operation for the biodiversity of our planet. With the publication of this strategy the zoo and aquaria community will show the world just how important a role it wants and intends to play in this great endeavour.

There is every reason to be proud of all these recent developments. However, there is no reason whatsoever for the zoo community to rest on its laurels! What has been accomplished so far – however much work it has cost – is not much more than an exploration of possibilities and potentials, a modelling of the components, and the construction of the framework. Having accomplished this, the real work still remains to be started. And the real work is a tremendous task! It requires the initiation, after careful choice, of many hundreds, maybe even a thousand or more, breeding programmes; the tightening of the world zoo network, involving over one thousand institutions; the transformation of what currently still is a rather loose assemblage of widely varied zoological institutions into a really effective, global conservation organization; the incorporation into this network of many thousands of zoo professionals to serve conservation goals; and the establishment of excellent contacts with other parts of the conservation community and with hundreds of local, national and international governments and organizations. All this seems almost to be an impossible task. Nevertheless it is the zoo community's obligation to the conservation of biodiversity to try to do the seemingly impossible.

The consequences to zoos of this obligation should not be taken lightly. It is beyond the scope of this paper to go into all the details of the far reaching changes that are required in the attitude of many zoos and their staffs. It seems appropriate, however, to make one final remark.

The keywords of the zoo community's contribution to the conservation of endangered species are **intensive management** and **cooperation**. It is the specific strength of the zoo world that under zoo conditions populations can be intensively managed to safeguard their genetic and demographic structures, which are crucial to their survival in captivity as well as in the wild. Intensive management can only be accomplished by intensive cooperation between the institutions involved. It is inherent that intensive management and intensive cooperation are labour intensive. Thus, above all, zoos must find ways to carry out all this work and to raise the budgetary means needed for it. If the zoo community fails in the attempt to enlarge the working potential of zoo staffs and that of the regional and global coordinating organizations, and if it fails in finding ways to finance this tremendous task, the magnificent edifice as constructed so laboriously over recent years – stone by stone – will remain partly empty, and hundreds, maybe thousands, of species will be denied its shelter.

REFERENCES

Ballou, J.D. (1990) *Capacity. Software to Establish Target Population Sizes for Managed Species*, National Zoological Park, Washington, DC.

Boer, L.E.M. de (1989a) Preservation of species in zoological and botanical gardens, in *EEP Co-ordinators' Manual* (ed. L.E.M. de Boer), Stichting Nationaal Onderzoek Dierentuinen, Amsterdam, Section 2, pp. 1–27.

Boer, L.E.M. de (1989b) Genetics and breeding programs; genetic guidelines and their background for EEP-co-ordinators, in *EEP Co-ordinators' Manual* (ed. L.E.M. de Boer), Stichting Nationaal Onderzoek Dierentuinen, Amsterdam, Section 4, 1–86.

Boer, L.E.M. de (1992) *Ex situ* propagation programmes as a contribution to the conservation of biodiversity, in *Conservation of Biodiversity for Sustainable Development* (eds O.T. Sandlund, K. Hindar and A.H.D. Brown), Scandinavian University Press, Oslo.

CBSG (1991) *Genome Resource Banking for Wild Species Conservation*, Captive Breeding Specialist Group, Apple Valley, MN.

CBSG (1992) Regional conservation coordinator committee; annual report 1992. Captive Breeding Specialist Group, Apple Valley, MN.

Clark, T.W., Backhouse, G.N. and Lacy, R.C. (1990) The population viability assessment workshop: a tool for threatened species management. *Endang. Spec. Update*, **8**, 1–5.

Flesness, N.R. and Mace, G.M. (1988) Population databases and zoological conservation. *Int. Zoo Yb.*, **27**, 42–9.

Foose, T.J. (1986) Riders of the last ark: the role of captive breeding in conservation strategies, in *The Last Extinctions* (eds L. Kaufman and D. Mallory), MIT Press, Cambridge, MA, pp. 14–65

Foose, T.J. and Ballou, J.D. (1988) Population management: theory and practice. *Int. Zoo Yb.*, **27**, 26–41.

Foose, T.J., Lande, R., Flesness, N.R. *et al.* (1986) Propagation plans. *Zoo Biol.*, 5, 139–46.

Foose, T.J., Seal, U.S. and Flesness, N.R. (1987) Captive propagation as a component of conservation strategies for endangered primates. *Monogr. Primatol.*, **9**, 263–99.

Hutchins, M., Wiese, R.J., Willis, K. and Becker, S. (eds) (1991) *AAZPA Annual Report on Conservation and Science 1990–91*, AAZPA, Bethesda, MD.

IUDZG/CBSG (1993) *The World Zoo Conservation Strategy*, Chicago Zoological Society, Brookfield, p. 122.

Jones, S.R. (1990) Overview of goals and activities of the IUCN Captive Breeding Specialist Group and International Species Information System. *Endang. Spec. Update*, **8** (1), 8–9.

Lacy, R.C. (1989) Analysis of founder representation in pedigrees: founder equivalents and founder genome equivalents. *Zoo Biol.*, **8**, 11–23.

Mace, G.M. and Lande, R. (1991) Assessing extinction threats: towards a re-evaluation of IUCN threatened species categories. *Conservation Biology*, **5** (2), 148–57.

Moore, H.D.M., Holt, W.V. and Mace, G.M. (eds) (1992) *Biotechnology and the Conservation of Genetic Diversity. Symp. Zool. Soc. London No. 64*, Claredon Press, Oxford.

Seal, U.S. (1986) Goals of captive propagation programmes for the conservation of endangered species. *Int. Zoo Yb.*, **24/25**, 174–9.

Seal, U.S. (1992) Tigers go global, finally. *Tiger Beat*, **5** (1), 1/16.

Seal, U.S., Ballou, J.D. and Padua, C.V. (1990). *Leontopithecus*: Population Viability Analysis Workshop Report. Captive Breeding Specialist Group, Apple Valley, MN.

16

Conservation Assessment and Management Plans (CAMPs) and Global Captive Action Plans (GCAPs)

U.S. Seal, T. J. Foose and S. Ellis
IUCN/SSC Captive Breeding Specialist Group, Minnesota, USA.

16.1 INTRODUCTION

Reduction and fragmentation of wildlife populations and habitats are occurring at rapid and accelerating rates. The results for an increasing number of taxa are small population sizes (i.e. a few tens to a few hundreds, or at best a few thousands) and isolated populations that are at risk of extinction. For such populations intensive management may become necessary for their survival and recovery. To an ever increasing extent, this intensive management includes, but is not limited to, captive breeding.

The problems for wildlife are so enormous that it is vital to apply the limited resources available for intensive management as efficiently and effectively as possible. Conservation Assessment and Management Plans (CAMPs) and their derivatives Global Captive Action Plans (GCAPs) are being developed to respond to this need. This paper

Creative Conservation: Interactive management of wild and captive animals.
Edited by P.J.S. Olney, G.M. Mace and A.T.C. Feistner.
Published in 1994 by Chapman & Hall, London.
ISBN 0 412 49570 8

describes the current state and development of these processes, which are still evolving.

16.2 CONSERVATION ASSESSMENT AND MANAGEMENT PLANS – CAMPs

CAMPs are intended to provide strategic guidance for the application of intensive management and information collection techniques to threatened taxa. CAMPs provide a rational and comprehensive means of assessing priorities for intensive management (including captive breeding) within the context of the broader conservation needs of threatened taxa.

CAMPs are conducted as collaborative ventures of the Captive Breeding Specialist Group (CBSG) of the Species Survival Commission (SSC) of the World Conservation Union (IUCN) with the other taxa-based Specialist Groups of the IUCN and Birdlife International, formerly International Council for Bird Preservation (ICBP), as well as with representatives of the Taxon Advisory Groups (TAGs) of the organized regional captive breeding programmes of the zoo/aquaria world. Hence, the CAMP process assembles a broad spectrum of expertise on captive and wild management of the taxa under review. Within the SSC of IUCN, the primary goal of the CBSG is to contribute to the development of holistic (i.e. integrating *in situ* and *ex situ*) and viable conservation strategies and action plans by the taxa-based SSC and ICBP Specialist Groups.

The CAMP process is initiated by preparing a briefing book for a subsequent workshop. The workshop participants develop assessments of risk of extinction and formulate recommendations for action for each species and/or subspecies, using a spreadsheet and data on the status of populations and habitats in the wild. These sheets also permit entry of the recommendations for intensive action. For this purpose, the process utilizes information from SSC Action Plans that may already have been formulated by the taxa-based Specialist Groups as well as additional data from experts on the taxa. Where such Action Plans do not yet exist, the CAMP process produces the necessary assessment of status and prospects to facilitate formulation of Global Action Plans for both *in situ* and *ex situ* efforts.

The CAMP process is also providing an opportunity to test the applicability of the Mace–Lande categories and criteria (Mace and Lande, 1991) for assessment of threat. The Mace–Lande system is being considered as a new process for assigning IUCN Categories of Threat to species, and is still under active development. The scheme attempts to assess threat in terms of likelihood of extinction within a specified

Table 16.1 Assessment of threat resulting from the CAMP Workshops

	Total taxa	*Critical*	*Endangered*	*Vulnerable*	*Safe*	*Unknown*	*Total threatened*
Parrots	428	25	36	78	228	61	139
Waterfowl	234	10	24	43	157	0	77
Asian hornbills	52	5	15	24	9	0	44
Primates	512	59	69	93	291	0	221
Felids	264	31	60	104	69	0	195
Canids, hyaenas	225	8	10	16	191	0	34
Antelope	395	9	21	46	87	232	76
Total (%)	2110	147 (7%)	235 (11%)	404 (19%)	1032 (49%)	293 (14%)	786

period of time. The proposed system defines three categories of threatened taxa as follows.

(a) Critical: 50% probability of extinction within five years or two generations, whichever is longer.
(b) Endangered: 20% probability of extinction within 20 years or 10 generations, whichever is longer.
(c) Vulnerable: 10% probability of extinction within 100 years.

Criteria are also proposed to estimate the probability of extinction of taxa based on information about the population size (total and effective), fragmentation, trends, and stochasticity for each category as well as conditions of the habitat now and in the future. Their purpose is to provide a system that is more objective and rational than previous schemes have been. The criteria are based on population viability theory (Gilpin and Soulé, 1986; Soulé, 1987a, b). Mace and Lande (1991) acknowledge that in most cases there will be insufficient data and imperfect models on which to base formal probabilistic analysis. For broader and cruder assessments they recommend the use of more qualitative criteria based on basic principles of population biology.

The CAMP process attempts to be as quantitative or numerate as possible for two major reasons:

(a) Action Plans (captive and wild) ultimately must establish numerical objectives for population sizes and distribution if they are to be viable;
(b) numbers provide for more objectivity, less ambiguity, more comparability, and better communication and hence cooperation.

There is a special attempt to estimate the total population size of each taxon. It is often very difficult, even agonizing, to be numerate because very often so few quantitative data on population sizes and distribution exist. However, with encouragement and mediation from workshop organizers, it is frequently possible to provide order-of-magnitude estimates, especially of whether the total population is greater or less than the numerical thresholds for the criteria relating to the three Mace–Lande categories of threat.

As a result of the CAMP process, most taxa are assigned to one of four categories: Critical, Endangered, Vulnerable, Safe. In assigning conservation priorities there is also an attempt to consider the taxonomic distinctiveness of each taxon, although this aspect of the process is still developing. Concerning taxonomy, the most conservative approach, relative to the preservation of biodiversity, is to attempt risk assessment and management recommendations initially in terms of the maximum distinction among possible 'subspecies' until taxonomic relationships are better elucidated. Splitting rather than lumping maximizes preserva-

Table 16.2 Intensive action recommendations from the CAMP Workshops

	Total taxa	*PHVA*	*More* in situ *management and protection*	*Research*	*Captive breeding*
Parrots	428	125 (29%)	175 (41%)	199 (47%)	170 (40%)
Waterfowl	234	92 (39%)	173 (74%)	166 (71%)	150 (64%)
Asian hornbills	52	35 (67%)	15 (29%)	50 (96%)	45 (87%)
Primates	512	136 (27%)	37 (7%)	192 (38%)	229 (45%)
Felids	264	30 (11%)	80 (30%)	120 (46%)	98 (37%)
Canids, hyaenas	225	14 (6%)	22 (10%)	47 (21%)	33 (15%)
Antelope	395	62 (16%)	111 (28%)	119 (30%)	138 (35%)
Total (%)	2110	494 (23%)	613 (29%)	893 (42%)	862 (41%)

tion of options and is generally advocated. Taxa can always be merged ('lumped') later if further information invalidates the distinctions, or if biological or logistical realities of sustaining viable populations precludes maintaining taxa as separate units for conservation.

The results and recommendations from a variety of Action Plans and CAMPs conducted so far are tabulated (Tables 16.1–16.3 and 16.5). Since many of these documents are still in draft form, assessments and recommendations may change, although we believe the general conclusions from these are robust (see summaries in CBSG, 1992). Table 16.1 shows the assessment of threat for various groups of taxa.

Based on these assessments, the CAMP process provides a set of recommendations including:

(a) Population and Habitat Viability Assessment (PHVA) and Conservation Management Plan Workshops;
(b) intensive protection and management in the wild;
(c) *in situ* and *ex situ* research;
(d) captive propagation programmes;
(e) genetic resource banking and application of reproductive technology.

Table 16.2 presents the results of the intensive management recommendations formulated in the CAMP Workshops conducted through mid-1992.

16.3 POPULATION AND HABITAT VIABILITY ASSESSMENTS – PHVAs

Population and Habitat Viability Assessments (PHVAs) are the enhanced versions of Population Viability Analyses (PVAs) that have evolved as the process has been applied by the CBSG. PHVAs use computer models:

(a) to explore extinction processes that operate on small and often fragmented populations of threatened taxa;
(b) to examine the probable consequences for the viability of the population under various management actions or inactions.

The models incorporate information on genetic and demographic characteristics of the population and on conditions in the environment to simulate probable fates (especially probability of extinction and loss of genetic variation) under a series of different scenarios (Clark, Backhouse and Lacy, 1991; Lindenmayer *et al.*, 1991). The primary computer program that has been used for the modelling to date is a stochastic population simulation model, VORTEX (Lacy, in press). VORTEX allows a range of scenarios under a variety of management regimes to be modelled. As a result it is possible to recommend management that maximizes the probability of survival or recovery of the population. The management actions may include establishment, enlargement, or more management of protected areas; poaching control; reintroduction or translocation; captive breeding; sustainable use programmes; and education efforts. PHVAs thus provide an important resource for development of comprehensive conservation and recovery programmes for threatened taxa. Moreover, while the PHVA process commences with a workshop, it normally continues as the results are reviewed and refined. The process frequently entails one or more follow-up workshops.

The PHVA process is still under development. In the past year, a major improvement has been the incorporation of more habitat information into the models (hence the term PHVA).

Because PHVAs are an important step in the development of an overall recovery programme for a threatened taxon, it has been SSC policy to conduct the workshops in a range state of the species in question, and only at the invitation of the appropriate wildlife agencies with management responsibility and authority.

The PHVA workshop provides population viability assessments for each population of a species or subspecies, as decided in arranging the workshop. The assessment for each species includes an in-depth analysis of information on the life history, population dynamics, ecology, and population history of the individual populations. Information on the demography, genetics, and environmental factors relevant to the status of each population, its risk of extinction under current management scenarios, and perceived threats is assembled in preparation for the PHVA, and for the individual populations before and during the workshop.

Relevant information includes data on age of first reproduction for males and females; inter-birth interval in the wild population; litter or clutch size; neonatal and first-year mortality; sex ratio at birth; juvenile survival to the age of first reproduction; adult sex ratio; breeding strategy (monogamous or polygynous in a season); adult mortality (by sex if available); population size; habitat carrying capacity and possible changes through time; environmental variables influencing either reproduction or mortality; potential catastrophic events and their frequency and possible severity in terms of effects upon reproduction or mortality in the year of occurrence; and dispersal and movement of animals between breeding groups.

An important feature of the workshops is the elicitation of expert information that may not be readily available in published form, yet which may be of decisive importance in understanding the behaviour of the species in the wild. This information will provide the basis for constructing simulation models of each population, which will, in a single model, evaluate the deterministic and stochastic effects and interactions of genetic, demographic, environmental, and catastrophic factors on population dynamics and extinction risks. The process of formulating information to put into the models requires that assumptions and the data available to support the assumptions be made explicit. This process tends to lead to consensus building on the biology of the species, which can serve as a basis for continuing discussion of management alternatives and adaptive management of the species or population as new information is obtained. In effect it provides a means for developing management programmes as scientific exercises, with continuing and timely evaluation of new information in a manner that is of benefit to adjusting management practices.

These workshop exercises are able to assist the formulation of management scenarios for the respective species, and evaluate their possible effects on reducing the risks of extinction. It is also possible through sensitivity analyses to search for factors whose manipulation may have the greatest effect on the survival and growth of the population. One can in effect rapidly explore a wide range of values for the parameters in the model to gain a picture of how the species might respond to changes in management. This approach may also be used to assist in evaluating the information contribution of proposed and ongoing research studies to the conservation management of the species.

Short reviews and summaries of new information on topics of importance to conservation management and recovery of the individual populations are also prepared during the workshop. Of particular interest are topics addressing:

(a) factors likely to have operated in the decline of the species or its failure to recover with management, and whether they are still important;
(b) the need for molecular taxonomic, genetic diversity, and parentage studies;
(c) the role of disease in the dynamics of the wild population, in potential reintroductions or translocations, and in the location and management of captive populations;
(d) the possible role of inbreeding in the dynamics and management of the captive and wild population(s);
(e) the potential uses of reproductive technology for the conservation of the species, whether through assisted reproduction or genome banking;
(f) techniques for monitoring the status of the population during the management manipulations to allow their evaluation and modification as new information is developed;

(g) the possible need for meta-population management for long-term survival of the species;
(h) formulation of quantitative genetic and demographic population goals for recovery of the species, and the level of management that will be needed to achieve and maintain those goals;
(i) cost estimates for each of the activities suggested for furthering conservation management of the species;
(j) local education programmes that can affect attitudes, leading to better understanding of and support for conservation.

16.4 *IN SITU* MANAGEMENT

Conservation strategies and Action Plans for threatened taxa must be based on viable populations (i.e. populations sufficiently large and well distributed that they are expected to survive stochastic risks as well as deterministic threats). These strategies and plans will also frequently require that management be undertaken, in addition to protection provided, for small populations.

Often the taxa may need to be managed as a meta-population, i.e. a system of disjunct sub-populations that are interactively managed with regulated interchanges among them, and interventions within them, to enhance survival.

16.5 RESEARCH

Three major kinds of conservation-related research are identified in the CAMP process to assist taxa under threat: taxonomic, survey work (especially to secure better estimates of population and habitat status), and husbandry. Table 16.3 presents results of these kinds of recommendations for the CAMPs conducted through mid-1992.

Table 16.3 Research recommendations from the CAMP Workshops

	Taxonomic	*Surveys*	*Husbandry*
Parrots	79	214	38
Waterfowl	94	150	30
Asian hornbills	36	3	4
Primates	135	156	48
Felids	101	120	39
Canids, hyaenas	38	45	5
Antelope	85	48	50
Total (%)	433	611	166

16.6 GLOBAL CAPTIVE ACTION PLANS – GCAPs

An important product of the CAMP process is a Global Captive Action Plan (GCAP) which attempts to provide a strategic overview and framework for captive breeding of the broad group of taxa of concern (i.e. an order, family, etc.). GCAPs provide strategic guidance for captive programmes at both the global and regional level in terms of captive breeding and other possible support (e.g. technical, financial) for *in situ* conservation. More specifically, GCAPs recommend which taxa are most in need of captive propagation and hence

(a) which taxa in captivity should remain there;
(b) which taxa not yet in captivity should be there;
(c) which taxa currently in captivity should no longer be maintained there.

The approximate scheme that has evolved for Global Captive Action Plans so far is given in Table 16.4, and Table 16.5 presents the captive

Table 16.4 The scheme for Global Captive Action Plans

Captive recommendation	*Level of captive programme*
90%/100 Years I	Population sufficient to preserve 90% of the average heterozygosity of the wild gene pool for 100 years, developed as soon as possible (1–5 years).
90%/100 Years II	Population sufficient to preserve 90% of the average heterozygosity of the wild gene pool for 100 years but developed more gradually (5–10 years).
Nucleus I	A captive nucleus (50–100 individuals) to always represent 98% of the wild gene pool. This type of programme will require periodic, but in most cases modest immigration/importation of individuals from the wild population to maintain this high level of genetic diversity in such a limited captive population. Reproductive technology will facilitate this strategy.
Nucleus II	A well-managed captive nucleus (25–100) for taxa not of conservation concern but present in captivity or otherwise of interest.
Elimination	Taxa are not of conservation concern and are not otherwise of interest. The population should be managed to extinction.

Table 16.5 Recommendations for captive breeding programmes from the CAMP Workshops

	90/100 I	*90/100 II*	*Nucleus I*	*Nucleus II*	*Pending research*	*Total (%)*
Parrots	24	7	69	40	30	170 (40%)
Waterfowl	12	32	37	62	7	150 (64%)
Asian hornbills	5	6	12	13	9	45 (87%)
Primates	77	41	40	71	0	229 (45%)
Felids	25	18	9	46	0	98 (37%)
Canids, hyaenas	6	0	18	8	0	33 (15%)
Antelope	34	22	19	63	0	138 (35%)
Total	183	126	204	300	46	862 (41%)

programmes recommended by the CAMPs that have already been conducted.

This system supposes that captive populations are an integral part of the meta-populations being managed by conservation strategies and Action Plans. Viable meta-populations for many species will need to include captive components. The IUCN Policy Statement on Captive Breeding (IUCN, 1987) recommends that in general captive propagation programmes be a component of conservation strategies for taxa whose wild population is below 1000 individuals. In these cases captive and wild populations should be interactively managed with interchanges of animals occurring as needed. There may be many problems with such interchanges including epidemiological risks, logistical difficulties, and financial limitations, etc. (Dixon, Chapter 10), but many of these can be resolved. Captive populations are support, not a substitute, for wild populations. This is the premise on which the proposals for nucleus populations are based. These would be small populations in captivity supplemented from the wild while natural populations are still large enough to fulfil this function without significant detriment (Safe, Vulnerable). This system would normally require the addition of one or two wild-caught individuals per generation to the captive nucleus. If and when the wild populations declined into a greater state of threat (i.e. Endangered, Critical), this supplementation would cease and the nucleus could be expanded into a full programme that ultimately could reinforce the wild populations.

The programme goals for 90%/100 Years I and II taxa are different from the general guidelines recommended for captive programmes in the past, i.e. 90% of genetic diversity for 200 years. A shorter time period is proposed because it buys time for more taxa that might be excluded from captive programmes if a longer time period (e.g. 200 years) is adopted, and it maintains more incentive to secure or restore viable populations *in situ*.

The GCAPs must confront the realities of limitation in captive habitat (space and other resources). The priorities for captive propagation must be reconciled by the potential or capacity of zoos and aquaria. Taxon Advisory Groups (TAGs) in many regions are now conducting surveys of the amount of captive space available. These surveys are rather sophisticated considering the captive ecologies and taxonomic affinities of the taxa, and zoogeographical themes of the institutions. Obviously the size of populations that can be maintained will be determined by the number of taxa for which programmes are developed. The regional TAGs will most accurately assess captive holding/exhibit space in their regions using surveys and censuses to supplement studbook databases, ISIS records, national or regional inventories, etc.

It is through the Regional Collection Plans and their Regional Breeding Programmes that the recommendations of the Global Captive Action Plans will realized. However, to maximize the efficiency and effectiveness of captive resources, regional programmes will need to be integrated and coordinated to form global programmes, i.e. the Global Animal Survival Plans (GASPs) (Figure 16.1). Programmes and masterplans for propagation and management now exceed 200 in the various regions of the zoo/aquaria world and development is in progress to form global programmes (GASPs) for at least a dozen taxa.

Any and all taxa that are maintained in captivity should be managed as populations. Hence, once taxa are selected for captive propagation, they should be managed by Regional (RCPP) and Global (GCPP or GASP) Captive Propagation Programmes. Therefore there should be studbooks, coordinators, masterplans, taxon advisory groups or other management provisions for these taxa. Moreover, animal spaces as well as the animals themselves should be managed. If zoos and aquaria are to respond to the great need for captive programmes, management will increasingly need to be more collective, i.e. more through Taxon Advisory Groups rather than individual taxon management and/or propagation committees.

While captive breeding programmes are emphasized in the GCAPs, the Plans also attempt:

(a) to identify where and how the captive community can assist with transfer of intensive management information and technology;
(b) to develop priorities for the limited financial support the captive community can provide for *in situ* conservation (e.g. adopt-a-sanctuary programmes).

16.7 SUMMARY AND CONCLUSIONS

The results of the initial CAMP/GCAP Workshops are first published as a review edition. These drafts are reviewed by distribution to 100–200 wildlife managers and regional captive programmes worldwide for comment; and at regional review sessions, utilizing local expertise with the taxonomic group in question. Thus CAMP/GCAP Workshops are not single events, although sometimes they are singular events. Instead, they are part of a continuing and evolving process of developing conservation and recovery plans for the taxa involved. The CAMP review process allows extraction of information from experts worldwide. CAMPs are continually evolving as new information becomes available and as global and regional situations and priorities shift. In nearly all cases, follow-up workshops will be required to consider particular issues in greater depth or on a regional basis. Moreover, some form of follow-up will always be necessary to monitor the implementation and effectiveness of the recommendations resulting from the workshop. In many cases a range of PHVA Workshops will result from the CAMP/GCAP Workshops. CAMPs are 'living' documents that will be continually reassessed and revised based upon new information and shifting needs.

The CAMP process is central to establishment of global priorities for intensive conservation action since CAMPs provide a global framework for intensive management in the wild and captivity. Wildlife Agencies and Regional Captive Breeding Programmes can use the CAMPs as guides as they develop their own action plans.

In conclusion, the result of the CAMP/GCAP process will be recommendations for intensive management of some kind, including captive breeding programmes, for thousands of species. This task is enormous, but so is the reality and the potential of the zoo and aquaria world.

REFERENCES

CBSG (1992) *CAMP/GCAP Summary Reports*, CBSG, Minnesota, USA.

Clark, T.W., Backhouse, G.N. and Lacy, R.C. (1991) The population viability assessment workshop. A tool for threatened species management and conservation. *Endangered Species Update*, **81**, 1–5.

Gilpin, M.E. and Soulé, M.E. (1986) Minimum viable populations: processes of species extinction, in *Conservation Biology: the Science of Scarcity and Diversity* (ed. M.E. Soulé), Sinauer, Sunderland, Massachusetts, pp. 19–34

Lacy, R.C. (in press) VORTEX: a computer simulation model for population viability analysis. *Wildlife Research*.

Lindenmayer, D.B., Thomas, V.C., Lacy, R.C., and Clark, T.W. (1991) Population viability analysis (PVA): the concept and its applications, with a case study of Leadbeater's possum, *Gymnobelideus leadbeateri*. Report to the Resource Assessment Commission. Department of Conservation and Environment, Melbourne.

Mace, G.M. and Lande, R. (1991) Assessing extinction threats: toward a reevaluation of IUCN threatened species categories. *Conservation Biology*, **51**, 148– 57.

Soulé, M.E. (ed.) (1987a) *Viable Populations for Conservation*, Cambridge University Press, Cambridge, UK.

Soulé, M.E. (1987b) Introduction and Where do we go from here? in *Viable Populations for Conservation* (ed. M.E. Soulé), Cambridge University Press, Cambridge, UK, pp. 1–10 and 175–84.

IUCN (1987) *The IUCN Policy Statement on Captive Breeding*, IUCN, Gland, Switzerland.

Part Three

Case Studies

17

Invertebrate propagation and re-establishment programmes: The conservation and education potential for zoos and related institutions

P. Pearce-Kelly
Zoological Society of London, UK.

17.1 INTRODUCTION

Zoos and related institutions such as aquaria and butterfly houses are becoming increasingly aware of the tremendous public interest that invertebrates generate (Robinson, 1991). This recognition is reflected in a growing number of impressive invertebrate exhibits, which in addition to providing visitors with a fascinating introduction to a diverse and important fauna (Rose, 1992), can also play a crucial role in highlighting the many threats facing the majority of the planet's invertebrate species. Examples of such exhibits are in Cincinnati Zoo (Ross, 1981), National Zoological Park, Washington, DC (Robinson, 1991), Melbourne Zoo (Toone, 1990), Tama Zoo, Tokyo (Yajima, 1991), Artis Zoo, Amsterdam (Veltman and Wilhelm, 1991), Berlin Zoo and Aquarium (Klos and Lange, 1986), London Zoo (Pearce-Kelly *et al.*, 1991), and a number of excellent butterfly houses (Collins, 1987; Toone, 1990) and aquaria throughout the world.

Creative Conservation: Interactive management of wild and captive animals.
Edited by P.J.S. Olney, G.M. Mace and A.T.C. Feistner.
Published in 1994 by Chapman & Hall, London.
ISBN 0 412 49570 8

If zoos are to realize their full potential for contributing towards invertebrate conservation, then their educational role needs to run in tandem with the development of propagation and establishment programmes for endangered species. The development of a number of such programmes over the last decade has served to prove the value of adopting this dual approach (Cooper, 1986; Pearce-Kelly *et al.*, 1991; Amaral, 1992; Barrett, 1992).

The educational, inspirational and entertainment value of well-designed invertebrate exhibits should not be underestimated. The characteristics of invertebrates can fascinate and educate at the same time. Their great diversity of colour, form and lifestyle, their abundance and the extraordinary biomass of so many species, their biological and economic importance, their interaction with plants and other animals (not least humans) and their essential role in most ecosystems are just a few of the biological facets which can be shown. An excellent synopsis of the value and need for invertebrate exhibits is given by Robinson (1991). The literature on invertebrates is enormous, but that on invertebrate exhibits is small and scattered. Two useful reviews are those by Morton (1991a) on butterfly and moth housing and display, and by Powell (1991) on the educational value of leaf-cutting ants (*Atta* and *Acromyrmex*).

A well-presented invertebrate exhibition will without doubt increase public attendance. Within one year of the opening of the Insect Ecological Land at Tama Zoo, Tokyo in 1988 there was a 20% increase in zoo attendance, and a survey before and after showed that the proportion of all zoo visitors who visited the insect exhibits rose in a year from 39% to almost 70% (Yajima, 1991).

In the case of the vast majority of the estimated 10–30 million invertebrate species on earth today, habitat preservation represents the only realistic conservation measure that can be taken to ensure their survival. Nevertheless, for many hundreds of fascinating and well-documented species it would be disastrous if the attitude were adopted that because captive breeding programmes can only ever save a small percentage of the total number of threatened species, there is little point in making the captive breeding option available for invertebrates. On the contrary, when deciding how best to allocate limited resources, invertebrates in many ways represent the most feasible, and certainly no less deserving, candidates for propagation and establishment programmes.

17.2 A RICH EXPERIENCE BASE

In addition to the often considerable expertise residing in zoos and aquaria that maintain invertebrate collections (Carlson, 1992; Sommer, 1992; Morgan, 1991, 1992), there is a greater pool of expertise outside the zoo world that can be drawn upon when developing invertebrate conservation programmes.

Universities and natural history museums have long incorporated invertebrates into a core area of their research and have contributed much to our present understanding of evolutionary biology, ecology, population dynamics, monitoring techniques and genetic research.

As a consequence of their intrinsic interest and their economic and medical importance, invertebrates have also been studied and cultured in a diverse range of research establishments, such as medical research units, biological control institutes and commercial breeding facilities, and also by many amateurs. Much of this research has necessitated the development of breeding and mass rearing techniques (Parsons, 1980; Pyle and Hughes, 1978; Gardiner, 1982; Friedrich, 1986; Wells and Coombes, 1987; Clark and Landford, 1991), dietary developments (Singh, 1977; Frye, 1986, 1991; Gardiner, 1978; Morton, 1981, 1991a; and the identification and treatment of invertebrate diseases (Cooper and Cunningham, 1991; Rivers, 1990, 1991; Williams, 1992).

The technical expertise built up by these academic and commercial organizations and by many dedicated amateurs, provides conservation bodies such as zoos with an array of potentially valuable partners with which to develop and refine management techniques for breeding endangered invertebrate species.

Governmental and NGO conservation bodies such as English Nature, the World Conservation Monitoring Centre (WCMC), the Species Survival Commission (SSC) of the World Conservation Union (IUCN), and World Wide Fund for Nature (WWF) have been effective in recognizing and drawing upon much of this expertise. In so doing, these organizations have successfully incorporated invertebrates into their mainstream monitoring and conservation programme work. To date, zoos have been slower to recognize and access this experience base. The establishment of the SSC Captive Breeding Specialist Group (CBSG) Invertebrate Task Force, and a number of regional and national Taxon Advisory Groups (TAGs) for invertebrates is, however, a welcome step in the right direction.

A wealth of evidence has been accumulated to suggest that, given the correct management regimes, many groups of invertebrates, by virtue of their small size, ease of transport and often spectacular reproductive potential, can be maintained in captivity as large populations, retaining a high degree of genetic variation (Morton, 1991b).

Research also indicates that many invertebrate groups are extremely well adapted to contend with periodic population drops, further enhancing the chances of maintaining healthy populations over time and ensuring successful establishments (Morton, 1991b).

For many invertebrate species, the level of funding and space required for establishing and maintaining large populations in captivity is much less than that required for most vertebrate species (Wilson, 1987).

Although invertebrates are often the first animals to suffer as a result of habitat disturbance and pollution (Erhardt and Thomas, 1991), making them valuable early indicators of habitat stress (Brown, 1991), their remarkable recovery powers mean that they also often have the best chance of being successfully re-established, once the habitat stress has been effectively removed (Morton, 1991b).

The habitat requirements and management considerations for a diverse range of invertebrate groups are well documented and understood (Kirby, 1992). This is particularly so in temperate regions (Fry and Lonsdale, 1991). Such detailed habitat knowledge and site management experience greatly enhances the feasibility of ensuring successful reintroductions.

Furthermore, and this is an important consideration, experience has shown that invertebrate breeding programmes are quite capable of holding their own in terms of attracting donor funding, media interest and public support. These combined factors make many invertebrate groups excellent candidates for incorporation into mainstream zoo conservation work.

A good example of an ongoing invertebrate breeding programme is the International Propagation Programme for the *Partula* tree snails (Tonge and Bloxham, 1991; Pearce-Kelly and Clarke, 1992). This conservation effort is one of the highest profile breeding programmes in the zoo world. Its success to date is in large part due to the high degree of institutional participation, involving zoos, universities and even botanical gardens. The resultant synergy of such a wide spectrum of institutional expertise and facilities has enabled the *Partula* programme to overcome an array of technical problems to a degree that would almost certainly not have been possible had there not been such cooperation.

Another attractive consideration in favour of invertebrate conservation programmes is the speed with which a breeding programme can be carried through to the release stage. A good example of such rapid programme development is English Nature's Recovery Programme for the British population of the field cricket *Gryllus campestris* (Whitten, 1990).

In the spring of 1992 a group of sub-adult *G. campestris* were collected from the remaining West Sussex colony and taken to London Zoo, where a subsequent F1 generation, numbering some 1000 crickets, was successfully produced. Within the space of a couple of months, 700 large F1 generation nymphs were released into selected sites during the summer of the same year (Pearce-Kelly, 1992). Experience gained from such fast turn-round programmes may well provide vertebrate recovery programmes, where the reintroduction phase may be many years away, with some valuable lessons and models.

Another excellent example of an invertebrate conservation programme benefiting from a collaboration of zoos, universities and government conservation agencies is the initiative for the American burying beetle (*Nicrophorus americanus*) (Amaral, 1992). A relatively rare example of a collaboration between zoos and dedicated amateurs is the conservation programme for the red-kneed bird-eating spider *Euathlus smithii* (Clarke, 1991).

17.3 SELECTING TARGET SPECIES

When determining which invertebrate species should be considered as priority candidates for conservation breeding programmes the same criteria that are used for identifying vertebrate candidates are equally pertinent. Considerations such as the rarity of a given species in the wild, its ecological and taxonomic importance, its research value, its ease of culture and the likely chances of eventual re-establishment are all important criteria.

A wealth of data, such as the *IUCN Invertebrate Red Data Book* (Wells, Pyle and Collins, 1983), the *1990 IUCN Red List of Threatened Animals* (IUCN, 1990), and the Nature Conservancy Council's report *Recovery* (Whitten, 1990), is available to help identify suitable target species. Island fauna, which include relatively large numbers of endemic species, are particularly susceptible to the effects of introduced aliens either as predators or as competitors, and to people-induced habitat changes. It is not surprising then that island fauna register high on the list of priority target species (Howarth and Ramsay, 1991).

17.4 THE REASON FOR THE SMALL NUMBER OF INVERTEBRATE BREEDING PROGRAMMES

Apart from a few notable exceptions (e.g. Malausa and Drescher, 1991), zoos and aquaria are the only institutions likely to provide the facilities and staff required for maintaining populations of non-commercial or non-research animal species. Though there are many invertebrates in aquaria world-wide there appear to be relatively few zoos yet to become involved with invertebrates, even in an exhibit capacity. It is not surprising, therefore, that the overall number of captive breeding and re-establishment programmes for invertebrates is low. There is surprisingly little information available on invertebrate exhibits outside aquaria.

In Great Britain a survey in 1987 of some 50 member zoos of the Federation showed that 25 kept invertebrates with a total of about 230 species (Hughes and Bennett, 1991); in 1986 there were 40 butterfly houses in the UK, of which in 1987 14 had 306 species of Lepidoptera, none being threatened in the wild (Collins, 1987). In Japan in 1990 there were 30 live insect exhibit facilities (Yajima, 1991).

The relatively few invertebrate breeding programmes that have been established, such as those for *Partula, Gryllus campestris, Nicrophorus americanus* and *Euathlus smithii*, serve to illustrate the potential (and difficulty) that exists in this area of conservation (Lees, 1989). In addition to aiding directly the target species concerned, the invertebrate conservation programmes that have been established serve as models which can be applied to a wide range of similar species under threat. For example, the management protocol that has been devised for the *Partula* snail propagation programme will be equally pertinent for a wide range of other endangered molluscs.

17.5 CONCLUSION

Invertebrates can provide perhaps the most viable and cost effective form of captive breeding and re-establishment work that zoos can conduct.

By joining forces with other organizations, such as universities and museums, zoos can access the experience base and establish the potential partners required to develop invertebrate breeding programmes.

With an estimated combined capacity of some 12 000 hectares, the world's zoos have been estimated to be able to support viable populations (i.e. 500-head) of around 800 land vertebrate species (Maier and Page, 1990). This figure is in stark contrast to the estimated potential for zoo-based invertebrate propagation programmes. If just 1% of the total world zoo space was devoted to housing invertebrate collections it can be calculated that some 15 000 invertebrate species, mainly Lepidoptera, could, in theory, be preserved as viable captive populations (Morton, 1991a).

At the eighteenth General Assembly of IUCN held in Perth, Australia in 1990, a resolution was adopted on the conservation of insects and other invertebrates urging action to 'strengthen invertebrate displays by zoos and butterfly houses linked to captive-breeding and re-establishment programmes (IUCN, 1991). It is to be hoped that with the recent formation of such bodies as the SSC/CBSG Invertebrate Task Force, the consolidation of international, regional and national links, and the increased cooperation between established centres of invertebrate expertise, ever more zoos and related institutions will realize their tremendous potential for playing a major role in preserving many of our planet's most remarkable animal species, and will allocate their resources accordingly. One per cent of the space on the 'global ark', together with the allocation of a handful of its 'crew', could increase our ability to conserve animal species by several orders of magnitude.

ACKNOWLEDGEMENTS

I am greatly indebted to the editors and my 'invertebrate colleagues' Dave Clarke, Victoria Silverton, Warren Spencer and Ajay Burlingham-Johnson.

REFERENCES

Amaral, M. (1992) Conservation of the endangered American Burying beetle *Nicrophorus americanus*, in *AAZPA/CAZPA Annual Conference Proceedings 1992*, American Association Zoological Parks and Aquariums, Wheeling, WV, pp. 509–520.

Barrett, P. (1992) Maintaining and breeding the common tree Weta *Hemideina crassidens* at Wellington Zoo. *Int. Zoo Yb.*, **31**, 30–6.

Brown, K.R. (1991) Conservation of neotropical environments: insects as indicators, in *The Conservation of Insects and their Habitats, 15th Symposium of the Royal Entomological Society of London* (eds N.M. Collins and J.A. Thomas), Academic Press, London, pp. 349–401.

Carlson, B.A. (1992) The potential of aquariums propagation of corals as a conservation measure, in *AAZPA/CAZPA Annual Conference Proceedings 1992*, pp. 370–377.

Clark, P.B. and Landford, A.D. (1991) Farming insects in Papua New Guinea. *Int. Zoo Yb.*, **30**, 127–31.

Clarke, D. (1991) Captive-breeding programme for the red-kneed bird-eating spider *Euathlus smithii* at London Zoo. *Int. Zoo Yb.*, **31**, 68–75.

Collins, N.M. (1987) *Butterfly Houses in Britain: The Conservation Implications*, IUCN, Cambridge.

Cooper, J.E. (1986) Captive breeding of invertebrates. *Int. Zoo Yb.*, **24/25**, 74–6.

Cooper, J.E. and Cunningham, A.A (1991) Pathological investigation of captive invertebrates. *Int. Zoo Yb.*, **30**, 137–43.

Erhardt, A. and Thomas, J.A. (1991) Lepidoptera as indicators of change in the semi-natural grasslands of lowland and upland Europe, in *The Conservation of Insects and their Habitats, 15th Symposium of the Royal Entomological Society of London* (eds N.M. Collins and J.A. Thomas), Academic Press, London, pp. 213–36.

Friedrich, E. (1986) *Breeding Butterflies and Moths: A Practical Handbook of British and European Species*, Harley Books, Colchester.

Fry, R. and Lonsdale, D. (1991) *Habitat Conservation for Insects – A Neglected Green Issue*, Amateur Entomol. Soc., **21**, Cravitz Printing Company, Brentwood, UK.

Frye, F.L. (1986) Care and feeding of invertebrates kept as pets or study animals, in *Zoo and Wild Animal Medicine*, 2nd edn (ed. M.E. Fowler), W.B. Saunders, Philadelphia, pp. 1039–54.

Frye, F.L. (1991) *Captive Invertebrates: A Guide to their Biology and Husbandry*, Robert E. Kreiger, Melbourne, Florida.

Gardiner, B.O.C. (1978) The preparation and use of artificial diets for the rearing of insects. *Entomologist's Rec. J. Var.*, **90**, 181–4, 267–70, 287–91.

Gardiner, B.O.C. (ed.) (1982) *A Silkmoth Rearer's Handbook*, The Amateur Entomol. Soc., Hanworth.

Hughes, D.G. and Bennett, P.M. (1991) Captive breeding and the conservation of invertebrates. *Int. Zoo Yb.*, **30**, 45–1.

Howarth, F.G. and Ramsay, G.W. (1991) The conservation of island insects and their habitats, in *The Conservation of Insects and their Habitats, 15th Symposium of the Royal Entomological Society of London* (eds N.M. Collins and J.A. Thomas), Academic Press, London, pp. 71–107.

IUCN (1990) *1990 IUCN Red List of Threatened Animals*, IUCN, Gland, Switzerland.

IUCN (1991) *IUCN 18th General Assembly Resolution and Recommendations*, IUCN, Gland, Switzerland.

Kirby, P. (1992) *Habitat Management for Invertebrates: A Practical Handbook*, Royal Society for the Protection of Birds, Sandy, UK.

Klos, H.G. and Lange, J. (1986) The modernisation of the Aquarium at Berlin Zoo. *Int. Zoo Yb.*, **24/25**, 322–32.

Lees, D. (1989) Practical considerations and techniques in the captive breeding of insects for conservation purposes. *The Entomologist*, **180** (1,2), 77–96.

Maier F. and Page. J. (1990) *Zoo: The Modern Ark*, Facts on File, Oxford.

Malausa, J.C. and Drescher, J. (1991) The project to rescue the Italian ground beetle *Chrysocarabus olympiae*. *Int. Zoo Yb.*, **30**, 75–9.

Morgan, R.C. (1991) Natural history, field collection and captive management of the Honey ant *Myrmecocystus mexicanus*. *Int. Zoo Yb.*, **30**, 108–17.

Morgan, R.C. (1992) Natural history, captive management and display of the Sunburst diving beetle *Thermonectus marmoratus*, in *AAZPA/CAZPA Annual Conference Proceedings 1992*, pp. 457–64.

Morton, A.C. (1981) Rearing butterflies on artificial diets. *J. Res. Lepid.*, **18**, 221–7.

Morton, A.C. (1991a) Captive breeding of butterflies and moths: I. Advances in equipment and techniques. *Int. Zoo Yb.*, **30**, 80–9.

Morton, A.C. (1991b) Captive breeding of butterflies and moths: II. Conserving genetic variation and managing biodiversity. *Int. Zoo Yb.*, **30**, 89–97.

Parsons, M. (1980) *Insect Farming and Trading Agency Farming Manual*, Insect Farming and Trading Agency, Bulolo, Papua New Guinea.

Pearce-Kelly, P.E. (1992) The feasibility and value of invertebrate captive breeding programmes, with special reference to the British field cricket *Gryllus campestris*, in *AAZPA/CAZPA Annual Conference Proceedings 1992*, pp. 505–8.

Pearce-Kelly, P. and Clarke, D. (eds) (1992) *Partula 1992*. Proceedings of the annual meeting of the *Partula* Propagation Group, Zoological Society of London, London.

Pearce-Kelly, P., Clarke, D., Robertson, M. and Andrews, C. (1991) The display, culture and conservation of invertebrates at London Zoo. *Int. Zoo Yb.*, **30**, 21–30.

Powell, R. (1991) The educational value of leaf-cutting ant colonies and their maintenance in captivity. *Int. Zoo Yb.*, **30**, 97–107.

Pyle, R.M. and Hughes, S.A. (1978) Conservation and utilisation of the insect resources of Papua New Guinea. Report to the Wildlife Branch, Dept. of Nat. Resources, Papua New Guinea.

Rivers, C.F. (1990) The control of diseases and pests of insects, in *The Management and Welfare of Invertebrates in Captivity* (ed. N.M. Collins), National Federation of Zoological Gardens, London, pp. 45–9.

Rivers, C.F. (1991) The control of diseases in insect culture. *Int. Zoo Yb.*, **30**, 131–7.

Robinson, M.H. (1991) Invertebrates: exhibiting the silent majority. *Int. Zoo Yb.*, **30**, 1–7.

Rose, R. (1992) Understanding the big picture: the importance of invertebrate education, in *AAZPA/CAZPA Annual Conference Proceedings 1992*, pp. 521–5.

Ross, J. (1981) The insect world at the Cincinnati Zoo. *Int. Zoo Yb.*, **21**, 36–41.

Sawyer, R.T. (1985) *Leech Biology and Behaviour*, Oxford University Press, Oxford.

Singh P. (1977) *Artificial Diets for Insects, Mites and Spiders*, IFI/Plenum Press, New York.

Sommer, F. (1992) Husbandry aspects of a jellyfish exhibit at the Monterey Bay Aquarium, in *AAZPA/CAZPA Annual Conference Proceedings 1992*, pp. 362–9.

Tonge, S. and Bloxham, Q. (1991) A review of the captive-breeding programme for Polynesian tree snails *Partula* spp. *Int. Zoo Yb.*, **30**, 51–9.

Toone, W.D. (1990) Butterfly exhibiting. *Int. Zoo Yb.*, **29**, 61–5.

Veltman, J. and Wilhelm W. (1991) Husbandry and display of the Jewel wasp *Ampulex compressa* and its potential value in destroying cockroaches. *Int. Zoo Yb.*, **30**, 118–26.

Wells, S.M. and Coombes W. (1987) The status and trade of the medicinal leech. *Traffic Bull.*, **8**, 64–9.

Wells, S.M., Pyle, R.M. and Collins N.M. (1983) *IUCN Invertebrate Red Data Book*, IUCN, Gland, Switzerland.

Whitten, A.J. (1990) *Recovery: A Proposed Programme for Britain's Protected Species*, CSD Report No. 1127, Nature Conservancy Council, Peterborough.

Willams, D.L. (1992) Studies in arachnid disease, in *Arachnida: Proceedings of a Symposium on Spiders and their Allies* (eds J.E. Cooper., P. Pearce-Kelly and D.L. Williams), Chiron Publications, Keighley, W. Yorks.

Wilson, E.O. (1987) The little things that run the world: the importance and conservation of invertebrates. *Conserv. Biol.*, **1**, 344–6.

Yajima, M. (1991) The Insect Ecological Land at Tama Zoo. *Int. Zoo Yb.*, **30**, 7–15.

18

Captive Breeding Programmes and their role in fish conservation

*C. Andrews**
National Aquarium in Baltimore, Maryland, USA
and
L. Kaufman
New England Aquarium, Boston, USA.

18.1 INTRODUCTION

According to Le Cren (1990), the biology of the vast majority of the 22 000 or so known fish species (which comprise 50% of all living vertebrates) is virtually unknown. Since available information suggests that in some parts of the world about one-third of the local fish species may be threatened, their conservation presents a massive challenge. The greatest number and variety of fish occur in the oldest freshwater lakes, wetlands and river systems of the tropics and subtropics, and in the waters surrounding coral reefs. Therefore single-species conservation is less relevant than the conservation of fish communities and their habitats, and the conservation of whole ecosystems must be the long-term goal of fish conservationists. This paper briefly summarizes information

* Corresponding author.

Creative Conservation: Interactive management of wild and captive animals.
Edited by P.J.S. Olney, G.M. Mace and A.T.C. Feistner.
Published in 1994 by Chapman & Hall, London.
ISBN 0 412 49570 8

on the importance and status of fish populations on a global basis, and also outlines the threats they face and their current conservation needs, particularly in relation to captive propagation efforts by zoos and aquaria.

18.2 IMPORTANCE

Since many fish live in a largely unknown and unseen environment their importance to humans is often overlooked. There are two overlapping justifications for the long-term conservation of fish (Nyman, 1991): economic factors and intrinsic factors.

18.2.1 Economic factors

Of the 90 million tonnes of fish which make up the annual global fisheries catch, 85% comes from the harvest of wild populations (the remainder is supplied by aquaculture), and fish are our only major food supply which is obtained principally from wild sources. In some areas, particularly certain tropical countries, fish are a vitally important source of available and/or inexpensive protein for the local populace.

Angling and ornamental fishkeeping are two popular pastimes enjoyed by many millions of people across the world; these hobbies promote a better understanding of the importance of aquatic ecosystems and have significant economic implications for the countries involved. Scuba-diving, snorkeling and similar water sports are also popular recreational activities, and are an important part of the tourist industry – on which some countries are heavily dependent.

Fish are the dominant aquatic vertebrates in the world's rivers, lakes and oceans, and as they vanish the integrity of whole ecosystems (and all of the ancillary resources they represent) are threatened. Fish are also used for ecological monitoring, as experimental animals and for pest control, and have contributed in a major way to the understanding of ecological principles and population genetics. Potential anti-cancer drugs have been identified from sharks, stingrays and various aquatic invertebrates. Therefore, as science advances (and as cultural values change), the potential value of all aquatic species must be emphasized.

18.2.2 Intrinsic factors

For centuries fish have played a significant part in art, in several religions and in our general appreciation of the natural world. As discussed by Nyman (1991), there is a need to move away from the dominant ethic in the developed world, which requires that species are commercially valuable before attracting significant conservation attention, and develop the more acceptable approach of long-term sustainable utilization of all natural resources (whatever their commercial value).

18.3 THREATS

As discussed by Andrews (1988), a number of human-related activities (either individually or acting in concert) are causing changes in natural fish populations at an alarming rate. These activities can be discussed under two broad headings:

(a) habitat alteration, including habitat destruction, pollution and the introduction of alien species;
(b) overexploitation including overfishing for food and over-collection for other purposes.

18.3.1 Habitat alteration

Habitat alteration is a major cause of changes in fish populations in both freshwater and marine environments. For example, dams may act as barriers to migrating fish, and artificial lakes and reservoirs may replace small streams or ponds, with consequent disruption to the endemic fauna. The widespread filling-in and/or draining of ponds, marshes and other small water bodies can have a significant effect on communities of smaller fish species, and such environments may also act as an important nursery ground for larger and perhaps commercially valuable species (Maitland, 1974).

The effect of pollution on aquatic ecosystems can be sudden and obvious (e.g. acute fish mortalities following the release of a toxic effluent), or more gradual and perhaps discrete (e.g. the effects of acid rain on upland fisheries in North America and Europe, or the sub-lethal effects of other pollutants on fish longevity, fecundity and resistance to diseases). Heated, nutrient-rich or sediment-laden effluents may change the chemical and/or physical characteristics of the receiving water body. Whilst toxic effluents are often assumed to exert their main effect on fish populations in downstream water courses, such effluents may also form a barrier to the migration of fish through the river system, and hence affect the viability of upstream populations (Maitland, 1974).

The introduction of alien fish species can also adversely affect local fish populations in a number of ways, including via competition, predation and hybridization.

18.3.2 Overexploitation

Subsistence fisheries have existed for a very long time, and have probably had very little impact on wild fish populations. However, the development and use of large-scale, mechanized fishing techniques in the 1950s and 1960s led to rapid overexploitation of fish stocks. Several of the commercially important North Atlantic and Pacific fish stocks declined dramatically as a result of overfishing. The sudden and drastic decline of once numerous species can also have a pronounced effect on other animals in the same ecosystem (e.g. the collapse of the Chilean and Peruvian anchoveta fishery in the 1970s and the decline in local sea birds). However, it is generally assumed that this type of overexploitation will not lead to the total extinction of a target species, as there will come a point when the fishing is no longer commercially viable and exploitation (fishing) will cease. This may not be the case for highly valued fish, such as some of the species for the ornamental fish trade – a thriving world-wide trade, the size and scope of which was reviewed by Andrews (1990). Although the majority of freshwater fish involved in the trade are captive-bred, significant numbers are still removed from the wild, and almost all the marine fish in the trade are wild-caught. There is now mounting concern that overexploitation may be causing the decline and even commercial extinction of a number of species which are popular in the trade; there is a need for greater monitoring and legislative controls (Andrews, 1990).

18.4 STATUS

According to the *1990 Red List of Threatened Animals* (IUCN, 1990), some 762 taxa of freshwater and marine fish are considered to be threatened or extinct. Since the vast majority of the listed species are freshwater fish, this represents about 9% of the 8500 or so species known to occur in freshwater habitats. However, available data from certain areas suggest that the actual situation may be far worse (Table 18.1). For example, in the African lakes of Barombi Mbo and Lake Victoria, large proportions of the fauna are endangered (Reid, 1990a; Kaufman, 1992; Witte *et al.*, 1992). In North America, Williams and Miller (1990) considered 292 (28%) of 1033 species of freshwater fish to be threatened or extinct, and in California 57% of the 113 species of native freshwater fish were thought to be threatened or in need of conservation action (Moyle and Williams, 1990). According to Almada-Villela (1990), 160 (32%) of the 500 or so species of Mexican freshwater fish may be threatened, with one-third of the threatened taxa in the desert regions of Sonora and Chihuahua. Further examples are cited in Table 18.1, and there is

Table 18.1 Some threatened freshwater fish faunas[a]

Location	*Estimated no. threatened/ total no.*	*Proportion of total (%)*	*Source*
British Isles	10/55	18.2	Maitland, 1974
Belgium	29/39	74.7	Bervoets *et al.*, 1990
Europe	98/200	49.0	Lelek, 1987
Mexico	160/500	32.0	Almada-Villela, 1990
California	64/113	57.0	Moyle and Williams, 1990
North America	292/1033	28.3	Williams and Miller, 1990
Australia	65/192	34.0	Pollard *et al.*, 1990
Waikako River, New Zealand	7/18	38.9	Swales, 1990
Sri Lanka	11/64	17.2	Evans, 1981
Peninsular Malaysia	145/382	38.0	Mohsin *et al.*, 1991
Singapore	19/53	35.8	Mohsin *et al*, 1991
Himalayan waters, Nepal	25/130	19.2	Shrestha, 1990
Lake Victoria	315/350+	90.0	Kaufman, 1992
Barombi Mbo	12/17	70.0	Reid, 1990a
Southern Africa	24/214	11.2	Skelton, 1990

[a]Definition of 'threatened' may vary: refer to source literature.

increasing (sometimes anecdotal or unpublished) information to suggest that the situation may be at least as critical in many other areas, including Madagascar, the Aral Sea, Philippines, Thailand and Hawaii.

Although the *1990 Red List of Threatened Animals* contains relatively few completely marine taxa, considerable concern is being expressed over a range of marine fish species, including certain sharks and rays (Gruber and Fowler, 1991), blue fin tuna (Anon., 1991) and some coral reef fish (Wood, 1985), and it has been suggested that limited distribution, slow attainment of maturity, intrinsic rareness and excessive exploitation may all be important factors in placing some marine fish species in jeopardy (American Fisheries Society, 1991).

18.5 CONSERVATION NEEDS

18.5.1 Preservation and management of habitats

Fish populations (not exclusively but especially those in freshwater) are in urgent need of conservation attention. As outlined by Nyman (1991), the long-term conservation of species must involve preservation of whole ecosystems and their constituent species at all trophic levels, and all conservation measures must have an evolutionary perspective. Protection of the world's aquatic biota depends on the fact that every ecosystem is interconnected and influenced (to a greater or lesser extent) by human activity. Since less than 3% of the world's surface area has protected status (with very few completely aquatic reserves), it is clear that wildlife conservation will depend upon multiple use of habitats outside formal reserves (Jones, 1991: Nyman, 1991). As outlined by Moyle and Sato (1991), it is likely to become necessary to promote the wise management of entire watersheds in certain areas, starting with those which are most pristine and have had least impact (as yet) from human activities.

18.5.2 The role of zoos and aquaria

As discussed by Kaufman (1987) and Andrews (1988), zoos and aquaria are well suited to help raise public awareness of aquatic conservation problems and to seek both popular and financial support for rational intervention and preventative or restorative efforts. Similarly, zoos and aquaria are now becoming more involved in breeding programmes for threatened fish

18.6 CAPTIVE BREEDING PROGRAMMES FOR THREATENED FISH

The Aquatic Section of the Captive Breeding Specialist Group (CBSG) of the World Conservation Union (IUCN) has identified three freshwater fish faunas for initial conservation attention: Lake Victoria fish, desert fish and Appalachian stream fish. Each of these faunas is characterized by moderate to extremely high levels of species richness, high levels of endemism and a large proportion of species which are threatened.

18.6.1 Lake Victoria fish

Severely affected by the introduction of the predatory Nile perch (*Lates niloticus*) and a number of other factors, most of the several hundred species of endemic fish (mainly haplochromine cichlids) are considered to be threatened or already extinct (Kaufman, 1992; Witte *et al.*, 1992). The captive breeding programme for Lake Victoria fish was recognized officially by the American Association of Zoological Parks and Aquariums (AAZPA) in December 1990 and has three primary goals (Kaufman, 1991). To:

(a) preserve in captivity 40 of the most endangered haplochromine cichlid species that are representative of the full range of the morphological diversity of the fauna, thus maintaining both research and reintroduction options for the future;
(b) establish captive populations of non-haplochromine food fish that are important in aquaculture;
(c) establish managed, productive populations of non-endangered haplochromine cichlids of long-term importance in research.

Chaired by the New England Aquarium (USA), with assistance from the Columbus Zoo Aquarium (USA), this project is being developed into a full AAZPA Species Survival Program (SSP), with a European coordinator for the captive breeding effort at the Amsterdam Zoo Aquarium (Holland). However, throughout the development of the Lake Victoria programme special emphasis has been placed on field conservation efforts and research. Known as the Lake Victoria Research and Conservation Programme, a group of international scientists is attempting to integrate the work currently underway in three of the lake's riparian countries (Kenya, Uganda and Tanzania), as well as the captive breeding efforts which are taking place in Europe and North America. High on the list of objectives is the establishment of a regional centre for aquatic conservation training in Kenya.

At present about 30 institutions are participating in the captive breeding efforts involving approximately 25 taxa of Lake Victoria cichlids. The taxa are from a number of different trophic groups, including piscivores, insectivores, epilithic grazers, mollusc crushers/shellers, paedophages and zooplanktivores (Kaufman, 1991). Several other trophic groups are, however, absent from the breeding programme (e.g. crab-eaters, scale-pickers) and the deeper water species are missing completely.

Overall the captive breeding programme for Lake Victoria cichlids presents a number of interesting challenges to those involved that may be of relevence to other fish breeding programmes. These include the following.

(a) Species selection, identification and availability

A large proportion of the lake's 350 (or more) endemic fish may be extinct or threatened, and new taxa are still being described. Founder individuals from certain taxa have proved difficult to obtain and, with a high degree of phenotypic plasticity and low genetic diversity in many of these fish, decisions must be made about which taxa should be preserved and at what taxonomic level.

(b) Recognition of individual fish

Breeding programmes frequently rely upon the marking of individual animals (which has been problematical with these fish) and the alternative of completely separating broods of the same species from different origins will become a significant drain on available resources.

(c) Reintroduction possibilities

Since one of the main threats which brought about the decline of the cichlid fauna (the Nile perch) cannot be removed from the lake, the likely success of any future introduction programme remains to be proven. Reintroduction may be more relevant to inshore, bottom-dwelling species than to others.

(d) Amateur hobbyists and the aquaria trade

Amateur hobbyists and the trade which they support are a significant resource in terms of tank space, manpower, publicity and even financial support. This is being explored in relation to the Lake Victoria programme and also a number of other fish conservation initiatives, although the involvement of hobbyists in captive breeding programmes for threatened fish will require careful management and coordination (Reid, 1990b).

18.6.2 Desert fish

Desert fish are being severely affected by a number of human-related activities, especially water extraction and pollution, on a global scale. The desert fish programme is being coordinated by the New York

Aquarium (USA) with assistance from the Dallas Aquarium (USA). Thus far a total of 14 North American public aquaria and museums have agreed to participate in the programme, which is under development into a full AAZPA SSP (P. Loiselle, pers. comm.). Attention is currently centred on a number of fish taxa from south-west USA and Mexico and a collaboration has begun between the New York Aquarium, the Dallas Aquarium and the Autonomous University of Nuevo Leon (Mexico). New York and Dallas Aquaria, aided by two corporate sponsors from the pet trade (Python Products, Milwaukee and Tetra Sales, USA), are supporting the Autonomous University's Desert Fishes Breeding Center. As part of the collaboration, two endangered Mexican endemics (*Cyprinodon alvarezi* and *Megupsilon apocus*) have been shipped to New York for captive breeding and the Dallas Aquarium has received stocks of an additional three species (*Cyprinodon fontinalis, C. nazas* and *Cyprinodon* 'Charco Azul'). The Dallas Aquarium is also working with *Cyprinodon eximius*, in conjunction with the Texas Department of Parks and Wildlife.

Eleven institutions are currenty displaying or holding the desert pupfish (*Cyprinodon macularis macularis*) and this particular subspecies is already receiving considerable attention from both state and federal agencies. Therefore it is likely that these animals will be replaced at all but two institutions by severely threatened Mexican *Cyprinodon* species. Five species of *Xiphophorus* are being managed by the New York Aquarium and four of these (*X. couchianus, X. gordoni, X. milleri* and *X. clemenciae*) are listed in the *1990 Red List* (IUCN, 1990). One of these (*X. couchianus*) is considered to be extinct in nature and has been maintained at the New York Aquarium since the late 1950s (Kallman, 1989). Stocks of this species have now also been dispersed to the Desert Fishes Breeding Center (above), the Columbus Zoo Aquarium and the Steinhart Aquarium (USA).

Additional taxa which have been proposed for involvement in the Desert Fishes SSP include 10 species of Mexican goodeids, eight of which are currently listed in IUCN (1990).

Although in its developmental stages, the desert fish programme shows great potential. If the anthropogenic threats can be removed from the discrete habitats of these fish, and the habitat subsequently secured and restored, reintroductions from carefully managed captive populations will have a significant impact on their conservation. However, too few founders and a lack of genetic diversity in the founding population is a familiar problem in captive breeding programmes for many endangered species, which may be further compounded by subsequent genetic drift. Whilst the genetic data from some electrophoretic examinations of endangered fish have indicated similarity between certain captive and wild populations, other studies have indicated a loss of rare alleles and reduced heterozygosity among captive populations (Nyman, 1991). Therefore, it is vital to take account of the genetic management of captive stocks of endangered fish and ensure that such management is carried out with a long-term evolutionary perspective (Meffe, 1990).

18.6.3 Appalachian stream fish

The streams and rivers of the south-eastern USA contain a large and diverse range of fish, some of which are naturally uncommon and/or suffering from the effects of habitat disturbance and destruction (Sheldon, 1988). To date only very preliminary efforts have been made to instigate the Appalachian stream fish as an AAZPA programme, which is likely to involve (among others) the recently opened Tennessee State Aquarium (USA). However, the US Fish and Wildlife Service, the University of Tennessee and a private non-profit corporation are involved in an active recovery programme and a restoration effort (including captive rearing and reintroduction) of the threatened smoky madtom (*Notorus baileyi*) and the yellowfin madtom (*N. flavipinnis*) (Tullock *et al.*, 1988; US Fish and Wildlife Service, 1990).

18.6.4 Other captive breeding projects for threatened fish

A number of other endangered fish taxa are the subject of captive breeding efforts, often at a local level and not involving zoos and aquaria to any great extent. These include Colorado River and New Mexican endemics (Johnson and Rinne, 1982; US Fish and Wildlife Service, 1990), *Hucho hucho* (Budihna and Ocvink, 1990), *Scleropages formosus* (Luxmoore, 1990), various threatened South African species (Skelton, 1990), *Oncorhynchus apache* and *O. gilae* (Rinne, 1990; US Fish and Wildlife Service, 1990), various threatened Australian species (Pollard *et al.*, 1990; Caughey, Hume and Wattam, 1990), *Garra barreimiae* (K. Banister, pers. comm.), and *Notopterus blanci*, *Pangasius gigas* and *P. jullieni* (T. Roberts, pers. comm.). In addition, several of the fish listed by IUCN (1990) are being bred commercially for the ornamental fish trade and/or by ornamental fish hobbyists (Andrews, 1990).

With an increasingly large number of candidate species, there is clearly a need to focus on species or faunal groups where there can be the greatest conservation impact. As noted by Stuart (1991), captive breeding programmes are only a small part of an overall conservation effort, but they can be extremely useful in drawing attention to and supporting *in situ* conservation. With this in mind it should be noted that 23 (27.4%) of the 84 federal recovery programmes for endangered freshwater fish in the USA include captive propagation as one of their components (US Fish and Wildlife Service, 1990). Unfortunately, zoos and aquaria are involved in very few of these projects. With their major focus on public awareness and education, many aquaria have the potential to catalyse genuine, community-based efforts to restore fish and their native habitats. The role of captive breeding programmes as a touchstone for longer-term environmental planning may be more important than the handful of species preserved through *ex situ* programmes.

All captive breeding programmes, especially those for fish where there are so many threatened forms, must have a clearly defined plan, goals and a time-line. In very few instances will it be possible to maintain threatened species in captivity in perpetuity. Simple logistics and artificial selection pressures (and consequent maladaption to the natural habitat of the species in question) are two concerns which apply particularly to fish. A number of factors have been suggested as important in selecting an endangered species for a captive breeding programme and the likelihood of a future reintroduction effort is often a prerequisite (Seal, 1986). Not only does this help to validate the breeding programme but it also helps support the need for continued habitat restoration and/or protection. Furthermore, aquaria should, wherever possible, incorporate elements of captive breeding programmes into their displays, thus publicizing the plight of aquatic habitats. With this in mind it is important that more 'flagship' species are selected, to draw attention to particular problem areas. However, many of the 'charismatic mega-fish' which are also endangered (e.g. arapaima [*Arapaima gigas*], white shark [*Carcharodon carcharias*]) are relatively large at maturity and their captive breeding presents a substantial challenge. Such species are best supported by *in situ* conservation efforts, or at least should have their breeding programmes sited adjacent to the animals' natural habitat (which is probably the best location for most large scale breeding efforts linked to reintroduction). Therefore, aquaria (in particular) may have to limit their activities to research on these species and/or concentrate their efforts on the captive breeding of the endangered but the apparently insignificant.

Both the Lake Victoria programme and the desert fish programme highlight the importance of collaboration among zoos, public aquaria and other organizations and individuals, and provide excellent examples of how aquaria can take a central coordinating role. Furthermore the AAZPA Fresh Water Fish and Marine Fish Taxon Advisory Groups are useful for instigating programmes at zoos and aquaria, but continued involvement with other outside interested parties is essential. This is particularly true with regard to species selection, ensuring adequate coordination between *in situ* and *ex situ* conservation, and in optimizing fund-raising efforts. With this in mind, the IUCN Fresh Water Fish Specialist Group is currently involved in seeking financial support for an Action Plan, which will assist in the direction of the conservation of fresh water fish on a global scale.

18.6.5 Captive breeding programmes for other fish

Aquaria generally house large, very diverse collections of fish and other aquatic animals and they have for the most part been able to continue to rely on specimens obtained from the wild for their exhibits. Since the sustainable utilization of natural resources is an important part of present-day conservation, and so long as fish are obtained in an environmentally acceptable fashion with their welfare requirements in mind, aquaria should be prepared to defend the continued removal of certain exhibit animals from the wild. However, it is clear that if present trends continue an increasing number of fish species (for legislative and/or political reasons and/or because of straightforward scarcity) will become increasingly difficult to obtain. The cost implications of developing and maintaining meaningful captive breeding programmes are significant, but to continue to support their mission of aquatic conservation aquaria need to be able to guarantee continued stocks of the exhibit animals around which their educational programmes are based. However, most aquaria are desperately short of off-show holding space and the resources to support large-scale breeding programmes. Resources that are available to address this critical issue are limited by design decisions made long in the past. As new aquaria are designed and constructed, and as existing facilities are renovated, the needs of conservation programmes must be incorporated into their design. Aquaria should, therefore, join together in collaborative breeding efforts and invest in additional facilities, and replace (where relevant) non-endangered species with endangered species in their exhibits and educational programmes.

18.7 CONCLUSIONS

Clearly freshwater and marine environments and the fish species they contain are in urgent need of conservation attention. Habitat restoration and/or protection is the main tool for preventing species loss, although captive breeding programmes (and the associated reintroduction efforts) can have a synergistic effect with *in situ* efforts. By combining research with collaborative breeding projects, and by integrating endangered (especially 'flagship') species into exhibits and other educational programmes, aquaria will continue to support and influence the aquatic conservation movement. However, this can only be achieved by providing more funding to relevant research projects and breeding programmes, with the added bonus that aquaria so involved can truly claim to be affecting fish conservation on a global basis.

REFERENCES

Almada-Villela, P.C. (1990) Status of threatened Mexican fishes. *Journal of Fish Biology*, **37A**, 197–9

American Fisheries Society (1991) Endangered finfish: a useful concept? *Fisheries*, **16** 23–6.

Andrews, C. (1988) The conservation of fish by zoos and aquaria, in *Proceedings of the 5th World Conference on Breeding Endangered Species in Captivity* (eds B.L. Dresser, R.W. Reece and E.J. Marurka), Cincinnati Zoo and Botanical Garden, pp. 9–20.

Andrews, C. (1990) The ornamental fish trade and fish conservation. *Journal of Fish Biology*, **37A**, 53–9.

Anon. (1991) WWF supports protection of bluefin tuna. *Focus*, **13**, 1, 6.

Bervoets, L., Coeck, J. and Verheyen, R.F. (1990) The value of lowland rivers for the conservation of rare fish in Flanders. *Journal of Fish Biology*, **37A**, 223–4.

Budihna, N. and Ocvink, A. (1990) Breeding and restocking of salmonoid fishes in Slovenia. *Journal of Fish Biology*, **37A**, 239–40.

Caughey, A., Hume, S. and Wattam, A. (1990) *Melanotaenia eachamensis* – history and management of the captive stocks. *Journal of the Australia New Guinea Fishes Association*, **6**, 241–7.

Evans, D. (1981) *Threatened Fresh Water Fish of Sri Lanka*. IUCN Publications, IUCN, Cambridge, UK.

Gruber, S.H. and Fowler, S. (1991) Elasmobranch conservation: the establishment of the IUCN Shark Specialist Group. *Aquatic Conservation*, **1**, 193–4.

IUCN (1990) *1990 IUCN Red List of Threatened Animals*, IUCN, Gland, Switzerland.

Johnson, J.E. and Rinne, J.N. (1982) The Endangered Species Act and southwest fishes. *Fisheries*, **7**, 2–72.

Jones, D.M. (1991) Introductory remarks, in *Beyond Captive Breeding: Reintroducing Endangered Mammals to the Wild, Symposia of the Zoological Society of London, No. 62*, (ed. J.H.W. Gipps), Clarendon Press, Oxford, pp. 3–5.

Kallman, K. (1989) Genetic diversity in small populations and their maintenance under captive conditions, *AAZPA 1989 Annual Conference Proceedings*, AAZPA, Wheeling, WV, pp. 154–62.

Kaufman, L. (1987) Caught between a reef and a hard place, *AAZPA 1987 Annual Conference Proceedings*, pp. 352–68.

Kaufman, L. (1991) Progress in the conservation of endemic fishes from Lake Victoria. *Today's Aquarist*, **4**, 3–8.

Kaufman, L. (1992) Lessons from Lake Victoria: catastrophic change in a species-rich fresh water ecosystem. *Bioscience*, 42(1), 846–58.

Le Cren, E.D. (1990) Rare fishes and their conservation. *Journal of Fish Biology*, **37A**, 1–3.

Lelek, A. (1987) *The Freshwater Fishes of Europe*, AULA–Verlag, Wiesbaden.

Luxmoore, R. (1990) *Trade and Captive Breeding of* Scleropages formosus *in Indonesia*. IUCN unpublished report, Cambridge, UK.

Maitland, P.S. (1974) The conservation of freshwater fishes in the British Isles. *Biological Conservation*, **6**, 7–14.

Meffe, G.K. (1990) Genetic approaches to conservation of rare fishes: examples from North American desert species. *Journal of Fish Biology*, **37A**, 105–12.

Mohsin, A.K. Mohammad and Ambak, M.A. (1991) *Freshwater Fishes of Peninsular Malaysia*, Penerbib Universiti, Pertanian, Malaysia.

Moyle, P.B. and Sato, G.M. (1991) On the design of preserves to protect native fishes, in *Battle Against Extinction: Native Fish Management in the American West* (eds W.L. Minckley and J.E. Deacon), University of Arizona Press.

Moyle, P.B. and Williams, J.E. (1990) Biodiversity loss in the temperate zone: decline of the native fish fauna of California. *Conservation Biology*, **4**, 275–84.

Nyman, L. (1991) *Conservation of Fresh Water Fish*, Fisheries Development Series 56, SWEDMAR/WWF, Sweden.

Pollard, D.A., Ingram, B.A., Harris, J.H. and Reynolds, L.F. (1990) Threatened fishes in Australia – an overview. *Journal of Fish Biology*, **37A**, 67–78.

Reid, G. (1990a) Threatened fishes of Barombi Mbo: a crater lake in Cameroon. *Journal of Fish Biology*, **37A**, 209–11.

Reid, G. (1990b) Captive breeding for the conservation of cichlid fishes. *Journal of Fish Biology*, **37A**, 157–66.

Rinne, J.N. (1990) Status, distribution, biology and conservation of two rare southeastern (USA) salmonids, the Apache trout, *Onchorhynchus apache* Miller, and the Gila trout, *O. gutae* Miller. *Journal of Fish Biology*, **37A**, 189–91.

Seal, U.S. (1986) Goals of captive propagation programmes for the conservation of endangered species. *Int. Zoo Yb.*, **24/25**, 174–9.

Sheldon, A.L. (1988) Conservation of stream fishes: patterns of diversity, rarity, and risk. *Conservation Biology*, **2**, 149–56.

Shrestha, T.K. (1990) Rare fishes of Himalayan waters of Nepal. *Journal of Fish Biology*, **37A**, 213–16.

Skelton, P.H. (1990) The conservation status of threatened fishes in southern Africa. *Journal of Fish Biology*, **37A**, 87–95.

Stuart, S.N. (1991) Reintroductions: to what extent are they needed?, in *Beyond Captive Breeding: Re-introducing Endangered Mammals to the Wild, Symposia of the Zoological Society of London, No. 62*, (ed. J.H.W. Gipps), Clarendon Press, Oxford, pp. 27–37.

Swales, S. (1990) Conservation status of large galaxids (Pisces: Galaxidae) in tributaries of the Waikako River, New Zealand. *Journal of Fish Biology*, **37A**, 217–18.

Tullock, J.H., Shute, J.R., Yarnell, J.W. and Etnier, D.A. (1988) *Captive Rearing of Endangered Fish Species*. Unpublished report, Aquatic Specialist Group, Knoxville, Tennessee.

US Fish and Wildlife Service (1990) *Report to Congress: Endangered and Threatened Species Recovery Programs*. US Department of Interior, Fish and Wildlife Service, Washington DC.

Williams, J.E. and Miller, R.R. (1990) Conservation status of the North American fish fauna in fresh water. *Journal of Fish Biology*, **37A**, 79–85.

Witte, F.T. Goldschmidt, J.H., Wanink, M.J.P. *et al.* (1992) The destruction of an endemic species flock: quantitative data on the decline of the haplochromine species from the Mwanza Gulf of Lake Victoria. *Environmental Biology of Fishes*, **34**, 1–28.

Wood, E. (1985) *Exploitation of Coral Reef Fishes for the Aquarium Trade*, Marine Conservation Society, Ross-on-Wye, UK.

19

The role of captive breeding in the conservation of Old World fruit bats

S. Mickleburgh
Fauna and Flora Preservation Society, London, UK
and
J.B. Carroll
Jersey Wildlife Preservation Trust, Channel Islands.

19.1 INTRODUCTION

Bats belong to the order Chiroptera. There are about 950 species worldwide, making up almost a quarter of all known mammal species. The order Chiroptera is divided into two major sub-orders; the Microchiroptera and the Megachiroptera (Old World fruit bats). The Megachiroptera are distinguished from the Microchiroptera by having a simple external ear forming an unbroken ring and by having a second finger that is relatively independent of the third finger and which usually bears a small claw. They do not possess a nose-leaf (often well-developed in the Microchiroptera) or tragus (a small structure inside the ear). They generally have large eyes, and sight and smell appear to be the major locational senses, in contrast to the Microchiroptera, which have small eyes. Echolocation, a method of orientation using ultrasonic sounds emitted through the mouth or nose, is universal among the Microchiroptera but is, with a few exceptions, unknown in the Megachiroptera.

Creative Conservation: Interactive management of wild and captive animals.
Edited by P.J.S. Olney, G.M. Mace and A.T.C. Feistner.
Published in 1994 by Chapman & Hall, London.
ISBN 0 412 49570 8

There are 41 genera of Megachiroptera containing a total of 161 species. They occur throughout the Old World tropics and subtropics from Africa through southern and South-East Asia to Australia and islands in the western Pacific. Many species occur on small islands or island groups.

19.2 THREATS

The main threats to fruit bats are deforestation, disturbance, hunting, conflict with commercial fruit growers, and tropical storms.

19.2.1 Deforestation

Many fruit bat species are dependent on primary forest for roosting and feeding, and are thus threatened by the large-scale destruction of rain forest in many tropical areas. Habitat loss has been cited as a major factor contributing to the declines in fruit bat populations (e.g. Wodzicki and Felten, 1975; Cheke and Dahl, 1981; Fujita and Tuttle, 1991). Many species, particularly those in mangrove swamps or lowland forest, have lost critical habitat, while others have lost vital food resources. The loss of tamarind trees (*Tamarindus indica*) has been identified as one factor responsible for the decline of the Rodrigues fruit bat (*Pteropus rodricensis*).

19.2.2 Disturbance

Fruit bats frequently roost in large, highly visible colonies. This makes them prone to disturbance by humans. Cave-roosting species are especially vulnerable, particularly where they are taken for food or share the roost sites with insect-eating species whose guano is collected by humans for use as fertilizer.

19.2.3 Hunting

In many areas of the world, fruit bats provide an important food source for local peoples, for example in Guam (Wiles, 1987a, b), Vanuatu (Chambers and Esrom, 1991), Samoa (Cox, 1983) and the Seychelles (Racey, 1979; Cheke and Dahl, 1981). In the past harvesting has been on a sustainable basis but more recently the use of firearms has led to serious declines in numbers (Wodzicki and Felten, 1975; Engbring, 1985; Wiles, 1987b). In the Pacific, a commercial trade in dead fruit bats has developed, centred on the island of Guam. In the period 1975–89 over 220 000 bats were imported into Guam from nearby islands to satisfy local and tourist demand (Wiles, 1992). The 1989 Convention on International Trade in Endangered Species of Wild Fauna and Flora (CITES) regulations were amended to curb much of this trade. However, bats are still imported, legally and illegally, into Guam, threatening a number of populations on surrounding Pacific islands.

19.2.4 Conflict with fruit growers

In areas of the world where commercial fruit-growing is a major industry (e.g. Israel, South Africa and Australia) there have been conflicts between bats and fruit growers. This has arisen because many cultivars are developed from wild species that are dependent upon bats for pollination and seed dispersal (van der Pijl, 1957; Marshall, 1983). Destructive control measures such as shooting are often still favoured. The conflict is particularly acute where fruit orchards have replaced primary or secondary forest habitat (Fleming and Robinson, 1987; Tidemann and Nelson, 1987).

19.2.5 Tropical storms

Species that occur on small islands are particularly vulnerable to the effects of tropical storms, which destroy habitat and food resources. Serious declines in populations after tropical storms have been recorded for a number of species (e.g. *Pteropus rodricensis* on Rodrigues [Jones, 1980; Carroll, 1984]; *P. samoensis* and *P. tonganus* on Samoa [E.D. Pierson, pers. comm.]; *P. rayneri* and *P. tonganus* in the Solomons [Flannery, 1989]).

19.3 IMPORTANCE OF FRUIT BATS

Fruit bats feed almost exclusively on plants, taking floral resources (nectar, pollen, petals and bracts), fruit and often the seeds themselves, and leaves (Marshall, 1985). They are very important pollinators and seed dispersers in tropical forests throughout the world (Marshall, 1983, 1985; Fleming, Breitwisch and Whitesides, 1987; Fleming, 1988; Cox *et al.*, 1991, 1992; Pierson and Rainey, 1992). Megachiroptera visiting flowers for food may effect pollination. This is known to be the case for 31 genera in 14 families (Marshall, 1985). Megachiropteran bats also feed upon at least 145 genera of fruit in 30 families widely distributed throughout the angiosperms (Marshall, 1985). Generally fruits are consumed when ripe. Large fruits must be consumed *in situ*, but smaller fruits may be carried away before being devoured and the seeds ejected through the mouth or anus. The distance a seed is carried will depend on its size and the size of the bat; tiny seeds which pass through the alimentary canal of a large bat will be carried furthest. Some species can travel up to 50 km each night to feed so that long-distance dispersal may sometimes occur. In the Philippines, an increased germination rate has been recorded for fig seeds (*Ficus chrysolepis*) taken from bat faecal masses (Utzurrum and Heideman, 1991). Because long-distance seed dispersal by fruit bats is primarily through faecal deposition, this makes bat dispersal even more effective than previously suggested.

On many oceanic islands, which have limited faunas, fruit bats are the only animals capable of carrying large-seeded fruits. In such ecosystems, fruit bats can be the single most important pollinators and seed-dispersers. In island ecosystems in the south-west Pacific, fruit bats are considered 'keystone species', because significant declines in forest regeneration rates and diversity would accompany their extinction (*Cox et al.*, 1991, 1992).

It has been shown that bats play a vital role in the regeneration of cleared areas. It has been estimated that bats contribute up to 95% of the seeds deposited in cleared, open areas (Thomas *et al.*, 1988).

Many of the plants that benefit from pollination or seed dispersal are economically important to man (Fujita and Tuttle, 1991; Wiles and Fujita, 1992). At least 443 products useful to man are derived from 163 plant species that rely to some degree on bats for pollination or seed dispersal (Fujita and Tuttle, 1991).

19.3.1 Conservation status of fruit bats

Mickleburgh, Hutson and Racey (1992) reviewed the conservation status of all Old World fruit bats. Table 19.1 summarizes the conservation gradings for all species.

19.3.2 Conservation options

There are a number of options available to help conserve threatened fruit bat populations. All of these are not mutually exclusively and ideally a species' survival is best guaranteed through a combination of options.

(a) Survey

For many species there is no information on current population status or trends. Surveys using standardized techniques need to be undertaken, particularly within protected areas.

(b) Habitat protection and management

Most fruit bat species are dependent to a large degree on forest, particularly primary forest. Protection of forest and the control of destruction are of the highest priority to ensure the survival of many

Table 19.1 Conservation gradings for Old World fruit bats (after Mickleburgh, Hutson and Racey, 1992)

Category[a]	*No. of species*
Extinct	3
Extinct (within last 50 years)	4
Endangered	14
Vulnerable	10
Rare	16
Indeterminate	7
No data	35
Not threatened	72
Total	161

[a]These are based on the criteria used by IUCN with the addition of a 'No Data' category for species where there is little or no information available to assess status.

species. In many cases the ultimate survival of endangered bat species rests with the Protected Areas System. The setting up of protected areas should be the goal of national conservation effort.

Forests can be managed with bats in mind through the maintenance of roosting trees (often large canopy trees) and a diverse understorey. Where caves are used as roosting sites, these can be managed to limit disturbance.

(c) Legislation

All *Pteropus* and *Acerodon* species are protected under the CITES regulations, although the level of protection varies. Other species receive varying degrees of protection at a local, national or international level. Legislation at all levels can be improved or implemented to ensure better protection for species, their habitats and roosts.

(d) Education

This is an almost universal need and can cover a variety of subjects. It can be general, emphasizing the role bats play in tropical ecosystems, or be aimed at specific groups such as fruit-growers, hunters or tourists.

(e) Captive breeding

Captive breeding is an invaluable tool to ensure the survival of critically endangered species. Ideally it should be undertaken before populations reach critical levels.

19.4 CAPTIVE BREEDING OF FRUIT BATS

Captive populations of at least 15 species of fruit bat are held at zoological institutions world-wide (Olney and Ellis, 1991). Most are non-threatened species. Some, such as the Egyptian fruit bat (*Rousettus aegyptiacus*) and the Indian fruit bat (*Pteropus giganteus*), have been successfully bred at a large number of institutions. Experience with such non-threatened species may provide invaluable information for current and future breeding programmes for endangered species. Mickleburgh, Hutson and Racey (1992) recommended that captive breeding programmes be considered for 15 species (Table 19.2). Three species were thought to be probably already extinct (two had not been seen for more than 50 years and one within the last 50 years) and for these captive breeding would be essential were they to be relocated in the wild.

In order to facilitate cooperation between zoos and to maximize effective use of cage space available for fruit bat breeding, regional Taxon Advisory Groups (TAGs) for fruit bats have been established in North America and the British Isles. These TAGs have been established under the auspices of the American Association of Zoological Parks and Aquariums and the Federation of Zoological Gardens of Great Britain and Ireland. Through liaison with field workers and the relevant SSC specialist groups, the Taxon Advisory Group seeks to prioritize species programmes for captive breeding and to ensure cooperation among zoos in order to maintain these priority species over the long term.

19.4.1 Recommending species for captive breeding

To date, the only successful endangered species breeding programme has involved the Rodrigues fruit bat, *Pteropus rodricensis*. A number of other schemes are underway or planned (Table 19.2). There are a number of criteria that should be considered when recommending species for captive breeding.

(a) Degree of threat

In general, the more threatened a species, the higher the priority for captive breeding. However, consideration should be given to species whose current status is more secure but whose future prospects are

Table 19.2 Captive-breeding programmes for Old World fruit bats (after Mickleburgh, Hutson and Racey, 1992)

Species	*Distribution*	*Category*[a]	*Captive breeding*
Acerodon jubatus	Philippines	E	Planned
Aproteles bulmerae	Papua New Guinea	E	Suggested
Nyctimene rabori	Philippines (Negros)	E	Planned
Pteralopex acrodonta	Fiji Islands (Taveuni)	E	Suggested
Pteralopex anceps	Bougainville, Solomon Islands (Choiseul)	E	Suggested
Pteralopex atrata	Solomon Islands (Guadalcanal, Santa Isabel)	E	Suggested
Pteropus leucopterus	Philippines	V	Planned
Pteropus livingstonii	Comoros	E	Underway
Pteropus pumilus	Philippines	V	Underway
Pteropus rodricensis	Rodrigues	E	Underway
Pteropus vampyrus lanensis	Philippines	V	Suggested
Pteropus voeltzkowi	Pemba	E	Planned
Acerodon lucifer[b]	Philippines (Panay)	Ex	
Dobsonia chapmani[b]	Philippines (Cebu, Negros)	Ex (?)	
Nyctimene sanctacrucis[b]	Solomon Islands	Ex	

[a]E = endangered; V = vulnerable; Ex = extinct.
[b]These species are thought to be extinct. Should they be relocated in the wild, a captive breeding programme would be an integral part of any management plan.

poor. Removal of individuals for a captive breeding programme from a larger population may be more acceptable than waiting until such action might jeopardize the survival of the wild population. In some instances, species that are thought to be extinct have been recommended for captive breeding. Should a supposedly extinct species be relocated in the wild (as has happened with Bulmer's fruit bat [*Aproteles bulmerae*]) its continued long-term survival may only be secured through captive breeding.

(b) Availability of suitable breeding institutions

The number of institutions with experience of breeding endangered species successfully is small. This will grow as knowledge is disseminated. The availability of expertise is vital to the success of any programme. Ideally, captive populations should be established within the range of the species with the involvement of local people. To date, this has rarely happened, although this situation will improve as more programmes are instigated.

(c) Control of threats

It is vital that the threats that currently face a species are controllable. A species cannot be released back into the wild if the factors that originally caused its decline are still operating. In some cases control of threats may not be possible and the only other option is to introduce the captive-bred species to another location.

(d) Potential for introduction/reintroduction

It is imperative that the potential for introduction/reintroduction be considered before a programme is started. In many cases the main threat is from habitat destruction. Unless there is a real chance of halting destruction or reclaiming habitat, captive breeding could be a fruitless exercise. While the ultimate aim is to reintroduce the species to its original habitat, introduction to a site outside the species' ecological range may be possible. This however raises ethical and ecological problems. Many fruit bats have very limited ranges, some occurring only on single islands. That single island may be the only possible location for a wild population, thus ruling out opportunities for introduction elsewhere. There is also the problem of the implications for indigenous fauna and flora.

All introduction or reintroduction programmes require strict scientific controls. The IUCN has produced guidelines for undertaking such programmes. These are discussed in detail by Mickleburgh, Hutson and Racey (1992).

19.4.2 *Pteropus rodricensis* – a case history

This species is endemic to the island of Rodrigues in the Indian Ocean, and in 1974 was described as probably the rarest bat in the world, with an estimated wild population of 75–80. In 1976 the Jersey Wildlife Preservation Trust in cooperation with the Mauritian government (which administers Rodrigues) initiated a captive breeding programme (reviewed by Carroll, 1988). Two founder colonies were established, one in Jersey and one in Mauritius. The Jersey colony increased without loss until 1980, although in subsequent years the number of neonatal and juvenile deaths increased. In recent years these have declined in frequency. In 1982 it was thought that density-dependent factors were increasing the incidence of juvenile mortality and the colony was divided (Carroll and Mace, 1988). Further colonies were established so that by 1991 about 250 bats were held in nine colonies in Jersey, Mauritius, England and the USA (Carroll and Barker, 1992).

The experience with *P. rodricensis* highlighted a number of potential problems with captive breeding programmes for fruit bats (Carroll and Mace, 1988). Many fruit bat species are gregarious. In *P. rodricensis* daytime social aggregations may be either harem groups with one male and several females, or mixed-sex subadult groups. At night new feeding groupings are formed. Harem males defend their roost sites with great vigilance while harem and other adult males control access to feeding stations. Thus a female may associate with and be mated by several different males in a 24-hour period. A recent study of *P. rodricensis* at JWPT by Young and Carroll (1989) showed that adult females were indeed promiscuous. They did, however, show marked fidelity to particular roost and feeding sites within the cage. As a result they did not encounter a great number of different males. Of the 11 study females, five mated with one male, four with two males and two with three males during the study period. This behaviour not only makes it impossible to assign paternity to offspring, it also means that considerable manipulation of social groups would be needed to ensure any particular male sired offspring. This would be costly in terms of cage space and would necessitate keeping the bats in unnatural social groups.

Maintaining genetic variability is crucial to any captive breeding programme. In general, the loss of genetic variation from any population depends on the effective population size and the number of generations over which the population is maintained (Carroll and Mace, 1988). It is suggested that for *P. rodricensis* the captive population should be divided into four distinct sub-populations. Within each sub-population there should be regular interchange of animals between each of the captive colonies, whereas movement of animals between sub-populations should be less frequent; only one individual per generation (Carroll and Mace, 1988). The animals moved should be females, as they integrate easily into a breeding situation and it is easier to monitor their breeding performance.

The lack of certainty of paternity is a serious hindrance in genetic management of the population. An attempt to determine paternity in *P. rodricensis* at JWPT was made using DNA fingerprinting techniques. Unfortunately, the fingerprints obtained were almost identical for most of the bats sampled and paternity could not be established (D. Ashworth, pers. comm.). Further attempts to assess paternity in the bats are being made at the Brookfield Zoo, Chicago and the Lubee Foundation, Florida (R. Phillips, pers. comm.; J. Seyjaget, pers. comm.). If data on male reproductive success can be obtained, better genetic modelling of the population and hence better long-term management of the population can be made.

Since 1976 the wild population on Rodrigues has increased. In 1990 it was estimated to be greater than 1000 individuals, but following a cyclone in February 1991 it had fallen to around 350 individuals by August of that year (K. Kelly, pers. comm.). The captive-bred population thus still provides an important safeguard against extinction as well as a reservoir of genetic variability.

19.4.3 Other captive breeding schemes

Livingstone's fruit bat (*Pteropus livingstonii*) is endemic to the Comoro Islands in the Indian Ocean. This species was abundant on the island of Anjouan in the nineteenth century but an expedition in 1989 estimated the population to be only around 60 individuals (Thorpe, 1989). In July 1990 a roost containing 60–120 individuals was discovered (Carroll and Thorpe, 1991) and a further roost with some 30 bats was discovered later that year but was abandoned in 1992 (K. Hunter, pers. comm.; S. Wray, pers. comm.). The major threats to the population are deforestation and underplanting of the remaining native forest. *P. livingstonii* appears to be unable to adapt to secondary, disturbed, or deforested habitats, unlike its sympatric cogener the Comoros lesser fruit bat (*P. seychellensis comorensis*). During 1992 an expedition from Bristol University carried out an ecological study of the two *Pteropus* species on Anjouan and, in collaboration with JWPT, attempted to capture some *P. livingstonii* for the initiation of a captive breeding programme. Six individuals were caught, five males and one female, which are now at JWPT. A further capture attempt is planned for 1993 to increase the number of bats in captivity and to equalize the sex ratio. The captive breeding programme is being carried out in collaboration with the Comorian government, which retains ownership of all the bats and of any that may be born in captivity in the future.

The Pemba fruit bat (*Pteropus voeltzkowi*) is endemic to the island of Pemba off the coast of Tanzania. Seehausen (1990) reported on a visit to the island in 1989 during which he managed to locate only three individuals. In late 1989 a single roost containing 150–200 bats was reported (H. Bentjee, pers. comm.). The main threats are from deforestation and possibly hunting. A captive breeding programme based at Phoenix Zoo is planned.

In the Philippines, two endangered fruit bat species, the little golden-mantled fruit bat (*Pteropus pumilus*) amd *P. leucopterus*, are currently held in a captive breeding establishment at CENTROP (Centre for Research in Tropical Ecology) based at the Department of Biology, Silliman University, Dumaguete City, Negros. Permission has been obtained to procure more individuals of six species, including the above two and the endangered golden-capped fruit bat (*Acerodon jubatus*) and Philippine tube-nosed bat (*Nyctimene rabori*), to enlarge this facility. This is being done in collaboration with DENR (Department of the Environment and Natural Resources) and the Lubee Foundation (W.L.R. Oliver, pers. comm.).

19.5 TRANSLOCATION

To date there have been no introductions or reintroductions of endangered fruit bats. It has been suggested that an introduction of *P. rodricensis* to an island outside the cyclone belt might be possible in the western Indian Ocean (Carroll, 1988). So far, no suitable sites have been identified. Controlling threats to populations is vital before captive-bred bats can be reintroduced. This is particularly difficult where the major threat is deforestation. Where threats cannot be controlled, and introduction to an alternative site is considered, there is the problem of availability of suitable areas. This is especially difficult with endemic species on small islands or island groups.

19.6 CONCLUSIONS

It is clear that it is possible to breed endangered fruit bats in captivity, although their survival after introduction or reintroduction has yet to be established. For a number of species, captive breeding may be the last chance to save them from extinction. In some cases, captive breeding of less threatened species may provide an opportunity to highlight their conservation problems in the wild and to prevent the population reaching the position where there is no alternative to a captive breeding programme. Captive breeding should be seen as one a number of tools that can be used to prevent the extinction of endangered species, and where possible it should be combined with education and improved management as a multifaceted approach to species conservation.

REFERENCES

Carroll, J.B. (1984) The conservation and wild status of the Rodrigues fruit bat (*Pteropus rodricensis*). *Myotis*, **21/22**, 148–54.

Carroll, J.B. (1988) The conservation programme for the Rodrigues fruit bat *Pteropus rodricensis*, in *Proceedings of the 5th World Conference on Breeding Endangered Species in Captivity* (eds B.L. Dresser, R.W. Reace and E.J. Maruska), Cincinnati Zoo, Cincinnati, pp. 457–76.

Carroll, J.B. and P. Barker, (1992) *Rodrigues fruit bat* Pteropus rodricensis *International Studbook*, Preliminary edition, Jersey Wildlife Preservation Trust, Jersey.

Carroll, J.B. and Mace G.M. (1988) Population management of the Rodrigues fruit bat *Pteropus rodricensis* in captivity. *Int. Zoo Yb.*, **27**, 70–8.

Carroll, J.B. and Thorpe, I.C. (1991) Conservation of Livingstone's fruit bat. A report on an expedition to the Comores in 1990. *Dodo*, **27**, 26–40.

Chambers, M.R. and Esrom, D. (1991) The fruit bats of Vanuatu. *Bat News*, **20**, 4–5.

Cheke, A.S. and Dahl, J.F. (1981) The status of bats on western Indian Ocean Islands with special reference to *Pteropus*. *Mammalia*, **45**, 205–38.

Cox, P.A. (1983). Observations on the natural history of Samoan bats. *Mammalia*, **47**, 519–23.

Cox, P.A., Elmqvist, T., Pierson, E.D. and Rainey W.E. (1991) Flying foxes as strong interactors in South Pacific island ecosystems: a conservation hypothesis. *Cons. Biol.*, **5**(4), 448–54.

Cox, P.A., Elmqvist, T. Pierson, E.D. and Rainey, W.E. (1992) Flying foxes as pollinators and seed dispersers in Pacific island ecosystems, in *Pacific Island Flying Foxes: Proceedings of an International Conservation Conference* (eds D.E. Wilson and G.L. Graham), US Fish and Wildlife Service Biological Report 90(23), US Department of the Interior, Fish and Wildlife Service, Washington DC, pp. 18–23.

Engbring, J. (1985) *A 1984 survey of the fruit bat (maga'lau) on Yap*. Report of the US Fish and Wildlife Service, Honolulu, Hawaii.

Flannery, T.F. (1989) Flying foxes in Melanesia: populations at risk. *Bats*, **7**(4), 5–7.

Fleming, P.J.S. and Robinson, D. (1987) Flying foxes (Chiroptera: Pteropodidae) on the north coast of New South Wales: damage to stonefruit crops and control methods. *Aust. Mamm.*, **10**, 143–7.

Fleming, T.H. (1988) *The Short-tailed Fruit Bat: A Study in Plant–Animal Interactions*, University of Chicago Press, Chicago.

Fleming, T.H., Breitwisch, R. and Whitesides, G.H. (1987) Patterns of tropical frugivore diversity. *Ann. Rev. Ecol. Syst.*, **18**, 91–109.

Fujita, M.S. and Tuttle, M.D. (1991) Flying foxes (Chiroptera: Pteropodidae): threatened animals of key ecological and economic importance. *Cons. Biol.*, **5**(4), 455–63. A complete list of plant and bat species, observation locations, specific bat–plant interactions, plant products and uses and full references can be obtained from Bat Conservation International, PO Box 162603, Austin TX 78716, USA.

Jones, C. (1980) *The Conservation of the Endemic Birds and Bats of Mauritius and Rodrigues (A Progress Report and Proposal for Future Activities)*. International Council for Bird Preservation, Mauritius and Washington.

Marshall, A.G. (1983) Bats, flowers, and fruit: evolutionary relationships in the Old World. *Biol. J. Linn. Soc.*, **20**, 115–35.

Marshall, A.G. (1985) Old World phytophagous bats (Megachiroptera) and their food plants: a survey. *Zool. J. Linn. Soc.*, **83**, 351–69.

Mickleburgh, S.P., Hutson, A.M. and Racey, P.A. (1992) *Old World Fruit Bats. An Action Plan for their Conservation*, IUCN, Gland, Switzerland.

Olney, P.J.S. and Ellis, P. (eds) (1991) *International Zoo Yearbook 30*, Zoological Society of London.

Pierson, E.D. and Rainey W.E. (1992) The biology of flying foxes of the genus *Pteropus*: a review, in *Pacific Island Flying Foxes: Proceedings of an International Conservation Conference* (eds D.E. Wilson and G.L. Graham), US Fish and Wildlife Service Biological Report 90(23), US Department of the Interior, Fish and Wildlife Service, Washington, DC, pp. 1–19.

Racey, P.A. (1979) Two bats in the Seychelles. *Oryx*, **15**, 148–52.

Seehausen, O., (1990) Vom Aussterben bedroht: Der Pemba Flughund. *Zool. Ges. f. Arten- und Populationsschutz*, **6**(1), 12–16.

Thomas, D.W., Cloutier, D., Provencher, M. and Houle, C. (1988) The shape of bird- and bat-generated seed shadows around a tropical fruiting tree. *Biotropica*, **20**, 347–8.

Thorpe, I. (1989) *University of East Anglia Comoro Islands Expedition 1989, Final report*. Unpublished Report, University of East Anglia, Norwich, UK.

Tidemann, C.R. and Nelson, J.E. (1987) Flying foxes (Chiroptera: Pteropodidae) and bananas: some interactions. *Aust. Mamm.*, **10**, 133–6.

Utzurrum, R.B. and Heideman, P.D. (1991) Differential ingestion of viable vs nonviable *Ficus* seeds by fruit bats. *Biotropica*, **23**(3), 311–12.

van der Pijl, L. (1957) The dispersal of plants by bats (chiropterchory). *Acta bot. neerl.*, **6**, 291–315.

Wiles, G.J. (1987a) The status of fruit bats on Guam. *Pacif. Sci.*, **41**, 148–57.

Wiles, G.J. (1987b) Current research and future management of Marianas fruit bats (Chiroptera; Pteropodidae) on Guam. *Aust. Mamm.*, **10**, 93–5.

Wiles, G.J. (1992) Recent trends in the fruit bat trade on Guam, in *Pacific Island Flying Foxes: Proceedings of an International Conservation Conference* (eds D.E. Wilson and G.L. Graham), US Fish and Wildlife Service Biological Report 90(23), US Department of the Interior, Fish and Wildlife Service, Washington, DC, pp. 53–60.

Wiles, G.J. and Fujita, M.S. (1992) Food plants and economic importance of flying foxes on Pacific islands, in *Pacific Island Flying Foxes: Proceedings of an International Conservation Conference* (eds D.E. Wilson and G.L. Graham), US Fish and Wildlife Service Biological Report 90(23), US Department of the Interior, Fish and Wildlife Service, Washington, DC, pp. 24–35.

Wodzicki, K. and Felten, H. (1975) The peka, or fruit bat (*Pteropus tonganus*) (Mammalia: Chiroptera), of Niue Island, South Pacific. *Pacif. Sci.*, **29**, 131–8.

Young, J.A. and Carroll, J.B. (1989) Male–female associations in captive Rodrigues fruit bats. *Dodo*, **26**, 48–60.

20

Captive breeding, reintroduction and the conservation of canids

J.R. Ginsberg
Zoological Society of London, UK.

20.1 INTRODUCTION

Captive breeding of wild animals may be pursued for utilitarian, aesthetic, educational, or scientific reasons: for consumptive use, either in meat production (Skinner, 1989) or the production of fur (Nowak *et al.*, 1987); to provide individuals for viewing in zoos or game parks without having to capture animals in the wild and further deplete natural populations (de Boer, 1992); or captive populations of wild animals can serve as latter-day arks, reinforcing and replenishing endangered wild populations (Anderson, 1986; Seal, 1986; Foose and Ballou, 1988; de Boer, 1992).

With an increasing rate of extinction (Wilson, 1988) the number of species requiring such captive management, or a combination of *ex situ* and *in situ* meta-population management, is sure to increase (Seal, 1989). Because we are limited financially in the number of large vertebrate species which can be bred in captivity (Conway, 1986), it is critical that the choice of which species are bred in captivity be made in a timely and considered manner (IUCN, 1987; Foose and Seal, 1992).

Great progress has been made in the last decade developing Species Survival Plans (SSPs), species studbooks and local (Animal Records Keeping System, ARKS) and global (International Species Information System, ISIS) databases for captive endangered vertebrate species (Fles-

Creative Conservation: Interactive management of wild and captive animals.
Edited by P.J.S. Olney, G.M. Mace and A.T.C. Feistner.
Published in 1994 by Chapman & Hall, London.
ISBN 0 412 49570 8

ness and Mace, 1988; de Boer, 1992). As of 1990, regional breeding programmes have been initiated for 172 species, the majority of them mammals (105) and birds (58) (de Boer, 1992). Despite these improvements in planning and management of single species, few studies have analysed, at the international level, zoo breeding effort and success either among or between higher taxa of captive animals (Ginsberg and Macdonald, 1990; Ralls, Ballou and Templeton, 1988).

Interspecific comparisons of which species are bred, and the survivorship of the young produced from these breeding attempts, are critical to future conservation efforts for several reasons. The rational allocation of breeding space and effort among subspecies of a particular species (Magquire and Lacy, 1990) is a specific example of a more general problem. If cage space, captive breeding research and money, are to be allocated appropriately, we need to have information on several variables: present allocation of cage space; some measure of current allocation of captive breeding effort, such as how many animals are being bred in captivity (number of litters and/or individuals bred) or how many zoos are actively breeding a species; and how captive breeding effort varies among species belonging to higher order taxonomic groups such as a family.

Comparative analyses also allow us to make generalizations about factors which may influence the success, or failure, of captive breeding programmes. In evolutionary studies, the appropriate level of comparison has been hotly debated (Harvey and Pagel, 1991). In zoo studies, the taxonomic level at which comparisons are made will provide qualitatively different information about how we set priorities. A comparison among phyla, for instance, would show quite clearly that entire groups of organisms (e.g. insects) are poorly represented in zoo collections and breeding programmes. Zoos, for the most part, breed and exhibit large vertebrates, particularly mammals (Conway, 1986). A more appropriate level at which to analyse the zoo community's recent experience in captive breeding, and its success or failure, would be among members of a single family (Foose and Seal, 1992). In this study I have limited myself to examining patterns of captive breeding among species within the family Canidae.

If captive breeding for eventual reintroduction is to be successful, we must:

(a) assess which species may require captive propagation;
(b) if these animals are not in captivity, find a source for these animals which does not compromise *in situ* populations;
(c) breed these species in captivity minimizing problems of inbreeding and selection for captive traits;
(d) successfully reintroduce these animals as quickly as possible.

Recent initiatives aim to set captive breeding priorities relative to conservation needs for a particular family Captive Animal Management Plans (CAMPs); Global Captive Animal Plans (CAPs); see Foose and Seal, 1992; Seal, Foose and Ellis, Chapter 16. These plans present detailed summaries of actions required; they do not, however, assess the tractability of their conclusions and recommendations. This paper will therefore address the following questions:

(a) what is the rate of success in captive breeding of endangered canids?
(b) are threatened species held in captivity and are they being bred?
(c) what is the success in reintroducing captive-bred canids to the wild?

20.2 METHODS

The analyses in this paper are based on data extracted from captive breeding records in the International Zoo Yearbooks (IZY) (e.g. Olney and Ellis, 1992). The IZY reports contain the only data available which provide long-term information on captive breeding in zoos. In this paper, annual breeding effort for a species is defined as the number of zoos reporting breeding of a particular species in a particular year. This measure, while ignoring specific efforts and breeding programmes, gives a rough measure of experience in captive breeding of a particular species. The measure of breeding success used is the percentage survival of pups born. This measure has been used in other studies to assess the success of captive breeding (Ralls, Harvey and Lyles, 1986; Ralls, Ballou and Tempelton, 1988). Poor survival of pups indicates that captive breeding cannot be sustained over a long period of time and may imply inadvertent selection for traits associated with captivity.

Captive breeding data reported in the IZY have limitations. One problem is that there is variation both in the accuracy and frequency of reporting. Of particular concern is the potential for variation in the reporting of juvenile mortality. For these analyses I make the following assumptions: there are no systematic biases in reporting that might influence inter-specific comparisons and accuracy of reporting does not change with year of reporting. This last assumption was tested by looking at mean mortality of juveniles across all species (Ginsberg and Macdonald, 1990). Since 1969, mean mortality has shown no statistically significant variation and these analyses are based only on data from the years 1970 to 1990.

Breeding results in the IZY are listed by zoological collection, not by litter. If several litters are bred in a single zoo in a single year, the data reported are the summed data of these litters. To simplify language, I refer to each entry in the IZY as a 'litter', recognizing that in a number of cases the data presented have been pooled from several litters. Such pooling will not affect mean litter mortality values reported, but will increase measured variance by reducing sample size.

The category of threat to which a particular canid species has been assigned was determined from the listings in Ginsberg and Macdonald (1990, p. 90) with the following caveat. As data on provenance of captive stock are often unavailable, the analyses were conducted at the level of the species, with the category of least threat for a sub-taxa being assigned to the entire species in question. This resulted in only a single species, the grey wolf (*Canis lupus*), being placed in a category of lower threat (Requiring No Immediate Protection) than may apply for particular subspecies such as the Mexican wolf. Species listed as 'Rare' are, for the purpose of this analysis, included in the category 'Vulnerable.'

20.2.1 Species bred in captivity

Breeding records for the Canidae, as reported in the IZY from 1970 to 1990, were divided into four, five-year periods (Figure 20.1), to allow an examination of the allocation of captive breeding effort (the number of litters reported) through time. Across two decades, no difference was seen in the number of litters of species falling into the categories 'Safe' and 'Status Unknown' (G^2 7.680, $P > 0.05$). In contrast, a significant difference in the number of litters reported for 'Endangered' and 'Vulnerable' species was seen ($G^2 = 12.7$, $P < 0.01$), with a marked increase in the years 1986–1990. While in the early 1970s 17% of all litters

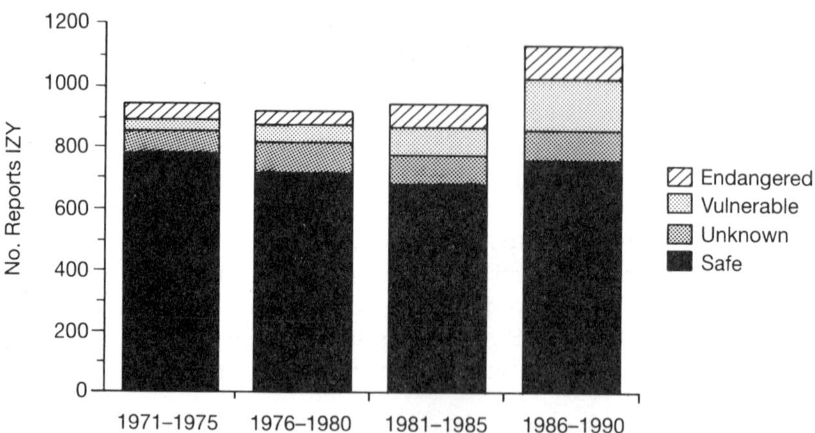

Figure 20.1 Number of reports of 32 canid species bred in captivity 1971–1990, by category of threat (data from International Zoo Yearbooks). In the late 1980s, there was a significant increase in the number of reports of breeding 'Endangered' and 'Vulnerable' species ($G^2 = 12.7$, $P \leqslant 0.01$).

reported were of Rare, Vulnerable or Unknown species, by the late 1980s, 38% of all litters bred fall into these categories.

From 1986 to 1990, however, three threatened species account for over 95% of the litters reported for species classified as Endangered or Vulnerable: maned wolf (*Chrysocyon brachyurus*), African wild dog (*Lycaon pictus*) and bush dog (*Speothos venaticus*) (Figure 20.2). During this time, two other species have at least five records (red wolf [*Canis rufus*], dhole [*Cuon alpinus*]), the grey zorro (*Dusicyon griseus*) is reported once, and two species (Ethiopian wolf [*Canis simensis*], island grey fox [*Urocyon littoralis*]) are not reported, or have not been bred during this time. The red wolf is poorly reported in the IZY, but none the less is being bred intensively. In 1990, the IZY lists three reports of litters of red wolves, but in the *1991 Red Wolf International Studbook* (Smith, 1992), individuals representing 10 litters born in 1990 are listed.

In the IZY from 1962 to 1990, 32 of the 34 canid species have been bred in captivity. The two species for which there are no records of

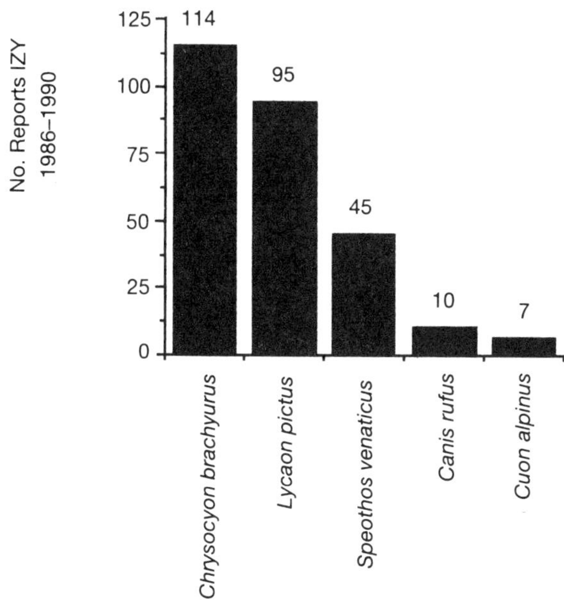

Figure 20.2 Number of litters reported for endangered and vulnerable canids. Despite the increase in the number of threatened canids being bred in captivity in the late 1980s, only three threatened canid species, the maned wolf, the African wild dog, and the bush dog, have been bred with any regularity. Most threatened species are neither bred in captivity nor held in zoological collections.

breeding in captivity are the Ethiopian wolf and the Tibetan fox (*Vulpes ferrilata*). Of the remaining 32 species bred in captivity, many have produced only a small number of litters. Of the 32 species, 14 have been bred, on average, in fewer than three collections per year. The remaining 18 species account for the great majority of all captive breeding of

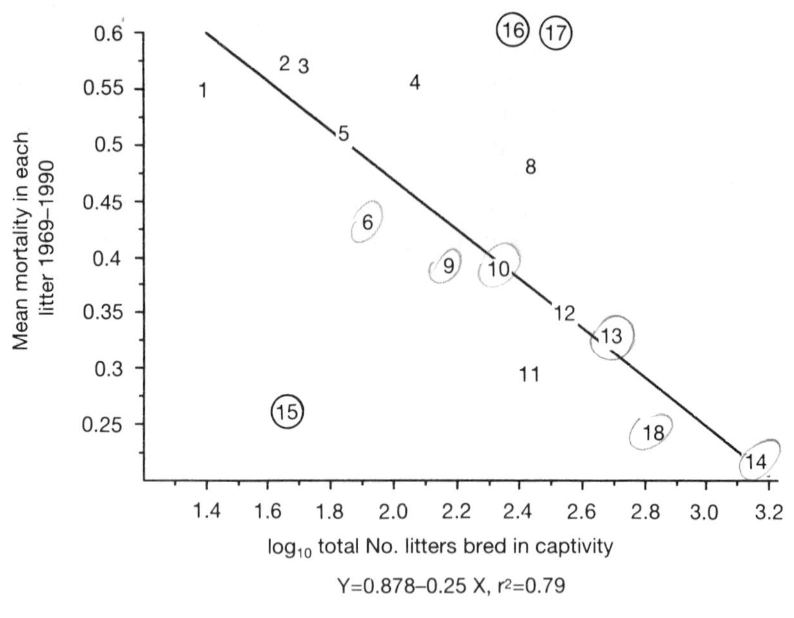

Key to species

1=Dhole	10=Golden jackal
2=Crab-eating zorro	11=Arctic fox
3=Bat-eared fox	12=Reccoon dog
4=Bush dog	13=Red fox
5=Grey fox	14=Grey wolf
6=Black-backed jackal	15=Kit/Swift fox
7=Corsac fox	16=Fennec fox
8=Maned wolf	17=African wild dog
9=Coyote	18=Dingo

○ =species excluded from analysis

Figure 20.3 Pup mortality in relation to number of captive-bred litters. Breeding success (low mortality of pups) is a log-linear function of the frequency with which a canid species has been bred in captivity (after Ginsberg and Macdonald, 1990). Extensive experience appears to be the variable most important to improving the survivorship of pups, explaining 77% of the variance in pup survivorship. Three outliers have been excluded as they exert significant leverage on the regression.

canid species (Figure 20.3). Seventy-five per cent of all captive breeding can be attributed to three species: the grey wolf, the dingo (*Canis familiaris dingo*), and the red fox (*Vulpes vulpes*). Only two threatened species, the African wild dog and the maned wolf, are among the species most frequently bred in captivity.

20.3 SUCCESS IN BREEDING THREATENED SPECIES

20.3.1 Results

There is a log-linear relationship between mean pup mortality per litter and the number of litters produced in captivity (Figure 20.3; $Y = 0.878 - 0.25\ X$, $r^2 = 0.79$). Three outliers were removed as they exert undue leverage on the regression (Sokal and Rohlf, 1981, p. 540). Two of the species excluded from the analysis show exceptionally high pup mortality given the number of litters which have been bred in captivity (African wild dog, fennec fox [*Fennceus zerda*]) and one species, the swift fox (*Vulpes velox*) has exceptionally low mortality despite relatively few individuals having been bred in zoos.

For three threatened species which have been bred in captivity extensively, patterns of mortality differ (Figure 20.4; Two-Way ANOVA, effect of species, $F = 5.3$, $P < 0.01$; effect of year, $F = 0.3$, $P > 0.05$; interaction $F = 1.0$, $P > 0.05$). In *Lycaon*, mortality has been relatively constant at around 55%, despite a large breeding effort. In *Chrysocyon*, efforts to improve the breeding programme for this species resulted in declining mortalities in the early 1980s (Ginsberg and Macdonald, 1990), but mortality has risen since. Of these three species, the one which has been bred least, *Speothos*, shows high and inconsistent variation in pup mortality, suggesting that variance may merely be a result of small sample size.

20.3.2 Discussion

The great majority of those species bred in captivity are relatively common. The majority of threatened taxa, or taxa whose status is uncertain, are either not bred in captivity, or are bred in captivity very rarely. Only three threatened species – the African wild dog, the bush dog, and the maned wolf – have been bred with any frequency in IZY reporting institutions. A fourth species, the red wolf, has also been bred relatively frequently, but only in recent years after a major effort brought on by extinction of the species in the wild (Waddell, 1992; Phillips *et al.*, in press).

Reported success in breeding canids is inconsistent among species. A single variable, experience, appears to account for much of the observed

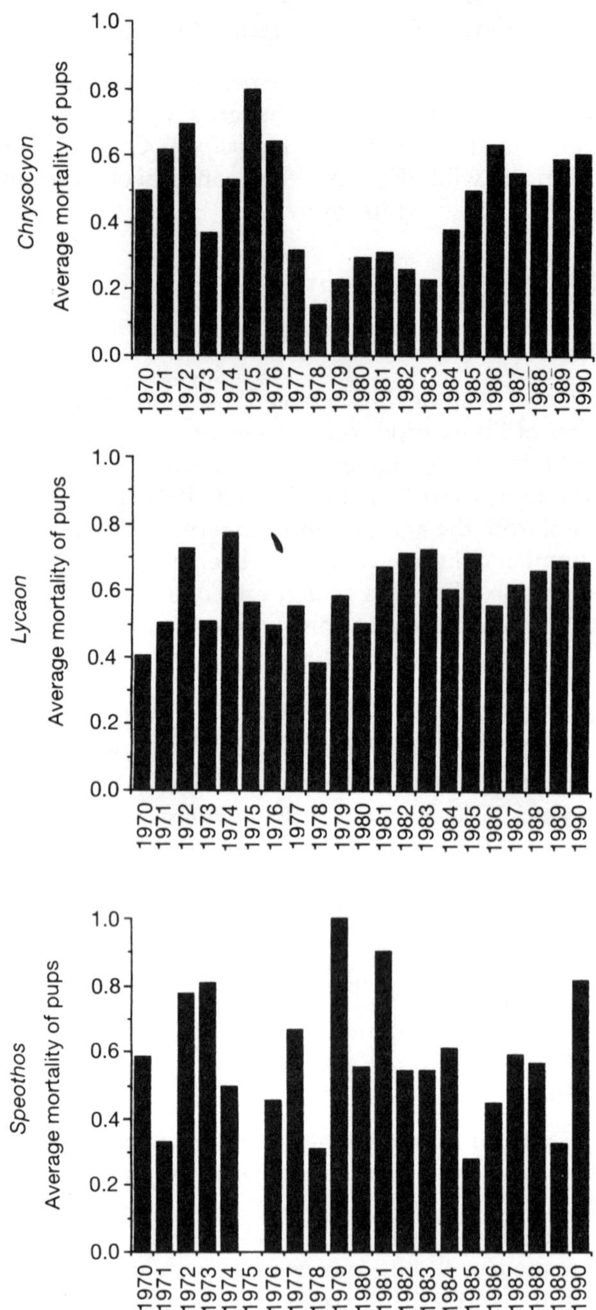

Figure 20.4 Patterns of annual pup mortality in the three threatened canid species bred extensively in captivity.

variation. The data indicate that experience must be gained on a species by species, or perhaps genus by genus, basis. A similar analysis of IZY data through 1985 (Ginsberg and Macdonald, 1990) found nearly identical results, with no significant difference in either the slope or intercept of the regression line and only slight change in the correlation coefficient (this study $r^2 = 0.79$; Ginsberg and Macdonald, 1990, $r^2 = 0.86$).

With improvements in captive breeding techniques and technology, it can be argued that intensive effort could reduce pup mortality without a 10-fold increase in the number of litters bred. Data on red wolf seem to support this hypothesis, while data on maned wolf do not. With fewer than 100 litters of red wolf born in captivity, we would expect a first year pup mortality of approximately 45% (see Figure 20.3). In the years 1989 and 1990, mortality averaged 33% (Smith, 1992), well below the expected. The red wolf, however, is either a hybrid of, or a species very closely related to, the coyote (*Canis latrans*) and the grey wolf (Wayne and Jenks, 1991). Pup mortality of the coyote (39%) is very similar to that of the red wolf, while pup mortality of grey wolves is under 25%, significantly lower than that of the red wolf.

In contrast to red wolf mortality, maned wolf pup mortality declined in the mid-1980s (Ginsberg and Macdonald, 1990) and has inexplicably increased in the late 1980s (Figure 20.4). This is despite intensive efforts at captive breeding of the species (Matern, 1991; Ginsberg and Macdonald, 1990). Even in good years, the captive population can only just sustain itself (e.g. 1990, 38% pup mortality, IZY; 42% studbook-listed mortality, Matern, 1991).

Unfortunately, problems remain even for those threatened taxa bred in captivity on a regular basis. African wild dog pup mortality is persistently high in many collections, although a few collections have had remarkable success in breeding these animals (e.g. reports from Pretoria, South Africa). The eastern and southern specimens of this species are considered to be separate subspecies (Girman *et al.*, in press), yet with a single exception, every African wild dog in captivity is of southern African origin (Brewer and Rhodes, 1992). Those populations in greatest danger of extinction are in eastern and western Africa (Ginsberg and Macdonald, in prep.). Success in breeding bush dogs is extremely variable and has led to several zoos abandoning breeding efforts (Ginsberg and Macdonald, 1990).

For species such as the African wild dog or Asian dhole, which experience high levels of juvenile mortality in captivity, large litter sizes mean that for many of those species already in captivity we can maintain a captive population without resorting to importing specimens from the wild. The effects of such selection, however, are unknown.

20.4 POTENTIAL FOR REINTRODUCTION

20.4.1 Results

When attempting the reintroduction of captive animals, a reasonable number of founder individuals and subsequent supplementation will improve the probability of establishing a self-sustaining population (Stanley Price, 1989; Griffith *et al.*, 1989; Pimm, Jones and Diamond, 1988). The size of captive stocks/meta-populations required for reintroduction programmes or successful captive breeding will vary according to the biology of a species (e.g. social structure, intrinsic rate of growth) and with assumptions one makes concerning starting levels of genetic variation and acceptable levels of loss of variation (heterozygosity) through time (Foose and Ballou, 1988).

Canids have relatively large litter sizes and can be expected to show relatively rapid rates of increase if mortality is controlled. Using a model which estimates the loss of heterozygosity as a function of effective population size, population growth, and generation time, (Ballou, 1989: CAPACITY V. 2.11), and standard estimates appropriate for canids of population growth (10%), founder number (30) and generation time (six years), the effects of social structure (N_e/N), projected loss of diversity and population size were examined (Figure 20.5). For social canids (wild dogs, wolves, dholes) which live in packs where only dominant indi-

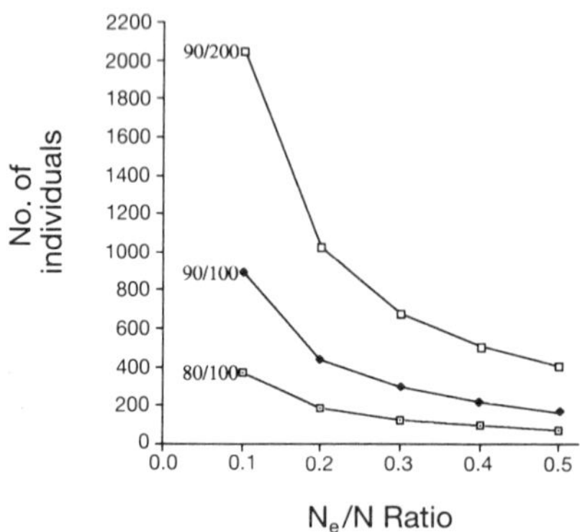

Figure 20.5 The relationship between population size, effective population ratio (N_e/N), and the preservation of genetic diversity (heterozygosity) through time for three scenarios: 80% of heterozygosity preserved after 100 years; 90%/100 years; and 90%/200 years. All populations assume 30 founders, a six-year generation time, and 10% growth rates.

viduals breed, we would expect a reduction in N_e/N, while for more asocial (most jackals, foxes), N_e/N ratios should be around 0.3–0.4 (data from Ginsberg and Macdonald, 1990). Roughly, a population of 200–600 is required to maintain 80–90% of the starting heterozygosity over 100–200 years.

Using the same model, the US Fish and Wildlife Service (1990) argue that to maintain 90% of the starting heterozygosity for 200 years, a red wolf recovery programme will require a meta-population of 540 individuals: 320 captive animals and 220 reintroduced wild animals. Population trends for the species (Smith, 1992; Waddell, 1992), suggest that captive populations are rapidly approaching this goal with over 170 individuals in the present meta-population. Wild populations, although small, have already reached 10% of the required number, with mortality in the first year after reintroduction of approximately 30% (M. Phillips, pers. comm.).

The history and success of a reintroduction for swift fox (*Vulpes velox*) in Canada are reviewed by Carbyn, Armbruster and Mamo (in press). Of particular interest are their findings concerning soft vs. hard release methods, and survivorship of captive-bred vs. wild-caught animals. Their data (Figure 20.6) show that wild-caught, translocated, hard-released animals survive better than hard-released captive-bred animals ($G^2 = 11.9$, $P < 0.001$). At one year following release, animals (captive and wild) which were introduced through soft release had a lower

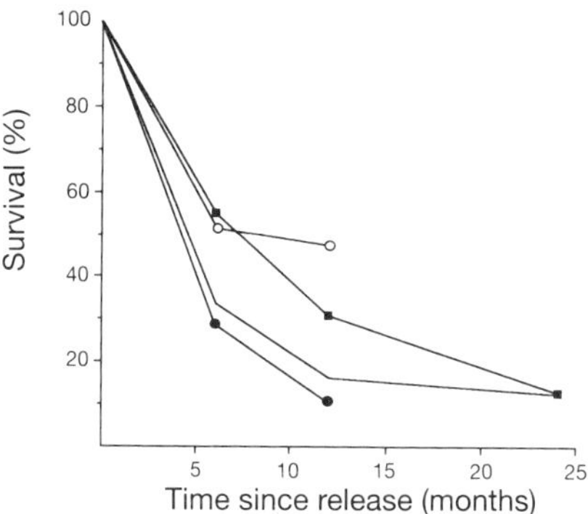

Figure 20.6 Survivorship of released kit foxes as a function of method of release and origin of released individuals. Wild-caught, translocated, hard-released animals survive better than hard-released captive-bred animals ($G^2 = 11.9$, $P < 0.001$). (Data from Carbyn, Armbruster and Mamo, in press.)
Key: ■ soft–all; ○ hard–wild; ● hard–captive; — hard–all.

survivorship (14/45, 31%) than hard-released wild-bred animals (9/19, 47%), but the difference was not significant (G^2 = 2.7, P = 0.09). Two years after release, no difference was observed in the survivorship of animals which were soft vs. hard released (hard: 6/45, 13%; soft, 5/43, 12%).

To investigate the effect of changing the size of starting populations, and to examine the effect of supplementation on population survival, population simulations were made of reintroductions using a packaged population dynamics program (VORTEX V. 4.1, Lacy, 1992). The model allows the user to examine how changing demography, and population size and structure affects the probability of extinction of single or multiple populations. In these simulations, a single population was modelled to determine the probability of extinction in a 100-year period. Each run of the model was simulated 100 times and data reported are for mean extinction probability at each 10-year interval.

Data on adult survivorship used in the model simulation were taken from Carbyn, Armbruster and Mamo (in press) on hard-released, wild-caught translocated animals. Other demographic parameters (e.g. litter size, juvenile survivorship, age at first breeding) were estimated using data from other sources (Ginsberg and Macdonald, 1990; O'Farrell, 1987). The model was run at each of three initial population sizes: 50, 100, and 500 individuals with even sex ratios at release. The model was also run for 50 and 100 individuals with supplementation of the population by 20 individuals per year every year for the first 10 years following initial release. Probability of survival was maximized by assuming that there was no effect of inbreeding and that breeding was density inde-

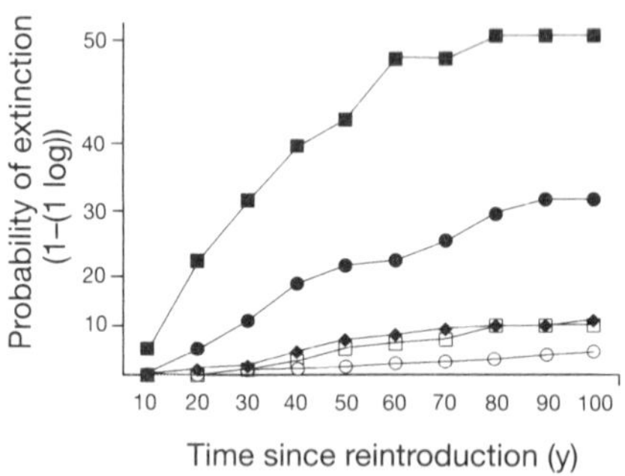

Figure 20.7 Results of a model examining effect of release number and method on probability of population extinction for swift fox. An increase in starting population size greatly reduces the probability of extinction. However, best results are achieved with relatively small initial release numbers (100SP, 50SP) followed by annual supplementation of 20 individuals/year for 10 years. Key: ■ N = 50; □ N = 50SP; ● N = 100; ○ N = 100SP; ◆ N = 500.

pendent. Carrying capacity was set at 1000 individuals. Results of the model simulation can be seen in Figure 20.7.

A single release of relatively few individuals (N = 50 or 100) has an extremely high probability of extinction due to demographic and environmental stochasticity (Figure 20.7), particularly in the first few decades following reintroduction. Increasing the size of the initial release reduces this probability, but even with 500 individuals released, many (10%) of the reintroduction attempts will fail. Supplementation acts to counteract the probability of extinction. There is no difference in extinction probability between a starting population of 50 individuals, supplemented by 20 individuals/year for 10 years and initial reintroduction of 500 individuals. An initial population of 100 individuals supplemented by 20 individuals/year for 10 years has the lowest probability of extinction of all models tested.

20.4.2 Discussion

Translocation of canids has been most successful when it has involved the accidental or intentional release of common species outside their historical range: European red foxes are common in Australia, and racoon dogs (*Nyctereutes procyonoides*) that escaped from fur-farming operations have established populations in eastern Europe (Ginsberg and Macdonald, 1990). Reintroduction of captive-bred African wild dogs has met with little success in Namibia (Scheepers, in prep.), but has been relatively successful in the Umfolozi Game Reserve in South Africa (Ginsberg and Macdonald, 1990). Initial efforts to reintroduce and translocate grey wolves were unsuccessful (Henshaw *et al.*, 1979; Weise *et al.*, 1975), but more recent efforts at translocation of wild animals from farm areas to areas in which wild wolves already exist were successful (Fritts, Paul and Mech, 1985).

The apparent success of the swift fox reintroduction programme in Canada, and the simple model generated using data from this study, suggest that reintroduction of once relatively common species to areas where they have been extirpated can be effective. The study (Carbyn, Armbruster and Mamo, in press) is important because, even for a species which is relatively common in many areas, it clarifies three important points:

(a) a large number of individuals (100–200) is required for such a programme if it is to have hope of long-term success;
(b) first year mortality following release is high (50–80%), higher than reported for other carnivores (Griffith *et al.*, 1989), even if measures are taken to soften the impact of reintroduction;

(c) translocated animals from other wild populations have a higher survivorship than captive-bred animals.

If captive-bred animals are used, first year mortality doubles, thus doubling the number of individuals required for a reintroduction programme to achieve the same effective starting population numbers in the first breeding season.

A study of golden lion tamarins also found higher survivorship of wild-born individuals; however sample sizes were small (Beck *et al.*, 1991: 4/6 wild born survived, 29/85 captive born survived). Furthermore, all tamarins received extensive post-reintroduction support in a variety of ways. Similar post-reintroduction support has been provided in other studies (red wolves: Phillips *et al.*, in press; Phillips, 1990) and has resulted in significantly lower first-year mortality than observed in the swift fox study. Clearly, any means of reducing mortality will improve the probability of success in reintroduction/translocation projects (Beck *et al.*, 1991; Stanley Price, 1989; Griffith *et al.*, 1989).

Data from the population model for swift fox suggest that for smaller, less social canids, somewhere between 200 and 500 individuals will be required to establish each wild population. The model also suggests that smaller initial reintroductions, followed by annual supplementation, will achieve a better result by dampening stochastic processes early in the reintroduction. In a simulation of oryx (*Oryx leucoryx*) populations, Stanley Price (1989) also found that supplementation can greatly improve probability of success of a reintroduction.

By reducing the intial population size of the reintroduction and using supplementation, programme managers can preserve their 'capital' (captive animals) while 'spending the interest' (annual growth). For a reintroduction in a single location, a captive population of 300 growing at 10% a year would provide 100 individuals for an initial 'inoculation' and 20 individuals/year supplementation. If after 10 years supplementation were stopped at the first site, the captive population could then grow to a sufficient size that a second reintroduction project at a different site using the same stock could then be attempted approximately 15–20 years after the beginning of the first project.

When reintroducing social carnivores effective population sizes and per capita rates of increase are smaller, hence larger numbers may be needed to ensure success. The need to acquire hunting skills, and more complex social behaviour, may exacerbate the difficulties observed in the swift fox programme when reintroducing captive-bred stock (e.g. *Lycaon* introductions in Namibia: Scheepers, in prep.). Reduced mortality of wild-caught individuals suggests that translocation from wild populations, rather than reintroduction of captive-bred animals, may be preferable for social canids; captive animals can form the founders for wild populations, as has been shown by the red wolf programme (Phillips *et al.*, in press).

Attempts to reintroduce or translocate endangered canids within their historical ranges are relatively uncommon (Ginsberg and Macdonald, 1990) and have met with little success. Early experiments with wolves (Henshaw *et al.*, 1979) and with wild dogs (Scheepers, in prep.) suggest that reintroduction of social carnivores is far from simple. Integrated meta-population management of an endangered species, such as the red wolf, is clearly possible. The reintroduction programme in the 477 km^2 Alligator River National Wildlife Refuge in North Carolina appears to be initially successful (Phillips and Parker, 1988; Phillips, 1990; Phillips *et al.*, in press) and, despite political opposition, reintroduction in the Great Smoky Mountains has occurred (Phillips *et al.*, in press). The recovery plan (US Fish and Wildlife Service, 1990), however, requires indefinite meta-population management for several reasons:

(a) the plan predicates a greater proportion of the population being held in captivity than in the wild;
(b) protected areas suitable for reintroduction are small and fragmented;
(c) political opposition and potential conflict requires close monitoring of the reintroduced population.

This results in an expensive programme (five-year projected budget of the field component and capital costs of captive breeding facility expansion, US$4.5 million; US Fish and Wildlife Service, 1990).

Reintroduced red wolf populations are unlikely to be self sustaining for two reasons. Firstly, adult mortality in the wild remains high (20–30%). Secondly no wild population is projected to exceed 100 individuals; for example, for a closely related species, the African wild dog, extensive modelling shows that small ($N < 350$) isolated populations have high probabilities of extinction (Mace *et al.*, in prep.) Thirdly, planned reintroduction sites are sufficiently disjunct that dispersal between sites is unlikely. These limitations are acknowledged by planners (US Fish and Wildlife Service, 1990) and reflect political/social realities rather than biological policy. None the less, they indicate the extent to which meta-population management will be required, and the costs of such management, if several relatively large, self-sustaining wild populations cannot be established.

For the red wolf costs are reduced and tractability improved, because all individuals are held in a single range-state country. For species which may require cross-border cooperation, and trans-continental meta-population management, costs and complications will be greater (Dixon, Chapter 10). Clearly, in regions where close management is difficult, or the expense is unlikely to be met, the only hope for large social canids is to establish and/or protect multiple wild populations. The potential for exchange among these populations by translocation of wild-born animals, whether to augment or re-establish populations which have been extirpated, needs to be investigated.

20.5 CONCLUSIONS

In the introduction, I posed three questions. The first question, What is our success in breeding endangered canids? has a simple answer. Our success in breeding canids is variable with species, with most of the variation being explained by the effort the zoo community has made in breeding any particular species. A coordinated effort may improve breeding success of species closely related to those with which we have extensive experience in captive breeding (e.g. red wolf, a close relative or hybrid of the grey wolf and coyote: Wayne and Jenks, 1991) or may not result in improved survivorship of young (maned wolf).

The answer to the second question, Are species of concern held in captivity and are they breeding? is a qualified 'No'. If our goal is to hold a collection of a particular taxon large enough to be self-sustaining and to provide adequate numbers for even a single reintroduction project, then we are clearly not meeting this goal for most threatened canids. This is particularly true if we expand the 'Threatened' category to include rare and unknown species such as the Sechuran zorro (*Dusicyon sechurae*), the pale fox (*Vulpes pallida*) or Blanford's fox (*Vulpes cana*).

A few species of concern are held in captive collections in numbers adequate to attempt reintroduction. As experience with the red wolf has shown, given time and sufficient funds, coordinated captive breeding/reintroduction programmes can be established. In most cases either a species is not held in captivity (e.g. island grey fox, Ethiopian wolf), the numbers in captivity, and presumably the numbers of founders, are too small (e.g. dhole, various rare fox species), or the sub-taxa of greatest conservation concern are not held in captive collections (African wild dog). Rectifying these deficiencies will require massive expansion or reallocation of cage space (Foose and Seal, 1992), and acquisition of breeding animals from threatened wild populations.

My third question, What is our success in reintroducing captive bred canids to the wild? is most difficult to answer. Both projects discussed (red wolf, swift fox) show that given adequate funds, appropriate habitat and adequate numbers of animals to reintroduce, success can be achieved. Preliminary confirmation of this success, however, takes decades not years (Beck *et al.*, Chapter 13). Translocation of wild caught canids appears to have a higher rate of success than reintroduction of captive-bred animals, suggesting that if such animals exist this may be a better way to establish new populations.

Conflicts between canids and people makes any reintroduction difficult. Reintroduction of grey wolves is controversial (Wilcove, 1987), even when naturally occurring populations are close to extinction (e.g. Italy: Boitani, pers. comm.). Where populations of a threatened canid are locally relatively common, the problems of protection are compounded (Fanshawe, Frame and Ginsberg, 1991; Townsend, 1988). Such antipathy can be a serious hurdle to reintroductions (Phillips *et al.*, in press).

Captive breeding and captive collections are clearly valuable tools for education. They also provide a temporary safe haven from human persecution, disease or genetic introgression through cross-species breeding (e.g. coyotes and red wolves, Phillips and Parker, 1988) or domestic dog breeds (dingo, Newsome and Corbett, 1985; grey wolf, Boitani, pers. comm.). Our relatively poor track record in the captive breeding of threatened canids, the uncertainty and political implications of removing threatened species from wild source populations, and the complexities and costs of reintroduction of captive-bred animals all suggest that we are a long way from being able to rely on captive breeding and reintroduction to reduce the extinction threat for most species.

ACKNOWLEDGEMENTS

Dr Lu Carbyn and his collaborators on the swift fox reintroduction programme in Canada, and various members of the red wolf reintroduction programme team including G. Henry, M. Phillips, W. Waddell, kindly provided unpublished manuscripts, reports, and material in press, Sally Gibbins assisted in data entry. G. Mace, M. Phillips, and L. Carbyn all provided comments which greatly improved the quality of the manuscript. Any errors in interpretation or presentation of data are entirely the fault of the author.

REFERENCES

Anderson, J.L. (1986) Restoring a wilderness: the reintroduction of wildlife to an African national park. *Int. Zoo Yb*, **24/25**, 192–9.

Ballou, J.D. (1989) *CAPACITY. Version 2.11*, National Zoological Park, Washington.

Beck, B.B., Kleiman, D.G., Dietz, J.M. *et al.* (1991) Losses and reproduction in reintroduced golden lion tamarins *Leontopithecus rosalia. Dodo, Journal of the Jersey Wildlife Preservation Trust*, **27**, 50–61.

Boer, L.E.M. de (1992) Current status of captive breeding programmes, in *Biotechnology and the Conservation of Genetic Diversity, Symposia Zoological Society of London No. 64*, (eds H.D.M. Moore, W.V. Holt and G.M. Mace), Clarendon Press, Oxford, pp. 5–16.

Brewer, B. and Rhodes, S. (1992) Lycaon pictus *African wild dog International Studbook*, Chicago Zoological Society, Chicago.

Carbyn, L.N., Armbruster, H.J. and Mamo, C. (in press) A review of the results of the swift fox reintroduction program in Canada, in *Restoration of Endangered Species* (eds M.L. Bowles and C.J. Whelan).

Conway, W.G. (1986) The practical difficulties and financial implications of endangered species breeding programmes. *Int. Zoo Yb*, **24/25**, 210–19.

Fanshawe, J.H., Frame, L.H. and Ginsberg, J.R. (1991) The wild dog – Africa's vanishing carnivore. *Oryx*, **25**, 137–46.

Flesness, N.R. and Mace, G.M. (1988) Population databases and zoological conservation. *Int. Zoo Yb*, **27**, 42–9.

Foose, T.J. and Ballou, J.D. (1988) Management of small populations. *Int. Zoo Yb*, **27**, 26–41.

Foose, T.J. and Seal, U.S. (1992) *Conservation Assessment and Management Plans, Global Captive Action Plans Summary Reports*, IUCN-CBSG, Minnesotta, USA.

Fritts, S.H., Paul, W.J. and Mech, L.D. (1985) Can relocated wolves survive? *Wildlife Society Bulletin*, **13**, 459–63.

Ginsberg, J.R. and Macdonald, D.W. (eds) (in prep.) *An Action Plan for the Conservation of the African Wild Dog*, Lycaon pictus, IUCN, Gland, Switzerland.

Ginsberg, J.R. and Macdonald, D.W. (1990) *The Canid Action Plan: An Action Plan for the Conservation of the World's Canids*, IUCN, Gland, Switzerland.

Girman, D.J., Kat, P.W., Mills, G. *et al.* (in press) A genetic and morphological analysis of the African wild dog (*Lycaon pictus*). *J. Heredity*.

Griffith, B., Scott, J.M., Carpenter, J.W. and C. Reed (1989) Translocation as a species conservation tool: status and strategy. *Science*, **245**, 477–80.

Harvey, P.H. and Pagel, M.D. (1991) *The Comparative Method in Evolutionary Biology*, Oxford University Press, Oxford.

Henshaw, R.E., Lockwood, R. Shideler, R. and Stephenson, R.D. (1979) Experimental release of captive wolves, in *The Behavior and Ecology of Wolves* (ed. E. Klinghammer), Garland STPM Press, New York, pp 319–45.

IUCN (1987) *The IUCN Policy Statement on Captive Breeding*, IUCN, Gland, Switzerland.

Lacy, R. (1992) *VORTEX Version 4.1*, Chicago Zoological Society, Chicago.

Mace, G., Ginsberg, J., Mills, G. *et al.* (in prep.) A population viability assessment of the African wild dog, in *An Action Plan for the Conservation of the African Painted Wolf*, Lycaon pictus (eds J.R. Ginsberg and D.W. Macdonald), IUCN, Gland, Switzerland.

Magquire, L.A. and Lacy, R.C. (1990) Allocating scarce resources for conservation of endangered sub-species: partitioning zoo space for tigers. *Conservation Biology*, **4**, 157–6.

Matern, B. (1991) *International Register and Studbook for the Maned Wolf* (Chrysocyon brachyurus) *1990*, Frankfurt Zoological Society, Frankfurt, Germany.

Newsome, A.E. and Corbett, L.K. (1985) The identity of the Dingo III: The incidence of Dingoes, dogs and hybrids in remote and unsettled regions of Australia. *Australian J. Zool.*, **33**, 363–75.

Nowak, M., Baker, J.A., Obbard, M.E. and Malloch, B. (1987) *Wild Furbearer Management and Conservation in North America*, Ministry of Natural Resources, Ontario.

O'Farrell, T.P. (1987) Kit Fox, in *Wild Furbearer Management and Conservation in North America* (eds M. Nowak, J.A. Baker, M.E. Obbard and B. Malloch), Ministry of Natural Resources, Ontario.

Olney, P. and Ellis, P.(1992) Species of wild animals bred in captivity during 1990 and multiple generation captive births. *Int. Zoo Yb*, **31**, 263–373.

Pimm, S.L., Jones, H.L. and Diamond, J.D. (1988) On the risks of extinction. *American Naturalist*, **132**, 757–85.

Phillips, M. (1990) The red wolf: recovery of an endangered species. *Endangered Species Update*, **8**, 79–81.

Phillips, M.K. and Parker, W.T. (1988) Red wolf recovery: progress report. *Conservation Biology*, **2**, 139–44.

Phillips, M.K., Smith, R., Henry, V.G. and Lucash, C. (in press) Red Wolf Recovery Program, in *Proceedings of the Second International Wolf Symposium* (ed. L. Carbyn), Canadian Wildlife Service, Ottawa.

Ralls, K., Ballou, J.D. and Templeton, A. (1988) Estimates of lethal equivalents and the costs of inbreeding in mammals. *Conservation Biology*, **2**, 185–93.

Ralls, K., Harvey, P.H. and Lyles, A.M. (1986) Inbreeding in natural populations of birds and mammals, in *Conservation Biology: The Science of Scarcity and Diversity* (ed. M. Soulé), Sinauer Associates, Sunderland, Mass.

Scheepers, L. (in prep). Attempts to reintroduce the African wild dog: a review, in Lycaon: *An Action Plan for the Conservation of the African Wild Dog* (eds J.R. Ginsberg and D.W. Macdonald), IUCN, Gland, Switzerland.

Seal, U.S. (1986) Goals of captive propagation programmes for the conservation of endangered species. *Int. Zoo Yb*, **24/25**, 174–9.

Seal, U.S. (1989) Introduction, in *Conservation Biology and the Black-footed Ferret*. (eds U.S. Seal, E.T. Thorne, M.A. Bogan and S.H. Anderson), Yale University Press, New Haven, pp. vi–xvii.

Skinner, J.D. (1989) Game ranching in Southern Africa, in *Wildlife Production Systems and Economic Utilization of Wild Ungulates* (eds R.J. Hudson, K.R. Drew and L.M. Baskin), Cambridge University Press, Cambridge, UK.

Smith, R. (1992) *1991 Red Wolf International Studbook*, Point Defiance Zoo, Washington, USA.

Stanley Price, M.R. (1989) *Animal Reintroductions: The Arabian Oryx in Oman*, Cambridge University Press, Cambridge, UK.

Sokal, R.R. and Rohlf, F.J. (1981) *Biometry*, 2nd edn, W.H. Freeman & Co, San Francisco.

Townsend, M.H.C. (1988) Glaring errors in wild dog story. *Zimbabwe Wildlife*, **53**, 19.

US Fish and Wildlife Service (1990) *Red Wolf Recovery Plan*, US Fish and Wildlife Service, Atlanta GA.

Waddell, W. (1992) *Red Wolf Newsletter*, **4**(1).

Wayne, R.K., and Jenks, S.N. (1991) Mitochondrial DNA analysis implying extensive hybridization of the endangered red wolf *Canis rufus*. *Nature*, **351**, 565–8.

Weise, T.F., Robinson, W.L., Hook, R.A. and Mech, L.D. (1975) An experimental translocation of the eastern timber wolf. *Audubon Conservation Report*, **5**, 1–28.

Wilcove, D.S. (1987) Recall to the wild: wolf reintroductions in Europe and North America. *Trends in Ecology and Evolution*, **2**, 146–7.

Wilson, E.O. (1988) The current state of biological diversity, in *Biodiversity* (ed. E. O. Wilson), National Academy Press, Washinton, DC.

21

The recovery of the angonoka (*Geochelone yniphora*) – an integrated approach to species conservation

L. Durrell
Jersey Wildlife Preservation Trust, Channel Islands
R. Rakotonindrina
Association Nationale pour la Gestion des Aires Protégées, Madagascar
D. Reid
Jersey Wildlife Preservation Trust, Channel Islands
and
J. Durbin
Durrell Institute of Conservation and Ecology, University of Kent, UK.

21.1 INTRODUCTION

No other country in the world has generated as much recent interest from conservation and development agencies as Madagascar. The island has few equals in terms of a people with a unique culture and a rich, diverse bio-resource on the one hand and extreme poverty and rapid environmental destruction on the other. Among the flora, endemism at the species level may be as high as 86%, and among the fauna, several

Creative Conservation: Interactive management of wild and captive animals.
Edited by P.J.S. Olney, G.M. Mace and A.T.C. Feistner.
Published in 1994 by Chapman & Hall, London.
ISBN 0 412 49570 8

groups rise to nearly 100%, e.g. mammals (except bats), reptiles and amphibians (IUCN/UNEP/WWF, 1987). But there is probably less than a fifth of the native woody plant cover remaining (Harcourt and Thornback, 1990), severe soil degradation is evident on three-quarters of the land surface (Le Bourdiec, 1972) and annual human population growth is higher than 3% (Kent and Haub, 1989).

Not surprisingly the faunal extinction rate in Madagascar within historical times is one of the highest in the world, and many more species are believed to be at risk at present. One of these is a tortoise, the angonoka (*Geochelone yniphora*). It is restricted to a small area in the Baly Bay region on the north-west coast, and numbers are less than 400 in the wild, according to the most recent estimate (Curl *et al.*, 1985). Its decline is linked primarily to the loss of its habitat, the dry forest/bamboo scrub mosaic that has given way to fire-created palm savanna, although collecting of adult tortoises by people and predation on eggs and young by the introduced bush pig (*Potamochoerus larvatus*) of Africa, are thought to be contributing factors (Durrell *et al.*, 1989). Although protected by Malagasy law from hunting and collecting (Curl *et al.*, 1985) and by the Convention on International Trade in Endangered Species (CITES) from international trade, the angonoka is the only one of the world's half dozen most threatened land tortoises not found in a protected area or a nature reserve (Juvik, 1991).

In 1985 the Tortoise Specialist Group of the World Conservation Union (IUCN) asked the Jersey Wildlife Preservation Trust (JWPT) to initiate a species recovery programme for the angonoka through JWPT's formal conservation agreement with the government of Madagascar. In 1986 this became the Angonoka Project, operated jointly by the Malagasy Waters and Forests Authority and the JWPT, and funded by a variety of international conservation organizations, notably the World Wide Fund For Nature. Of the 31 people who have worked on the project, 22 have been Malagasy.

The Angonoka Project has shown that it is possible to work simultaneously towards the two main objectives of a species recovery programme: to reverse the decline in absolute numbers and to maintain self-sustaining wild populations. By adopting an integrated strategy from the outset, with diverse project disciplines combining with local interests to produce mutual benefit to tortoises and to humans, both aims are served. Indeed, the first objective will generate interest and funding to achieve the second.

21.2 BREEDING THE ANGONOKA IN CAPTIVITY

The first stage of the project was to establish a captive breeding programme. In 1986 eight angonoka, already held in captivity elsewhere in

Madagascar, were brought to the Forestry Station at Ampijoroa, about 150 km from Baly Bay and in the same phytogeographical domain (Curl, 1986). With the exception of one hatching at the Honolulu Zoo (Juvik, Meier and McKeown, 1991), no angonoka had bred before in captivity, but by the end of 1991 the Ampijoroa animals had produced 62 surviving young. Between 1987 and 1991 there were 132 apparently viable eggs laid and 66 hatchings, giving a hatch success of 50% and a survival success of 95%, the figures improving each season thanks to techniques developed at Ampijoroa (Reid, 1991a, 1991b; Reid, Durrell and Rakotobearison, 1989). Research on health, diet, breeding performance and hatchling development is on-going and is vital to the continued success of the breeding programme and to the formulation of protocols for the eventual release programme.

21.3 RESEARCH ON WILD ANGONOKA

The breeding station has served as home base for the planning and execution of eight field trips to the Baly Bay region to assess the distribution and status of the angonoka and its habitat and to begin studies on its ecology and behaviour (Curl, 1986; Reid, 1987, 1989, 1990a, 1990b, 1991c, 1991d; J. Juvik and R. Kiester, pers.comm.).

The Baly Bay region is characterized by a long dry season and a climax vegetation of lowland, dry, deciduous tropical forest, said to represent the most endangered of tropical ecosystems (Janzen, 1988). Biodiversity here has been little studied, but is likely to be high, given the high levels in other western Malagasy forests (Langrand, 1990). The forest is interspersed with bamboo and scrub bush, which is typical angonoka habitat, and palm savanna.

The project is developing a regional habitat database utilizing both ground surveys and remote sensing techniques, i.e. multi-spectral satellite imagery, and Space Shuttle and low altitude colour-infra-red photography, for which the required facilities have been made available by the US Forest Service. The database, which is being designed to integrate with other geographical information systems (GISs) currently under development for Madagascar, will provide high-resolution identification and discrimination of actual and potential angonoka habitat, and facilitate future detailed monitoring of regional environmental change (Juvik, 1991). Data are now available to analyse the landscape dynamics of the region over the last 40 years (R. Kiester, pers.comm.).

Low altitude overflights in January 1992 revealed degradation of most vegetation types to the east of Baly Bay, except on Cape Sada, and large areas of untouched forest interspersed with the bamboo scrub habitat favoured by the angonoka in the Belambo area, which is to the west of

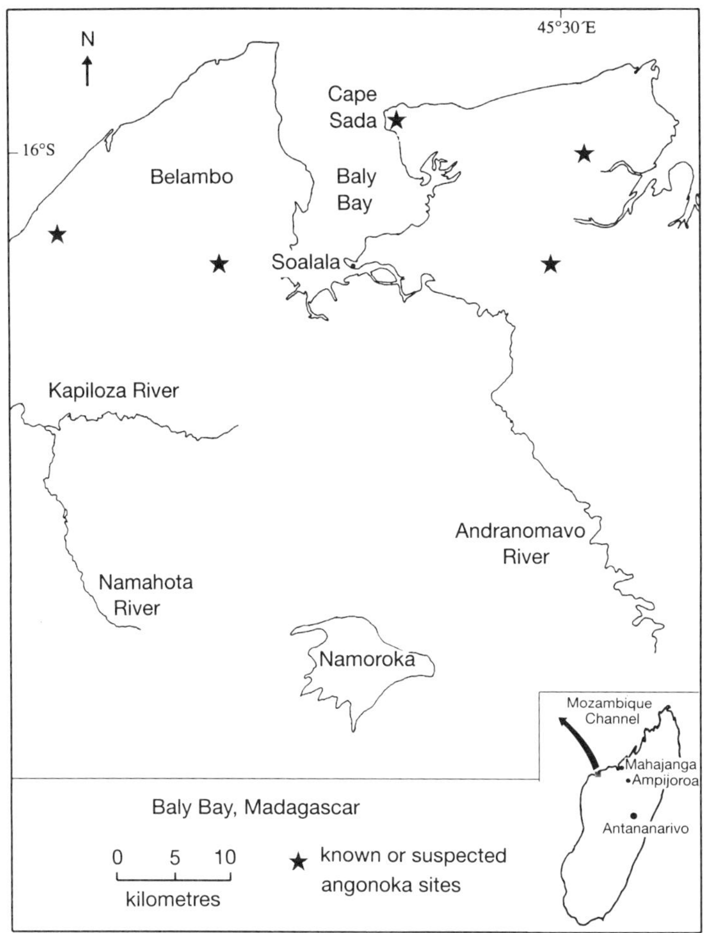

Figure 21.1 The angonoka sites around Baly Bay, Madagascar.

Baly Bay (Figure 21.1). At Namoroka, about 30 km south of the Bay, there is largely intact forest, which lies on spectacular limestone karst formations and is dominated by xerophytes. Extensive mangrove forests occur in areas around the coasts (J. Juvik and R. Kiester, pers.comm.)

The field work on the ground to date has revealed mixed fortunes with respect to the angonoka's distribution: the probable demise of one population, but the discovery of at least 35 km^2 of angonoka habitat in the far west. Heavy bush pig activity in the region has been reaffirmed, but with no direct evidence of predation on angonoka. On Cape Sada itself, the angonoka population is estimated at 30–36 individuals, skewing towards females, and there is about 150 ha of suitable habitat which could support about 300 angonoka. In January 1992 transects were cut on Cape Sada and radio transmitters were attached to eight tortoises, which will enable later researchers to gather quantitative data on angonoka seasonal movement and habitat use. A two-year field study in the Baly Bay region, scheduled to begin in 1993, will amass more detailed information on angonoka ecology and behaviour, and will assess the degree and the relative importance of the various factors which are believed to threaten the species.

21.4 SOCIAL STUDIES

Information on angonoka biology obtained at Ampijoroa and around Baly Bay is obviously vital to the planning of protective measures for wild angonoka, but equally important are efforts to bring the people of the region into the planning process.

Since 1990 three trips have been undertaken to study socio-economics in the area to the east of the Bay, with particular reference to the role of local people in the successful maintenance of a potential protected area for the angonoka (Durbin and Rakotoniaina, 1990, 1991).

The only town and administrative centre of the region is Soalala, lying on the south-eastern rim of Baly Bay and with a population of less than 1000. Most people live in small, dispersed and mobile communities, health and education services are minimal and there is little immigration from other parts of Madagascar. Near the coasts people depend primarily on the sea, and fishing provides a modest income. People living inland are more dependent on rice and cattle. Everyone plants other crops for home consumption, but on a small scale due to particularly heavy depredations in the region by the bush pigs.

As sparsely populated as the region is, environmental degradation is well advanced in some areas. People fell trees to build houses and boats and to fence their gardens against bush pigs. They burn grasslands frequently in the erroneous belief that pasture quality will be improved, but the fires often rage out of control. Several forest patches in the region have been totally destroyed by fire in the last 10 years. As hardwood forests become degraded or disappear, people are turning to mangrove for construction material, although it is not as durable. Recent discussions between villagers and project personnel indicate local concern about the effects of these activities in the long term. The villagers remark on the poor quality of pasture after burning, on the distance to the forests and the effort it now takes to gather traditional forest products, the lack of surface water and the sanding up of rice fields and lakes. Deforestation inland, resulting in increased silt loads in the rivers, and disturbance among the mangroves themselves are likely to be having a negative effect on the richness of the coastal ecosystems, including the nascent fishing industry.

The local people do not eat the angonoka, but until recently, national law notwithstanding, it was collected as a pet for aesthetic reasons and in the belief that it wards off sickness in humans and domestic poultry. The tortoise is part of the culture of the region, and people look upon its decline with regret. Therefore the education aspect of the project has been well received.

21.5 EDUCATION AND TRAINING

Visits by local schools and other groups to the breeding station are encouraged, and an educational poster has been produced for distribution around Ampijoroa and Baly Bay.

In 1991 two environmental awareness campaigns were mounted in the area to the east of the Bay, the ground having been prepared by the socio-economic studies. They were undertaken jointly by the WWF Education Programme and the Angonoka Project. Classes centred around the angonoka as a symbol of environmental health and a better quality of life. They revealed that the local people:

(a) are genuinely concerned for the angonoka, once they know that it is unique to the region;
(b) understand the benefits of rational management, especially the control of fire and bush pigs, and are interested in learning and implementing management techniques;
(c) want to improve both subsistence and market agriculture, with the specific aims of obtaining better health and education services;

(Rajafetra and Rabodomalala, 1991; Rabodomalala, Rajafetra and Andrianalinirina, 1991.) An immediate result of the local education effort was that people donated their pets to the Ampijoroa breeding station, bringing the number of adult tortoises there to 20.

As for professional training, the Malagasy involved in angonoka husbandry and field biology have received instruction on site. In 1989 the Malagasy head of the breeding station, Germain Rakotobearison, came to Jersey for three months to study tortoise husbandry at the International Training Centre for the Conservation and Captive Breeding of Endangered Species.

21.6 DISCUSSION AND PLANS FOR THE FUTURE

The Angonoka Project employed a multi-disciplinary strategy early in its development, and this has produced, for modest financial outlay, the broad information base required to make intelligent plans to safeguard the species from extinction. The Project has, however, evolved into something bigger than an effort to save a tortoise. Among the local people and the project participants, it has inspired a regional vision for development in north-west Madagascar, integrally based on a programme of sound environmental management and protection of biodiversity, in line with the Environmental Action Plan recently adopted by the government of Madagascar. Thus the Angonoka Project is evolving into the Baly Bay Project.

The JWPT will continue the work on the angonoka and its habitat, and other agencies will be brought in to cover the wider aspects of the project, such as natural resource management, e.g. a local bush pig products marketing scheme.

Cape Sada will be developed as a 'small management unit' for the study and protection of wild angonoka and as a site for the introduction of captive-bred and, possibly, other angonoka from non-viable wild populations. Perhaps Cape Sada will warrant legally protected status, or, more desirably, will require only local agreement for its protection from fire, cattle grazing and other human use.

Meanwhile, work in the Belambo area will begin and develop much as it has for the eastern side of the bay. The global importance of the lowland, dry, deciduous forests merits the establishment of a field station, which will accommodate national and foreign scientists and their students. All activities will be done with a view to assessing the suitability of Belambo as a prescribed area for the protection and maintenance of biodiversity and natural resources. Although the Namoroka area is already legally protected, virtually nothing is known of its biodiversity and natural resources, and therefore Namoroka is included in the plans for the future.

What has been learned from the Angonoka Project can be applied to many other species recovery programmes. The lessons are these:

(a) from the outset, involve the people of the region concerned;
(b) assemble a multi-disciplinary team, composed of nationals and non-nationals;
(c) find the positive ecological connection between the species concerned and the lives of the people, disseminate this information and retain this focus.

One of the most popular proposals for the Baly Bay Project is that the town of Soalala, as well as serving as the coordination centre for the project, will become the site of a second breeding unit for angonoka, working closely with the well-established breeding centre at Ampijoroa, and a holding facility for tortoises destined for release to the wild. Thus the project as a whole will come full circle and remind participants and visitors alike of the origins of the effort to improve the quality of life in north-west Madagascar.

ACKNOWLEDGEMENTS

The species recovery programme for the angonoka has been made possible by the support and cooperation of the Government of Madagascar, particularly the Ministry of Animal Production and Waters and Forests (now the Ministry of Waters and Forests). In spite of the enormous difficulties facing it today as one of the world's poorest developing countries, Madagascar has adopted a far-sighted Environmental Action Plan and is in the process of implementing it. The authors of this paper and other participants and supporters of the Angonoka Project wish to record here their gratitude to the Government of Madagascar and present their best wishes for the future.

The Angonoka Project has received help from numerous institutions worldwide, to whom sincere thanks are offered: funding has been provided by Arcadia Foundation (USA), Beckett Foundation (Denmark), Conservation International (USA), Herpetological Conservation Trust (UK), Jersey Wildlife Preservation Trust, New York Zoological Society, People's Trust for Endangered Species (UK), Wildlife Preservation Trust International (USA), World Wide Fund For Nature-International and World Wildlife Fund-US; goods and/or services have been provided by Air Madagascar, British Airways, Durrell Institute of Conservation and Ecology, Honda Company, IUCN/SSC Tortoise and Freshwater Turtle Specialist Group, Marine Fisheries Service (Madagascar), Shell Exploration and Development (Madagascar), SOTEMA (Madagascar), Toyota Motor Corporation, Turtle Graphics, WWF Education Programme (Madagascar), WWF Protected Areas Programme (Madagascar) and the communities of Soalala, Marotia and Antanandava (Madagascar), as well as by JWPT and the Malagasy Waters and Forests Authority.

The authors are most grateful to Phillip Coffey and Karen McDonnell for preparing the map of the Baly Bay region.

REFERENCES

Curl, D.A. (1986) A recovery programme for the Madagascan Plowshare tortoise *Astrochelys* (= *Geochelone*) *yniphora*. *Dodo*, **23**, 68–79.

Curl, D.A., Scoones, I.C., Guy, M.K. and Rakotoarisoa, G. (1985) The Madagascan tortoise *Geochelone yniphora*: current status and distribution. *Biological Conservation*, **34**, 35–54.

Durbin, J. and Rakotoniaina, L.J. (1990) Project Angonoka: the influence of local people on the recommendations for a reserve in the Soalala region. Unpublished report to Jersey Wildlife Preservation Trust 19/11/90.

Durbin, J. and Rakotoniaina, L.J. (1991) Project Angonoka: local people and conservation in the Soalala region. Unpublished report to Jersey Wildlife Preservation Trust 5/9/91.

Durrell, L., Groombridge, B., Tonge, S. and Bloxam, Q. (1989) *Geochelone yniphora* ploughshare tortoise, plowshare tortoise, angulated tortoise, angonoka, in *The Conservation Biology of Tortoises* (eds I.R. Swingland and M.W. Klemens), Occasional Papers of the IUCN Species Survival Commission No. 5, IUCN, Gland, pp. 99–102.

Harcourt, C. and Thornback, J. (1990) *Lemurs of Madagaascar and the Comores. The IUCN Red Data Book,* IUCN Gland, Switzerland and Cambridge, UK.

IUCN/UNEP/WWF (1987) *Madagascar, An Environmental Profile* (ed. M.D. Jenkins), IUCN, Gland, Switzerland and Cambridge, UK.

Janzen, D.H. (1988) Tropical dry forests: the most endangered tropical ecosystem, in *Biodiversity* (ed. E.O. Wilson), National Academy Press, Washington, DC, pp. 130–7.

Juvik, J.O. (1991) Conservation strategies for fragmented dry forest ecosytems in western Madagascar, with special reference to protection of *Geochelone yniphora* habitat in the Baly Bay region. Unpublished project proposal to IUCN/SSC Tortoise and Freshwater Turtle Specialist Group 5/91.

Juvik, J.O., Meier, D.E. and McKeown, S. (1991) Captive husbandry and conservation of the Madagascar Ploughshare tortoise *Geochelone yniphora,* in *Proceedings of the First International Symposium on Turtles and Tortoises: Conservation and Captive Husbandry* (eds K.R. Beaman, F. Caporaso, S. McKeown and M.D. Graff), Chapman University, California, pp. 127–37.

Kent, M.M. and Haub, C. (1989) *World Populations Data Sheet,* Population Reference Bureau, Washington, DC.

Langrand, O. (1990) *Guide to the Birds of Madagascar,* Yale University Press, New Haven and London.

Le Bourdiec, P. (1972) Acceleraated erosion and soil degradation, in *Biogeography and Ecology of Madagascar* (eds. R. Battistini and G. Richard-Vindard), Dr W. Junk B. V. Publishers, The Hague, pp. 227–259.

Rabodomalala, A., Rajafetra, V. and Andrianalinirina, Y. (1991) Rapport sur la campagne de sensibilisation pour la préservation de l'angonoka et la protection de l'environnement à Tanandava et Marotia (Fivondronana de Soalala). Unpublished report to World Wide Fund For Nature, 4/11/91.

Rajafetra, V. and Rabodomalala, A. (1991) Rapport du séminaire de sensibilisation sur la protection de l'environnement et la sauvegarde de l'angonoka. Unpublished report to World Wide Fund For Nature, 3/5/91.

Reid, D. (1987) Project report July 1987. Unpublished report to Jersey Wildlife Preservation Trust.

Reid, D. (1989) Field report. Unpublished report to Jersey Wildlife Preservation Trust, 25/1/89.

Reid, D. (1990a) Field report. Unpublished report to Jersey Wildlife Preservation Trust, 15/2/90.

Reid, D. (1990b) Field report. Unpublished report to Jersey Wildlife Preservation Trust, 13/9/90.

Reid, D. (1991a) Project report January 1990 – June 1991. Unpublished report to Jersey Wildlife Preservation Trust.

Reid, D. (1991b) Project report December 1991. Unpublished report to Jersey Wildlife Preservation Trust.

Reid, D. (1991c) Field report. Unpublished report to Jersey Wildlife Preservation Trust, 5/2/91.

Reid, D. (1991d) Field report. Unpublished report to Jersey Wildlife Preservation Trust, 1/11/91.

Reid, D. Durrell, L. and Rakotobearison, G. (1989) The captive breeding project for the angonoka *Geochelone yniphora* in Madagascar. *Dodo*, **26**, 34–48.

22

Is the Hawaiian goose (*Branta sandvicensis*) saved from extinction?

J.M. Black
The Wildfowl and Wetlands Trust, Slimbridge, UK
and
P.C. Banko[a]
Wildlife Science Group, College of Forest Resources, University of Washington, Seattle, USA.

22.1 INTRODUCTION

In this paper we answer the question in the title with a negative, not yet, not in the past or at current levels of management. We provide evidence that past efforts to save the Hawaiian goose (*Branta sandvicensis*), commonly known as nene, have effectively prolonged the extinction process, but potent limiting factors in its environment are still active and inhibiting recovery. As early as 1864, naturalists warned that the nene was on the brink of extinction (Baldwin, 1945) but, through the intensive rearing and release programmes some 500 nene are currently living in the wild on three of the larger Hawaiian islands, Hawaii, Maui and Kauai (Black *et al.* 1991). When restocking began in 1960, fewer than

[a]Present affiliation: US Fish and Wildlife Service, Patuxent Wildlife Research Center, Maryland, USA.

Creative Conservation: Interactive management of wild and captive animals.
Edited by P.J.S. Olney, G.M. Mace and A.T.C. Feistner.
Published in 1994 by Chapman & Hall, London.
ISBN 0 412 49570 8

30 nene remained only on the island of Hawaii. In 1962, the first set of nene was released on Maui and, in 1982, a hurricane liberated 12 captive nene from a private collection on Kauai.

In the early phases of the recovery programme, managers removed the main limiting factor which affected the demise of this most tame of waterfowl by legislating a hunting ban, and setting aside several hundred acres of volcanic montane scrubland as sanctuaries (see Kear and Berger, 1980). Stone *et al.* (1983), Morin and Walker (1986) and Hoshide, Price and Katahira (1990) outlined a further set of potential limiting factors which may explain the low productivity and survival of the captive-reared nene that have been released; over 2100 individuals have been liberated between 1960 and 1990 (Black *et al.*, 1991). The potential problems include inbreeding depression, loss of adaptive skills, disease, poaching, road kills, dietary deficiencies and predation from introduced mammals. We tested the prevalence of a number of these factors by conducting a natural-style experiment where we monitored the breeding success of individuals (with the same genetic, behavioural and captive backgrounds as the now wild-type birds) in large enclosures that were erected in habitats where others have been released. We predicted that if the birds in the enclosures bred with the same or better success as those in the wild, then genetics and appropriate behaviour may not be limiting; rather, diet and/or predation would be. The enclosed areas kept predators out and we lessened potential feeding deficiencies by adding readily accessible food. To identify further the area which needs most urgent management action, we recorded all occurrences of predation throughout the study period.

Using demographic data, including numbers of females that hatch eggs, the proportion of successful broods and mortality rates, we modelled the likelihood of population growth or decline in three scenarios for the three sub-populations of nene.

(a) The status quo scenario – data from the released wild birds and without further supplementations from captive stocks.
(b) A repeat effort scenario – data from the wild birds (as above) but with a large-scale release programme mirroring the effort of the previous 30 years.
(c) An optimal management scenario – data from the enclosure experiment and no further supplementations.

22.2 METHODS

22.2.1 Study sites

Extending from Kauai in the north to Hawaii in the south, the islands decrease in age and increase in size, elevation, habitat availability and habitat heterogeneity (Figure 22.1). The Hawaiian goose (Figure 22.2) on Kauai occupies rich coastal marshes and lowland pastures which are irrigated during dry periods. Seasonally moist, montane, volcanic shrub-grassland above about 1800 m elevation is the primary habitat of five wild populations on Hawaii, the Keauhou, Keauhou II, Kahuku, and Kipuka Ainahou sanctuaries, and one area on Maui, Haleakala National Park (HALE). The Hawaii Volcanoes National Park (HAVO) population utilizes a variety of vegetation types between about 300 and

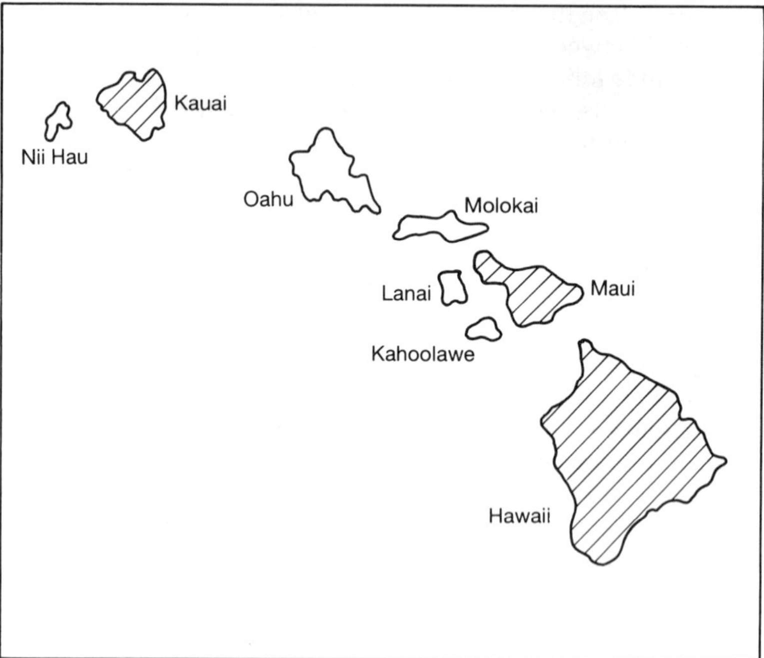

Figure 22.1 The larger islands of the Hawaii. The shaded islands are where Hawaiian geese are currently found.

1300 m elevation. Although wild populations on Maui and Hawaii generally occupy habitats dominated by native vegetation, all utilize alien plants, especially introduced pasture grasses, golf course grasses and road-side grasses.

We also studied captive nene that were placed in 10 enclosures at HAVO ranging from 0.1 to 1.5 ha in size and extending from near sea level to 1220 m elevation. Commercial poultry ration and fresh drinking water supplemented the naturally occurring vegetation in the pens.

All populations were exposed to varying levels of introduced predators, including mongooses (*Herpestes auropunctatus*), feral cats (*Felis cattus*) and dogs (*Canis familiaris*), although mongooses do not occur on Kauai. During the study predators were controlled by trapping around the enclosures, but not in the wild areas.

Figure 22.2 A two-year-old Hawaiian goose, commonly known as nene (pronounced neyney), with a large coded plastic leg-band (AJ).

22.2.2 Nests in the wild and enclosures

During 1978–81, nests were visited at approximately weekly intervals to determine the progress and fate of clutches. Mongooses were considered to have disturbed nests if the appearance of broken eggs was similar to chicken eggs fed experimentally to two captive mongooses and/or if eggs were away from nests; mongooses were considered responsible for the predation events if tooth marks were found on the bones of partially eaten carcasses.

22.2.3 Predation study

To identify predators and the incidence and rate of predation we placed a series of chicken egg clutches (each with three or four chicken eggs and spaced 48–83 m apart) along an elevational gradient. Four low elevation sites (213–762 m), two mid-elevation sites (1128 and 1219 m), and two high elevation sites (1950 and 2042 m) were selected in nene breeding areas in HAVO and HALE, and a total of 606 clutches were monitored in 1979 and 1980 from November through May (approximately the nene breeding season). To determine directly the identity of egg predators, 61 of the 606 artificial clutches were placed inside live-traps having 18 × 18 cm entrances. Similar traps, baited with commercial cat food, were maintained outside HAVO pens (five traps/enclosure).

22.2.4 VORTEX simulation of stochastic population change

We used the April 1991 version of the VORTEX program designed by Bob Lacy and James Grier (Grier and Barclay, 1988; Lacy, Flesness and Seal, 1989; Lindenmayer *et al.*, 1991) with the help of the Captive Breeding Specialist Group of the World Conservation Union (IUCN). The programme incorporates a number of randomly occurring events as well as environmental variation.

VORTEX simulates birth and death processes and the transmission of genes through the generations by generating random numbers to determine whether each animal lives or dies, whether each adult female produces broods of size 0 (hatches no eggs), or one, two, etc. goslings during each year, and which of the two alleles at a genetic locus are transmitted from each parent to each offspring. Fecundity is assumed to be independent of age (after an animal reaches reproductive age, set at 2–25 years for the nene) and sex-specific mortality rates were specified for each age class (see below). We specified the mating system as monogamous for nene, although 19% of the pair-units at Slimbridge (N = 45) were 2M:1F trios in 1990 and two wild trios were reported by

Banko (1988): one 1M:2F and one 2M:1F. Due to the occurrences of trios we assumed that 98% of males sired offspring in any breeding period. The sex ratio was set at 1M:1F because the available data were variable; current ringing data 26M:26F for Maui, 385M:373F for Slimbridge, and 52M:64F for HAVO.

The computer program simulates and tracks the fate of each population, and outputs summary statistics on the probability of population extinction over specified time intervals, the mean time to extinction of those simulated populations that went extinct, the mean size of populations not yet extinct and the levels of genetic variation remaining in any extant populations. Each animal in the initial population is assigned two unique alleles at some hypothetical genetic locus. The model can include the presence of inbreeding depression – expressed in the model as a loss of viability in inbred animals. We chose to include this feature because of the possibility that a catastrophe could reduce the already small population to such a level that inbreeding could become a problem. Kear and Berger (1980) argue that the Slimbridge stock went through a temporary bottleneck leading to low productivity but this was revived after two wild-type males were incorporated. Due to the high degree of band sharing in the captive stocks (E.H. Rave, pers. comm.) we included the possibility that a lethal gene could be expressed. An example is that three of 62 eggs from the Olinda stock showed some deformity in 1990 (F. Duvall, pers. comm.).

VORTEX also models catastrophes, the extreme of environmental variation, as events that occur with some specified probability and reduce survival and reproduction for one year. We included four potential catastrophes which we assumed had the following effects. Firstly, volcanoes, 1 in 100 years, where violent eruptions of gases, ash and other pyroclastic material disrupt breeding for 5% of the population that year and kill 5% more. This was based on the facts that the flocking season lasts for 2 months (17% of a year) and that the largest flocks, at current levels, are about 30 (N. Santos and T. Telfer, unpublished data). In the other months, flock size is limited to breeding units (usually two birds), so a localized eruption will effect fewer birds. Two phreatic eruptions of Kilauea have occurred during the past 200 years (Heliker, 1989). Volcanic eruptions are not a problem for the Maui and Kauai flocks. Secondly, tsunami, 1 in 20 years, where the severity values are the same as for volcanic eruptions for the Hawaii flock but on Kauai the value of 30% mortality was used. A tsunami destroyed 31 of H. Shipman's 42 semi-captive nene on Hawaii in 1946. Tsunami are not a problem to the nene on Maui. Thirdly, hurricanes, 1 in 200 years, where all nesting attempts are destroyed in that year but all adults survive. Fourthly, disease epidemics, 1 in 50 years, were attributed with the same severity values as volcanic eruptions. We decided to write a disease possibility in because although infrequent it can cause massive mortality in waterfowl populations (Owen and Black, 1990). However, since nene currently rarely congregate in flocks larger than 30 birds, the potential effect has been scaled down, to the severity values as volcanic eruptions.

VORTEX also allows the user to supplement the population from captive stocks. In the Repeat Effort Scenario we released birds at the same rate as in the previous 30 years of the restocking programme (on Hawaii and Maui) to simulate what would happen in the future if we were to repeat our efforts. We ran each scenario 1000 times for a 200-year period with the carrying capacity truncation set at 2000 individuals.

22.3 RESULTS

22.3.1 Predation

Nest success (at least one egg hatched) in enclosures was nearly twice as high as in the wild (Table 22.1; Banko, 1992): 83% of 30 enclosure nests vs. 44% of 70 wild nests. Predation by mongooses, the leading cause of nest failure in the wild, resulted in the loss of 28 of 70 nests (40%); they destroyed 24 wild clutches and killed four females at the nest. Predation in the enclosures was rare; only one clutch in 39 nests (3%).

Trapping and the chicken egg placements also demonstrated that mongooses were the most serious nest predators in all habitats. Artificial clutches (N = 606) were disturbed or destroyed by mongooses in 54% of all cases; 43% of all clutches were disturbed within 30 days (nene incubation period). Clutches at lower elevations were destroyed most rapidly; 79% of low elevation clutches (N = 168), 71% of mid-elevation clutches (N = 54) and 20% of high elevation clutches (N = 384) were destroyed after 30 days (Chi square = 203.9, df = 2, $P < 0.001$, also see Stone, Hoshide and Banko 1983). Forty-six mongooses were caught at 61 live-traps baited with chicken eggs. During 1975–80, 745 mongooses and 11 feral cats were trapped at enclosures, thus illustrating the high density of these introduced predators in low and mid-elevation habitats.

22.3.2 Simulations

(a) Values for the model

Table 22.2 lists the demographic values that changed in the various scenarios. The value for first year mortality (between hatching and 12 months) was based on two data sets. We assessed that only one gosling fledged from 56 eggs that hatched, and Hoshide, Price and Katahira (1990) reported that 48 fledglings came from 144 hatchlings. This yields

Table 22.1 Breeding demographics and breeding success of nene in the enclosures at HAVO

	1975	*1976*	*1977*	*1978*	*1979*	*1980*	*1981*
Total population[a]	2	16	21	29	14	17	20
Pairs capable of breeding	1	6	5	8	5	7	6
Pairs breeding	1	5	5	5	2	4	4
Percentage of pairs breeding	100	83	100	62	40	57	67
Percentage of total population breeding	100	62	48	34	29	47	40
Nest attempts[b]	1	5	7	5	2	5	5
Successful nests[c]	1	5	6	4	1	4	4
% Nest success	100	100	86	80	50	80	80

[a]Includes juveniles (<2 years) and adults. Juveniles were fledglings produced 1 year earlier; in reality they fledged from the enclosures before their first year. Not included in annual totals were free-flying individuals that voluntarily nested in enclosures or associated with captives; individuals from other captive flocks that were too old (>/= 15 years); or those which had insufficient time to adjust to new conditions before the breeding season (<1 year).

[b]Includes first nests and renest attempts.

[c]At least one egg hatched.

Table 22.2 Demographic data used in three computer simulations: A = status quo scenario; B = repeat effort scenario; C = optimal management scenario

	Hawaii	*Maui*	*Kauai*	*Source*[a]
Starting pop. size	340	180	37	a,b,c
Number released				
A and C	0	0	0	
B	1560	460	–	a,d
Maximum brood size				
A and B	4	4	6	c
C	6	6	–	c
Average brood size				
A and B	2.56	1.89	3.33	c,e
C	3.00	3.00	–	
Percentage of no. hatching				
A and B	79.1	77.4	83.0	f
C	69.1	69.1	–	
First year mortality				
A and B	75.5	75.5	27.4	f
C	55.0	55.0	–	

[a]a = Black *et al.* (1991). b = Natividad Hodges (1991). c = The Nene Recovery Team (unpublished data). d = Releases occurred in years 1–30 on Hawaii and in years 1–19 on Maui. Equal numbers of both sexes were released. A pair of two- and three-year olds accompanied each batch of one-year old releases to enhance survival (see Marshall and Black, 1992). e = Banko (1988); Banko (1992). f = Values calculated in the text.

a value of 75.5% mortality in the first year, the value attributed to the Hawaii and Maui flocks in the Status Quo and Repeat Effort Scenarios. In the enclosures, first year mortality was only 12.5% (30 of 241 hatched goslings) so we chose a more realistic figure of 55% for the Optimal Managment Scenario, the same value as the steadily increasing population of barnacle geese (*Branta leucopsis*) (Owen 1982; Prop, van Eerden and Drent, 1984). In the second and 3+ years we attributed a 14% and 5% mortality rate, respectively, in all scenarios (Banko and Elder, 1989, unpublished data). We chose to make male mortality rates slightly lower (as for other geese: Owen, 1982); age 2 = 13% and age 3+ = 4%. The variation in each of these mortality values was set at ±10%.

The values chosen for the proportion of nene that succeed in breeding attempts (hatching at least one egg) were calculated using the following data on numbers of non-breeders and failed breeders.

Hawaii: based on data from Banko (1992) and using the proportion of paired and unpaired birds in the population, 85% of the 340 Hawaii nene were capable of breeding and 51–62% of these attempted to breed.

This means between 147–179 individuals or 73–89 pairs tried to breed each year. One or more eggs hatched in 44% of 70 nests. This means that 32–39 nests hatched one or more eggs (mean of 35.5 broods), and this involves 64–78 individuals. Therefore, 262–276 nene, or a mean of 79.1% of the population, did not successfully hatch eggs. We set the value for environmental variation (EV) at 18.5% based on the extreme proportions of breeders in the population; 44% in 1978 and 54% in 1979 (Banko, 1992); EV = (0.44/0.54)−1×100.

Maui: based on Banko (1992), 81% of the Maui flock (N = 180) were capable of breeding and 24–76% of these attempted to bred. This means between 35–111 individuals or 18–56 pairs bred and 44% of these hatched one or more eggs, yielding 8–25 successful nests or 16–50 successful individuals. Therefore, 96–130 nene or a mean of 77.4% of the population did not hatch eggs. We set EV at 68.3% based on the extreme proportions of breeders in the flock; 24% in 1979 and 76% in 1981. In addition, due to the high elevation and location of Haleakala, entire cohorts face cold, wet conditions in some years, thus causing early mortality. For example, in 1989 it snowed and hailed for 12 days during the hatching or brooding period (C. Natividad Hodges, pers. comm.).

Kauai: based on T. Telfer (unpublished data), 86% of the Kauai flock (N = 37) were capable of breeding, so eight birds or four pairs bred (24%) and 73% of these succeeded in hatching one or more eggs. Therefore, 31 birds or 83.7% of the population did not hatch eggs. We set EV at 28% based on the extreme proportions of breeders in the flock; 21.6% in 1990 and 30% in 1987.

The enclosure value: including juveniles produced the previous year, 51% of the captive flock was capable of breeding and 40–100% (mean = 73%) of these attempted to breed (Table 22.1). Five of 30 (17%) nesting attempts failed. Thus, in the model 69.1% of the population did not hatch eggs, when the mean value used for non-breeding was 63%. We set EV at 54.2% based on the extreme proportions of breeders in the flock; 28.6% in 1979 and 62.5% in 1976.

(b) Status Quo Scenario

In this model we found that the Kauai flock was the only one to survive, although in only 91% of the simulations (Figure 22.3; Table 22.3). Numbers on Kauai rose throughout the 200-year period, peaking at 1400 (18 SE) individuals. The number of alleles in the Kauai flock was preserved quite well over time; 37 birds and 74 alleles at the start and 23 alleles preserved (31%) in 200 years. In contrast, only 41% and 2% of Hawaii and Maui populations, respectively, survived the 200-year period. Even in these surviving cases numbers were approaching extinc-

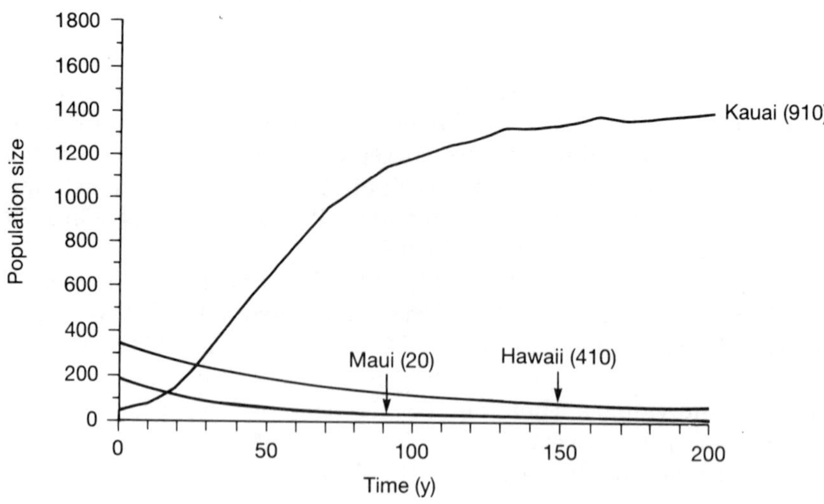

Figure 22.3 A model showing population change in the Status Quo Scenario over a 200-year period without further supplementations. The curves are based on the means of the 1000 simulations for each population. The figures in parentheses indicate the number of simulations when the population did not go extinct. The arrows indicate the mean year of extinction for the unsuccessful simulations (see Table 22.3).

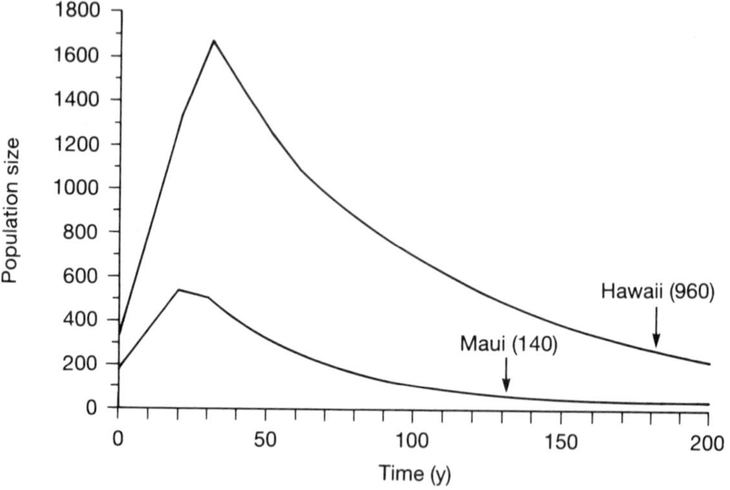

Figure 22.4 A model showing population change in the Repeat Effort Scenario over a 200-year period with initial supplementations. The curves are based on the means of the 1000 simulations for each population. The figures in parentheses indicate the number of simulations when the population did not go extinct. The arrows indicate the mean year of extinction for the unsuccessful simulations (see Table 22.3).

tion. The populations on Hawaii and Maui declined throughout the period (Maui at a faster rate than Hawaii) and on average went extinct by year 149 (1 SE) and 91 (1 SE), respectively.

(c) Repeat Effort Scenario

In this scenario 1560 nene were released over a 30-year period on Hawaii and 460 nene were released over 19 years on Maui, which closely matches the previous release efforts (Black, Duvall and Hoshide, 1991). Annual releases included 25:25 (female:male) one-year-old geese together with a 1:1 two-year-old pair. With these additional releases, the livelihood of the Hawaii and Maui flocks was prolonged for 33 and 41 years (the mean time till extinction), and increased final population size (at the end of the 200 year period) by 171 and 33 individuals, respectively (Figure 22.4; Table 22.3). However, in all cases, numbers were declining and extinction was approaching.

On both islands numbers peaked in the year of the final releases and thereafter declined. This repeated release effort was more successful on Hawaii than on Maui (final populations of 234 vs.44, respectively).

(d) Optimal Management Scenario

In this model the values for numbers successfully hatching, brood sizes and survival reflected a situation where predation was reduced and feeding opportunity was enhanced. The values we chose were based on real situations within the enclosure experiments (Table 22.1). In this enhanced environment both the Hawaii and Maui flocks quickly flourished to self-sustaining levels and approached the level of carrying capacity truncation at 2000 individuals (Figure 22.5; Table 22.3). By the 40th and 50th year the Hawaii and Maui populations, respectively, rose to above 1600 individuals and continued to rise at a slower rate for the remaining period.

We checked the effect of each of the three parameters in turn (Table 22.2); only when all three were incorporated did the population reach a self-sustaining level. The preservation of original alleles was 18% for Hawaii and 27% for Maui birds (Table 22.3).

22.4 DISCUSSION

The situation which has arisen on Kauai gives grounds for optimism. Our model shows that even with status quo demographic data the Kauai flock grew toward carrying capacity in 91% of the simulations. Recent counts support this result; 12 birds in 1982, 27 in 1987 and 75 in 1992 (T. Telfer, pers. comm.). However, the population went extinct in the other 9% of the simulations. It could be argued that only a 100% chance of

Table 22.3 Summary statistics of three scenarios for the three nene populations: in each case 1000 simulations were run and the carrying capacity truncation was set at 2000 individuals

	Status quo			*Repeat effort*		*Optimal mangt.*	
	Kauai	*Hawaii*	*Maui*	*Hawaii*	*Maui*	*Hawaii*	*Maui*
Starting population	37	340	180	340	180	340	180
Final population[a]	1400	63	11	234	44	1631	1673
Probability of extinction (%)	9	59	98	4	86	1	0
Mean years to extinction	64	149	91	182	132	115	–
Growth rate (r)[b]	0.05	−0.02	−0.04	−0.02	−0.04	0.08	0.09
Mean lambda	1.08	0.98	0.96	0.98	0.96	1.10	1.11
Final heterozygosity							
expected	0.86	0.78	0.68	0.94	0.80	0.98	0.98
observed	0.88	0.87	0.84	0.98	0.90	0.99	0.99
Final number of alleles	23	12	6	51	14	119	98

[a]Mean of the extant populations after 200 years.
[b]Mean growth rate in years with supplementation, prior to carrying capacity truncation.

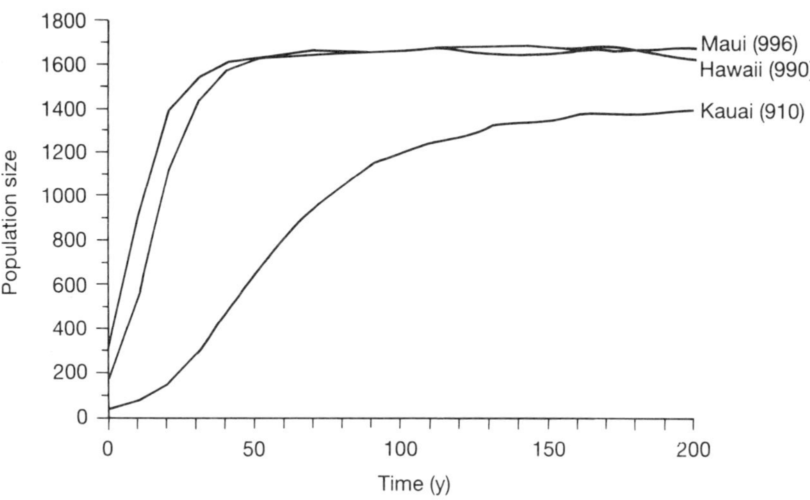

Figure 22.5 A model showing population change in the Optimal Management Scenario over a 200-year period without further supplementations. The curves are based on the means of the 1000 simulations for each population. The Kauai curve is from the Status Quo Scenario (Figure 22.2) for comparison. The figures in parentheses indicate the number of simulations when the population did not go extinct, which was nearly 100% in this model.

success is acceptable and that further management is needed. On Kauai habitat suitability and predators are not apparently limiting the nene. Instead, the genetic managment of the flock could be improved; the flock was founded by three breeding pairs which may themselves have been related. The Nene Recovery Team has realized the potential significance of the Kauai population and has begun a programme to increase the genetic diversity. Seven birds with different origins were released at a refuge on the northern coast of the island in 1991, and further releases are planned until the full component of genetic lineages have been included on the island – a positive step towards a healthier population (see Tomlinson *et al.*, 1991).

The finding that the median time to extinction for the nene on Hawaii was increased by 31 years after a further 30 year release programme indicates that the extinction process is being held at bay by the captive breeding and release scheme. The modelling shows that a further period of releases on Hawaii maintained numbers above 200 individuals, and that the population survived in 96% of the simulations. This suggests that reintroduction as a management tool may continue to be effective in prolonging extinction on Hawaii, but it will not enable the population to achieve self-sustaining status. Actual counts at HAVO reveal that numbers have apparently declined by 15% since 1989 (159 then and 135 in 1991) (Katahira, pers. comm.).

In retrospect, therefore, it appears that management efforts in the first 30 years of the nene recovery programme have prolonged extinction of the original 30 nene that remained in the uplands of Hawaii in the 1950s. The Repeat Effort Scenario for the Maui population, on the other hand, was not so successful as on Hawaii; in only 14% of the simulations did the populations survive. In this model we used a large value for environmental variation due to the annual variation in the proportion of non-breeders (suggesting that breeding condition is sometimes not achieved), and to the harsh weather conditions during the breeding season, which can limit the survival of young goslings. In spite of this, recent counts from this area have been consistent at about 150 individuals in recent years (Natividad Hodges, 1991; Black *et al.*, 1991). This stability may be due to the large-scale predator control programme that was instigated in the Haleakala Crater in 1981 (see below).

Our enclosure and non-enclosure comparison shows that when mongooses and feral cats were removed from current nene habitats, then and only then could the population become self-sustaining. The Optimal Management Scenario model showed that after the first 10 years under an enhanced management both the Hawaii and Maui flocks tripled in size. This indicates that large-scale predator control on behalf of the nene could be an effective conservation action. Managers in Haleakala National Park, Maui, where about 35% of wild nene live, have begun such a programme. They work a series of about 400 live-traps throughout the crater. Managers on Hawaii are seeking support to enable them to increase their predator control effort to similar levels.

We have ruled out the possibility that inbreeding depression is limiting productivity and survivability in nene. The model shows that, even when including the inbreeding depression feature, numbers can increase to safe levels. The data from the enclosures show that nene do have the necessary skills to establish partnerships, copulate, incubate and rear offspring. However, our model does not reflect how naïve captive-bred nene cope once released. Marshall and Black (1992) showed that the early experiences of nene, whether reared by parents, foster parents or without parents, can affect their social development, reactions to predators and ability to integrate with adult nene after release. Managers and aviculturalists in Hawaii have acted on these findings and are attempting to offer nene more appropriate early experiences prior to release.

In addition, an intensive series of field studies has begun to assess the current feeding and energetic constraints of nene in different habitats (Black, 1990; Black *et al.*, 1991). Banko (1988) showed that the nene in enclosures had larger clutches, took fewer and shorter nest recesses and slept for shorter periods than those nesting in the wild. He attributed these differences to the fact that supplemental feed was available to the geese in the enclosures. By enhancing the feeding opportunities for nene the number of potential breeders, successful breeders and surviving offspring will increase. Providing predation is kept to a minimum, these revived demographic parameters will enable the nene to reach a self-sustaining level.

In conclusion, the Hawaiian goose is once again in danger of extinction. However, we predict that when management efforts are enhanced to include an appropriate predator control effort the nene flocks in the Hawaiian state will be able to thrive. When predation levels are kept to a minimum, as on Kauai, managers will look forward to a new set of challenges as numbers increase and individuals disperse.

ACKNOWLEDGEMENTS

This study was funded by the National Park Service and The Wildfowl & Wetlands Trust, with contributions from DOFAW (Honolulu), CBSG and British Airways Assisting Nature Conservation. We thank Janet Kear, Georgina Mace and Jim Jacobi for constructive comments on the manuscript. We thank Dave Manuwal, Chuck Stone and Dan Taylor for their support during the field work and all other Nene Recovery Team members for their recent contributions. Ulie Seal, Fern Duvall, Ann Marshall and Mike Bruford gave us their expert assistance during the modelling process.

REFERENCES

Baldwin, P.H. (1945) The Hawaiian goose, its distribution and reduction in numbers. *Condor* **47**, 27–37.

Banko, P.C. (1988) Breeding biology and conservation of the nene, Hawaiian goose (*Nesochen sandvicensis*). University of Washington, Seattle, Ph.D. Thesis.

Banko, P.C. (1992) Constraints on wild nene productivity. *Wildfowl*, **44**, 99–106.

Banko, W.E. and Elder, W.H. (1989) Population histories: species accounts, subgrassland birds: Hawaiian goose, nene. Unpublished Report, University of Hawaii.

Black, J.M. (1990) The nene recovery initiative. Report to The Wildfowl & Wetlands Trust, Slimbridge, UK.

Black, J.M., Duvall, F., Hoshide, H. *et al.* (1991) The current status of the Hawaiian goose and its recovery programme. *Wildfowl*, **42**, 149–154.

Grier, J.W. and J.H. Barclay (1988) Dynamics of founder populations established by reintroduction, in *Peregrine Falcon Populations: Their Management and Recovery* (eds T.J. Cade, J.H. Enderson, C.G. Thelander, and C.M. White), The Peregrine Fund, Boise, Idaho, pp. 689–701.

Heliker, C. (1989) Volcanic origins of the Hawaiian Islands, in *Conservation Biology in Hawaii* (eds C.P. Stone and D.B. Stone), University of Hawaii Press, Honolulu, pp. 11–15.

Hoshide, H., Price, A.J. and Katahira, L. (1990) A progress report on Nene *Branta sandvicensis* in Volcanoes National Park, 1974–1989. *Wildfowl*, **41**, 152–5.
Kear, J. and Berger, A.J. (1980) *The Hawaiian Goose: An Experiment in Conservation*. Poyser, Calton.
Lacy, R.C., Flesness, N.R., and Seal, U.S. (1989) Puerto Rican parrot population viability analysis. Report to the U.S. Fish and Wildlife Service. Captive Breeding Specialist Group, Species Survival Commission, IUCN, Apple Valley, Minnesota.
Lindenmayer, D.B., Thomas, V.C., Lacy, R.C. and Clark, T.W. (1991) Population Viability Analysis (PVA): the concept and its applications, with a case study of Leadbeater's Possum, *Gymnobelideus leadbeateri* McCoy. Report to the Forest and Timber Inquiry (Resource Assessment Commission), Canberra, Australia.
Marshall, A. and Black, J.M. (1992) The effect of rearing experience on subsequent behaviour traits in captive-reared Hawaiian geese: implications for the re-introduction programme. *Bird Conservation International*, **2**, 131–47.
Morin, M. and Walker, R.L. (1986) The nene restoration plan. Unpublished document, Hawaii DOFAW/DLNR, Honolulu.
Natividad Hodges, C. (1991) Survey of Nene *Nesochen sandvicensis* at Haleakala National Park, 1988 through 1990. *'Elepaio* **51**, 38–9.
Owen, M. (1982) Population dynamics of Svalbard barnacle geese, 1970–1980. The rate, pattern and causes of mortality as determined by individual marking. *Aquila*, **89**, 229–47.
Owen, M. and Black, J.M. (1990) *Waterfowl Ecology*, Blackie Publ., Glasgow.
Prop, J., van Eerden, M.R. and Drent, R. (1984) Reproductive success of the barnacle goose *Branta leucopsis* in relation to food exploitation on the breeding grounds, western Spitsbergen. *Norsk Polarinstitutt Skrifter*, **181**, 87–117.
Stone, C.P., Walker, R.L., Scott, J.M. *et al.* (1983) Hawaiian goose management and research: where do we go from here? *'Elepaio*, **44**, 11–15.
Stone, C.P., Hoshide, H. and Banko, P.C. (1983) Productivity, mortality and movements of Nene in the Kau desert, Hawaii Volcanoes National Park, 1981–1982. *Pacific Science*, **38**, 301–11.
Tomlinson, C., Mace, G.M., Black, J.M. and Hewston, N. (1991) Improving the management of a highly inbred species: the case of the white-winged wood duck *Cairina scutulata* in captivity. *Wildfowl*, **42**, 123–33.

23

The extinction in the wild and reintroduction of the California condor (*Gymnogyps californianus*)

W.D. Toone
San Diego Wild Animal Park California, USA
and
M.P. Wallace
Los Angeles Zoo, California, USA.

23.1 INTRODUCTION

Fossil records for the genus *Gymnogyps* date well into the late Pleistocene. As with most avian species, the fossil records for the condor are probably far from complete but they indicate that the species ranged as far south as Nuevo Leon, Mexico, along the west coast of North America into the Pacific Northwest to Canada, across the southern United States to Florida and up the east coast as far as New York (Brodkorb, 1964; Emslie, 1987). Much of the California condors' (*Gymnogyps californianus*) range reduction took place along with the Pleistocene extinctions of North American mega-fauna (Emslie, 1987).

The first written account of the California condor appeared in 1602 (de la Ascension, 1928) and at that time the condors' range was probably limited to western areas of North America from Canada to Baja California (Koford, 1953). Though legally protected since the 1900s, the decline of the population had been watched with a certain sense of helplessness

Creative Conservation: Interactive management of wild and captive animals.
Edited by P.J.S. Olney, G.M. Mace and A.T.C. Feistner.
Published in 1994 by Chapman & Hall, London.
ISBN 0 412 49570 8

stemming primarily from a poor understanding of the sources of condor mortality. Beginning with the work of Koford, the biology of condors has attracted the attention of an increasing number of biologists. Nevertheless, the vast and rugged terrain made the discovery and recovery of a dead condor a rarity, shrouding the causes of death in mystery and leading only to speculation as to the reasons for the precipitous decline.

Historically, the use of the rodenticide 1080, loss of habitat, shooting, and disturbance have been implicated as the primary causes of decline (Wilbur, 1978).

In spite of observations of condors feeding on 1080 poisoned ground squirrels, however, there is no evidence that the birds were harmed. Though there are no documented cases of condor losses due to 1080, the number of overall recovered carcasses is very small and certainly does not provide a complete picture of condor mortality.

Considering the historical range of the condor, the nearly pristine conditions of its nesting habitat, and the fact that its foraging habitat has changed little over the last 100 years, it is difficult to place the full weight of the decline on a loss of habitat.

Disturbance has often been suggested as a major cause of decline. Several organizations suggested that if people entered condor habitat the birds would abandon it, if a nest site was disturbed the pair would never breed again and that if a condor was handled it would die. Though abundant evidence to the contrary has existed for years, the political clout of these accusations had a profound impact on the recovery of the condor. W.L. Finley and H.T. Bohlman often entered condor habitat and nests to document the chicks (Finley, 1908). In spite of frequent disturbance of this type which continued in the work of Koford in the 1930s, nesting success of study nests continued to be the same as that of undisturbed pairs (Snyder and Snyder, 1989). The 'disturbance' of pairs during the intensive research period between 1982 and 1986, when six eggs were removed from wild nests for the captive program, did nothing other than to cause the predicted increase in the reproductive rate.

Shooting, for target and collection of museum specimens, has certainly had an impact, though its relative importance in the decline of the condor cannot be quantified. Records indicate numerous collections for museums (Wilbur, 1978). Random shootings are not as easily evaluated but they certainly occurred and, one would guess, with some frequency. One of the four dead condors subjected to a post-mortem in the 1980s had been shot (Janssen *et al.*, 1986). Although several authors report many condors being shot, during the 1940s Koford estimated that at that time only one condor per year was lost to shooting (Wilbur, 1978). Extensive public education from the 1970s and 1980s to the present will, hopefully, contribute significantly to reducing this threat in the future.

23.2 EXTINCTION IN THE WILD

In 1978 the California Condor Recovery Team was officially recognized and the recovery plan that they prepared under the auspices of the US Fish and Wildlife Service was adopted. It called for a three-pronged approach including:

(a) habitat studies, acquisition, and protection;
(b) radio-telemetry studies with wild birds;
(c) establishment of a captive population to produce birds for reintroduction.

It was not until 1982 that an effective means to accurately census the population and identify individuals was developed. A programme of photographing condors in flight was initiated and individuals were identified by following feather loss and replacement during the molt (Snyder and Johnson, 1985a,b). Though this system allowed for the identification of birds and therefore increased awareness of disappearances or additions to the population, it was limited in its use for understanding movements and it did nothing to aid in the recovery of dead or dying birds. To follow condors during release experiments in Peru, a wing or patagial transmitter was developed and tested on 47 wild and 12 released Andean condors between 1980 and 1984 (Wallace and Temple, 1987). Late in 1982 radio-telemetry became a reality in California with two condors being fitted with transmitters and released. Ultimately a total of nine wild California condors flew with radio-transmitters.

Because of the development and use of photo-census and radio-telemetry, accurate population records were kept from 1983 to 1987, when the last wild bird was captured. During that period, at least 15 wild condors perished. Of these birds only four were recovered, two of them with the aid of radio-telemetry (Snyder and Snyder, 1989). Though the numbers are far too small to make definitive statements with regard to the overall decline of the population, data on the causes of mortality at least began to help produce possible scenarios. There were three mortalities resulting from lead poisoning as a result of ingested lead, and one from cyanide poisoning from an M-44 'coyote getter'. In one case of lead poisoning it was confirmed to have been from the ingestion of a bullet fragment (Wiemeyer *et al.*, 1988). Several large mammals which are frequently hunted using lead shot or bullets are present in condor habitat and it is easy to imagine carcasses of unretrieved but mortally wounded game animals as a likely source of lead for a carrion eating species.

In general, it has been the conclusion of recent investigators that loss of habitat is not the immediate threat to the future of the condor (Snyder and Snyder, 1989; Wallace, 1991). Food sources seem to be abundant.

It is likely, however, that the feeding behaviour of the condor may place it at risk, by their flying into areas which hunters frequent, increasing the possibility of being shot and feeding on food contaminated with lead. The bird depends on sight to locate its food and therefore seldom feeds in the more remote and mountainous areas where the dense cover of chaparral hides any carrion. Instead they are drawn to the rolling open foothills of the central valley, where both ranching and hunting are more practical.

The final blow to the wild population of condors came during the winter of 1984–1985. At the end of the 1984 breeding season four wild pairs were manipulated to increase egg production, several chicks were collected and distributed to the zoos and a fifth pair was discovered. The total wild population was 15 birds but by the spring of 1985, it was clear that catastrophic losses had occurred. By the summer it was finally accepted by most involved in the programme that six birds, 40% of the wild population, had been lost. Missing birds included mates of four of the remaining five pairs. Ultimately agreement was reached that the wild population was too small to sustain itself without immediate supplementation from the captive population. Due to the age and the genetic and sexual composition of the captive population, that support was not considered wise. The loss of AC-3, the female of the last remaining pair, to lead poisoning in January 1986, made it apparent that all the wild birds should be caught and placed in temporary custody in the two zoos which had established condor facilities – the Los Angeles Zoo and the San Diego Wild Animal Park. The last wild condor, AC-9, was captured on Sunday, 19 April 1987 and joined 26 others in captivity. A large chapter in the history of the condor was brought to a close, and for the short term the future of the California condor lay completely in the hands of two zoos.

23.3 CAPTIVE HISTORY

There are a variety of records of the California condor in captivity. Birds were kept prior to European settlement when native Americans (Chumash) along the coast of California occasionally had condors either as pets or for ceremonial use. After European settlement, condors were kept by locals, including 'Ben Butler' kept by F.H. Holmes, and 'General', originally kept by W.L. Finley and later transferred to the New York Zoo (Wilbur, 1978). A condor was taken by the California Department of Fish and Game and placed on exhibit in the Golden Gate Park in San Francisco around 1917 (Wilbur, 1978). There are numerous other accounts of captive condors, but with the exception of those at the National Zoo, Washington, DC, birds were apparently never housed together and attempts to breed California condors were not made until the 1980s. Dixon (1924) reported egg-laying by a 12-year-old female at the National Zoo, housed with other condors, but post-mortem examinations later revealed that all birds in the collection were females. Extensive behavioural records were not kept on the California condors at the National Zoo, but in Los Angeles and San Diego experience seems to indicate that cathartid vultures are behaviourally quite conservative. The egg-laying reported by Dixon may closely parallel the regular production of eggs laid by a female–female pair of American black vultures (*Coragyps atratus*) at the San Diego Wild Animal Park. This pair externally appeared to be entirely normal, copulating regularly, laying a two-egg clutch and taking turns incubating. As with the California condors at the National Zoo, this mate choice was probably driven by the lack of males.

More productive efforts were made with the closely related Andean condor (*Vultur gryphus*) which though also threatened was much more common in zoos, and could act as a husbandry model. The first Andean condors to breed in captivity did so as early as 1840 at the London Zoo, and during the 1940s at the San Diego Zoo K.C. Lint experienced great success in condor breeding. His work documented the occurrence of annual nesting (when a chick or egg is removed or lost in the first year), artificial incubation and hand-rearing. These successes ultimately led Director Belle Benchley and K.C. Lint to propose, in 1952, a captive breeding programme for California condors. The proposal called for the collection of a pair of juvenile condors to be used for breeding in captivity, and their young to be eventually released to augment the dwindling wild population. Though a permit was issued for this project, it was never completed because of active opposition primarily on the part of the National Audubon Society, which resulted in legislation preventing any type of interference, including most types of intensive field research. Later, the Patuxent Wildlife Research Center, the Bronx Zoo, and the San Diego Zoo would carry out more intensive work with the Andean condor, thoroughly defining artificial incubation parameters (Kuehler and Witman, 1988), puppet-rearing techniques (Bruning, 1983; Toone and Risser, 1988), diet and many facets of their medical care.

Though one young bird, a male called Topa Topa, had been found in a weakened condition and brought into captivity at the Los Angeles Zoo in 1967, the recent efforts to bring further birds into captivity did not begin until 1982 with the emergency collection of a two-to-three month-old chick that was being neglected by its parents.

Permission was granted in 1983 to collect the first eggs from the nests of wild condors for captive incubation. From 1983 to 1985 15 eggs were collected from wild pairs and incubated at the San Diego Zoo and San

Diego Wild Animal Park. An overall hatchability of 86.7% and a survivability of 92.3% was achieved (Kuehler and Witman, 1988; Kuehler *et al.*, 1991).

These successes were soon followed by the successful breeding of the first California condor in captivity in 1988 (Toone and Risser, 1988). Forty-nine California condors have now been bred (to 1993) at the Los Angeles Zoo and the San Diego Wild Animal Park.

23.4 ANDEAN CONDORS AS SURROGATES

Perhaps the most important contribution to the recovery programme has been the use of Andean condors as surrogate species. Work with Andean condors has for many years served as a model for the California condor programme. Incubation parameters and husbandry techniques now successfully applied to California condors were first developed with Andean condors (Bruning, 1983; Cade, 1986).

Radiotelemetry transmitters for the wing of condors, and techniques for tracking were also developed using Andean condors, as well as trapping and field handling methods prior to their use on California condors (Wallace and Temple, 1987).

Anticipating success with captive propagation of California condors, Andeans were used to develop methods for releasing captive-reared condors to the wild. Based on successful experiments with releasing two smaller cathartid vultures, *Coragyps atratus and Cathartes aura*, in Florida (Wallace and Temple, 1983), 11 Andean condors were released in northern Peru. These experiments, conducted between 1980 and 1984, outlined the basic husbandry and field techniques necessary for successful condor releases to occur (Wallace and Temple, 1987). However, the difference in climate, topography, carrion availability and human activity level are drastically different between the area where Andean condors were released in Peru and the former range of California condors where releases would be expected to occur.

In 1987 we felt that four to five years would be required to sufficiently build the captive flock to a level that releases could take place. During that interim period, Andean condors were again used in the recovery effort to tailor condor release methods developed in Peru to the conditions that exist in the southern California mountains. Between 1988 and 1991, 13 first-year female Andeans were released from three different sites, allowing the programme to 'fine tune' release methods, develop means to circumvent new logistical problems in the field and train a field crew for California condor release (Wallace, 1991).

23.5 RELEASE OF CALIFORNIA CONDORS

Three criteria for the release of California condors were established:

(a) founder birds must have 96% of their genes represented in offspring before any future offspring would be eligible;
(b) the birds had to be physically and behaviourally suitable for release;
(c) at least three birds meeting these criteria had to be available.

Due to extreme financial and political concerns the Recovery Team agreed to amend the final criterion so that the release numbers could be filled by the addition of Andean condors as social cohorts.

The third criterion was considered purely for the social needs of the birds; they simply are less stressed and are more easily managed if released in a group of three or more. The captive programmes production of 12 young in 1991 raised founder representation in the captive flock sufficiently to consider the release of two of the offspring. To meet the criterion for a large group size, two female Andeans were also included in the release contingent. All four birds were within two months of age, and were raised together at the Los Angeles Zoo throughout their five-month fledging period. Using puppets, from behind one-way glass, and video cameras, complete isolation from human contact was maintained during the captivity period.

On October 10, 1991, all four condors, two Andeans, and two Californians were flown by helicopter to a remote release pen in the Los Padres National Forest where they acclimatized for three months to the sights and sounds of the area in which they would be released. On January 14 1992, all four birds were released. Although the actual release was uneventful, inclement weather prolonged their adjustment period. Soon, however, they were exploring up to 20 miles from the release area in different directions. During one of these flights Xewe, a California condor, was shot at three times by a man with a .22 calibre rifle. The bird was unharmed and the individuals were apprehended and are now being prosecuted. The incident points to the need for better education about the condor programme and hunting laws in general. On October 6 1992, Chocuyens, the other California condor, died from ingesting ethylene glycol, a toxin found in car antifreeze. The tragic loss added to our list a new threat to condors in the wild. The two Andean condors were captured, and as with the previous birds in the program, were re-released in areas of Colombia and Venezuela to build a population in areas where the species had become rare or extinct (Lieberman *et al.*, 1991).

Six more California condors were released on December 1 1992, from those 12 condors produced in captivity during the breeding season. The single free-flying condor, Xewe, quickly joined the others. All seven birds have so far been adjusting well to their new environment over the last four months without incident.

As of this writing the released birds continue to do well, feeding on provided carcasses to minimize the opportunity for accidental poisoning.

23.6 DISCUSSION

No discussion of a captive programme would be complete without some comment on public education. Clearly the efforts of the San Diego Zoo were blocked in 1952 because the proposals did not then fit the public perception of what the roles of a zoo are. In spite of the changes in zoo philosophy over the last 30 years, and the efforts by zoos to establish themselves as breeders and guardians of endangered species, the aversion to zoo involvement nearly cost us the species. This aversion may persist because captive breeding is not always, in and of itself, a defensible justification for zoo involvement in conservation efforts. Public education must also be an integral part of any recovery programme. Our zoo visitors and the mass media are the best outlets available to us. Though efforts with the media have been extensive, early political pressures have prevented the exhibition of the California condor, and this has limited our ability to communicate the important lessons of the condor recovery. Interviews with the media, regardless of the statements made in the interviews, often result in articles designed to dramatize a personal effort, expose controversy within the conversation effort, or emphasize the emotional drama of the successes and failures. Rarely is there a clear statement concerining the ultimate causes of environmental degradation, pressures of a growing human population, lack of concern for recycling resources, habitat protection or other environmental issues. Zoo literature must also include these issues. Unfortunately an audience deprived of environmental education about the condor programme because of the prevalence of myths about the species sensitivity and supposed delicate nature, is an audience unaware, and therefore unable to intelligently support the ongoing efforts of recovery.

It would be beyond comprehension that a zoo should provide animals for release to the wild without ensuring that all possible had been done within the environment to provide for the future well-being of the released individuals. The implication here for evolving zoos is that they should also be active in habitat conservation as political advocates, land managers and perhaps land owners.

REFERENCES

Brodkorb, P. (1964) Catalogue of fossil birds. Part 2 (Anseriformes through Galliformes). *Bull. Florida State Mus. Biol. Sci.*, **8(3)**, 195–335.

Bruning, D.F. (1983) Breeding the Andean condor *Vultur gryphys* at the New York Zoological Park. *Int. Zoo Yb.*, **23**, 11–14.

Cade, T.J. (1986) Propagating diurnal raptors in captivity: a review. *Int. Zoo Yb.*, **24/25**, 1–20.

de la Ascension, A. (1928) Father Antonio de la Ascension's account of the voyage of Sebastion Vizcaino. *Calif. Hist. Soc. Q.*, **7(4)**, 295–394.

Dixon, J. (1924) California Condors breed in captivity. *Condor*, **26**, 192.

Emslie, S.D. (1987) Age and diet of fossil California condors in Grand Canyon, Arizona. *Science*, **237**, 768–70.

Finley, W.L. (1908) Life history of the California Condor. III. Home life of the condors. *Condor*, **10**, 59–65.

Janssen, D.L., Oosterhuis, J.E., Allen, J.L. *et al.* (1986) Lead poisoning in free range California condors. *Journ. Am. Vet. Med. Ass.*, **189(9)**, 1115–17.

Koford, C.B. (1953) The California condor. *National Audubon Soc. Res. Rep.* **4**, 1–54.

Kuehler, C.M., Sterner, D.J., Jones, D.S. *et al.* (1991) Report on captive hatches of California condors *Gymnogyps californianus*: 1983–1990. *Zoo Biol.* **10(1)**, 65–8.

Kuehler, C.M. and Witman, P.N. (1988) Artificial incubation of California condors *Gymnogyps californianus* eggs removed from the wild. *Zoo Biol.*, **7(2)**, 123–32.

Lieberman, A., Wiley, J., Rodriquez, J.V. and Paez, J.M. (1991) First experimental reintroduction of captive reared Andean condors *Vultur gryphus* into Colombia, S.A., in *American Assoc. Zoological Parks and Aquariums Annual Conf. Proc. 1991*, AAZPA, Wheeling, WV, pp. 129–31.

Snyder, N.F.R. and Johnson, E.V. (1985a) Photographic censusing of the 1982–1983 California condor population. *Condor*, **87**, 1–13.

Synder, N.F.R. and Johnson, E.V. (1985b) Photos key to condor census. *Outdoor Calif.*, **46**, 22–5.

Snyder, N.F.R. and Snyder, H.A. (1989) Biology and conservation of the California condor, in *Current Ornithology vol. 6* (ed. D.M. Powell), Plenum Press, New York.

Toone, W.D. and Risser, A.C. (1988) Captive management of the California condor *Gymnogyps californianus. Int. Zoo Yb.*, **27**, 50–8.

Wallace. M.P. (1991) Methods and strategies for releasing California condors to the wild, in *American Assoc. Zoological Parks and Aquariums Annual Conf. Proc. 1991*, AAZPA, Wheeling, WV, pp. 121–8.

Wallace, M.P. and Temple, S.A. (1983) An evaluation of techniques for releasing hand-reared vultures to the wild, in *Vulture Biology and Management* (eds S.R. Wilbur and J.A. Jackson), University of Calif. Press, Berkeley and Los Angeles, pp. 400–23.

Wallace M.P. and Temple, S.A. (1987) Releasing captive reared Andean Condors to the wild. *J. Wildl. Mgmt.* **51**, 541–50.

Wiemeyer, S.N., Scott, J.M., Anderson, M.P. *et al.* (1988) Environmental contaminants in California Condors. *J. Wildl. Mgmt.*, **52(2)**, 238–47.

Wilbur, S.R. (1978) The California Condor, 1966–1976: a look at its past and future. *North American Fauna*, **72**, 1–136.

24

The captive breeding and conservation programme of the Bali starling (*Leucopsar rothschildi*)

B. van Balen
ICBP Indonesia Programme, Indonesia
and
V.H. Gepak
Kebun Binatang Surabaya, Indonesia.

24.1 INTRODUCTION

The Bali starling (*Leucopsar rothschildi*) is a strikingly beautiful silky-white bird with black tips to the flight feathers and blue naked skin around the eyes. It was first described and placed in a monotypic genus by Stresemann in 1912, a year after he discovered it in the dry lowland forest along the coast of north-west Bali (Stresemann, 1912). Since its discovery the numbers have declined and its distribution has receded. In the 1920s it occupied roughly 30 000 ha of uninhabited land (Paardt, 1926; Plessen, 1926; Helvoort, 1990), but with the progressive conversion of forest to agricultural land, by the late 1980s its range had shrunk to less than 4000 ha and the population was restricted to a small part of the Bali Barat National Park in the north-west of the island (Helvoort, 1990). In the last 20 years the decline in numbers has been accelerated by trapping for the international pet trade and an increased demand from aviculturists. By 1990 the total wild population was estimated to

Creative Conservation: Interactive management of wild and captive animals.
Edited by P.J.S. Olney, G.M. Mace and A.T.C. Feistner.
Published in 1994 by Chapman & Hall, London.
ISBN 0 412 49570 8

be as low as 13 (Balen and Soetawidjaya, 1991). The Bali starling has been included in the IUCN Red Data Book since 1966, in the Endangered category, and international trade is prohibited under the Convention on International Trade in Endangered Species (CITES). Since 1970 the species has had absolute protection under Indonesian law.

In 1983 the Indonesian government, represented by the Directorate-General of Forest Protection and Nature Conservation (PHPA) formally requested the International Council for Bird Preservation (ICBP) to draw up and put into action a conservation project for the Bali starling. The implementation of this project was preceded by a feasibility and preparation period of 1983–1986 (Helvoort, Soetawidjaja and Hartojo, 1986), and by 1987 PHPA, ICBP, the American Association of Zoological Parks and Aquariums (AAZPA) and the Jersey Wildlife Preservation Trust (JWPT) had produced a comprehensive five-year cooperative programme.

The agreed overall aim was to restore a viable and self-sustaining population in Bali Barat National Park. The objectives included:

(a) monitoring and protecting the birds in the wild;
(b) establishing a captive breading programme in Indonesia with input from other captive breeding programmes in America, Jersey and elsewhere;
(c) restocking the wild population;
(d) promoting public awareness.

A further three-year plan was agreed in 1992 which continued the original aims and expanded the objectives to include;

(a) stopping the illegal capture of birds;
(b) reducing the demand for wild-caught birds;
(c) establishing new populations within the species' dispersive range from captive stock
(d) continuing to promote an awareness of the cultural and aesthetic value of conserving the Bali starling in the wild;
(e) undertaking management-orientated studies of the behaviour and ecology of the species;
(f) developing the capability of the Bali Barat National Park to be self-sufficient in conserving the species.

24.2 CAPTIVE STOCK WORLD-WIDE

Fortunately there has been for some time a relatively large captive population in zoos, bird collections and private collections world-wide. This population has been estimated to be in excess of 700 individuals (Helvoort, 1990), but only recently has there been an attempt to cooperatively manage parts of this scattered population. Poor record-keeping and uncontrolled breeding has made any analyses and management difficult. There are two regional studbooks which do provide usable data. One, the American studbook, which is under the auspices of AAZPA and one of their Species Survival Plans (SSP), recorded as of July 1992, 381 birds in 68 participating institutions (Seibels, 1992). The other, which is under the auspices of the Federation of Zoological Gardens of Great Britain and Ireland, registered as at the end of 1991, 110 birds in 20 insititutions (Fisher, 1992). In Europe the European Association of Zoos and Aquaria (EAZA) has approved the setting up of a coordinated breeding programme (EEP).

24.3 CAPTIVE BREEDING PROGRAMME IN INDONESIA

24.3.1 Breeding stock

In August–November 1987 the breeding facility already in existence in Surabaya Zoo in east Java was renovated. This facility comprised 29 aviaries with 16 Bali starlings. In November 1987 the captive population was increased with the addition of 37 birds donated by zoos and private collections in the USA and by Jersey Wildlife Preservation Trust. Five of these birds, most of them over 10 years old, died shortly after arrival. In 1992 the breeding stock in Surabaya Zoo consisted of 44 birds, with 18 birds (11 male, seven female) born before 1985, 10 birds (four male, six female) born in 1985 or later, and 16 birds (six male, eight female, two unknown) of unknown age, but all born before 1987.

24.3.2 Husbandry

A number of publications on breeding Bali starlings have appeared in the last two decades (Taynton and Jeggo, 1988; Partington *et al.*, 1989; see also bibliographies in Seal (1990) and Seibels (1991). Husbandry used in Surabaya Zoo followed the guidelines given by American zoos (Seibels, 1991) and a brief account only is given here.

The breeding aviaries for single pairs were at least 2.5 m high, 2.5 m wide and 4 m deep, well planted with low shrubs and small trees. Breeding results during the first season were disappointing and measures to enhance productivity were taken:

(a) aviaries were screened in order to avoid interaction with starlings in adjoining breeding units;
(b) in 1989 the old nest-boxes were replaced by boxes that followed a design widely used in the USA (Seibels, 1991);
(c) disturbance from visitors to the zoo was decreased by closing off the immediate surroundings of the aviaries;

(d) birds with poor breeding performance were repaired;
(e) in addition to dry food pellets, fresh local fruits (papaw, bananas) and live food (mealworms, ant pupae) were provided.

24.3.3 Breeding results

Egg hatching during the year was satisfactory. Mortality, however, after hatching remained high and to date an average of only six to nine birds reach maturity each year. The introduction of new nest-boxes in August 1989 resulted in some increase in the number of clutches, but the chick mortality stayed high. To date 39 birds have been successfully raised. Figure 24.1 shows the breeding results for 1988–1992. As three pairs that were amongst the most productive of the breeding stock, were stolen in March 1991, the period April 1991 to July 1992 has been omitted from the graph. The stolen birds were retrieved in the second half of 1992.

24.3.4 Studbooks

In order to manage the captive population scientifically, and in particular to minimize inbreeding, carefully maintained studbooks are essential. American and British birds are already registered, and regional studbooks for birds in Indonesia and in Europe are being prepared. Other studbooks should be prepared for birds held in Japan and Singapore.

Taynton and Jeggo (1988) found evidence of increased chick mortality with higher inbreeding levels in Jersey birds, and Helvoort (1990) reported an inbreeding depression in the American population leading to a reduction in fertility. A recent study (Thohari *et al.*, 1991) indicated an extremely low heterozygosity of certain blood protein types in the captive population held in Indonesia, with as yet unknown implications for the species. The introduction of wild-caught Bali starlings, of which a fair number are still in private hands, in Java and Bali would diversify the captive gene pool.

24.3.5 Minimal viable population

It has been tentatively suggested that for a species' long-term survival a minimum effective population of 500 individuals is needed (Franklin, 1980). More specifically a recent population viability analysis (PVA) (Seal, 1990) considered that to be viable, 100 birds in the wild and another 1000 in captivity were desirable, with these two populations being managed as a single meta-population. These numbers are based on little empirical data and their feasibility (especially with regard to the wild population) is doubtful, but it is clear that more birds will decrease

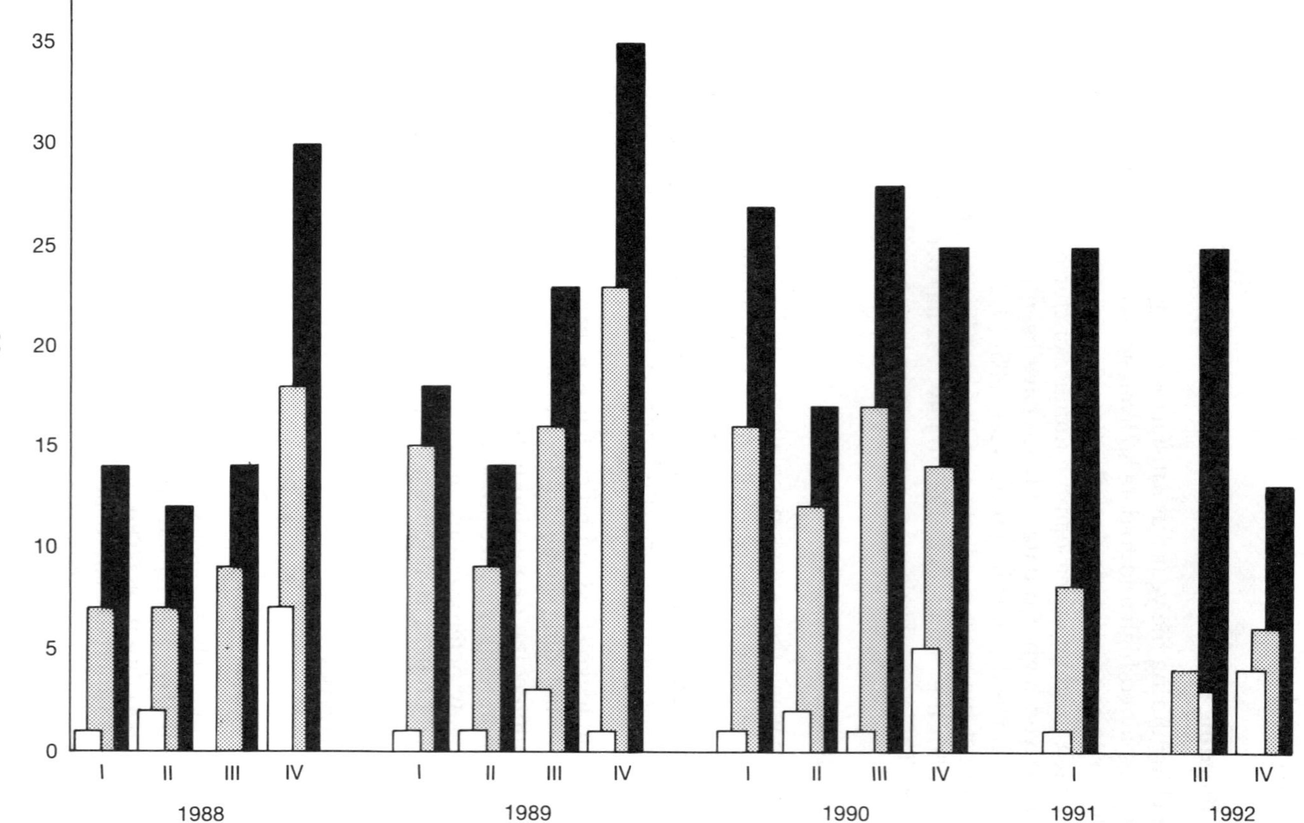

Figure 24.1 Breeding results of the Bali Starling Propagation Centre at the Surabaya Zoo, 1988–1992; reliable figures for the period were not available (I: January–March, II: April–June, III: July–September, IV: October–December; solid bars: number of eggs laid; grey bars: hatchlings, white bars: successfully fledged birds).

the risk of genetic deterioration and extinction. The zoos and private collections in the USA can together house only a restricted number of birds. Those registered in the studbook are closely managed and monitored, as are those in the British studbook, but there are many in the widely scattered world captive population which are outside any managed breeding programme. Coordination amongst collections is necessary to increase the effective size and viability of the captive breeding programme.

Of the 44 birds at present in the Surabaya zoo only four are successful breeders. Recombination of the pairs, especially among the groups of different origin, is essential to enhance productivity and increase the effective breeding population. The extension of the Indonesian propagation programme to other sites would increase the size of the Indonesian captive population and the number of birds for release. To this end, birds obtained under a one-time amnesty campaign from local private owners in exchange for captive-bred birds brought over from the USA (in June 1992, 17 birds were brought over for this purpose), and those handed over directly with the help of local PHPA officers, are now being registered. To date more than 80 birds have been registered and fitted with transponders. In 1993 they will be placed in the new additional captive breeding centres, or be released into the wild, if considered appropriate.

24.3.6 Release programme

The building of a Pre-release Training Centre (PTC) in the Bali Barat National Park was completed in June 1988. The unit comprises 10 aviaries each 5 × 3 × 2 m and follows the design of the Captive Propagation Centre in Surabaya zoo. The PTC is located in an off-public site, with restricted access for interested visitors. The aviaries are sufficiently isolated to reduce any habituation to humans, including the bird keeper.

In autumn 1987, when less than 50 birds survived in the wild, a release was felt to be justified. In July 1988 the first group of three birds from the Surabaya zoo was accomodated in the PTC. To adjust the birds to their future environment, they received six week's training which focused on developing skills for foraging in the wild, retaining fear of humans, and gradually accustoming the birds to the boxes in which they were to be transported to the release site. In the field, the birds were released in turns during the first week, in order to maintain the birds close to the release site – the caged birds always attracted the ones already released. This attempt resulted in one known casualty and the disappearance of the other two birds. The extremely dry conditions and strong wind at the time of release, the birds unfamiliarity with the area, and the location of the site distant from any known Bali starling roosting area may have been contributory factors which caused the failure (Helvoort, pers. comm.)

During the Bali starling PVA workshop, held in Bogor, Indonesia, and attended by an international group of conservationists, aviculturists, and other experts (see Seal, 1990), it was decided that a second attempt to release captive-bred birds into the wild population should be undertaken as soon as possible.

Accordingly, in April 1988, eight captive-bred birds were brought over from Surabaya Zoo to Bali. One bird died shortly afterwards, probably due to stress, and another was considered unfit for release. The birds were given various kinds of wild fruits which were known to have been eaten by Bali starlings. They readily took various arthropds, including scorpions and millipeds, and small reptiles, that entered their cages. They showed instinctive reactions towards raptors flying over. In early 1990, two birds confiscated in east Java were added to the PTC group; one of these was considered for release, but the other was assumed to be unfit for release as its malformed bill suggested it had been hand-raised.

In April 1990 another six captive-bred birds were transported from Surabaya Zoo to Bali Barat and housed in the PTC. The second attempt was planned for a location in the far north-east edge of the Prapat Agung Peninsula in Bali Barat, a short distance form the Teluk Kelor guard house. Here wild Bali starlings were known to roost regularly. A two-compartment simple cage (2 × 2 × 2.5 m) was built on the site to serve as training accommodation. On the day the birds were transported to the release site, transponders were inserted, and colour rings were attached. Special heavy-duty rings designed for the Bali starling conservation programme were attached to the two confiscated birds, whilst the other birds had their metal zoo rings. In order to tell the released birds from the wild birds during at least the first weeks, the breast feathers of the birds to be released were dyed red with rhodamine B. Two coloured plastic spiral rings were attached to all the birds each coded with a unique combination of numbers. On April 15 1990 the first four birds were released from the cage where they had been housed during the previous two days. On April 17, three birds were released, followed by two, two, one and one on each consecutive day. Birds unfit for release stayed in the cage to decoy released birds back to the cage, where food and water were provided during the first weeks. The wild-caught confiscated bird that was released with the captive-bred birds, and which was expected to act as a sort of guide was a disappointment: immediately after its reluctant take off, it flew away in a direct line and was never seen again.

Daily monitoring of the starlings by telescope (20–60x) from a hide near the release site was maintained during the first weeks, where food pellets, fruits and drinking water were provided. Acceptance of the wild by the released birds went smoothly and soon mixed foraging, communal anting and roosting flocks could be seen. One pair was formed within one week and the wild bird would follow its partner close to the food and drinking water container. The observation of several birds around the site, but not at the drinking place soon after release, suggested that acclimatization was rapid in some cases.

Reading of the ring codes became increasingly difficult, as the birds became more wary in the process of adaptation, making the success rate hard to assess. This was aggravated by unexpectedly high poaching pressure near the sites. Within one month one of the released birds, detectable by its transponder, was rediscovered in the hands of a local bird dealer, and an unknown number of other birds may have been trapped. In early October 1990, however, six months after the release, a marked bird was identified about 8 km from the release site. It was observed copulating with a wild bird (in the *kapok* plantation enclave along the main road that cuts through the National Park), but disappeared soon after. In November the same year, another released bird was rediscovered, paired to a wild bird. In January the following year this pair successfully raised three young in a tree hole not far from the release site.

A major decline in numbers of Bali starlings was found during the pre-breeding census of 1990. Even with the 13 released birds, numbers had dropped to some 15 birds largely due to poaching. The following breeding season guarding of the park was increased, but no releases took place, primarily because sufficient were not available from the Surabaya breeding facility. Poaching appeared to be better controlled, though still going on, and the post-breeding censuses completed in June 1991 and June 1992 showed about 35 and 55 birds, respectively (Figure 24.2). Eight occupied nests were located in the Teluk Kelor/Batu Gondang area. Only natural nest holes were used, and again the nestboxes provided a few years ago were ignored.

24.3.7 Disease

Though a quarantine period in the PTC is common practice before release, and the birds kept in the zoo are examined regularly, there is still a considerable risk of disease transmission. The incidence of atoxoplasmosis in captive Bali starlings in American zoos (Partington *et al.*, 1989) is especially worrying. Following a discussion paper prepared by PHPA and ICBP, an AAZPA team came from the USA in August 1992 to examine the birds held in Indonesian zoos and in the PTC in Bali. A medical quarantine protocol for all birds to be released and for birds

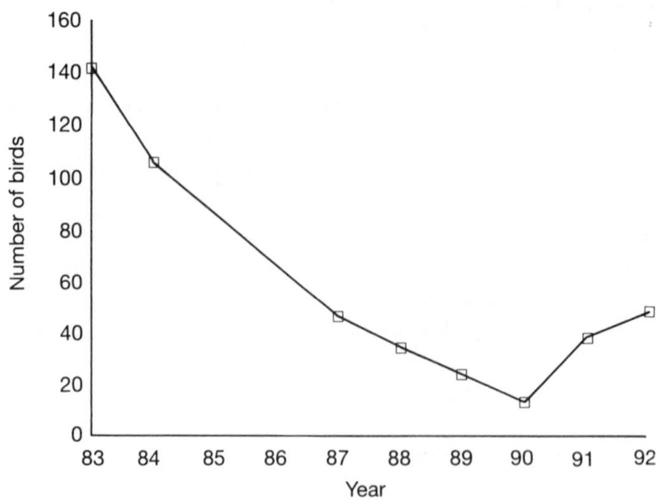

Figure 24.2 Bali starling numbers in the Bali Barat National Park, 1983–1992 (after van Balen and Jepson, 1992).

held in captive breeding centres has since been developed (Appendix 4 of Balen and Jepson, 1992).

To further reduce the risk of disease transmission, release in the future will be in former, but now empty, Bali starling habitat. This reintroduction, as opposed to restocking, will involve rather different and more elaborate release techniques, as no resident guides will be available. An intensive field study is being prepared by ICBP and AAZPA, aimed at collecting data on behaviour and breeding success of released birds (M. Collins, pers. comm.). A possible release site has been identified on the island of Menjangan, pending more information on diseases in the captive population and full control of poaching in the area. Furthermore, the use of radio-telemetry is being considered and preliminary tests on captive starlings has had promising results (Elbin *et al.*, 1991).

24.4 CONCLUSION

Habitat availability in the present National Park as a limiting factor on the recovery of the Bali starting is currently being investigated by the project. There may not be enough suitable habitat in the national park to support more than a three- or fourfold increase in the present population of 55 Bali starlings. Even if attained, this figure would be far below any number suggested for a viable population. Any continuation of the release programme will have to take this into account and the conversion of the plantation enclaves that exist in the Bali Barat National Park into Bali starling habitat must keep pace with an increasing number of starlings.

Considerable time and effort has been put into the captive-breeding programme, but to date its success in terms of contribution to the conservation of the Bali starling has been limited. The recovery of the wild population following the improvement in protection shows that other techniques can be possibly more immediately efficient. However, the potentially deleterious consequences of inbreeding cannot be discounted and the introduction of new genes is justified. Further releases are planned for 1993 and feasibility studies are now being carried out. To avoid disturbance of the present wild population, sites in Bali starling habitat other than the previous ones will be selected.

ACKNOWLEDGEMENTS

The Bali Starling Project Phases 1–3 would not have been possible without the assistance and commitment of a large number of people and institutions. The Project is managed by ICBP and financed by AAZPA and the New York Zoological Society, Liz Claiborne/Art Ortenberg Foundation, and JWPT. The PHPA head offices, the Bali Barat National Park and Surabaya Zoo were the local partners in the implementation of the Project. In particular, M. Noer Soetawidjaja, Slamet Suparto and Made Rasma c.s. were most closely involved in the release programme. Thanks are forwarded to the former Bali Starling Project officer, Bas van Helvoort, for the discussions and support provided during the early stages of the project, and to Paul Jepson and Professor H. Prins for their valuable comments on an early draft of the paper.

REFERENCES

Balen B. van and Jepson, P. (1992) *Bali Starling Project: Activity Report, January-August 1992*, ICBP Indonesian Programme, Bogor.

Balen, B. van and Soetawidjaja, M.N. (1991) Bali Starling Project: Interim Report October–December 1990. Internal Document. ICBP Indonesian Programme, Bogor.

Elbin, S.B., Burger, J., Koontz, F.W. and Bruning, D. (1991) *Preliminary evaluation of radio-transmitter attachment methods for captive and reintroduced Bali Mynahs*. Poster presentation at the Am. Ornith Union meeting, Montreal.

Fisher, I.J. (1992) *The Bali Starling Regional Studbook 31 Dec. 1991*, Zool. Soc. of London, London.

Frankin, I.R. (1980) Evolutionary change in small populations, in *Conservation Biology* (eds M.E. Soulé and B.A. Wilcox), Sinauer Associates, Sunderland, MA, pp. 135–49.

Helvoort, B.E. van (1990) The Bali Starling *Leucopsar rothschildi* Stresemann 1912;

its current status and need for conservation. *ASEAN Workshop on Wildlife Res. and Manag.* PHPA, Bogor, pp. 115–31.

Helvoort, B.E. van, Soetawidjaja, M.N. and Hartojo, p. (1986) *The Bali Starling* (Leucopsar rothschildi): *A Case for Wild and Captive Breeding*, ICBP, Cambridge UK.

Paardt, Th. van der (1926) Manoek Putih: *Leucopsar rothschildi. De Tropische Natuur.*, **15**, 169–73.

Partington, C.J., Gardiner, C.H., Fritz, D. *et al.* (1989) Atoxoplasmosis in Bali mynahs *(Leucopsar rothschildi). J. Zoo and Wildlife Medicine*, **20**, 328–35.

Plessen, V. von (1926) Verbreitung und Lebensweise von *Leucopsar rothschildi* Stres. *Ornith. Monatsb.*, **34**, 71–3.

Seal, U.S. (ed.) (1990) *Bali Starling* Leucopsar rothschildi: *Viability Analysis and Species Survival Plan, Workshop Report*, CBSG/IUCN, MN.

Seibels, R.E. (1991) *1990 Regional Studbook for the Bali Mynah* (Leucopsar rothschildi), Riverbanks Zoological Park, Columbia.

Seibels, R.E. (1992) *Bali Mynah*, in AAZPA Annual Report Conservation and Research Report (1991–1992) (eds R.S. Weise, M. Hutchins, K. Willis and S. Becker), AAZPA, Bethesd, Maryland, pp. 177–8.

Stresemann, E. (1912) Description of a new genus and a new species of bird from the Dutch East Islands. *Bull. B.O.C.*, **31**, 4–6.

Taynton, K and Jeggo, D. (1988) Factors affecting breeding success of Rothschild's Mynah (*Leucopsar rothschildi*) at the Jersey Wildlife Preservation Trust. *Dodo, J. Jersey Wildl. Preserv. Trust*, **25**, 66–76.

Thohari, M., Masyud, B., Mansjoer, S.S. *et al.* (1991) Comparative study on blood protein polymorphism of captive Bali Starling *(Leucopsar rothschildi)* from Indonesia, the United States and England. *Media Konservasi*, **3**, 1–10.

25

An experimental reintroduction programme for brush-tailed phascogales (*Phascogale tapoatafa*): the interface between captivity and the wild

T.R. Soderquist
Monash University, Victoria, Australia
and
M. Serena,
Healesville Sanctuary, Victoria, Australia and Chicago Zoological Society, Illinois, USA.

25.1 INTRODUCTION

The tuan or brush-tailed phascogale (*Phascogale tapoatafa*) is a carnivorous marsupial belonging to the family Dasyuridae. Adult phascogales weigh 110–310 g and are primarily aboreal in their habits. Like many of the smaller Australian temperate zone marsupials, phascogales breed only once a year, in May–July during the austral winter. Females are monoestrous and can accommodate up to eight young in the pouch; the size of litters has been reported to vary from three to eight both in captivity and in the wild (Cuttle, 1982; Halley, 1992). Young are firmly

Creative Conservation: Interactive management of wild and captive animals.
Edited by P.J.S. Olney, G.M. Mace and A.T.C. Feistner.
Published in 1994 by Chapman & Hall, London.
ISBN 0 412 49570 8

Figure 25.1 Distribution of *Phascogale tapoatafa* in Australia. Shaded areas represent former distribution; hatched areas represent approximate current range.

attached to the teat until the age of seven weeks and are thereafter deposited in a nest when their mother is active. As in the related genus *Antechinus*, wild male phascogales characteristically die soon after their first breeding season at the age of about one year (Cuttle, 1982).

Phascogales currently are found at low densities in areas of dry open sclerophyll forest or woodland in the Northern Territory and the states of Victoria, New South Wales, Queensland and Western Australia (Figure 25.1). Since European settlement a large proportion of suitable phascogale habitat has been cleared for agriculture or grazing by domestic livestock, contributing to a major decline in the species' range in Victoria (Figure 25.2). Much of the remaining forest habitat occupied by phascogales in Victoria has been highly fragmented, isolating small populations in restricted enclaves. For example, Chiltern State Park, an isolated forest remnant in north-eastern Victoria, supports an estimated 35–50 breeding females in 4200 ha of high quality habitat (Soderquist, unpublished data).

Phascogales have been maintained in captivity at Healesville Sanctuary since the 1930s. However, no concerted effort was made to breed the species regularly or in large numbers until 1990. At that time, the decision was taken to assess and improve breeding protocols, so captive-bred juveniles would be available for an experimental reintroduction programme, jointly sponsored by Healesville Sanctuary and the Chicago Zoological Society.

Figure 25.2 Distribution of *Phascogale tapoatafa* in Victoria. Shaded areas represent the extent of the historical distribution; hatched areas represent the current range. The location of reintroduction sites in this study is identified by a star. (Based on Atlas of Victorian Wildlife, 1992.)

25.2 EXPERIMENTAL REINTRODUCTION

25.2.1 Criteria

The selection of *P. tapoatafa* as an appropriate subject for experimental reintroduction studies was predicated on the following criteria being met.

(a) Phascogales are sufficiently rare that any information bearing on optimum reintroduction protocols is inherently valuable in helping to conserve the species. In addition to developing appropriate captive breeding and reintroduction techniques for phascogales, a major objective of the experimental programme was to obtain information on the site-specific causes of phascogale mortality in areas of forest habitat from which the animals had disappeared.

(b) Phascogales still occur in sufficient numbers that data can be obtained reasonably readily from wild populations as the basis for evaluating how well released animals cope with their environment.

(c) The animals are large enough to carry a reasonably long-lived and powerful radio-transmitter collar for post-release monitoring purposes.

(d) A suitable release site could be identified, namely an area of forest in Gippsland, in south-eastern Victoria, where phascogales are known to have occurred as recently as the 1960s (Atlas of Victorian Wildlife, 1992). The habitat at this site appeared to be comparable to areas in central Victoria where phascogales persist, both in terms of tree species composition and the presence of mature trees (potential nest sites).

(e) Captive husbandry techniques for the species were initiated and have been improved at Healesville Sanctuary, and suitable captive breeding stock was available (Halley, 1992).

25.2.2 Information from the wild

Information on the ecology of wild phascogales was central to the development of pre-release and release protocols for captive-bred individuals. For example, based on radio-tracking and spotlight observations it was known that juveniles first venture outside the immediate vicinity of their nursery nest at the age of 15 to 16 weeks, are weaned at 19 to 21 weeks, and typically disperse from the maternal home range by 36 weeks (Soderquist, unpublished data). Foraging by both juveniles and adults is primarily arboreal and involves gleaning insects on and under bark. Activity is nocturnal, with the daylight hours spent in a small tree hollow or similar nest site.

25.2.3 Captive breeding and management protocols

Captive breeding protocols for phascogales were developed based on the assumption that animals were most likely to reproduce successfully if the mating system in captivity mimicked that in the wild. From radio-tracking and direct observational data, it was known that breeding females occupy mutually exclusive home ranges which overlap with those of one to several males (Soderquist, unpublished data). Females are much more sedentary than males, and breeding appears to be promiscuous with no evidence of prolonged mate guarding by males. Salient features of the corresponding captive breeding protocol included maintaining females individually in separate enclosures and introducing males to female enclosures rather than vice versa. Females were paired as close to the anticipated onset of oestrus as possible in order to minimize potential premature harassment by males. In 1990, four females were each paired with a single male, while in 1991 a 'round-robin' system was employed in which individual males were rotated every few days among five females. Both protocols were found to be satisfactory, with all females producing litters in both years.

In order to provide captive-bred phascogales with a range of experiences relevant to life in the wild, juveniles were raised in large (3–4 m wide × 10–12 m long) weldmesh enclosures that were mostly unroofed and hence exposed to natural weather conditions. Enclosures were planted with large grass tussocks and furnished with wooden nest-boxes mounted 1.5 m above the ground and a variety of vertical logs and tree limbs to provide appropriate substrates for climbing. For several weeks before the time of release, mealworms were hidden under bark or in holes drilled in the logs and branches in order to encourage captive-bred animals to forage in a typical wild manner; other foods that were routinely provided included crickets, moths, fly pupae, whole dead mice and two-day-old chicks (small vertebrates are occasionally preyed upon in the wild). Young phascogales were also observed feeding on free-flying insects (beetles and moths) that entered the enclosures.

25.2.4 Release

Young phascogales were released at the age of 30–39 weeks. Animals were initially released from their home nest-boxes, which were transported into the field and mounted on large trees about 6 m above the ground. In order to discourage animals from exiting the boxes before their normal nocturnal activity period, cardboard was taped over the entrance. Phascogales typically responded by remaining inside the nest-box until dusk and then rapidly chewing a hole in the cardboard so they were free shortly after dark. After release, animals were located regularly by means of radio-telemetry both during the day and at night. Phascogales that dispersed out of range of a hand-held receiver were located again by tracking their signals from an aeroplane. Direct observation of nocturnal behaviour was also undertaken regularly for periods of 30–60 min with the assistance of a dim red spotlight and binoculars. All released animals received supplemental food twice a week, usually in the form of mealworms left in paper cups attached to trees at nest sites.

25.2.5 Evaluation

From this project's inception, it was considered essential that success in both the captive breeding and release programmes be evaluated by comparison with relevant parameters in the wild, so as to provide an objective basis for assessing whether specific protocols were appropriate and effective. In the case of the breeding programme, the key variables were conception rate and mean litter size at weaning. Evaluation of success in the release programme was much more complex. Major variables considered in the first year included mortality rate, relative ability of animals to locate secure nest sites, range of foraging behaviours utilized and rate of foraging success, relative ability of animals to navigate successfully in the forest environment, and the extent of long-range dispersal from the release site resulting in phascogales being effectively lost from the newly established population.

The site of the initial 1991 release of four males and four females comprised a tract of prime Gippsland lowland forest which has been fragmented by agricultural clearing. The results of this release demonstrated that captive-bred phascogales were adept at foraging, and, that they typically selected suitable nest sites. The animals were also remarkably good at finding their way through the forest, often returning to preferred nest sites from foraging areas many hundreds of metres away. However, losses due to predation from ground predators (introduced red foxes and feral cats) were high; this initially was thought to be due to high predator densities in adjoining agricultural land. In consequence, the second 1991 release of six males and six females was undertaken in a nearby area of essentially continuous foothill forest habitat. The second group of animals proved to be as capable of finding food and nest sites as the first group; however, they were also subject to very heavy predation by ground predators. In addition, four males were observed to disperse relatively long distances (up to 8 km) from the second release site.

In parallel with the Gippsland release program, two litters of captive-raised wild orphans (comprising five males and three females) were reintroduced to their former maternal home ranges in central Victoria. This work was undertaken to enable a comparison of post-release results in areas where wild phascogales persist with sites where the species had vanished. Both litters were the offspring of radio-collared females, and were hand-reared or fostered to captive females (Hunter, 1991) from the ages of 14 and nine weeks, respectively. In brief, the survival rate of the captive-reared orphans in central Victoria was much better than that of the young animals released in Gippsland, indicating that a high level of predation is not an inevitable outcome when captive-bred phascogales are released, and suggesting that introduced predators may be more abundant in areas from which phascogales have recently disappeared than in areas where the species remains. Males released into established populations in central Victoria also showed less inclination to disperse from the vicinity of the release site than Gippsland males, thereby supporting the prediction that male phascogales prefer to establish home ranges in areas already containing females.

25.2.6 Protocol modification

After considering all of the above results, protocols for the 1992 Gippsland release programme were modified as follows. First, the area selected for reintroduction was baited intensively for introduced mammalian predators (using buried meat baits) beginning about 10 weeks prior to the first phascogales being released. Surveys for fox and cat tracks carried out along dusty road margins in the release area indicated that the baiting technique was successful in greatly reducing foxes but not cats. Second, male phascogales were not released until females had been introduced to and occupied a site for a minimum of two weeks. In brief, survival of the nine males and 18 females released in 1992 was substantially improved relative to the previous year. Of 13 animals known to have died by the beginning of the breeding season, five deaths were attributed to feral cats, five to raptors, two to reptiles (large varanid lizards) and one to a motor vehicle. In addition, the extent of male dispersal from the release area decreased substantially relative to that observed in 1991. A minimum of two females, at least one of which successfully weaned a litter, survived to the lactational period.

Research undertaken on phascogales in captivity was important in providing information needed to implement the release programme and/or enhance understanding of the natural behaviour and ecology of the species. For example, captive growth curves (e.g. head width, tibia length and body mass) were generated by measuring juveniles at scheduled intervals from the beginning of pouch life to weaning. While such data may not accurately predict the growth of wild juveniles through the entire course of their development, measured differences in wild and captive growth rates may also be of value in assessing the degree to which reproduction constitutes an energetic drain on wild females (Serena and Soderquist, 1988). Genetic research based on DNA fingerprinting techniques was undertaken in part to identify the paternity of offspring arising from the round-robin breeding protocol, and in part to investigate aspects of phascogale reproduction that would be more difficult to address in the wild, e.g. whether litters of mixed paternity can occur. Finally, overnight consumption of various foods (meat mix, boiled egg, cheese, whole dead mice, two-day-old chickens, mealworms, crickets, moths and fly pupae) by adult females was carefully monitored over a 10-week period in order to help decide which food type(s) would best function as a post-release supplement. Information on total overnight food consumption was also required to develop guidelines for a baiting programme designed to control foxes and cats but not be hazardous to phascogales and other native carnivores.

25.3 SUMMARY

In summary, it is our experience that captive breeding and reintroduction programmes benefit greatly from the use of extensive prior data on the ecology and behaviour of a species. Unfortunately, the amount of time, effort and resources required to generate detailed ecological information about rare species of fauna is usually considerable. Zoos have a valuable role to play in this regard, both in identifying those species which are particularly high-priority candidates for future reintroduction studies, and helping to support relevant field programmes financially and otherwise.

By the same token, reintroduction studies clearly have significant potential to function as experiments providing insight into the factors causing the decline of threatened fauna. Opportunities also exist for collecting data in the context of captive breeding programmes that both contribute to the design of effective reintroduction protocols and improve understanding of the ecology of rare species. In practice, development of effective, integrated programmes of captive breeding and conservation in the field will most obviously be expedited by zoos hiring trained scientific staff to communicate with external researchers and wildlife managers and co-ordinate their work with in-house activities.

REFERENCES

Atlas of Victorian Wildlife (1992) Unpublished database maintained by Victorian Department of Conservation and Natural Resources, Heidelberg, Victoria.

Cuttle, P. (1982) Life history strategy of the dasyurid marsupial *Phascogale tapoatafa*, in *Carnivorous marsupials* (ed. M. Archer), Royal Zoological Society of New South Wales, Sydney, pp. 13–22.

Halley, M. (1992) Maintenance and captive breeding of the brush-tailed phascogale *Phascogale tapoatafa* at Healesville Sanctuary. *Int. Zoo Yb.*, **31**, 71–8.

Hunter, S. (1991) Fostering tuans. *Thylacinus*, **16** (1), 10.

Serena, M. and Soderquist, T.R. (1988) Growth and development of pouch young of wild and captive *Dasyurus geoffroii* (Marsupialia: Dasyuridae). *Aust. J. Zool.*, **36**, 533–43.

26

Coordinating conservation for the drill (*Mandrillus leucophaeus*): endangered in forest and zoo

E.L. Gadsby and P.D. Jenkins Jr
Pandrillus, Nigeria
and
*A.T.C. Feistner**
Jersey Wildlife Preservation Trust, Channel Islands.

26.1 INTRODUCTION

Drills (*Mandrillus leucophaeus*) and mandrills (*Mandrillus sphinx*) are the only members of their genus, are among the most impressive of the primates, yet among the least known. They have contiguous ranges in the Lower Guinean forest block of west-central Africa. The mandrill is found south of the Sanaga River in Cameroon, Rio Muni, Gabon, and parts of Congo. The drill is found only between the Sanaga River in Cameroon and the Cross River in Nigeria, and on the island of Bioko (Grubb, 1973; Oates, 1986; Blom *et al.*, 1992) (Figure 26.1).

* Corresponding author.

Creative Conservation: Interactive management of wild and captive animals.
Edited by P.J.S. Olney, G.M. Mace and A.T.C. Feistner.
Published in 1994 by Chapman & Hall, London.
ISBN 0 412 49570 8

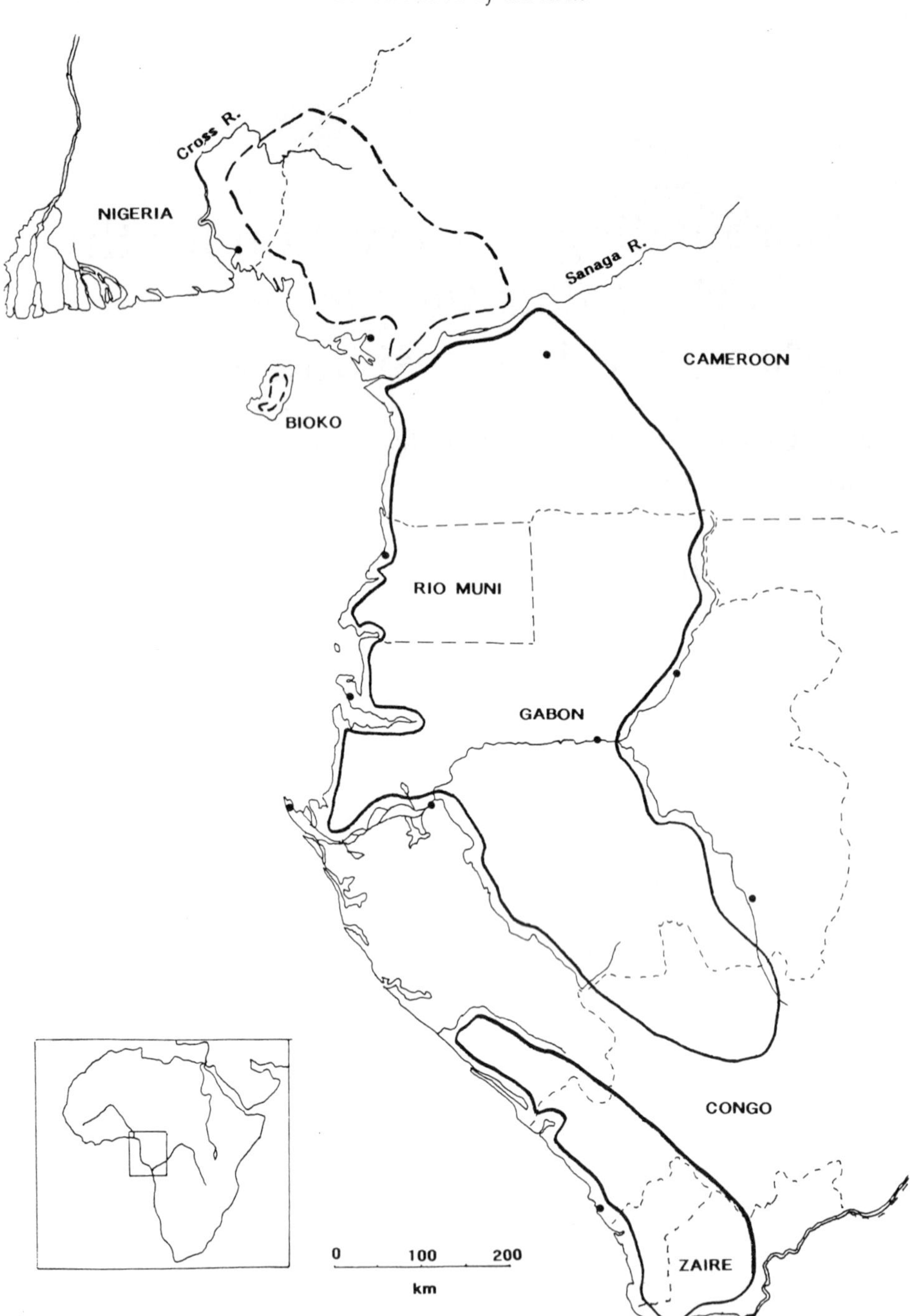

Figure 26.1 Distribution of drills (*Mandrillus leucophaeus*) (– –) and mandrills (*M. sphinx*) (—) in west central Africa. - - - - international boundary.

Both species are large, sexually dimorphic (females 12–15 kg, males 20–40 kg), short-tailed, possess sternal scent glands, and have a rich vocal repertoire (Gartlan, 1970; Hill, 1970; Jouventin, 1975; Kudo, 1987; Feistner, 1991). They are readily distinguished by the pattern of colour on the face. In drills the face is moon-shaped, black, and has a single ridge either side of the muzzle. In mandrills the prominent muzzle is fluted and is red and blue. There are also differences in rump colour. Both species appear very similar in their behaviour and ecology. They are mainly terrestrial, and feed largely on fruit, herbaceous vegetation, and invertebrates (Gartlan, 1970; Jouventin, 1975; Hoshino, 1985; Lahm, 1986). They live in groups of 20–40 individuals, which may contain a single fully mature adult male. Solitary males are also encountered. Social organization appears flexible and in both species groups may split to form foraging sub-groups, or coalesce at certain times of year to form super-groups of 100–300 individuals. They have large home ranges of up to 40 km^2, travel widely each day, and live at low density (Struhsaker, 1969; Gartlan, 1970; Gartlan and Struhsaker, 1972; Jouventin, 1975; Hoshino *et al.*, 1984). All these factors combine to make them difficult to census or study. However, there are strong indications that their numbers are declining rapidly. This is particularly serious for the drill, which has the smaller range.

The drill has been given the highest conservation priority rating by the World Conservation Union (IUCN), is classified as Endangered and is Appendix 1 listed on The Convention on International Trade in Endangered Species CITES (Oates, 1986; Lee, Thornback and Bennett, 1988). This paper describes the problems facing drills in their natural habitats and in the small captive population, the initiatives which have resulted from a recognition of the drill's critical situation, and outlines the advantages of developing an *in situ* facility for drill conservation.

26.2 PROBLEMS FACING WILD DRILLS

No long-term ecological studies have yet been conducted on drills, and the species has suffered from a low international profile, both scientific and popular. Thus, although highly endangered, the conservation problems of drills were long overlooked. The drill has the most limited range of any large African primate, and is thus especially vulnerable to habitat disturbance and destruction. Drills appear to have been eliminated from significant portions of their historical range. Their remaining, highly fragmented, range is about 40 000 km^2 (an area smaller than Switzerland), but only about half the area now contains habitat usable by drills.

The mainland habitat is split up into at least 11 islands, some as small as 100 km^2, which are isolated by barriers to drill migration such as major roads and substantial areas of human settlement and cultivation

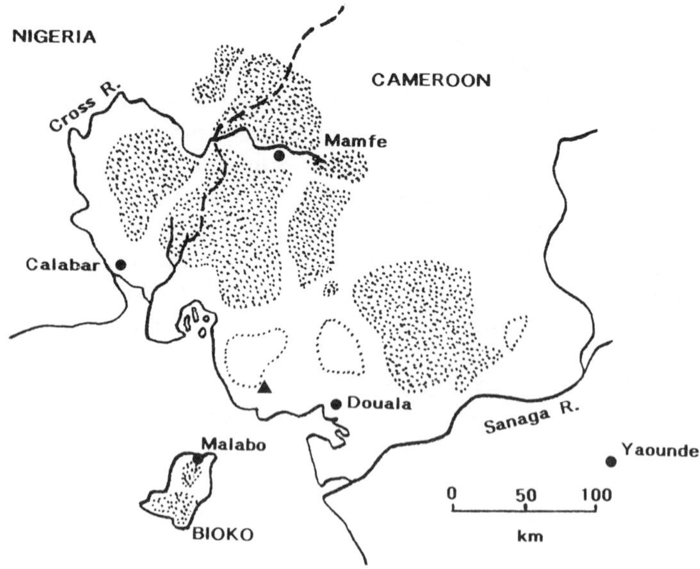

Figure 26.2 Fragmentation of known drill habitat. Habitat islands in Nigeria and Bioko confirmed, those in Cameroon are based on preliminary results of survey in progress. ● Towns; - - - - international frontier, ⁝⁝⁝ forest areas inhabited by drills; ▲ Mt Cameroon.

(Figure 26.2). On Bioko, drills are divided into two populations, isolated by similar barriers (J.R. Gil Gomez, pers. comm.). Within each habitat island, sub-populations are becoming increasingly isolated by the construction of minor roads, expanding cultivation and settlement, logging activities and local hunting pressure. As development continues these minor barriers are growing, becoming major barriers which will eventually effectively isolate the small and decreasing sub-populations of drills into yet smaller and more fragmented habitat islands. These small, isolated sub-populations are even more vulnerable to disturbance and hunting pressure, thus leading to an 'extinction vortex'.

Within remaining habitats hunting for bushmeat is the greatest threat to the species' survival. Hunting is virtually unchecked in all known habitats and is presently increasing in at least two areas, and probably more. Drills are hunted indiscriminately with 12-gauge shotguns and are especially vulnerable to local methods in which dogs are trained to hold a group 'at bay' in small trees. Thus immobilized, one or more hunters can massacre the animals, shooting-gallery fashion. Reports of eight to 10 drills shot in a single encounter are common and instances of 20 or more have been confirmed. In this way whole groups are devastated in a single afternoon. Reproduction cannot compensate for such heavy losses.

All mammals have traditionally been hunted as a source of protein for subsistence in this part of Africa. However, in the past 10–15 years the nature of hunting has changed from a subsistence activity to a commercial one, albeit illegal. At the same time, the introduction and increasing availability of modern weapons has boosted kill-rates to unprecedented and unsustainable levels. Although drills are regarded as one of the most preferred bushmeats, the vast majority of hunters now sell carcasses for the cash income rather than use them to feed their families. Highly organized commercial hunting operations are becoming common, particularly in parts of Cameroon. Professional hunters establish semi-permanent camps in the forest and smoke carcasses for routine vehicle pick-up by bushmeat traders. These then sell the meat in urban centres and provide the hunters with supplies and ammunition, which may include superior imported weapons.

The relentless persecution reduces group size, lowers density, increasingly isolates groups of drills, and may also be affecting behavioural and ecological strategies. It is becoming apparent that the formation of super-groups, which may play a crucial role in transfer of individuals, and thus genetic material, is occurring with decreasing frequency.

26.2.1 Current conservation initiatives for wild drills

Drills are protected on paper by the Endangered Species Decree in Nigeria and by some hunting and firearms restrictions in Cameroon and Equatorial Guinea. They are listed in Class B of the African Convention and thus in theory can only be hunted, captured, killed, or collected under authorization issued by the competent authority. Nigeria and Cameroon (though not yet Equatorial Guinea) are signatories to CITES, on which the drill is Appendix 1 listed. However, none of these governments has the resources or trained personnel to enforce their laws, and the public is often unaware of their existence.

In Nigeria, over 50% of remaining drill habitat has been gazetted to the Cross River National Park (3720 km^2). In Cameroon, the Korup National Park in Southwest Province has an area of 1259 km^2 (Figure 26.3). Portions of these two parks are contiguous along 17 km of border. Two protected areas of about 525 km^2 and 700 km^2 have been decreed for Bioko (Fa, 1991; 1992). The establishment of protected areas such as these is a great stride forward for the conservation of forest habitats. However, simply preventing forest destruction does not in itself ensure the continuing existence of drills and other wildlife. Despite the establishment of large internationally financed conservation projects for both

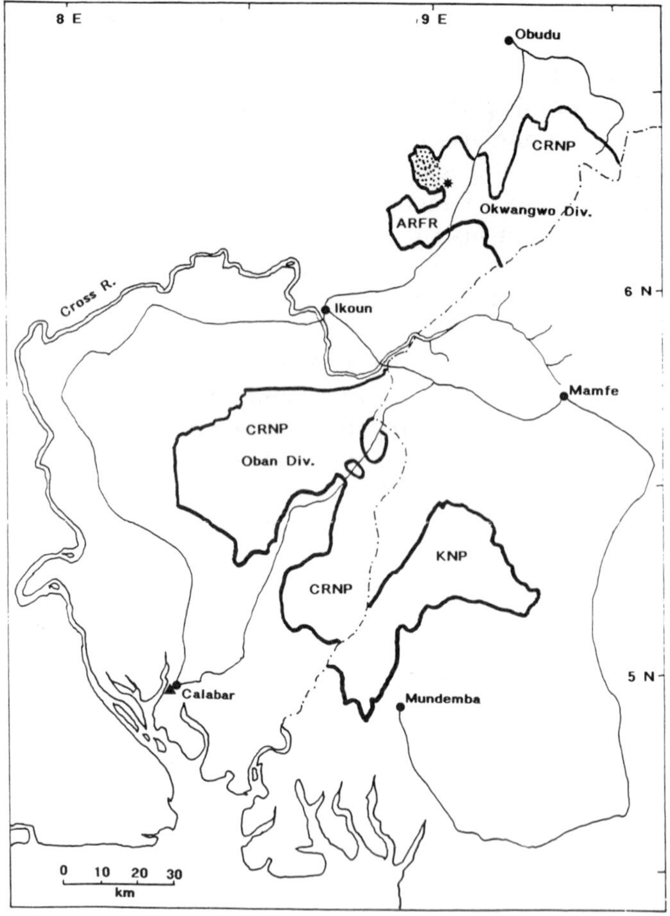

Figure 26.3 Map showing Cross River National Park (CRNP), Nigeria; Korup National Park (KNP), Cameroon; Afi River Forest Reserve (ARFR), and ⁝⁝⁝ proposed Afi River Wildlife Sanctuary, Nigeria; ▲ present location of interim facility at Drill Ranch; ★ proposed site of permanent facility – the Drill Rehabilitation and Breeding Centre; –·–·– international frontier.

national parks, no **effective** protection for their wildlife yet exists, and hunting levels have remained constant or even increased within park boundaries. Even where protected areas exist, drills are vulnerable to intense predation around the edges. Because of their low overall density, this means that populations can be easily 'drained' out of even very large areas. Outside the parks, hunting and habitat destruction continues unchecked.

In order to assess the status of the drill in the wild, a survey of mainland drill populations has been initiated by two of the authors (E.L.G and P.D.J) to define the species' present distribution, to examine the role of hunting in the species' decline, to identify the principal threats to each remaining habitat area, and to estimate the long-term viability of sub-populations. This work is carried out by ground-checking habitat areas against existing maps, and by intensive interviews with experienced local hunters, and is coordinated with the respective government departments on an ongoing basis. The survey in Nigeria was completed in 1990 and the Cameroon work will be finished in 1993 (Gadsby, 1990). Based on results, practical recommendations for a comprehensive conservation programme for the drill will be drawn up. Survey work has also been undertaken on Bioko (Butynski and Koster, 1990; Schaaf, Butynski and Hearn, 1990; Schaaf, Struhsaker and Hearn, 1992).

26.3 PROBLEMS FACING CAPTIVE DRILLS

Unfortunately it is also realistic to regard drills as endangered in captivity. The total world population currently recorded in the International Studbook is about 65 individuals (Böer, 1992). Additionally, the breeding record is extremely poor. For example, the last successful drill birth in the USA occurred in 1982. The situation is a little better in Europe, with reproduction occurring regularly at two institutions and sporadically in others; however since births are matched by deaths of older animals the population is not growing. This can be attributed to the fact that drills have been scattered amongst the world's zoos and many are kept in pairs or singly. Many captive drills have been housed for long periods in sub-standard accommodation and are behaviourally inadequate. Many exhibit socialization problems and some do not even recognize conspecifics as suitable breeding partners.

26.3.1 Current conservation initiatives for captive drills

In response to the conservation priority given to drills in the wild and the recognition that reproduction in captivity was declining, various initiatives have been taken to enhance the role zoos can play in the continued survival of this threatened species. International and regional studbooks (North America, Europe, Japan) have been compiled. In the USA a Species Survival Plan (SSP) has been developed and five zoos there have attempted to set up drill propagation groups for the 8.14 drills i.e. 8 males, 14 females. A European Endangered Species Programme (EEP) has been established in Europe to coordinate reproduction among the 9.15 drills held there. In Japan, efforts are underway to integrate the 5.4 drills held into international programmes. These initiatives have resulted in:

(a) an increase in the transfer of information and drills between institutions, e.g. drills have been moved within the SSP and EEP and four drills have gone from EEP to SSP;
(b) an increase in the number of studies with captive drills, on behaviour (e.g. all USA zoos are using the same ethogram and data collection protocols), on environmental enrichment for drills, and on reproduction (urine samples are collected from female drills for hormonal analysis and semen samples from males for possible artificial insemination);
(c) an increase in general awareness and public education about drill conservation problems and a willingness to raise funds for both *in situ* and *ex situ* work.

(Böer, 1987a,b, 1988, 1989; Cox, 1987a,b, 1991; Desmond, Laule and McNary, 1987; Hearn, Weikel and Schaaf, 1988; DuBois, 1991.)

In the process of survey work in Nigeria and Cameroon, various initiatives with African zoos holding drills are being explored, and improvements are being made where possible. The two Cameroonian zoos hold 3.3 drills. In one, 2.2 animals that were kept singly have now been re-housed together. In Nigeria over 10 drills are kept at the seven zoos so far investigated, but numerous zoos have not yet been visited. The possibility of developing within-country coordination for breeding drills in African zoos is currently being discussed.

However, even with these initiatives the legacy of decades of lack of interest and mismanagement means that the prospects for drills in captivity are likely to improve only slowly.

26.3.2 The role of *in situ* captive breeding

There is a third initiative, which has the potential to contribute significantly to the conservation of drills both in captivity and in the wild. This is the establishment of a rehabiliation and captive breeding centre in a country with a wild drill population. There are numerous advantages to this strategy, which can be summarized as follows.

(a) It could result in the establishment of a large, cohesive, reproducing drill group. A precedent for this has been set by the establishment of a mandrill group in Gabon, which grew from 14 founders to a group of 38 individuals in six years by births within the group (Feistner, 1989; Feistner, Cooper and Evans, 1992) (Figure 26.4). Given the similarity of drills to mandrills, and the advantages of keeping animals in enclosures of natural vegetation under their natural climatic régime, it seems likely that drill breeding could be equally successful (Feistner, 1990, 1992).

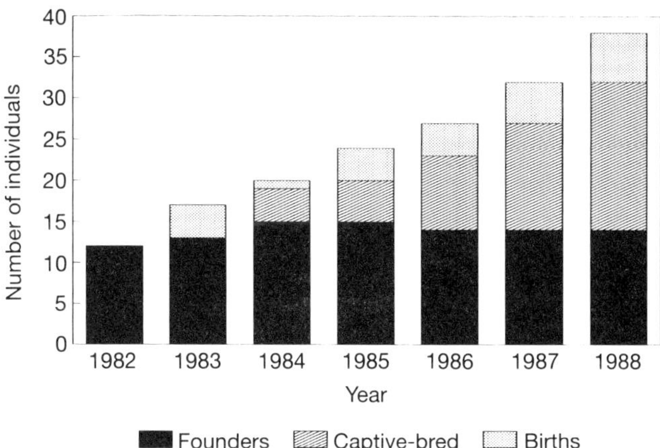

Figure 26.4 The growth of the semifree-ranging mandrill group at CIRMF, Gabon 1982–1988.

(b) The rescue and rehabilitation of drills currently illegally kept in villages would not only improve the life of those drills but also provide founder stock and increase the current captive gene pool.
(c) Drills could be provided on loan to EEP/SSP-participating zoos to enhance their captive breeding programmes, allowing sub-population management of captive drills and integration of the SSP and EEP with *in situ* work.
(d) A drill group in naturalistic semi-free-ranging conditions would represent an important resource for research, especially for studies of behaviour and reproduction.
(e) Such a group could act as a focus for local conservation education, since few people have seen a live drill or realize they are special to their country and endangered.
(f) Such a centre would provide training and employment locally thus directly benefiting local people.
(g) The establishment of such a drill group creates the possibility for the future of reintroducing drills to the wild in areas from which they have become exterminated.

The potential benefits outlined above have been recognized, and efforts are now underway to realize them by establishing such a facility in Nigeria.

26.4 THE DRILL REHABILITATION AND BREEDING CENTRE, NIGERIA

Considering the poor history and status of the world-wide captive drill population, drill orphans seen in villages during the course of the

Nigerian survey were thought to be a valuable resource wasted, both from a captive breeding standpoint and, more importantly, from a local conservation perspective. It was thus decided, in collaboration with the Cross River State Department of Parks and Wildlife, to rescue and rehabilitate these animals to form a founder group for an *in situ* captive breeding programme.

The first drill was acquired in 1988, but the programme began in earnest in early 1991 with a commitment to seek out and rescue all captive drills in Cross River State. This effort proved immensely productive and by the end of that year the founder group numbered 12 drills. Today, the Drill Rehabilitation and Breeding Centre maintains 8.6 drills (Table 26.1), the world's largest captive group, at an interim facility, Drill Ranch, in the Cross River State capital of Calabar. The drills are officially owned by the Nigerian government, with Gadsby and Jenkins as custodians.

Illegally kept drills were either donated freely to the project by local citizens, or confiscated by the Department of Parks and Wildlife. All were orphaned when hunters shot their nursing mothers and had been living in captivity without conspecifics, often enduring indescribably poor conditions without proper nutrition and health care. Most were acquired as infants, but some had survived their captivity over three years. They are now rehabilitated to the extent that they live together as an evolving and dynamic social group. While at one time infants had to be hand-reared, that task is increasingly being performed by adoptive juvenile and adolescent members of the group.

The project has been largely self-financed. Operating costs are defrayed by in-country donations from individuals and businesses in the form of cash, animal foodstuffs, enclosure furnishings, building materials and other goods. A local group of Nigerian medical professionals provide dental, X-ray, lab-testing and diagnostic services to the drills free of charge. Nearly US$4000 was raised by the Drill SSP in 1991; other, smaller contributions were also received from abroad.

It is pertinent to note the importance of developing the local support base during the project's critical early stages. Residents and businesses in a drill habitat area not only become aware of local conservation issues through the captives, but are participating directly in species preservation. This support is nurtured by public visits to Drill Ranch, the distribution of a newsletter and other types of promotion. The newsletter is written for popular appeal and is used to maintain contact, for example, with villagers who have donated drills, and with businesses or individuals who have donated goods, cash or services, and to credit publicly those people for their contributions. The local media have provided positive coverage to raise the programme's profile state-wide.

The project is designed to be a cooperative effort between the

Table 26.1 Current population of orphaned wild-born drills (*Mandrillus leucophaeus*) at Drill Ranch, Calabar, Nigeria

Studbk No.	*House*	*House name*	*Sex*	*Est. D.o.B*[a]	*Area of origin*	*Date acquired*	*How acquired*
1042	001	Calabar	F	Sep 88	E. Oban Hills	02 Nov 88	Donation
1044	002	Basi-Boje	M	Oct 89	Afi Massif	01 Apr 90	Donation
1043	003	Billy O'Ban	M	1986	Unknown	01 Jun 90	Confiscation
	004	Scarlett Ebe	F	Oct 90	Ikpan block	26 Feb 91	Confiscation
	005	Jules	M	1985	Unknown		Cared for locally
	006	Alhaja	F	Sep 90	Unknown	30 Apr 91	Confiscation
	007	Kola	F	1989	Unknown	25 Jul 91	Donation
	008	Antifon	M	Oct 90	W. Oban Hills	20 Sep 91	Donation
	009	Eliza Bemi	F	Dec 90	Okwangwo/Mbe	17 Oct 91	Confiscation
	010	Ekki Ikang	M	1987	Unknown	08 Nov 91	Confiscation
	011	Glory	F	Oct 91	Unknown	05 Dec 91	Confiscation
	012	Kebi	M	1989	W. Oban Hills	19 Dec 91	Confiscation
	013	Tchiko	M	Mar 90	Unknown	10 Mar 92	Donation
1041	014	Moka Boi	M	1989	Bioko	24 Mar 92	Donation

[a]Estimated date of birth.

Nigerian government, international non-governmental organizations, and contributing SSP and EEP institutions. Project development has benefited from close association with the Cross River National Park Project, which is being established by the government with assistance from the World Wide Fund for Nature. While the drill centre does not receive financial assistance from the park project or the government, it does receive logistical and moral support.

26.5 FUTURE DEVELOPMENTS FOR *IN SITU* DRILL CONSERVATION

As major funding is secured, construction of the drills' permanent facility in the forest will begin, and relocation is planned for mid-1994. The drill centre will occupy about 1 km^2, including the enclosures and various support structures. The main enclosure will contain several hectares of naturally forested former drill habitat. The drills will roam, forage, and sleep naturally within the enclosure, although their diet will be supplemented. The expected site lies in the Afi River Forest Reserve, directly west of the northern division of the Cross River National Park (see Figure 26.3).

The Afi Reserve is an important habitat area containing the north-western-most populations of drills and gorillas in Africa, as well as other endangered species such as red-eared guenons (*Cercopithecus erythrotis*), chimpanzees and elephants (Harcourt, Stewart and Inahoro, 1989; Oates *et al.*, 1990). The reserve is currently under heavy pressure from logging and hunting activities. It was excluded from the national park, but by locating the drill facility here a conservation initiative will be established which will hopefully secure the future of the area. Within the Afi Reserve are village enclaves and it is intended to site the drill centre on an unused parcel of these community-owned forest lands. Leasing the land on a long-term basis directly from the villagers would bring them immediate and recurrent financial benefit, and stimulate a sense of pride and responsibility for drills within the surrounding communities. During preliminary discussions the villagers and their chiefs have demonstrated great enthusiasm for the centre's location and development on their tribal lands.

Establishment of the captive breeding centre is the first phase of a larger programme for *in situ* drill conservation. It is hoped that the drill centre will serve as a catalyst for the creation of an active conservation project in adjacent drill habitat. The north-west corner of the Afi Reserve is a spectacular massif of about 120 km^2, an important watershed, and too rugged for commercial logging. We feel that the Drill Rehabilitation and Breeding Centre could be the ideal vehicle with which to launch a village-based anti-poaching and conservation education programme for the massif, and to work with the state government to realize its previously recommended regazettement as a wildlife sanctuary. Creation of the sanctuary, comprising about 25% of the reserve, would provide a core area for wildlife populations which might then allow sustainable hunting of non-threatened species in surrounding areas.

The essence of the conservation project would be the recruitment and training of a ranger corps of former hunters familiar with the massif, its hunting patterns, and animal movements – thus employing those men most economically affected by the sanctuary's restrictions. Hunting and trapping of any animal would be prohibited within the sanctuary and the enforcement of existing laws encouraged outside its boundaries. Staff from the drill centre would regularly visit the villages to develop public awareness and support using written materials, posters, public meetings, classroom projects and other methods. Public awareness and conservation education is a major goal of the captive breeding and conservation programme and is fundamental to its success and to the long-term survival of drills and other endangered species in their natural habitat.

A tourism scheme could be created, offering controlled viewing of the captive group and multi-day treks into the Sanctuary. This plan would build upon a preliminary effort initiated by the villagers themselves in recent years. While trekkers may not see a gorilla, drill, or other large mammal, all could enjoy the rain forest adventure and this speaks highly of the scenic quality of the Afi Massif. Visitors' fees and employment of local men as porters and guides could create a sustainable, if small, community industry. As an added attraction to the nearby national park, visitors would be able to view a group of one of the world's rarest primates in a naturalistic setting and/or trek out for a possible glimpse of a wild drill, gorilla, or chimpanzee. Outside attention and support, especially from foreigners, often reinforces local enthusiasm for, and understanding of, wildlife preservation.

26.6 CONCLUSIONS – INTEGRATED AND COORDINATED CONSERVATION FOR WILD AND CAPTIVE DRILLS

Through the initiatives described above the captive drills themselves will serve as the impetus and focus for the primary objective – that of ensuring the long-term survival of the species in its natural habitat. An opportunity is at hand for zoos world-wide to work not just with each other, but with a country in which the species occurs, and to create a viable captive breeding population, and simultaneously to assist the long-term survival of drills and sympatric species in their natural habitats in Africa.

ACKNOWLEDGEMENTS

We would like to thank Clement O. Ebin, Director, Cross River State Department of Parks and Wildlife and Project Coordinator of the Cross River National Park Project for his valued assistance and support of all the work in Nigeria. Mr Ebin is also a co-founder of the Drill Rehabilitation and Breeding Centre. We are especially grateful for the guidance of John Oates and Julian Caldecott. The Nigerian survey received grants from Wildlife Conservation International and World Wildlife Fund-US and was supported in part by work for the World Wide Fund for Nature-UK. All activities in Nigeria are carried out for the Ministry of Agriculture. In Cameroon we would like to thank John Banser of the Institute of Animal Research MESIRES, under whose auspices the survey is being conducted; also Augustin Bokwe, Director of Wildlife, Ministry of Tourism; Beryl Fofung and Stephen Gartlan, WWF-Cameroon; and Hans and Tracey Hockey, Yaounde. The drill survey in Cameroon is the Fauna and Flora Preservation Society Project No. 90/1/1 and we are grateful to the society, and Amanda Hillier and Roger Wilson for their patience, support, and confidence in our work.

The captive drills benefit from the efforts of many, but we would especially like to thank Daphne Onderdonk and Daniel Lende, volunteers at Drill Ranch, for their care of the drills while we resume the survey, and Aniekan Effiong, Shelagh Heard, Jurgen Maresch, Ernest Nwufoh, and Zena Tooze who have been instrumental in the project's development on the ground. We would like to thank the project's many local sponsors and friends; Dietrich Schaaf of Zoo Atlanta for organizing timely support from the drill SSP; and Nick Ashton-Jones, Cross River National Park Project Manager, and family have given us unflagging help and encouragement. Thanks to Louise and Paul Davies who provide essential logistical support in London.

REFERENCES

Blom, A., Alers, M.P.T., Feistner, A.T.C. *et al.* (1992) Notes on the current status and distribution of primates in Gabon. *Oryx*, **26**, 223–34.

Böer, M. (1987a) *International Studbook for the Drill*, Zoologischer Garten Hannover, Germany.

Böer, M. (1987b) Beobachtungen zur Fortpflanzung und zum Verhalten des Drill (*Mandrillus leucophaeus*, Ritgen, 1824) im Zoo Hannover. *Zeitschrift für Säugetierkunde*, **52**, 265–81.

Böer, M. (1989) Beschreibung einer jungtierentwicklung beim drill (*Papio* [*Mandrillus*] *leucophaeus*) inklusive einiger bemerkungen zu seiner verbreitung. *Zoologischer Garten*, **5–6**, 331–44.

Böer, M. (1989) Möglichkeiten und voraussetzungen einer erhaltungszucht des drill (*Mandrillus leucophaeus*) in Europa. *Haltung und Zucht von Primaten*, **8**, 35–9.

Böer, M. (1992) Drill (*Mandrillus* (= *Papio leucophaeus*) EEP Annual Report 1991, in *EEP Yearbook 1991/1992 Including the Proceedings of the 9th EEP Conference, Edinburgh 6–8 July 1992* (eds Brouwer, K., Smits, S. and de Boer, L.E.M.), EAZA/EEP Executive Office, Amsterdam, pp. 60–1.

Butynski, T.M. and Koster, S.H. (1990) The status and conservation of forests and primates on Bioko Island (Fernando Poo), Equatorial Guinea. Unpublished report to WWF (US) and the Chicago Zoological Society.

Cox, C. (1987a) Increasing the likelihood of reproduction among drills. *Annual Proceedings of the American Association of Zoological Parks and Aquaria*, AAZPA, Wheeling, WV, pp. 425–34.

Cox, C. (1987b) Social behavior and reproductive status of drills (*Mandrillus leucophaeus*). *Regional Proceedings of the American Association of Zoological Parks and Aquaria*, AAZPA, Wheeling, WV, pp. 321–8.

Cox, C. (1991) Progress on the drill front. *Zoo View*, **25**, 19.

Desmond, T., Laule, G. and McNary, J. (1987) A training project to enhance positive social interactions in a community of drills (*Mandrillus leucophaeus*). *Regional Proceedings of the American Association of Zoological Parks and Aquaria*, AAZPA, Wheeling, WV, pp. 352–358.

DuBois, T. (1991) Behavioural enrichment. *Zoo View*, **25**, 9–11.

Fa, J. (1991) *Guinea Ecuatorial – Conservacion y Manejo Sostenible de los Ecosistemas Forestales*, IUCN, Gland, Switzerland, and Cambridge, UK.

Fa, J. (1992) Conservation in Equatorial Guinea. *Oryx*, **26**, 87–94.

Feistner, A.T.C. (1989) The behaviour of a social group of mandrills *(Mandrillus sphinx)*. University of Stirling, Ph.D. thesis.

Feistner, A.T.C. (1990) Reproductive parameters in a semifree-ranging group of mandrills, in *Baboons: Behaviour and Ecology, Use and Care* (eds de Mello, M.T., Whiten, A. and Byrne, R.W.), University of Brasilia, Brasilia, pp. 77–88.

Feistner, A.T.C. (1991) Scent marking in mandrills (*Mandrillus sphinx*). *Folia Primatologica*, **57**, 42–7.

Feistner, A.T.C. (1992) Aspects of reproduction of female mandrills (*Mandrillus sphinx*). *Int. Zoo Yb*, **31**, 170–8.

Feistner, A.T.C., Cooper, R.W. and Evans, S. (1992) The establishment and reproduction of a group of semifree-ranging mandrills. *Zoo Biology*, **11**, 385–95.

Gadsby, E.L. (1990) The status and distribution of the drill, *Mandrillus leucophaeus*, in Nigeria. Unpublished report to the Nigerian Government, WCI, and WWF-US.

Gartlan, S.J. (1970) Preliminary notes on the ecology and behavior of the drill, in *Old World Monkeys, Evolution, Systematica and Behavior* (eds Napier, J.R. and Napier, P.H.), Academic Press, New York, pp. 445–80.

Gartlan, J.S. and Struhsaker, T.T. (1972) Polyspecific associations and niche separation of rain-forest anthropoids in Cameroon, West Africa. *Journal of Zoology, London*, **168**, 221–66.

Grubb, P. (1973) Distribution, divergence and speciation of the drill and mandrill. *Folia Primatologica*, **20**, 161–77.

Harcourt, A.H., Stewart, K.J. and Inahoro, I.M. (1989) Gorilla quest in Nigeria. *Oryx*, **23**, 7–13.

Hearn, G.W., Weikel, E.C. and Schaaf, C.D. (1988) A preliminary ethogram and study of social behavior in captive drills, *Mandrillus leucophaeus*. *Primate Report*, **19**, 11–17.

Hill, W.C.O. (1970) *Primates Comparative Anatomy and Taxonomy VIII Cynopithecinae Papio, Mandrillus, Theropithecus*, Edinburgh University Press, Edinburgh.

Hoshino, J. (1985) Feeding ecology of mandrills (*Mandrillus sphinx*) in Campo Animal Reserve, Cameroon. *Primates*, **26**, 248–73.

Hoshino, J., Mori, A., Kudo, H. and Kawai, M. (1984) Preliminary report on the grouping of mandrills (*Mandrillus sphinx*) in Cameroon. *Primates*, **25**, 295–307.
Jouventin, P. (1975) Observations sur la socio-écologie du mandrill. *La Terre et la Vie*, **29**, 493–532.
Lahm, S. (1986) Diet and habitat preference of *Mandrillus sphinx* in Gabon: implications of foraging strategy. *American Journal of Primatology*, **11**, 9–26.
Lee, P.C., Thornback, J. and Bennett, E.L. (1988) *Threatened Primates of Africa. The IUCN Red Data Book*, IUCN, Gland, Switzerland.
Kudo, H. (1987) The study of vocal communication of wild mandrills in Cameroon in relation to their social structure. *Primates*, **28**, 289–308.
Oates, J.F. (1986) *Action Plan for African Primate Conservation: 1986–90*. IUCN/SSC Primate Specialist Group, New York.
Oates, J.F., White, D., Gadsby, E.L. and Bisong, P.O. (1990) Conservation of gorillas and other species. Appendix 1 *Cross River National Park (Okwangwo Division) Feasibility Study*. WWF, Gland, Switzerland.
Schaaf, C.D., Butynski, T.M. and Hearn, G.W. (1990) The Drill (*Mandrillus leucophaeus*) and Other Primates in the Gran Caldera Volcanica de Luba. Results of a Survey Conducted March 7–22, 1990. Unpublished report to the Government of the Republic of Equatorial Guinea.
Schaaf, C.D., Struhsaker, T.T. and Hearn, G.W. (1992) Recommendations for Biological Conservation Areas on the Island of Bioko, Equatorial Guinea. Unpublished report to the Government of the Republic of Equatorial Guinea.
Struhsaker, T.T. (1969) Correlates of ecology and social organization among African cercopithecines. *Folia Primatologica*, **11**, 80–118.

27

Reintroduction of the black-footed ferret (*Mustela nigripes*)

B. Miller
Universidad Nacional Autonoma de Mexico, Mexico
D. Biggins and L. Hanebury
US Fish and Wildlife Service, Fort Collins, Colorado, USA
and
A. Vargas
US Fish and Wildlife Service, University of Wyoming, Wyoming, USA.

27.1 INTRODUCTION

The black-footed ferret (*Mustela nigripes*) (Figure 27.1) is a small, secretive, nocturnal member of the family Mustelidae (Hall, 1981; Honacki, Kurman and Koeppl, 1982). Ferrets have an obligate dependence on the prairie dog (*Cynomys* spp.) community, utilizing the prairie dog for food and its burrows for shelter (Campbell *et al.*, 1987).

By 1960, government eradication programmes had reduced the original geographical distribution of prairie dogs by 98% (Marsh, 1984), and the resulting habitat fragmentation left patches that were probably too small, or too widely separated, to sustain viable populations of ferrets (Brussard and Gilpin, 1989). After the 1985 collapse of the last known

Creative Conservation: Interactive management of wild and captive animals.
Edited by P.J.S. Olney, G.M. Mace and A.T.C. Feistner.
Published in 1994 by Chapman & Hall, London.
ISBN 0 412 49570 8

Figure 27.1 Black-footed ferret photographed in the wild at Meeteetse, Wyoming, 1983. (Dean Biggins.)

wild black-footed ferret population, located near Meeteetse, Wyoming, the few remaining ferrets were taken into captivity (see Weinberg, 1986; May, 1986; Clark and Harvey, 1988; Miller *et al.*, 1988; Thorne and Williams, 1988; Clark, 1989; Seal *et al.*, 1989 for accounts of biology and/or policy of captive breeding).

Captive breeding, under the direction of the Wyoming Game and Fish Department in cooperation with the US Fish and Wildlife Service, has now produced large numbers of ferrets. The captive population has grown from 18 in the spring of 1987 to 349 animals located at seven institutions in the fall of 1992, and it is very likely that 100 or more ferrets could be available for reintroduction each year (Cassidy, 1992). The guidance of the World Conservation Union (IUCN) Captive Breeding Specialist Group (CBSG) was particularly important in the inception and success of the captive breeding programme. Breeding animals in captivity, however, is only part of recovery. Those captive animals must also be re-established in native habitat.

Re-establishment in the wild is not an easy task. Effects of captivity, such as relaxed selection pressure or artificial selection, can impose a wide variety of behavioural and physiological responses between and within species (Kleiman, 1989; Derrickson and Snyder, 1992). The development of any complex behaviour pattern is the result of extensive interaction between genetics and the animal's experience (Polsky, 1975). Instinctive behaviours may not develop effectively in an unnatural environment, and learned behaviours may require the proper stimuli at the appropriate stage of maturation (Beck *et al.*, Chapter 13). Indeed, some behaviours may require repeated cues throughout juvenile development if they are to be performed efficiently as adults (Gossow, 1970). There is also evidence that early experience can alter brain size and other cerebral measures which potentially affect behaviour later in life (Greenough and Juraska, 1979; Rosenzweigh, 1979).

Captive environments can potentially affect behavioural formation and, consequently, reintroductory success. Because little is known about effects of captivity on survival traits or about optimal release techniques, reintroductory success can be maximized by using experimental methods and a statistically valid design. The experimental process should be based on open scientific discussion, and there should be an unbiased periodic review process that incorporates accountability for actions during all stages of planning and implementing the programme.

Most of the following discussion involves release strategies. That does not mean that we do not recognize the importance of policy and education (Reading, Clark and Kellert, 1992). Indeed, without addressing those two issues, the best scientific design will still ultimately fail. The following discussion merely reflects our research role in the recovery of the black-footed ferret.

27.2 REINTRODUCTION RESEARCH

It is imperative that reintroductions are carefully designed. To gain knowledge and experience with ferret releases, we conducted two years' (1989–1990) of research with captive-raised Siberian ferrets, (*Mustela eversmanni*) the closest relative of the black-footed ferret.

This research investigated development of survival skills and was augmented by test releases of neutered animals (Biggins *et al.*, 1990; Miller *et al.*, 1990a; 1990b; 1992; Biggins *et al.*, in review a). The experiments compared effects of different captive environments on survival skills, compared effects of differential release methods on survival, and refined the radio-telemetry monitoring system that would be used on black-footed ferrets. In addition, one of us (A.V.) is conducting behavioural experiments with black-footed ferrets raised in different captive environments.

We first analysed the innate and learned aspects of predator avoidance with surrogate Siberian ferrets captive-raised at the Conservation and Research Center of the National Zoological Park, Washington, DC (Miller *et al.*, 1990a). We tested avoidance responses to artificial terrestrial and avian predators with two, three, and four-month-old Siberian ferrets. Siberian ferrets reacted innately to predator models at four months of age, and that response was enhanced after a single aversive experience (Miller *et al.*, 1990a). The learning response was heightened when the models were replaced by a live dog; Siberian ferrets remembered that experience for at least two weeks (Miller *et al.*, 1992).

In addition to avoiding predators, newly released black-footed ferrets must locate and kill prairie dogs, their primary source of food. Black-footed ferrets raised in an enriched cage (the same size but more hiding places, concrete blocks, etc.) killed hamsters more effectively than black-footed ferrets raised in standard cages (A. Vargas, in prep.). Cage-raised Siberian ferrets killed mice (Miller *et al.*, 1992) and prairie dogs more efficiently when they had previous experience.

Locating prey is an equally important part of the predatory sequence. Cage-raised fishers (*Martes pennanti*) killed porcupines (*Erithizon dorsatum*) at the first opportunity in captivity, but starved after release into the wild, presumably because they did not know how to locate prey (Kelly, 1977). Cage-raised Siberian ferrets released into an enclosed 200 m^2 mock prairie dog colony (arena) located burrows with food progressively faster as they matured (Miller *et al.*, 1990b). They also located food in a burrow plugged with dirt, which simulated winter conditions when white-tailed prairie dogs (*Cynomys leucurus*) hibernate (Miller *et al.*, 1990b).

Although cage-raised Siberian ferrets located food in the enclosed prairie dog arena, they spent excessive time on the surface, which would expose them to predators in the wild (Miller *et al.*, 1990b). In contrast, captive-raised Siberian ferrets with prairie dog burrow experience settled more quickly into burrows in a novel situation than cage-raised Siberian ferrets without burrow experience (Miller *et al.*, 1992).

In 1989 and 1990 these results were tested during releases of neutered captive-raised Siberian ferrets (Biggins *et al.*, 1990; Biggins *et al.*, in review a). In 1990, one group was raised in cages similar to the captive conditions of black-footed ferrets (Biggins *et al.*, in review a). That group (n = 14) had no previous experience with burrows or live prey but was released onto the prairie dog colony from elevated cages with supplemental feeding provided at the cage. This was similar to the black-footed ferret reintroduction plan proposed by the Wyoming Game and Fish Department (Thorne and Oakleaf, 1990). The other group was raised in cages until the age of 3.5 months, then introduced into two 200 m^2 enclosed prairie dog arenas where they lived in prairie dog burrows and were exposed to live prey for one month. One arena was free of predators (n = 10), but Siberian ferrets in the second arena (n = 15) were occasionally harassed by a domestic dog. Arena animals were released to the prairie dog colony without the use of elevated cages.

Because of simultaneous experiments during development, all captive-raised Siberian ferrets had extensive contact with humans and novel situations. Arena animals spent developmentally sensitive periods in cages before their exposure to the enclosed semi-natural prairie dog colonies. Young Siberian ferrets were cross-fostered between litters to provide equal genetic representation between treatments. Both groups were released onto prairie dog colonies at 4.5 months of age.

A third group of five wild-caught Siberian ferrets (*M. e. dauricus*) translocated from China was also neutered and released onto the same prairie dog colonies (Biggins *et al.*, in review a). All Siberian ferrets were monitored with radio-telemetry to develop information on movements and survivorship.

Captive-raised Siberian ferrets with arena pre-release conditioning exhibited behavioural traits similar to wild, translocated Siberian ferrets; Siberian ferrets without pre-release conditioning that were introduced to a prairie dog colony via an elevated release cage behaved differently from wild-caught and arena pre-conditioned groups (Biggins *et al.*, in review a). Most animals stayed on the prairie dog town, but Siberian ferrets with arena pre-release conditioning moved more rapidly and covered greater distances on the surface. Loitering above ground makes an individual susceptible to coyotes (*Canis latrans*) and owls (*Bubo virginianus*). Ferrets remaining in the same area attracted badgers (*Taxidae taxus*), particularly when they were supplementally fed prairie dogs. The only animals that were verified as having killed in the wild had pre-release conditioning with prey (Biggins *et al.*, in review a).

Although no significant differences in survival were observed, survival ranged from 0.2 to 21 days for cage-raised animals, 0.4 to 33.3 days for arena pre-conditioned animals, and 4.3 to 41.9 days for wild-caught animals (Biggins *et al.*, in review a). Known causes of mortality included coyotes (n = 21), badgers (n = 8), owls (n = 1) and starvation (n = 3).

Based on the apparent benefits of minimal pre-release conditioning in small arenas, it was recommended that a sample of the first black-footed ferret reintroduction be raised from birth in larger enclosed prairie dog arenas (US Fish and Wildlife Service, 1990; Miller *et al.*, 1990a; Beck and Miller, in press). The US Army offered to donate several 400 m^2 warehouses for the arenas.

In 1991, six years after the first black-footed ferrets were taken into captivity, that initial reintroduction took place. All 49 animals were raised in cages and liberated via on-site elevated release cages where they were supplementally fed. The reintroduction occurred under the direction of the Wyoming Game and Fish Department with cooperation from the US Fish and Wildlife Service, US Bureau of Land Management, SSP, and, importantly, private landowners. The release site was the Shirley Basin area of south-eastern Wyoming. Thirty-seven animals were monitored by radio-telemetry during the period of release in September and October (Biggins *et al.*, in review b). There was additional monitoring by occasional spotlighting and snow tracking.

Twenty-seven of 37 black-footed ferrets (73%) abandoned the release cage area within 30 hours (Biggins *et al.*, in review b). About half of the telemetry-monitored black-footed ferrets rapidly dispersed from the release area, moving 4.1–17.1 km (Biggins *et al.*, in review b). Known causes of mortality included coyotes (n = 5), badger (n = 1), injury (n = 2) and starvation (n = 2) (injured and starving animals were rescued). According to observation and scat analysis, some ferrets killed prairie dogs (Biggins *et al.*, in review b). Four black-footed ferrets survived the winter and two females produced litters (Anon, 1992).

Although release design prevented comparison of many variables that influence survival, litter effect was significant. Fate of an animal was the same as its litter-mate in 77% of all possible comparisons, but non-sibling pairs shared the same fate only 51% of the time (Biggins *et al*, in review b). The work of A. Vargas will soon shed more light on differential captive conditions and their effects on survival of black-footed ferrets. Attempts to correlate success with time of release, place of release, human disturbance, and weather, were inconclusive (Biggins *et al.*, in review b).

In the fall of 1992, there was a second reintroduction of 90 black-footed ferrets into the same area of Wyoming, but this time 17 of those 90 animals were raised in small outdoor arenas (approximately 35 m^2) with access to prairie dog burrows and live prairie dogs. The other 73 were raised in the standard cages. After one month, the post-release survival seemed superior for the group raised in the outdoor arenas (A. Vargas, in prep.). As a result, six males and six females have now been moved to the 400 m^2 enclosed prairie dog arenas donated by the US Army, and their offspring are scheduled to be released in the 1993 black-footed ferret reintroduction.

27.3 DISCUSSION AND RECOMMENDATIONS

Our recommendations are by no means comprehensive. They discuss only information presented earlier in this paper.

27.3.1 Test the effects of different captive environments including a semi-natural environment

Our data indicated that pre-release conditioning in semi-natural conditions provided benefits to captive-raised animals scheduled for release to the wild. By raising young ferrets in large enclosed prairie dog colonies, the captive environment may provide stimuli more similar to those encountered by developing juveniles in nature. Behavioural traits do not exist in isolation. Individual behaviours must be expressed in the context of a host of simultaneous behaviours necessary for survival; the expression of an individual behaviour must be adjusted according to each situation that the animal faces (Fentress, 1978; Caro, 1989). A semi-natural pre-release environment may allow animals to develop responses to a variety of ecologically relevant stimuli. The approach seems safe when developmental processes are incompletely understood, and may provide a holistic pre-release preparation for captive-raised animals.

27.3.2 Test more than one release strategy

The Black-footed Ferret Recovery Plan (US Fish and Wildlife Service, 1988) suggested that the first three black-footed ferret releases experimentally test variables that could affect success. The results of those early tests could then be incorporated into an operational technique for later large-scale releases.

By comparing reasonable, but widely different, release strategies, the fastest road to recovery will not be overlooked. The most promising strategy can be refined further with comparative tests. The comparative process avoids testing a single unsuccessful technique for several years, or even settling on a partially successful method without knowing whether there were more efficient procedures. Comparing a strategy one year to a different strategy the following year, without considering inter-year variability, is an inferior scientific design.

27.3.3 Monitor animals effectively

Early reintroductions should be intensively monitored to determine causes of mortality, movements of released animals, and life history attributes that will assist in guiding future releases. Accurate records should be kept on the fate of reintroduced animals and resulting offspring.

27.3.4 Encourage and stimulate communication with outside expertise through open discussion and publication of techniques and results

Data should be accessible in the peer-reviewed scientific literature. Accessibility of information stimulates the creative input necessary for solving the complex problems of reintroduction. The CBSG provided valuable insight when black-footed ferrets entered captivity. With their guidance, a number of research projects by mustelid experts, reproductive physiologists, and nutritionists led to rapid advances in ferret captive breeding.

27.3.5 Design goals carefully

Judging success of early reintroductions is difficult, and the definition should be carefully chosen. Expense is a favourite of accountants, but it is much easier to see cost than value. A cheaper technique that establishes fewer animals may not be thrifty over time. Financial costs cannot be adequately assessed until survival rates are established.

Defining success of early releases solely by survival can be misleading because high mortality is likely during the first reintroductions. Analysis of behavioural traits may provide an understanding of how a captive-raised animal responds to a natural environment and contribute clues for improving techniques and increasing survival.

Knowledge gained toward an optimum release technique may be the most important goal during initial releases. High mortality is not a failure unless biologists do not learn enough to increase survival in future reintroductions. The comparative process of a well-designed release best offers that opportunity to learn.

In conclusion, careful planning with a scientific approach to management of reintroductions will offer the most efficient path toward recovery, and we strongly recommend a scientific progression toward an operational rearing and release tactic that uses the comparative process. We are optimistic that the comparative process can speed progress toward recovery of the black-footed ferret.

ACKNOWLEDGEMENTS

The Conservation and Research Center of the National Zoo and the US Army Pueblo Depot supplied facilities and donated time and materials. Funding sources included: Wildlife Preservation Trust International, the USFWS National Ecology Research Center, USFWS Wyoming Cooperative Research Unit, USFWS Region 6 Enhancement, Smithsonian Institution, National Zoological Park, Friends of the National Zoo, Chevron USA, National Fish and Wildlife Foundation, Ferret Fanciers of Louden County, Wyoming Game and Fish Department, and Brookfield Zoo.

Many biologists and technicians contributed to nocturnal collection of radio-telemetry data and raising young Siberian ferrets in captivity. John Oldemeyer, Ben Beck, Larry Shanks, and Mike Hutchins reviewed the manuscript.

REFERENCES

Anon (1992) The Drumming Post, non-game newsletter of the Wyoming Game and Fish Department, **5**(3), 1.

Beck, B.B., Kleiman, D.G., Castro, I. *et al.* (in press) Preparation of the golden lion tamarin for release into the wild, in *A Case Study for Conservation Biology: The Golden Lion Tamarin* (ed D.G. Kleiman), Smithsonian Institution Press, Washington, DC.

Beck, B.B. and Miller, B. (in press) Implications from the golden lion tamarin for the reintroduction of the black-footed ferret, in *Proceedings of the Black-footed Ferret Reintroduction Workshop*, Laramie Wyoming, March 29–31, 1990 (ed. E.T. Thorne), Wyoming Game and Fish Department, Cheyenne.

Biggins, D.E., Hanebury, L.R., Miller, B.J. and Powell, R.A. (1990) Release of Siberian polecats on a prairie dog colony. *70th Annual Meeting of the American Society of Mammologists*, June 9–13, 1990, Frostburg, Maryland.

Biggins, D.E., Hanebury, L.R., Miller, B.J. *et al.* (in review a) Release of Siberian ferrets (*Mustela eversmanni*) to facilitate reintroduction of black-footed ferrets (*Mustela nigripes*). *Wildlife Society Bulletin*.

Biggins, D.E., Miller, B.J. and Hanebury, L.R. (in review b). First reintroduction of black-footed ferrets (*Mustela nigripes*). *Dodo, Journal of the Jersey Wildlife Preservation Trust.*

Brussard, P.F. and Gilpin, M.E. (1989) Demographic and genetic problems associated with small population size, with special reference to the black-footed ferret, in *Conservation Biology of the Black-footed Ferret* (eds U.S. Seal, E.T. Thorne, S.H. Anderson and M. Bogan), Yale University Press, New Haven, Connecticut.

Campbell, T.M., Clark, T.W., Richardson, L. *et al.* (1987) Food habits of the Wyoming black-footed ferrets. *American Midland Naturalist*, **117**, 208–10.

Caro, T. (1989) Missing links in predator and anti-predator behavior. *Trends in Ecology and Evolution*, **4**, 333–4.

Cassidy, D. (1992) Unpublished presentation at the black-footed ferret Intersvate Working Group Meeting, December 2–3, 1992.

Clark, T.W. (1989) Conservation biology of the black-footed ferret (*Mustela nigripes*). *Wildlife Preservation Trust Special Scientific Report* No 3, Philadelphia, Pennsylvania.

Clark, T.W. and Harvey, A.H. (1988) Implementing endangered species recovery policy: Learning as we go? *Endangered Species Update*, **5**, 35–42.

Derrickson, S.R. and Snyder, N.F.R. (1992) Potentials and limits of captive-breeding in parrot conservation, in *New World Parrots in Crisis: Solutions from Conservation Biology* (eds S.R. Beissinger and N.F.R. Snyder), Smithsonian Institution Press, Washington, DC, pp. 133–63.

Fentress, J.C. (1978) Conflict and context in sexual behavior, in *Biological Determinants of Sexual Behavior* (ed. J.B. Hutchinson), John Wiley & Sons, New York, pp. 579–614.

Gossow, H. (1970) Vergleichende verhaltensstudien an Marderartigen I. uber Lautausserungen und zum Beuterhalten. *Zeitschrift fur Tierpsychologie*, **27**, 405–80.

Greenough, W.T. and Juraska, J.M. (1979) Experience induced changes in brain fine structure: their behavioral implications, in *Development and Evolution of Brain Size: Behavioral Implications* (eds M.E. Hahn, C. Jensen and B.C. Dudek), Academic Press, New York, pp. 263–94.

Hall, E.R. (1981) *The Mammals of North America*, John Wiley & Sons, New York.

Honacki, J.H., Kurman, K.E. and Koeppl, J.W. (1982) *Mammal Species of the World*, Allen Press and Associates and Systematic Collections, Lawrence, Kansas.

Kelly, G.M. (1977) Fisher (*Martes pennanti*) biology in the White Mountain National Forest and adjacent areas. University of Massachusetts, Ph.D. dissertation.

Kleiman, D.G. (1989) Reintroduction of captive mammals for conservation. *Bioscience*, **39**, 152–61.

Marsh, R.E. (1984) Ground squirrels, prairie dogs and marmots as pests on rangeland, in *Proceedings of the Conference for Organization and Practice of Vertebrate Pest Control*, August 30–September 3, 1982, Hampshire, England, ICI Plant Protection Division, Fernherst, England, pp 195–208.

May, R.M. (1986) The black-footed ferret: a cautionary tale. *Nature*, **320**, 13–14.

Miller, B.J., Anderson, S.H., DonCarlos, M. and Thorne, E.T. (1988) Biology of the endangered black-footed ferret and the role of captive propagation in its conservation. *Canadian Journal of Zoology*, **66**, 765–73.

Miller, B., Biggins, D., Wemmer, C. *et al.* (1990a) Development of survival skills in captive-raised Siberian polecats (*Mustela eversmanni*) II: predator avoidance. *Journal of Ethology*, **8**, 95–104

Miller, B., Biggins, D., Wemmer, C. *et al.*, (1990b) Development of survival skills in captive-raised Siberian polecats (*Mustela eversmanni*) I: locating prey. *Journal of Ethology*, **8**, 89–94.

Miller, B., Biggins, D., Hanebury, L. *et al.* (1992) Rehabilitation of a species: the black-footed ferret (*Mustela nigripes*), in *Wildlife Rehabilitation Vol. 9* (ed. D.R. Ludwig), Burgess Printing, Edina Minnesota, pp. 183–92.

Polsky, R.H. (1975) Developmental factors in mammalian predation. *Behavioral Biology*, **15**, 353–82.

Reading, R.P., Clark, T.W. and Kellert, S.R. (1992) Towards an endangered species reintroduction paradigm. *Endangered Species Update*, **8**, 1–4.

Rosenzweigh, M.R. (1979) Responsiveness of brain size to individual experience: behavioral and evolutionary implications, in *Development and Evolution of Brain Size: Behavioral Implications* (eds M.E. Hahn, C. Jensen and B.C. Dudek), Academic Press, New York, pp. 263–94.

Seal, U.S., Thorne, E.T., Anderson, S.H. and Bogan, M. (eds) (1989) *Conservation of the Black-footed Ferret*, Yale University Press, New Haven, Connecticut.

Thorne, E. and Oakleaf, B. (1990) Unpublished presentation at the Black-footed Ferret Reintroduction Workshop, March 29–31, 1990.

Thorne, E.T. and Williams, E.S. (1988) Disease and endangered species: the black-footed ferret as a recent example. *Conservation Biology*, **21**, 66–74.

US Fish and Wildlife Service (1988) Black-footed ferret recovery plan. US Fish and Wildlife Service, Denver Colorado.

US Fish and Wildlife Service (1990) Unpublished Study Plan, US Fish and Wildlife Service.

Vargas, A. (in prep) Behavioral preparation of captive-raised black-footed ferrets for reintroduction. University of Wyoming, PhD dissertation.

Weinberg, D. (1986) Decline and fall of the black-footed ferret. *Natural History Magazine*, **95**, 62–9.

Part Four

Regional Approaches

28

Threatened endemic mammals of the Philippines: an integrated approach to the management of wild and captive populations

W.L.R. Oliver
IUCN/SSC Pigs and Peccaries Specialist Group, and Deer Specialist Group, Norwich, UK.

28.1 INTRODUCTION

The Philippines archipelago, a vast array of more than 7100 islands, has an exceptionally rich fauna. Among the native land mammals at least 175 species of nine orders (Insectivora, Scandentia, Dermoptera, Chiroptera, Primates, Pholidota, Rodentia, Carnivora and Artiodactyla) occur in the Philippines. This compares, for example, to the 105 species of six orders known from Madagascar, which has nearly twice the land area (Ingle and Heaney, 1992). About 120 (69%) species (and many more subspecies) of Philippine mammals are endemic. This is an unusually high percentage, higher in fact than any other biogeographical province in the Indo-Malayan Realm, itself one of the richest and most distinct of the world's biogeographical regions (Heaney, Gonzales and Alcala, 1987; MacKinnon and MacKinnon, 1986).

Part of the reason for the richness of the Philippine fauna is that the country, being bisected by 'Wallace's Line' (Figure 28.1), includes both 'Sundaic' and 'Wallacean' elements. Thus, Balabac, Palawan and the

Creative Conservation: Interactive management of wild and captive animals.
Edited by P.J.S. Olney, G.M. Mace and A.T.C. Feistner.
Published in 1994 by Chapman & Hall, London.
ISBN 0 412 49570 8

Calamian Islands (or the 'Palawan Faunal Region'), comprise the easternmost extension of the Sunda Shelf, while the rest of the archipelago (the Philippine biogeographical province *sensu stricto*) constitutes the northernmost part of the Wallacean Sub-region. The latter area is itself divided by deep water channels (> 120 m) into at least four major centres of endemism or late Pleistocene islands – namely: 'Mindoro', the 'Luzon Faunal Region' (Luzon and associated islands), the 'Mindanao

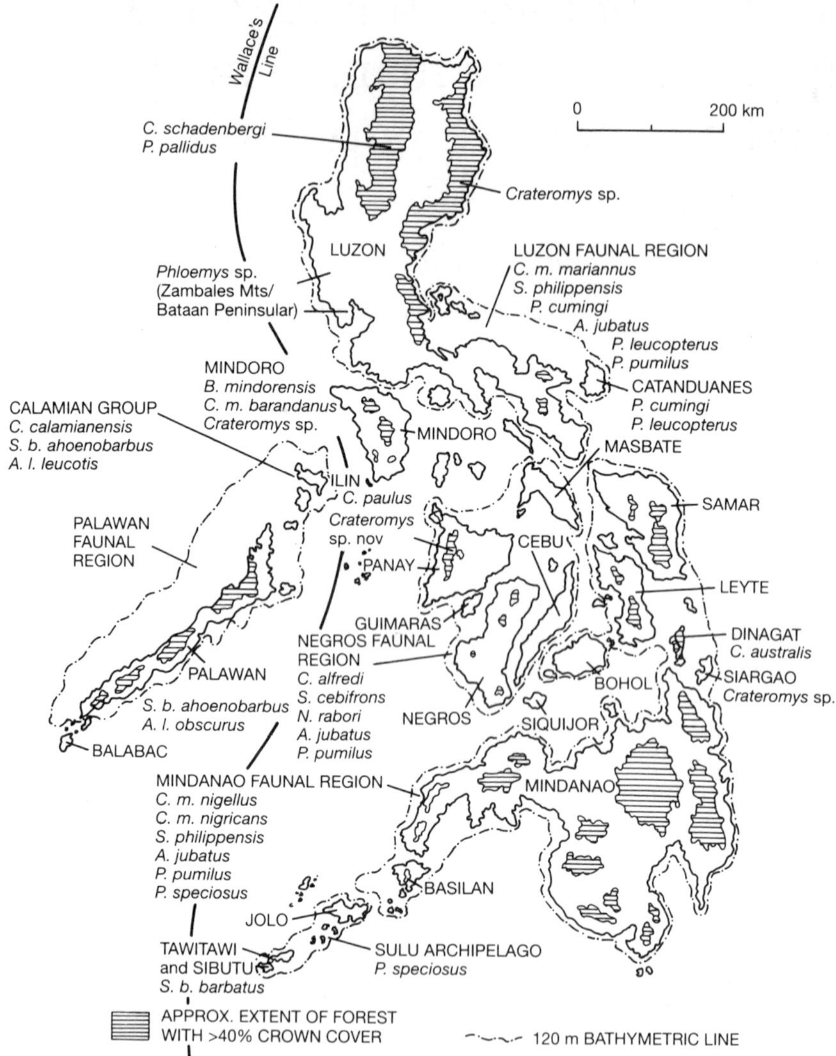

Figure 28.1 Distribution of taxa mentioned in the text in relation to the principal faunal regions/late Pleiotocene islands of the Philippines, as defined by the 120 m bathymetric line. (Modified from Heaney, 1986.)

Faunal Region' (Samar, Leyte, Mindanao and associated islands), and the West Visayan or 'Negros–Panay Faunal Region' (Masbate, Panay, Guimaras, Negros and Cebu) (Heaney, 1986; Figure 28.1).

However, while these divisions are basic to our understanding of the distributional relationships of the Philippine mammals and other fauna, the biology and ecology of the overwhelming majority of these mammals remains very poorly known. Many species are known only from their type localities, and many of the smaller islands have never been surveyed. When surveys have been done they have often resulted in the discovery of new species. Indeed, with the exception of two highly speciose genera of murine rodents (*Oryzomys* and *Akodon*), more new species of mammals have been described from the Philippines during the last few decades than from the whole of South America (S. Goodwin, pers. comm.). Over the past 20 years, for example, six new fruit bats have been described over the world-wide distribution of the family, of which three are from the Philippines (Heaney and Heideman, 1987). Two more recently collected species are awaiting formal description (S. Goodman, pers. comm.; C. Ross, pers. comm.).

A spate of new descriptions in the 1980s included two new cloud rats, *Crateromys australis* (Musser, Heaney and Rabor, 1985) from Dinagat Island, off Mindanao, and *C. paulus* (Musser and Gordon, 1981) from Ilin island, off Mindoro. Both of these species are known only from the holotype specimens. The *C. paulus* specimen was collected in 1953, but the species was almost certainly extinct in the type and only known locality by the time it was described. A recent survey of Ilin revealed that the island had been virtually deforested for 30 years, and none of the local people could recall seeing any of these large, highly distinct rodents since that time (Pritchard, 1989).

A rather similar situation exists for the naked-backed fruit bat (*Dobsonia chapmani*) which was known only from the islands of Negros and Cebu, where little forest remains. This species was first described in 1952, but it has not been recorded since 1964, despite systematic searches by trained investigators in the 1980s (Heaney and Heideman, 1987). Recent faunal surveys on the neighbouring island of Panay have failed to locate another endemic fruit bat, *Acerodon lucifer*, which has not been recorded since the original series of specimens was obtained in 1888.

The principal threats to the fauna of the Philippines are the burgeoning human population (now numbering about 67 million people) and continued deforestation. At least 94% of the total land area of approximately 300 000 km^2 was originally covered by tropical forests. However, by 1988 satellite imagery revealed that only about 21% natural forest cover remained (Forest Development Bureau, 1988). These problems are exacerbated by the disproportionate extent of habitat loss in some of the

Table 28.1 Distribution of taxa of priority concern[a]

Species/subspecies[b]	*Distribution*	*Comment*
Tamaraw* *Bubalus mindorensis*	Mindoro	Pop'n. *c.* 350?
Spotted deer* *Cervus alfredi*	Panay; Negros	Extinct over most of range, status critical
Calamian deer *C. calamianensis*	Busuanga; Culion (Calamian Is)	Very limited range, inc. on Calavit
Mindoro rusa *C. mariannus barandanus*	Mindoro	Status uncertain, but limited range
Lowland Mindanao rusa *C. mariannus nigellus*	Mindanao	Status uncertain, but certainly depleted
Visayan warty pig* *Sus cebifrons*	Panay; Negros	As for *C. alfredi*, but poss. more resilient
Panay cloud rat* *Crateromys* sp. *nov.*	Panay	Very limited range, probably endangered
Ilin Is. cloud rat* *Crateromys paulus*	Ilin Is.	Extinct on Ilin, but may occur on Mindoro
Naked-backed fruit bat *Dobsonia chapmani*	Cebu; Negros	Probably extinct
Tube-nosed fruit bat* *Nyctimene rabori*	Negros; Sibuyan Is	Rare over very limited range
Philippine nectar bat *Eonycteris robusta*	East Philippines	Widespread, but very uncommon
Panay flying fox* *Acerodon lucifer*	Panay	Probably extinct
Giant golden-mantled flying fox, *A. jubatus*	East Philippines	Rare, with patchy distribution
White-eared flying fox *A. l. leucotis*	Busuanga (Calamian Is)	Very limited range, no recent reports
White-winged flying fox *Pteropus leucopterus*	East Philippines	Rare, very patchy distribution
Little golden-mantled flying fox, *P. pumilus*	East Philippines	Limited distribution, (but locally common?)

[a]For further details see Heaney, Gonzales and Alcala (1987).
[b]A number of endemic rodents and other small mammals are excluded although potentially threatened through loss of habitat over their highly restricted known or presumed ranges.
* = high priority species.

faunistically most important regions (notably the Western Visayas), and by intense hunting pressure on all larger or commercially valuable species. As indicated in Table 28.1, the most threatened Philippine endemic mammals also have limited ranges, and several are single island endemics. All are potentially seriously threatened, some critically so. The Protected Areas and Wildlife Bureau (PAWB) of the Department of Environment and Natural Resources (DENR) is attempting to curb the rate of habitat destruction and illegal trade in wildlife species, but it is seriously hampered by lack of resources and the prevailing political instability in parts of the country.

For all of these reasons the Philippines has been identified by the Species Survival Commission (SSC) of the World Conservation Union (IUCN) as one of the world's highest priority areas for conservation concern (S. Stuart, pers. comm.), though to date it has received little attention from the conservation community. The country is already facing a major ecological crisis, and many other extinctions seem inevitable unless urgent action is taken to address this situation.

28.2 CURRENT CONSERVATION PROJECTS AND THE *CERVUS ALFREDI* PROTOCOL

In recent years, concerns about the declining status of particular endemic taxa have resulted in field status surveys and the formulation of conservation recommendations for the tamaraw (*Bubalus mindorensis*) (Cox and Woodford, 1990; Callo, 1991); Calamian deer (*Cervus calamianensis*) (Villamor, 1989); Philippine spotted deer (*C. alfredi*) (Cox, 1987a,b; Oliver, Cox and Dolar, 1991); the Visayan warty pig (*Sus cebifrons*) (Cox, 1987a; Oliver, Cox and Groves, in press), and the cloud rats (*Crateromys* spp. and *Phloeomys* spp.) (Oliver and Cox, 1990).

Of these, the tamaraw is widely regarded as one of the principal focal species for conservation interest in the Philippines. The species has increased in numbers from an estimated wild population of about 200 individuals when the Tamaraw Conservation Programme (TCP) was initiated by the Philippine Government in 1979, to a current estimated population of about 350 individuals; though these figures are not based on any systematic surveys throughout the species' range. Moreover, wild tamaraw remain poorly protected, and the majority of resources allocated for the TCP have been invested in the local captive and semi-captive components of this programme. However, with numerous losses and no successful rearings so far, these efforts have proved unsuccessful (for details see Cox and Woodford, 1990; Callo, 1991).

Rather more success was achieved in the captive breeding of the Calamian deer at the Ecosystems Research and Development Bureau (ERDB) at Los Banos, Luzon, though this 10 year project was 'terminated' in 1991, and the remaining stock of eight (2.6) barely exceeds the number of founders originally obtained for this project (Villamor, 1989). However, the overall survival prospects of this species have improved following the creation of the Calauit Game Reserve and Wildlife Sanctuary (3 760 ha) in 1976, where an estimated original population of about 35 individuals had increased to about 550 individuals by the end of 1991 (J. Gapuz, pers. comm.). Moreover, animals taken from this population have been used for reintroduction projects on at least three smaller islands in this group. Changes in land usage and hunting pressure have eliminated Calamian deer over much of the rest of their (highly restricted) range on Busuanga and Culion, though it is doubtful if the species is in any immediate danger of extinction on either of these islands (Oliver and Villamor, 1992).

In contrast, the Philippine spotted deer (*C. alfredi*) and the Visayan warty pig (*S. cebifrons*) have already been eliminated from several much larger islands, and both are undoubtedly seriously endangered and declining throughout their small remaining range. Their plight was first revealed during a field status survey organized by the IUCN/SSC Pigs and Peccaries Specialist Group (PPSG) and the Zoological Society for the Conservation of Species and Populations (ZSCSP) in 1985. Both species are endemic to the west-central Visayan Islands or Negros–Panay Faunal Region. However, the 1985 survey confirmed that deforestation in this region had been disproportionately severe, and that both species had been extirpated from over 95% of their former range, including all of Cebu, Guimaras and (probably) Masbate, all of which have been virtually deforested (Cox, 1987a). As a result, both species are now known to survive only on Panay and Negros, where they are still subject to intense hunting pressure in each of the few isolated patches of remaining forest on these islands (Cox, 1987a; Oliver, Cox and Dolar, 1991).

These findings resulted in the formulation of two principal recommendations, which were agreed in consultation with the DENR. These were to establish a new national park, the (proposed) Panay Mountains National Park (of about 40 000 ha), to protect the sole surviving population of both species on Panay, and to set up an international cooperative breeding programme for the spotted deer, which was considered more immediately threatened than the wild pig. The spotted deer, being the largest and one of the most 'attractive' endemic forms, was also deemed an appropriate 'flagship' for increased conservation interest in the region (see also Chapter 2). In 1987, with funding assistance from the West Berlin Zoo, the first of these recommendations was implemented by means of a faunal survey and the development of a preliminary management plan for the proposed national park (Cox, 1987b). The Park is expected to be officially gazetted in the near future (W. Dee, pers.

comm.). In 1990, with funding assistance from the Mulhouse and San Diego Zoos, the captive breeding programme was initiated under the aegis of a Memorandum of Agreement (MOA) between DENR and the Mulhouse Zoo, France. This was the first such international agreement for a species' conservation project in the Philippines. Under the terms of the MOA, two *in situ* (one each on Panay and Negros for animals from each of these islands) and one *ex situ* (in Mulhouse Zoo, of Negros stock) breeding groups were established from animals of known origin. All of these animals were being held locally by private owners, most of whom donated their 'pets' to this project. A total of 14 (6.8) animals were acquired in this way during the first year, and a further 7 (4.3) have been obtained since that time. Clearly, one of the many advantages of establishing these *in situ* facilities is that they can also serve as 'rescue centres', thereby enabling additional founders to be incorporated into the breeding programme on an opportunistic basis, as well as facilitating local research and conservation–education projects.

Of the total of 21 (10.11) founders obtained to date, three (1.2) proved to be in extremely poor condition and have since died. However, four (2.2) have been born thus far, bringing the May 1992 total to 22 (11.11). These animals, and all future additional wild-caught or captive-bred individuals, will collectively comprise the World Herd, and all animals will be maintained on breeding loan from the Philippine Government. A stock register or studbook is being kept by Mulhouse Zoo, which will also be largely responsible for the coordination and development of this programme. In addition, a Philippine Spotted Deer Conservation Committee, comprising a small number of Filipino and foreign members and co-chaired by the directors of the PAWB and the Mulhouse Zoo, has been established to oversee and advise on these developments. Perhaps most importantly, this Committee is also empowered under the terms of the MOA to: 'negotiate the donation of funds for related conservation activities in the Philippines, from any outside institution which may become directly involved in this programme in the future through its receipt of animals on breeding loan'. By this means, it is hoped to generate periodic, but potentially substantial, funding support for the further development of this programme and for other conservation projects in the West Visayas.

28.3 FOLLOW-UP ACTIVITIES AND NEW PROJECTS

In line with the protocol established during the development of the Spotted Deer Conservation Programme, a new series of species' projects was initiated earlier this year, with funding support from the Fauna and Flora Preservation Society, the Lubee Foundation, and San Diego and

Mulhouse Zoos. These projects are intended to address a variety of follow-up activities and outstanding recommendations arising from the Philippine spotted deer and warty pig projects, a recent cloud rat survey (Oliver and Cox, 1990), and two new initiatives for the Calamian deer and for the fruit bats *Nyctimene, Acerodon* and *Pteropus* spp. These projects also represent a refinement of the priorities for the threatened mammals of the Philippines identified by IUCN/SSC in an outline proposal of July 1990, entitled 'Conservation and Recovery of Threatened Species in the Philippines' (IUCN/SSC, 1990). The only notable omission from the IUCN/SSC listing of species' conservation priorities, the Visayan warty pig, has been added as a high priority for conservation action following recent clarification of its taxonomy and distribution, which reveal that it is both more distinct and more endangered than formerly supposed (Oliver, Cox and Groves, in press.).

Briefly, these activities include the following.

(a) Production and distribution of 2500 spotted deer conservation–education posters (printed in three local languages); organization of a one-day workshop (entitled: 'The Philippine Spotted Deer Conservation Programme and Related Conservation Priorities in the West and Central Visayas') for local DENR officers, decision makers and land managers; and convening of the inaugaural meeting of the Philippine Spotted Deer Conservation Committee. (Both meetings were held at Silliman University in Negros, in March 1992.)

(b) The repair and improvement of the spotted deer enclosures at Silliman University; acquisition of additional founders; and development of a second 'rescue and breeding centre' for the spotted deer in Negros Occidental, from whence most (illegally) wild-caught animals now originate.

(c) A field status survey of the Calamian Islands (Busuanga, Calauit, Culion and Coron) to assess current status of the Calamian deer, bearded pig (*S. b. ahoenbarbus*) and white-eared fruit bat (*A. l. leucotis*); and development of recommendations for the enhanced future protection of representative populations and habitats of these species (Oliver and Villamor, 1992).

(d) Formulation and implementation of a MOA for the development of a international cooperative breeding programme for *C. calamianensis*, and other related (*in situ*) conservation activities, between DENR and the Zoological Society of San Diego.

(e) Formulation of relevant agreements, and construction of enclosures at Silliman University for initiation of a captive breeding programme for the Visayan warty pig; production and distribution of a conservation–education poster featuring the Philippine wild pigs; and the collection and export of blood samples from captive Philippine pigs, of several representative taxa, for cytogenetic analyses (de Haan *et al.*, in press).

(f) Formulation of a tripartite MOA between DENR, Silliman University and the Lubee Foundation, USA for development of international cooperative breeding programmes for (in the first instance) four species of fruit bats (i.e. *N. rabori, A. jubatus, P. leucopterus* and *P. pumilus*); extension of existing facilities for the captive management of fruit bats at Silliman University, including the construction of three large flight cages; provision of equipment and funds for status surveys and the capture and management of founder stocks of these species.

It is hoped that the finalization of the aforementioned agreements, and the implementation of other outstanding recommendations, including the acquisition of sufficient (i.e. 12 or more) founder stocks of each of the above species will be completed within one to two years, when several expeditions to selected locations are being planned in conjunction with a number of Philippine institutions and relevant local researchers. During this period it is also hoped to implement some of the recommendations from the 1990 cloud rat survey, including the investigation of reports and collection of voucher specimens of the (as yet undescribed) populations of these animals in the Zambales Mts/Bataan Peninsular (Luzon) and Mindoro and Siargao Islands. All of these populations are thought likely to be threatened and each may be a new taxon (Oliver and Cox, 1990). Several other conservation–education posters, depicting these and other threatened endemic species, are also in the planning stages.

28.4 DISCUSSION

As indicated in Table 28.1, the most threatened Philippine endemic mammals have very limited ranges, and several are single island endemics. Although loss of habitat is the single greatest threat to these faunas, there are important regional differences in the extent of deforestation, so considerable overlap in the ranges of some of the most threatened species is to be expected. Consequently, projects focused on different taxa can be conducted concurrently in many instances, and proportionately more resources directed towards the development of conservation strategies in priority locations. This facilitates a more coherent approach to the conservation needs of these animals, without diluting resources directed towards particular prioritized taxa. This consideration also obtains with respect to captive breeding programmes, **provided** that those breeding programmes are part of an integrated species' conservation strategy, rather than an isolated objective. Given that funding for species' conservation projects is often difficult to obtain from mainstream conservation funding agencies, and that species are the stock-in-trade of zoos, any zoo purporting a commitment to species' conservation should accept a responsibility to assist the development of conservation initiatives in the countries of origin of those species. In the protocol established for the Philippine spotted deer, captive breeding is part of a wider strategy which embraces a variety of other fundamentally important activities, including habitat protection, field research and conservation–education. Moreover, each of these activities has been formulated in response to recommendations arising from preliminary field work, rather than the other way around, as frequently happens in zoo-based initiatives. In addition, mechanisms are built into the terms of the MOA with the Philippine government, which are intended to ensure that the captive breeding programme is not only structured to meet the long-term management interests of this species, but that the successful development of this breeding programme will generate resources for a variety of in-country conservation activities from an increasing number of breeding centres outside the Philippines. If this can be achieved, perhaps another perceptual barrier between captive and wild may be set aside, and the respective approaches to the management of these populations may become more congruent.

REFERENCES

Callo, R.A. (1991) The tamaraw population: increasing or decreasing? *Canopy Int.*, **16** (4), 4, 9.

Cox, C.R. (1987a) The Philippines spotted deer and the Visayan warty pig. *Oryx*, **21** (1), 37–42.

Cox, C.R. (1987b) A preliminary survey of the proposed Panay Mountains National Park. Unpublished Report to the Zoologischer Garten, Berlin.

Cox, C.R. and Woodford, M. (1990) A technical evaluation of the Philippine Tamaraw Conservation Programme. Unpublished Report to DENR, IUCN/SSC and the Zoological Society of London.

Forest Development Bureau (1988) *Natural Forest Resources of the Philippines, Philippine – German Forest Resources Inventory Project*. Forest Management Bureau, Dept. of the Environment and Natural Resources, Manila.

Haan, N.A. de, Bosma, A.A., Macdonald, A.A. and Oliver, W.L.R. (in press) A species of wild pig in the Philippines with a type of centric fusion new to *Sus: 13/16*, in *Proc. 10th European Colloquium on Cytogenetics of Domestic Animals*, Utrecht.

Heaney, L.R. (1986) Biogeography of mammals of S.E. Asia: estimates of rates of colonisation, extinction and speciation. *Biol. J. Linnean Soc.*, **28,** 127–65.

Heaney, L.R., Gonzales, P.C. and Alcala, A.C. (1987) An annotated checklist of the taxonomic and conservation status of land mammals in the Philippines. *Silliman J.* **34** (1–4), 32–66.

Heaney, L.R. and Heideman, P.D. (1987) Philippine fruit bats, endangered and extinct. *Bats*, **5,** 3–5.

Ingle, N.R. and Heaney, L.R. (1992) A key to the bats of the Philippine Islands. *Fieldiana: Zoology*, No. 69.

IUCN/SSC (1990) Conservation and recovery of threatened species in the Philippines. Unpublished project proposal, IUCN, Gland, Switzerland.

MacKinnon, J. and MacKinnon, K. (1986) *Review of the Protected Areas System in the Indo-Malayan Realm*, IUCN/UNEP, Cambridge, UK and Gland, Switzerland.

Musser, G.G. and Gordon, L.K. (1981) A new species of *Crateromys* (Muridae) from the Philippines. *J. Mamm.*, **62** (3), 513–25.

Musser, G.G., Heaney, L.R. and Rabor, D.S. (1985) Philippine rats: a new species of *Crateromys* from Dinagat Island. *Amer. Mus. Novitates.*, **2821,** 1–25.

Oliver, W.L.R. and Cox, C.R. (1990) Distribution and status of the Philippine cloud rats. Unpublished report to DENR and the Zoological Society of London.

Oliver, W.L.R., Cox, C.R. and Dolar, L. (1991) The Philippine spotted deer, *Cervus alfredi*, conservation project. *Oryx*, **25** (4), 199–205.

Oliver, W.L.R., Cox, C.R. and Groves C.P. (in press) The Philippine warty pigs, *Sus philippensis* and *S. cebifrons*, in *Pigs, Peccaries and Hippos. IUCN/SSC Action Plan for the Suiformes* (ed. W. Oliver), IUCN, Cambridge, UK, and Gland Switzerland.

Oliver, W.L.R. and Villamor, C.I. (1992) The distribution and status of the Calamanian deer *Cervus* (= *Axis*) *calamianensis*, and the Palawan bearded pig, *Sus barbartus ahoenobarbus*, in the Calamanian Islands, Palawan Province. Unpublished report to DENR, Zoological Society of San Diego and the Fauna and Flora Preservation Society.

Pritchard, J.S. (1989) Ilin Island cloud rat extinct? *Oryx*, **23** (3), 126.

Villamor, C.I. (1989): Breeding of Calamian deer (*Axis calamianensis*) in captivity. *Syvatrop, Philippines Forestry Res. J.*, **12** (1–2), 49–60.

29

Interface between captive and wild populations of New Zealand fauna

P. Garland
Orana Park Wildlife Trust, New Zealand
and
D. Butler
Department of Conservation, New Zealand.

29.1 WHY IS NEW ZEALAND'S FAUNA NOW SO ENDANGERED?

New Zealand's isolation from other major land masses, since the breakup of the great southern continent of Gondwanaland, has led to the separate evolution of species with spectacular adaptations – the kiwi (*Apteryx* spp.) with its sense of smell, the huia (*Heteralocha acutirostris*) with its different feeding pattern for male and female and the kea (*Nestor notabilis*), the only alpine parrot in the world.

The absence of large placental land mammals (and only two species of bats) has allowed New Zealand's fauna to respond to the varied ecological opportunities in a landscape and climate that fluctuates from sub-tropical rain forests to alpine glacial mountains. The richness and diversity of New Zealand's endemic fauna is clearly remarkable for such a small land mass (269 000 km^2), with two land mammals, 133 birds, 40 reptiles, three amphibians (species and subspecies), both richness and diversity rivalling the world's tropical rain forest for uniqueness.

Creative Conservation: Interactive management of wild and captive animals.
Edited by P.J.S. Olney, G.M. Mace and A.T.C. Feistner.
Published in 1994 by Chapman & Hall, London.
ISBN 0 412 49570 8

The arrival of humans – first Polynesian people over 1000 years ago and then European settlers, 150 years ago – has dramatically changed this ecological balance and led to a sharp decline and then extinction for many species. While habitat destruction has played a significant role in this decline (in the early 1800s 66% of New Zealand was forested but now only 22%), a much more severe and continuing threat was created through the introduction of many exotic mammals. Between 1851 and 1910 more than 34 exotic species were introduced. While the browsing animals, deer, possums, goats, pigs, etc., created obvious modification to the landscape, and destruction of the plant life, exerting subtle effects as competitors of native species, it has been the small mammalian predators, such as the rat, stoat and ferret, that have brought about the most disastrous results. It is estimated that at least 50 species have suffered extinction or serious decline as a result of their introduction.

New Zealand is now in the unenviable position of having the greatest number of endangered species per capita of any country in the world. Of the 94 endemic land bird (full) species known to have occurred in New Zealand, 35 (37%) became extinct in the Polynesian era and a further 11 (18.6%) became extinct in the European era. Of the 48 remaining endemic land bird (full) species, 23 (48%) are threatened (Bell, 1986). A truly frightening scenario.

29.2 WHY AREN'T MORE NEW ZEALAND SPECIES EXTINCT?

New Zealand is fortunate in having one of the largest national park systems per area of land mass in the world, with 30% of New Zealand's forest, lakes and mountains having protected status. Started in 1887, these national parks provide extensive areas of forest, mountains and rivers suitable for wildlife habitation.

Fortunately, many parts of New Zealand still remain relatively inaccessible, particularly in the alpine areas of South Island and central North Island. These areas remained unsuitable for farming, logging, etc., and hence were left relatively unscathed until protected.

Along the coastline there are over 600 offshore islands greater than 1 ha in area (Daugherty *et al.*, 1990), some of which have remained predator-free, and which serve as last-ditch refuges for many at-risk species.

Wildlife protection laws are extensive in New Zealand, with most native species receiving full protection under the Wildlife Act 1953. The export of New Zealand fauna has always been prohibited without permit and few species have ever entered the pet trade. Exploitation for commercial purposes has also been minimal.

29.3 HOW IS NEW ZEALAND WILDLIFE'S RECOVERY TAKING PLACE?

A wide range of conservation-based field programmes have been in effect since the New Zealand Wildlife Service was established in 1947 (now the Department of Conservation), and these include enforcement of wildlife protection laws, public education, protection and management of habitat, and direct and intensive species management both *in situ* and *ex situ*. In the wildlife recovery planning for many New Zealand endangered species, two practical applications have been used very successfully.

29.3.1 Use of offshore islands as conservation reserves

This technique has been the most successful of all methods employed in New Zealand to provide recovery assistance for threatened species. The movement of individuals to islands free of mammalian predators has been sufficient to improve markedly the status of several species. The North Island saddleback (*Philesturnus carunculatus rufusater*) had become restricted to a single island, Hen Island (Tarauga), by late last century, but it is now established on at least seven islands following transfers from the 1960s onwards (Lovegrove, 1992).

However, in many cases a transfer alone has not been sufficient and additional intensive management has been needed, as shown by two of New Zealand's most famous threatened species. The recovery of the Chatham Island black robin (*Petroica traversi*) began with the transfer of the remaining seven birds, but the population declined further to five before the innovative cross-fostering programme of Don Merton and his colleagues led to its restoration (Merton, 1990). This programme involved manipulation of nest contents, transfers of eggs and chicks, placing eggs in nest-boxes and supplementary feeding – a wide range of interventionist techniques.

Similarly the transfer of kakapo (*Strigops habroptilus*) from Stewart Island to the predator-free islands of Little Barrier, Codfish and Maud did not in itself lead to any increase in numbers. Successful breeding only occurred on Little Barrier in 1990/91 following two seasons of supplementary feeding, eight years after the arrival of the first birds on the island. An intensive research programme was needed to design feeding hoppers that kiore (*Rattus exulans*) could not get access to, and then to train kakapo to use them. Currently (April 1992) kakapo are nesting on Codfish Island in the absence of supportive management, but indications are that shortage of food and scavenging and predation by kiore are having a negative effect on their ability to rear chicks.

These two examples show the value of applying techniques developed through intensive management breeding (similar to those used in captive breeding) to free-ranging animals in the wild, perhaps leading to the ideal combination. However for practical or logistical reasons, this approach may not be possible in many situations, and captive breeding *ex situ* may be appropriate.

29.3.2 Captive breeding in support of wild populations

Captive breeding for New Zealand native fauna species began in the late 1950s and over recent years has produced substantial results in aiding recovery of threatened or at-risk species.

(a) Government-supported breeding facilities

The Department of Conservation (DOC) maintains the National Wildlife Centre at Mt Bruce which undertook much of the pioneering work with captive breeding of native birds. The DOC has also developed purpose-built facilities for rearing and reintroduction of takahe (*Notornis mantelli*) at Burwood Bush, Te Anau, and black stilt (*Himantopus novaezealandiae*) at Twizel. Both programmes have resulted in released birds breeding in the wild.

(b) Zoos and wildlife parks

Both zoos and wildlife parks have been instrumental in pioneering and reinforcing captive breeding programmes for New Zealand native species, many with spectacular results (Butler, 1992).

The Otorohanga Zoological Society, with its establishment of the National Kiwi Centre in the 1970s, has produced in-depth and quality information on kiwi biology, physiology, reproduction, etc., as well as building up an extensive breeding record for other native birds.

Wellington Zoo has developed effective captive breeding and advocacy programmes for the giant weta (*Deinacrida* spp.), using these large insects to focus attention on invertebrate conservation (Barrett, 1992).

Auckland Zoo through incubation and rearing for kea and kaka (*Nestor meridionalis*) has gained valuable knowledge to be applied to the captive breeding of kakapo. Three kakapo chicks were moved to the zoo from Codfish Island in April/May 1992 to capitalize on this expertise.

Orana Park has developed species management programmes involving the breeding of the New Zealand pigeon (*Hemiphaga novaeseelandiae*) as an analogue for the endangered Chatham Island pigeon (*Hemiphaga novaeseelandiae chathamensis*) (Blanchard, 1992), which numbers less than 50 birds and South Island skinks (*Leiolopisma* spp.).

(c) Private breeders

Private breeders have also contributed significantly to captive breeding programmes associated with recovery plans and this is particularly so with the waterfowl breeders who are members of the Ducks Unlimited organization. To date they have reared over 1200 brown teal (*Anas aucklandica*) for release and re-establishment to the wild. Some private organizations and breeders have a considerable advantage over city-based urban zoos since they are not open to the public and have available space, as can be seen by the extensive wetland development at facilities like the Isaac Wildlife Trust.

29.3.3 Coordination of captive and wild management

The establishment of species recovery plans through the DOC is a multi-disciplined approach to assessing and rectifying the threats facing many New Zealand endangered species. Recovery plans are developed and implemented by a recovery group made up of experts from a wide range of backgrounds. They serve to integrate *in situ* and *ex situ* elements into a comprehensive strategy maximizing the use of all the expertise and resources available. The implementation of captive management programmes is incorporated as a major factor in many plans. At least five have been approved, kakapo (Powlesland, 1989), blue duck (*Hymenolaimus malacorhychos*) (Williams, 1988), North Island kakako (*Calleanas cinerea wilsoni*) (Rasch, 1991), yellow-eyed penguin (*Megadyptes antipodes*) (Department of Conservation, 1991) and kiwis (Butler and McLennan, 1991).

The DOC identifies captive breeding priorities in the following order:

(a) where captive breeding can be used to supplement the wild population or establish new ones;
(b) where captive animals can provide research findings which will assist the wild population;
(c) where captive stock can promote public education fostering greater support for the conservation of wild populations;
(d) as an insurance policy against further decline or extinction of wild populations.

Cooperative planning by all parties involved in captive breeding is now well developed to pool resources towards work on priority species. Collectively, all holders of New Zealand native birds (zoos, private breeders, etc.) have amalgamated to form a New Zealand Taxon Advisory Group (TAG) and meet annually to discuss options and opportunities and to develop species management planning. The New Zealand Bird TAG operates as an advisory group under the Australasian Species

Management Programme (ASMP). Breeders of reptiles, amphibians and invertebrates will become organized to meet under other TAGs in the future. Collective captive breeding for conservation is being coordinated through this network and all of the major captive breeding programmes are appointing species management coordinators. Both DOC and the major zoos operate computer-based Single Population Analysis and Records Keeping System (SPARKS) programmes for species management.

More recently new techniques and planning options developed by the Captive Breeding Specialist Group (CBSG) of the Species Survival Commission (SSC) of the World Conservation Union (IUCN) have been introduced, resulting in a number of specific workshops on threatened species. The Population and Viability Analysis Workshop (PVA) for kea and kaka held at Orana Park in December 1991 proved a valuable tool in reducing the risks of extinction. This assessment and action plan process is expected to be developed further and a Conservation Assessment and Management Plan CAMP Workshop is planned for New Zealand penguins in August, 1992.

The *Captive Breeding Index of New Zealand Native Fauna* (Garland and Butler, 1991) gives a good account of captive breeding programmes currently being undertaken in New Zealand and highlights further research opportunities.

29.4 HOW SUCCESSFUL ARE WE AT RE-ESTABLISHING CAPTIVE-BRED SPECIES IN THE WILD?

In terms of volume, some of our achievements can be described as high-input, particularly the release of 1200 brown teal to the wild. However population surveys and radio-telemetry have shown that relatively few of these teal are surviving to breed, due to continued predation and poor habitat. Release of takahe, on the other hand, looks likely to lead to the development of a new wild population, as the factors that caused earlier declines are no longer operating to the same extent. Several other programmes involving the propagation and reintroduction of threatened species, e.g. black stilt and blue duck, are in early stages but look promising.

We expect that as captive breeding techniques improve and our knowledge of reintroduction procedures are fine-tuned, we should become better at successful reintroduction programmes, but the initial threats causing the species' decline in the first place must have been removed or at least reduced. Hence the failure to date of the brown teal reintroduction programme. The greater public awareness of conservation programmes, particularly those involving captive breeding, has increased rapidly New Zealand's expectations for the re-establishment of much of our native fauna. Increased support, through sponsorship, fundraising, etc., is now being achieved as a result and inevitably this will benefit the reintroduction of threatened species.

29.5 WHAT IS THE FUTURE FOR NEW ZEALAND'S NATIVE FAUNA?

There are currently 30 species recovery plans in draft form or complete and this is expected to grow to 50 in the next five years. The ultimate success of these recovery plans, e.g. the securement of stable, genetically viable populations in the wild, will depend greatly on our ability to provide the resources necessary, not only to secure the current and re-establish new populations, but to reduce those threats which remain. In broad terms, habitat availability may still be sufficient to provide the base habitat requirements of most species, but it is the threat of mammalian predation and competition which offers the greatest challenge. New Zealand has achieved much in wildlife conservation in the last 25 years, but it will be at least another 25 years before we are able to say that we have saved the remnants of what was a diverse and unique fauna.

REFERENCES

Barrett, P. (1992) Maintaining and breeding the Common tree weta *Hemideina crassidens* at Wellington Zoo. *Int. Zoo Yb.*, **31**, 30–6.

Bell, B.D. (1986) *The Conservation Status of New Zealand Wildlife. New Zealand Wildlife Series*, Occasional publ. 12, Wellington.

Blanchard, B. (1992) Hand-rearing a New Zealand pigeon *Hemiphaga n. novaeseelandiae* at the Wellington Zoological Gardens. *Int. Zoo Yb.*, **31**, 49–53.

Butler, D.J. (1992) The role of zoos in the captive breeding of New Zealand's threatened fauna. *Int. Zoo Yb.*, **31**, 4–9.

Butler, D.J. and McLennan, J.A. (1991) *Kiwi Recovery Plan. Threatened Species Unit Recovery Plan*, Ser. No. 2, Department of Conservation, Wellington.

Daugherty, C.H., Towns, D.R., Atkinson, I.A.E. and Gibbs, G.W. (1990) The significance of the biological resources of New Zealand islands for ecological restoration, in *Ecological Restoration of New Zealand Islands* (eds D.R. Towns, C.H. Daugherty and I.A.E. Atkinson), Conservation Sciences Publication No.2, Department of Conservation, Wellington.

Department of Conservation (1991) *Yellow-eyed Penguin* (Megadyptes antipodes) *Species Conservation Strategy (1990–1995)* (Revised), Department of Conservation, Dunedin.

Garland, P.D. and Butler, D. (1991) *Captive Breeding Index of New Zealand Native Fauna*, Orana Park Wildlife Trust, Christchurch.

Lovegrove, T.G. (1992) The effects of introduced predators on the Saddleback (*Philesturnus carunculatus*), and implications for management. University of Auckland Ph.D. Thesis.

Merton, D.V. (1990) The Chatham Island black robin. *Forest and Bird*, **21**(3), 14–19.

Powlesland, R.G. (1989) *Kakapo Recovery Plan 1989–1994*, Department of Conservation, Wellington.

Rasch, G. (1991) *Recovery Plan for North Island Kokako. Threatened Species Unit Recovery Plan*, Ser. No. 1, Department of Conservation, Wellington.

Williams, M.J. (1988) *Conservation Strategy for Blue Duck. Science and Research Internal Report* No. 30, Department of Conservation, Wellington.

30

The potential for captive breeding programmes in Venezuela – efforts between zoos, government and non-governmental organizations

P. Trebbau, M. Diaz and E. Mujica
Fundación Nacional de Parques Zoológicos y Acuarios, Venezuela.

30.1 INTRODUCTION

The geographical location of Venezuela makes it a privileged Neotropical country. It is rich in natural resources and biodiversity, a consequence of more than seven different climatic zones in an area of 916 445km^2. These rich resources are vulnerable to the considerable impact of development, which results in habitat destruction and species extinction.

30.1.1 Protected areas

Two decades ago, the realization of the effect of these pressures on Venezuela's natural resources led to politicians favouring the creation of extensive protected areas, including national parks and natural monuments. Currently 15% of the national territory is under some kind

Creative Conservation: Interactive management of wild and captive animals.
Edited by P.J.S. Olney, G.M. Mace and A.T.C. Feistner.
Published in 1994 by Chapman & Hall, London.
ISBN 0 412 49570 8

of governmental protection, more than in any other South American country (González and Febres, 1992). However, the same problems that affect many countries under development also affect Venezuela and its protected areas. In particular, lack of effective guards and equipment make it almost impossible to protect the ecosystems within the protected areas.

Venezuela has a governmental Institute of Parks (Instituto Nacional de Parques – INPARQUES) that manages national parks as well as the administrative and legal aspects of a few recreational parks and zoos (Amend, 1991). In 1991 a new initiative was established to strengthen the system of Venezuelan national parks. Two programmes were developed: technical training programmes for rangers, technicians, management personnel, and field staff of the national parks service; and research training and ecological monitoring of each park. This second programme has as its goal the generation of an ecological database for each park which can be used to guide planning and decision-making processes for the parks' management. These programmes, which have been developed through multi-lateral cooperation, are of fundamental importance, both strategically and economically, to the national parks service, and represent the first of their kind in South America.

The organizations participating in these programmes are: INPARQUES, Econatura, Wildlife Conservation International, and the European Community. INPARQUES is an autonomous governmental agency of the Ministry of the Environment, and is responsible for the protection, planning, improvement and management of the national park system. Econatura is a Venezuelan non-governmental organization (NGO), whose main activity is the promotion of professional training in the conservation of natural resources. Wildlife Conservation International (WCI), a division of the New York Zoological Society (NYZS), is in charge of the society's international programmes. The NYZS is the oldest NGO in the USA working internationally in the field of conservation, and it has been supporting programmes in Venezuela since the 1920s. The Commission of the European Community (EC) represents European interests in integration and development. Funding for international cooperation in projects such as those described above is obtained through the EC's environmental programme.

One of the biggest problems in the national parks is the illegal hunting pressure on their wildlife. From 1985 to 1987 Silva and Strahl (1987) conducted a study on the ecology and conservation of curassows (Cracidae) in nine national parks in the northern region. They highlighted the need for an educational programme targeted at hunters (Silva and Strahl, 1989). As a result of this, in 1989 a programme about hunting was started in the northern Venezuelan parks. Education, vigilance, and ecotourism were explored as possible solutions to the illegal hunting

(Silva and Pellegrini, 1991, 1992). INPARQUES and WCI provided official backing and financial support.

The initial evaluation of the programme indicated the following: 87% of the surveyed hunters were illegally hunting for sport; hunting pressure was more intense on species initially more abundant in any given area; the mean take was 22 game pieces/hunter/year; most game hunted was small to medium size and there was no selectivity; and poachers operated within park boundaries mainly during weekends and holidays.

From these data the main objectives of the study were to teach people how to hunt legally outside the parks, and to explain the importance of conserving national parks. Silva and Pellegrini (1991, 1992) emphasized the unfairness of prosecuting people ignorant about the legal aspects of hunting, when the main fault is governmental lack of effective publication and dissemination of information on environmental and legal issues in the country.

Solutions were investigated through a three-pronged approach.

(a) Education – three-week extension courses were provided, which gave information about legal aspects of hunting and also emphasized the importance of and reasons for conserving national parks. Participants were evaluated and certificates were provided as proof of successful participation.
(b) Vigilance – hunting clubs were formed to serve as *ad honorem* guards together with officially assigned park guards from INPARQUES and other National Guard bodies.
(c) Ecotourism – proposals to employ hunters as nature guides to provide a monetary alternative to hunting were investigated.

Results from four national parks targeted as a first step indicated that not one hunter who had attended the programme had relapsed to illegal hunting within the national parks. This good news indicated that the programme could be effective. Efforts must now continue to cover every national park on a long-term basis.

30.1.2 The role of captive breeding

Properly managed protected areas provide possible sites for reintroduction programmes. Venezuela is probably one of the most stable countries in the Neotropics and has considerable potential for developing conservation programmes. However, the minor role that Venezuelan zoos have played in captive breeding programmes is not encouraging. Unfortunately, up to now the main objective of zoos has been to provide a recreational resource. This has been due mainly to their uncoordinated and unstable governing boards which have not developed any medium-

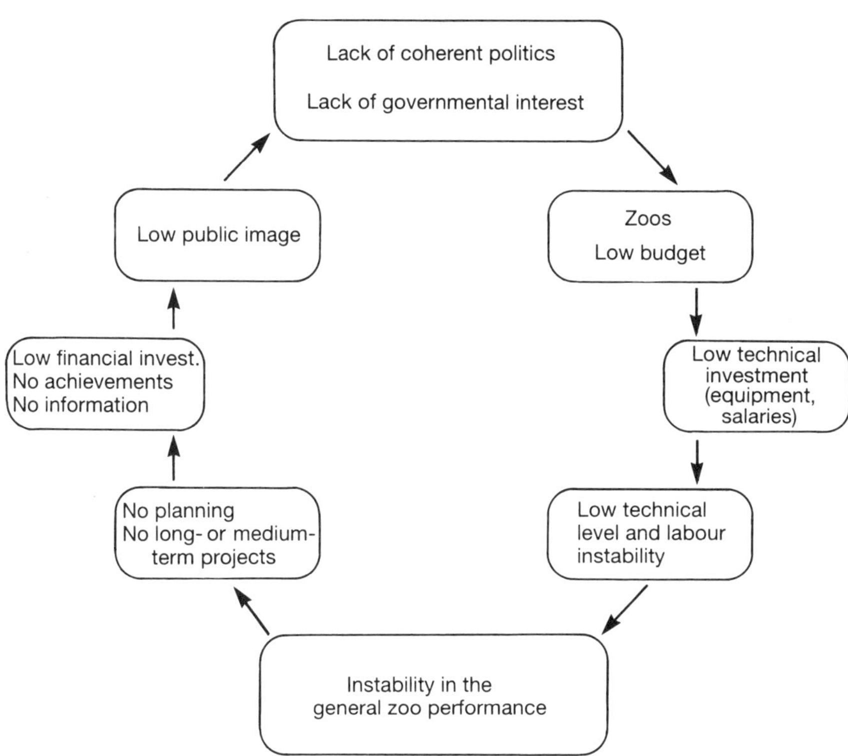

Figure 30.1 The negative feed-back circle of problems experienced by Venezuelan zoos which recent initiatives are helping to break.

or long-term breeding projects (Diaz and Barraza, 1992). Hence the zoo community's performance has been based on local issues, rather than framed in terms of national and international perspectives (Figure 30.1).

The few independent efforts which have resulted in breeding success are little known, both nationally and internationally. The lack of information and coordination has resulted in leading Venezuelan zoological institutions performing below internationally acceptable levels, with few exceptions.

Recently, however, a National Zoo Foundation, the Fundación Nacional de Parques Zoológicos y Acuarios (FUNPZA), has been created to work as a coordinating body and to develop coherent policies for expanding the role of Venezuelan zoos in conservation, and to assist with breeding programmes. FUNPZA works with a governmental institution that controls the legal aspects of holding, transferring, or hunting wildlife.

As the existing uncoordinated situation is being improved, international support and respect for Venezuelan captive breeding programmes and professionals has been growing. As a brief example three major captive breeding and reintroduction programmes will be mentioned.

30.2 CAPTIVE BREEDING AND REINTRODUCTION PROGRAMMES

30.2.1 Orinoco crocodile (*Crocodylus intermedius*)

In 1978 a global programme was started to rescue the Orinoco crocodile (*Crocodylus intermedius*). In Venezuela this programme had been coordinated by an NGO, the National Foundation for the Defence of Nature (Fundación Nacional para la Defensa de la Naturaleza – FUDENA) since its inception. The programme was developed with support from the Crocodile Specialist Group (CSG) of the Species Survival Commission (SSC) of the World Conservation Union (IUCN), a national advisory group coordinated by FUDENA, and the Ministry of the Environment. The programme is divided into two main projects – the study of wild populations, and the development of captive breeding centres. In 1984, surveys resulted in two protected areas for crocodiles being assigned in Los Llanos, Apure State. Three centres now hold adults for breeding purposes and raise juveniles collected from the wild. The aim of the centres is to rear the animals up to one or one-and-a-half years old until they reach at least a metre in length, and then release them into the protected areas. This second project is further subdivided (Arteaga, 1992): Venezuelan members of the CSG have developed plans for 1993–1995 for improving the breeding success at each centre; projects are ongoing to encourage private landowners to develop captive breeding centres in collaboration with the CSG who provide technical advice on improving husbandry techniques; reintroduction and follow-up projects using radio-telemetry are being instigated. Additionally, since the threatened status of the species in the wild is largely due to human misconceptions and hunting pressure for skins, education of local people is regarded as a priority, along with incorporation of local people into the conservation programmes. Adequate policing of the newly created protected areas in Los Llanos has not yet been accomplished due to a guard force compromised by lack of training and equipment, and lack of communication in the area.

30.2.2 Andean condor (*Vultur gryphus*)

A reintroduction programme for the Andean condor (*Vultur gryphus*) is being developed in Venezuela for the initial release of five birds (2.3) coming from the collaborative project between San Diego and Los Angeles Zoos. The chosen release area is located in El Valle Mifasí,

Páramo Piedras Blancas, Mérida State, and is part of the northern range of the Andes mountains. The programme is coordinated and funded in Venezuela by a private bank, Banco Andino, with the official support of INPARQUES (Gil and Pinto, 1991). The release platforms, quarantine headquarters, and biological station are almost complete, and the birds arrived in December 1992. The construction of a captive breeding centre in the area is planned soon after this initial phase, and more widespread breeding of this species in Venezuelan zoos is the long-term aim.

30.2.3 Spectacled bear (*Tremarctos ornatus*)

A collaborative programme for the spectacled bear (*Tremarctos ornatus*) between the Parque Zoologico y Ambiental Gustavo Rivera and the Lincoln Park Zoo started in 1991. The aim was to breed spectacled bears in Venezuela in accord with the recommendations of the international studbook. An agreement between Banco Andino and INPARQUES will lead, in the near future, to the creation of a captive breeding centre in Los Andes. The ultimate goal of this project is to reintroduce captive-bred bears to national parks in nearby areas (Gil and Pinto, 1991). This plan is in its early phases and would benefit from reinforcement from national and international institutions, to ensure the establishment of the centre and the training of personnel to carry out the breeding and reintroduction programmes.

These three projects, which involve collaboration between the government, NGOs, private individuals and banks, and national and international organizations, indicate that there is considerable potential in Venezuela for developing conservation programmes, both through captive breeding in special centres and zoos, and through protection in the wild by the creation of protected areas. For Venezuela to realize this potential, collaboration and support from the international conservation community is likely to be needed.

30.3 OTHER CONSERVATION PROGRAMMES

There are several other conservation programmes which are being developed.

30.3.1 Sea turtle conservation in national parks

The first sea turtle conservation effort in national parks was started in 1976 in the Archipelago Los Roques National Park under the Fundación Cientifica Los Roques programme for the evaluation of sea turtle nesting, hatchery, and head-starting of *Caretta caretta*, *Eretmochelys imbricata*, and *Chelonia mydas*. In 1990 an evaluation of sea turtle nesting beaches in different national parks was started through the wildlife programme of the National Parks Authority. Together with inventories undertaken in 1987 and 1988 by FUNDENA, this information resulted in the formulation of special zoning regulations to protect sea turtle nesting sites. Other goals of the project include running courses for park rangers on sea turtle biology, monitoring and conservation techniques, and the provision of information on the endangered sea turtles to local people within and near the protected areas (Guada and Vernet, 1992).

30.3.2 Programmes and projects of PROVITA

PROVITA is an NGO active in conservation and is running the following programmes.

(a) Spectacled bear programme

This includes four different projects involving environmental education for local people in habitat areas, training of young professionals, and ethnozoological studies in the Sierra Nevada National Park, Mérida State.

(b) Psittacidae programme

This includes projects for the conservation of insular parrots and parakeets: conservation of the Margarita parrot (*Amazona barbadensis*) through management and biological studies and education programmes on Margarita Island; studies of *Amazona barbadensis* on La Blanquilla Island; and captive care of confiscated birds (for subsequent release), protection of breeding areas, and design and creation of protected areas for this parrot. Additionally, studies of the present status and conservation of *Aratinga acuticaudata neoxena* in La Aresting National Park are underway.

(c) Endangered Venezuelan species programme

Under the supervision of 300 specialists and 43 specialized councillors, a preliminary list of Venezuelan threatened species has been drawn up. This will form the basis for the publication of a Red Data Book for Venezuela, which will help zoos establish priorities for captive breeding programmes.

(c) Peninsula de Paria National Park Project

PROVITA has undertaken a sociological study to establish an education programme in this national park. The next step will be the provision of a multi-disciplinary programme to offer farming alternatives to the local population.

Other projects include education workshops for conservation, an environmental education programme for the conservation of the northern helmeted curassow (*Pauxi pauxi*) in Yacambu National Park, and conservation of the oil bird (*Steatornis caripensis*) in El Guachero National Park, Monagas State.

30.3.3 Breeding programmes in zoos

Species likely to be selected for priority breeding programmes in Venezuelan zoos and aquaria are the following.
Mammals: spectacled bear (*Tremarctos ornatus*), giant otter (*Pteronura brasiliensis*), Caribbean manatee (*Trichechus manatus*), giant armadillo (*Priodontes giganteus*), giant anteater (*Myrmecophaga tridactyla*), little spotted cat (*Felis tigrina*) and black-headed uakari (*Cacajao melanocephalus*). Birds: northern helmeted curassow (*Pauxi pauxi*), horned screamer (*Anhima cornuta*), red siskin (*Carduelis cuculatta*), harpy eagle (*Harpia harpia*) and Andean condor (*Vultur gryphus*). Reptiles: Orinoco crocodile (*Crocodylus intermedius*) and American crocodile (*Crocodylus acutus*).

Las Delicias Zoo together with Miguel Romero Antoni Zoo are planning to breed harpy eagles (*Harpia harpia*). The UNELLEZ Zoo is planning to start a breeding colony of giant otter (*Pteronura brasiliensis*).

30.4 SUMMARY

Much progress is being made in developing conservation programmes in Venezuela, through fruitful collaboration between national and international institutions. Protection of areas of habitat is being improved and education programmes are teaching the benefits of conservation. Practical help to local people through employment in conservation programmes and the offering of alternative technologies is also being expanded. Further effort is needed to support the few breeding centres now working and to enlarge the possibilities for Venezuelan zoos and aquaria to contribute to national and international conservation efforts.

ACKNOWLEDGEMENTS

We would like to thank the following individuals for their useful information about captive breeding and reintroduction programmes in the country. Edgar Yerena (INPARQUES-Spectacled Bear Programme), José Lorenzo Silva (WCI-NYZS), Hedelvy Guada (INPARQUES), Alfredo Arteaga and Ana De Lucca (FUDENA), Maria Rosa Cuestas (Andean Condor–Spectacled Bear Programme). Franklin Rojas (PRO-

VITA). Anna Feistner and David Waugh provided useful comments on the manuscript.

REFERENCES

Amend, S. (1991) *Parque Nacional El Avila. Parques Nacionales y Conservación Ambiental*, Editorial Torino, Caracas.

Arteaga, A. (1992) Programa de conservación del Caimán del Orinoco (*Crocodylus intermedius*): Proyectos para consolidar su recuperación. FUDENA, Caracas.

Diaz, M. and Barraza, L. (1992) Bridging the gap – Latin American zoos and the world zoo community. *Int. Zoo News*, **234**, 11–13.

Gil, G. and Pinto, F. (1991) El Cóndor de los Andes: mensajero olridado. *Gaceta Ecologica*, **1**, 6–7.

González, C. and Febres, M.A. (coord.) (1992) Areas Naturales Protegidas de Venezuela. *Serie Aspectos Conceptuales y Metodológicos, DGSPOA/ACM/01*, Venezuela.

Guada, H. and Vernet, P. (1992) The sea turtle conservation in the National Parks of Venezuela. Poster presented in the 'International Workshop in Sea Turtle Conservation and Biology', Jenkill Island, USA.

Silva, J.L. and Strahl, S. (1987) Participación venezolana en el Proyecto 'Ecología conductual y conservación de la familia Cracidae en Venezuela' Unpublished report to FUDENA and WCI/NYZS, Caracas.

Silva, J.L. and Strahl, S. (1989) Necesidad de un proyecto de educación para cazadores furtivos en Venezuela. Unpublished report to FUDENA and WCI/NYZS, Caracas.

Silva, J.L. and Pellegrini, N. (1991) Proyectos de educación para cazadores furtivos en los Parques Nacionales San Esteban y Henry Pittier. Unpublished report to FUDENA and WCI/NYZS, Caracas.

Silva, J.L. and Pellegrini, N. (1992) Proyectos de educación para cazadores furtivos en los Parques Nacionales Terepaima y Yacambú. Unpublished report to FUDENA and WCI/NYZS, Caracas.

31

Species conservation priorities in Vietnam and the potential role of zoos

H.J. Adler
Allwetterzoo, Münster, Germany
and
R. Wirth
IUCN/SSC, Germany.

31.1 INTRODUCTION

Vietnam has been identified as one of the highest priority areas for conservation action by the Species Survival Commission (SSC) of the World Conservation Union (IUCN) because of the high percentage of threatened endemics which occur in the country (Stuart and Wirth, 1990). Species diversity and abundance has been adversely affected by bombing, bulldozing, defoliation and application of napalm during decades of war (MacKinnon, 1985) and since the termination of hostilities in 1975 pressure on the country's wildlife resources has further increased due to shifting cultivation, timber extraction and hunting caused by the rapidly growing human population.

Vietnam was once almost completely forest-covered but today only about 20% of the country's total land area supports natural vegetation, and less than 10% of this remains undisturbed (MacKinnon, 1985). In

Creative Conservation: Interactive management of wild and captive animals.
Edited by P.J.S. Olney, G.M. Mace and A.T.C. Feistner.
Published in 1994 by Chapman & Hall, London.
ISBN 0 412 49570 8

addition to long-term projects, several species and subspecies will need 'emergency intervention' if they are to survive, even in the short term.

The Ministry of Forestry (MoF) which is responsible for nature and species conservation, and as such also has responsibility for managing the two major zoos in the country, is willing to address these problems (Ministry of Forestry, 1985). However the depressed economic situation in Vietnam prevents the MoF from initiating most of the activities needed without foreign aid. Other important factors are the absence of a sufficient number of trained personnel and an improper understanding of nature conservation priorities. This multitude of problems is further complicated by a convoluted bureaucratic structure which occasionally prevents funds allocated for protected area development actually being disbursed to the sector.

The Vietnamese government is well aware of the important role which foreign nature conservation organizations can play in helping to alleviate these problems, and a wide variety of opportunities exist for zoos to contribute significantly to species conservation in Vietnam through the provision of financial, technical and scientific support. In this paper we identify several important species which are critically endangered in Vietnam, and discuss the conservation activities in which zoos can participate to ensure their long-term survival

31.2 BIRDS

31.2.1 Pheasants

Vietnam has a diverse avifauna which includes many endemics. These have been the target of several surveys by Birdlife International, formerly the International Council for Bird Preservation (ICBP), the most recent of which took place in 1991 (Eames and Robson, 1992).

These surveys revealed that the conservation status of most of Vietnam's endemic birds gives great cause for concern.

Apart from the fact that zoos could adopt *in situ* conservation activities for particular species, such as the black-hooded laughing thrush (*Garrulax milleti*) which was only rediscovered in 1991 and which has not yet any officially protected area in its now very restricted remaining range, zoos' assistance would be particularly important for several *Lophura* pheasants.

Of these Edwards' pheasant (*Lophura edwardsi*) deserves special mention. Despite concerted efforts by the ICBP survey teams, no evidence has been found of any surviving Edwards' pheasant populations and this species is now extinct in the wild.

Fortunately, a captive population of several hundred individuals exists in zoos and private collections throughout the world. As far as is known, all these birds derive from a founder population of 15, imported by Jean Delacour to France some 60 years ago. Unfortunately, however, during its long captive history the species has never been subject to any coordinated management and may have suffered significant inbreeding, loss of genetic variation, and selection. While we cannot nullify past mistakes there is now an urgent need to give priority attention to this critically endangered species. As a basis for a correctly structured recovery effort a PVA Workshop, cooperatively organized by ICBP, the Captive Breeding Specialist Group (CBSG), the World Pheasant Association and Antwerp Zoo took place in Belgium in February 1993. The outcome of this workshop suggests that there is a need for many zoos to provide holding space for this neglected species and to allocate funding for research, which should include DNA fingerprinting.

Of the closely related Vietnamese pheasant (*Lophura hatinhensis*) only three specimens exist in Hanoi Zoo as part of a cooperative captive breeding effort between the Centre For Natural Resources Management and Environmental Studies (CRES) (whose director Professor Vo Quy described the species) and the Brehm Fund for International Bird Conservation (Adler, 1992a).

No pure imperial pheasants (*Lophura imperialis*) are at present held in captivity. Both *L. hatinhensis* and *L. imperialis* still occur in the wild, but their continued survival is very uncertain. The exact status and distribution of both species needs further research, and should results indicate that captive breeding is needed, the credibility of the captive breeding community would obviously be enhanced if the problems affecting the captive stock of Edwards' pheasant were to be properly addressed.

31.3 MAMMALS

31.3.1 Kouprey

The kouprey (*Bos sauveli*) has been the subject of several surveys financed by the Kouprey Conservation Trust under the management of Dr Lee Simmons, Henry Doorly Zoo, Omaha (Cox and Duc, 1990). However, no evidence has been found to suggest that *B. sauveli* still occurs in Vietnam, and additional field surveys to locate the species are required before specific conservation recommendations can be formulated.

31.3.2 Tonkin sika deer

A project which has proceeded relatively successfully is the Tonkin Sika Deer (*Cervus nippon pseudaxis*) Conservation Programme. This subspecies is extinct in the wild and probably only survives pure bred in the Deer Farm of Cuc Phuong National Park and as a small population of 90–100 animals in zoos. The programme was initiated by the authors and combines captive breeding with financial and scientific support for *in situ* conservation (Adler and Wirth, 1992). Cooperating agencies are the Ministry of Forestry and Cuc Phuong National Park in Vietnam, as well as Whipsnade Wild Animal Park, Poznan Zoo, Mulhouse Zoo and Zürich Deer Park. The four zoos have received a number of Tonkin sika deer on breeding loan from the Vietnamese government in an attempt to increase genetic variability of existing captive stock in Europe. At the same time these zoos have provided support to improve the management infrastructure at Cuc Phuong, and have developed a strategy for eventual reintroduction of the deer to the park. A European Endangered Species Programme (EEP) for Tonkin sika deer is now in place and is coordinated by Klaus Rudloff of Tierpark Berlin, and an international studbook is proposed. A Memorandum of Agreement which included the whole EEP would be a further beneficial step.

31.3.3 Owston's palm civet

One of the least known viverrids until recently, the endangered Owston's palm civet (*Chrotogale ownstoni*) has now received considerable attention thanks to the activities of the IUCN Mustelid, Viverrid and Procyonid Specialist Group. Efforts by the Institute of Ecology and Biological Resources (IEBR), under the leadership of Professor Dang Huy Huynh, to study and conserve the species have received a boost with funds made available through the Cuc Phuong Project of Frankfurt Zoological Society (Adler, 1991a). As part of this project Frankfurt Zoo participates in a captive breeding and research programme. Likewise, the Frankfurt Zoological Society and the Jersey Wildlife Preservation Trust have supported the training of Vietnamese biologist Nguyen Xuan Dang in Jersey.

31.3.4 Concolor gibbons

The concolor gibbon occurs in the extreme south of China and in Indo China, east of the Mekong River. Until recently most taxonomists recognized at least six subspecies of concolor gibbon but it has now been suggested that these animals should be separated into two distinct species: the concolor gibbon (*Hylobates concolor*) which lacks white cheek whiskers in the black colour phase, and the white-cheeked gibbon (*Hylobates leucogenys*) which is characterized by cheek whiskers which are white or pale red. One subspecies of concolor gibbon (*H. c. concolor*) and three subspecies of the white-cheeked gibbon (*H. l. leucogenys, H. l. siki and H. l. gabriellae*) occur in Vietnam. All are threatened by habitat loss and hunting. Specimens of each subspecies of the white-cheeked group of gibbons are held in captivity. The most endangered of these (*H. l. siki*) has a founder population of four and a total captive population of only six animals. A studbook, research programme on taxonomy and coordination effort is directed by Jean-Marc Lernould of Mulhouse Zoo. The total captive population of *H. c. concolor* comprises only a single male animal. The International Studbook on all *H. concolor* individuals in captivity has recently been published (Lernould, 1993), and also contains useful data on taxonomy, geographical distribution, genetics and status in the wild.

In addition to habitat loss and hunting for medical purposes, gibbons in Vietnam are very much threatened by trapping of juveniles for the pet markets of the major Vietnamese cities. These gibbons are usually obtained by shooting the mothers, and surveys have noted a great scarcity of females in certain areas as a result. Very few of the captured gibbon babies survive for any length of time, not even those regularly acquired by Saigon Zoo (Adler, 1991b).

There is an urgent need for a concerted effort by the zoo community to develop a strategy with the Vietnamese government which permits legal action against this destructive and cruel trade. As part of this effort zoos could provide funds and expertise for the proper maintenance of confiscated gibbons. Some of the confiscated gibbons could then be made available on a non-commerical basis if and when required for a captive breeding programme.

31.3.5 Tonkin snubnose monkey

In early 1992 a survey, carried out by Radoslaw Ratajszczak and funded by the Fauna and Flora Preservation Society and the World Wide Fund for Nature, confirmed that the Tonkin snubnose monkey (*Rhinopithecus avunculus*) still occurs in Vietnam (R. Ratajszczak, pers. comm.). However, it was also confirmed that the species is seriously endangered, as probably less than 300 animals survive, distributed in four forest patches, the largest of which is no more than 20 km^2 in extent. It was reported that at least 17 snubnoses had been killed by hunters between January and March 1992 and at this rate the species will probably be extinct within three years. Only an immediate major conservation effort can still save it. The authors are hopeful that the Vietnamese government will establish and police a new protected area for the Tonkin snubnose monkey as a result of a planned meeting in Hanoi in spring 1993.

31.3.6 Langurs

Langurs of the *francoisi* complex seem to be less problematic in captivity than many other leaf-eating primates. The population of Francois' langurs (*Trachypithecus (f.) francoisi*) in American zoos is increasing (for 1990 the *International Zoo Yearbook* lists 19.33 animals in 9 collections). Cyndi Kuehler of San Diego has investigated and documented all aspects of their captive maintenance and an unofficial studbook exists. These efforts are important as Francois' langur is now almost extinct in Vietnam (Ratajszczak, Cox and Duc, 1989). The fact that the last remaining individuals are subject to poaching for export to China indicates that very few now also survive in China.

The status of the related Delacour's langur (*Trachypithecus (f.) delacouri*) is even more critical. In fact, together with the Tonkin snubnose it shares the dubious distinction of being one of Asia's most endangered primates (Eudey, 1987). A few dozen at best now survive in Cuc Phuong National Park, the only protected area within its range (Adler, 1991c). An unknown, but certainly very low, number of animals occur in a few forest fragments along the Laotian border with Vietanm. Both inside and outside the national park they are heavily hunted and their extinction seems imminent.

The Cuc Phuong Project of the Frankfurt Zoological Society, which aims at the protection of the local remnant population of Delacour's langur, was implemented by Dr Richard Faust and the authors in 1991 (Adler and Peter, 1991). Despite some success it soon became evident that the permanent presence of a Project Executant will be necessary in Cuc Phuong to achieve the planned goals (Adler, 1992b). It is hoped that more institutions will follow the positive example of Frankfurt Zoological Society in providing rapid objective support.

It is also of great urgency to obtain the necessary funds for a survey of areas outside Cuc Phuong where some Delacour's langurs may still survive. Considering the rapid rate of forest destruction and the severe and uncontrolled hunting pressure, one must realistically assume that there will be little hope for Delacour's langurs in unprotected forest patches other than incorporating them into a coordinated captive breeding programme.

Captive breeding may not immediately be needed for two other forms of *T. francoisi*, namely the Cat Ba langur (*T. f. poliocephalus*) and the black langur (*T. f. hatinhensis*). The former has a small, but apparantly stable population and prospects for conservation action in the wild are reasonably good. *T. f. hatinhensis*, which was feared extinct, has just been rediscovered and the status of the population is not yet known.

REFERENCES

Adler, H.J. (1991a) Conservation program of *Chrotogale owstoni* Thomas, 1912: some first results. *Mustelid and Viverrid Conservation. The Newsletter of the IUCN MVSG*, **4**, 8.

Adler, H.J. (1991b) The crested gibbon, *Hylobates (Nomascus) concolor* Harlan, 1826, in the Nam Cat Tien Reserve, Southern Vietnam. *Primate Report*, **29** 59–64.

Adler, H.J. (1991c) On the situation of the Delacour's langur (*Trachypithecus f. delacouri*) in the north of Vietnam – proposal for a survey and conservation project. *Primate Report*, **31**, 6–7.

Adler, H.J. (1992a) Vietnam – Fasan im Zoo Hanoi! *Tropische Vögel/Trochilus*, **13**, 30–1.

Adler, H.J. (1992b) Cuc Phuong National Park, Nordvietnam/Projekt 1129/91 der Zoologischen Gesellschaft Frankfurt von 1858 – ein Situationsbericht. *Mittlg. der Zoolog. Gesellschaft f. Populations- u. Artenschutz, München*, **8** (1), 1–5.

Adler, H.J. and Peter, W.P. (1991) *Cuc Phuong National Park Conservation Program, North Vietnam. Project Proposal*, Zool. Soc. Frankfurt, Frankfurt/Main.

Adler, H.J. and Wirth, R. (1992) Das Erhaltungsprojekt für den Vietnamsika. *Mittlg. der Zool. Gesellschaft f. Populations – u. Artenschutz, München*, **8** (2), 1–5.

Cox, R. and Ha Dinh Duc (1990) *Survey for Kouprey in the Yok Don Nature Reserve, Dak Lac Province, Vietnam*. Unpublished Report to the Kouprey Conservation Trust.

Eames, J.C. and Robson, C.R. (1992) Vietnam Forest Project – Forest Bird Survey 1991. *ICBP Study Report No. 51*, ICBP, Cambridge.

Eudey, A.A. (1987) *Action Plan for Asian Primate Conservation*, IUCN Primate Specialist Group, Riverside, California.

Lernould, J.M. (1993) *International Studbook Hylobates concolor 1990*, Parc Zoologique et Botanique de la Mulhouse, France.

MacKinnon, J. (1985) *Review of the Protected Area System and Species Conservation Needs of Vietnam*. WWF, Gland, Switzerland.

Ministry of Forestry (1985) *Forest Preserves in Vietnam*. Ministry of Forestry, Hanoi.

Ratajszczak, R., Cox, R. and Ha Dihn Duc (1989) *A Preliminary Survey of Primates in North Vietnam*, WWF Project 3869, WWF, Gland, Switzerland.

Stuart, S. and Wirth, R. (1990) *Preliminary Project Proposal – Conservation and Recovery of Threatened Species in Vietnam*, IUCN, Gland, Switzerland.

Index

Note: page numbers in **bold** type refer to **figures**, page numbers in *italic* type refer to *tables*.